Three-Dimensional Television, Video, and Display Technologies

Springer
Berlin
Heidelberg
New York
Barcelona
Hong Kong
London
Milan
Paris
Tokyo

Bahram Javidi   Fumio Okano

Three-Dimensional Television, Video, and Display Technologies

With 352 Figures

Springer

Bahram Javidi
Electrical and Computer Engineering Department
University of Connecticut
Storrs, CT 06269, USA
E-mail: bahram@engr.uconn.edu

Fumio Okano
Science & Technical Research Laboratories
NHK Japan Broadcasting Corp.
1-10-11 Kinuta, Setagaya-ku
Tokyo 157-8510, Japan
E-mail: okano@strl.nhk.or.jp

Library of Congress Cataloging-in-Publication Data

Three-dimensional television, video and display technology / Bahram Javidi, Fumio Okano [editors].
p. cm.
Includes bibliographical references and index.
ISBN 3540435492 (alk. paper)
1. Three-dimensional display systems. 2. Television display systems. 3. Imaging systems.
I. Javidi, Bahram. II. Okano, Fumio.

TK7882.I6 T474 2002
621.3815'422–dc21 2002070539

ISBN 3-540-43549-2 Springer-Verlag Berlin Heidelberg New York

Springer-Verlag Berlin Heidelberg New York
a member of BertelsmannSpringer Science+Business Media GmbH

http://www.springer.de

Printed in Germany

Typesetting: Data conversion by LE-TEX Jelonek, Schmidt & Vöckler GbR, Leipzig
Cover design: *design & production*, Heidelberg

Printed on acid-free paper SPIN 10792942 56/3141/yl 5 4 3 2 1 0

For Bethany, Ariana, Darius, and Vida

Preface

Three-dimensional (3-D) imaging and display are among the most important applications of information systems with a broad range of applications in computer display, TV, video, robotics, metrology, reconnaissance, and medicine. Developing technology to display high-quality 3-D scenes, especially those with motion, is a very challenging problem. Attempts were made to produce stereoscopic effects in paintings even before photographic technology was developed. Ideal 3-D images, especially moving ones, are among the goals of future technologies because of their diverse and significant applications. Science fiction motion pictures frequently use 3-D images; for instance, the Star Wars movie features a scene in which Princess Leia's message is conveyed as a 3-D image.

The first attempts to generate moving stereoscopic images were for movie production. Video technology, as it continues to advance, is increasingly being exploited to produce moving pictures with 3-D effects. In recent years, virtual-reality applications are attracting strong interest, and many experimental systems have been reported. Three-dimensional moving-picture systems are increasingly in demand for various applications in entertainment, industry, and medicine. Some of them have already matured to a practical level. However, higher-quality 3-D image systems are generally more difficult to implement.

With the rapid advances in electronic/optoelectronic hardware and software technology, these 3-D-imaging techniques can now be implemented with commercially available components and can be used for many commercial applications. A number of experimental 3-D moving picture systems have been proposed and developed. Depending on the characteristics of the reproduced 3-D images, they can be roughly classified into three types: binocular, multiview, and spatial-image reproduction. Each of these is a complex technology that combines optical, display, and image-capture techniques. These technologies are multidisciplinary and incorporate knowledge from the fields of optics, electronics, signal processing, and psychological science.

The chapters of this book are contributed by leaders in the field of 3-D TV and display technology and discuss both theory and applications of these techniques. Engineering examples, numerical simulations, and experimental results illustrate the theoretical concepts. We believe that this book will be very useful for students, engineers, scientists, and managers who are interested in this important field. There is a large community that does R&D on imaging systems and is interested in learning about the advances in the rapidly evolving field of 3-D TV and display. This book addresses these needs by providing recent advances in 3-D display and TV and by emphasizing the development of new applications.

We are grateful to the contributors, most of whom are our friends, for their outstanding contributions. We would like to thank the Springer editor Dr. Hans Koelsch for his support of the project, and the Springer production staff, particularly Ms. Jacqueline Lenz for her assistance with the preparation of this manuscript.

Storrs, Connecticut
Tokyo, Japan
May 2002

Bahram Javidi
Fumio Okano

Contents

Part I Capture, Display, and Transmission System

Contributors

Kiyoharu Aizawa
Graduate School
of Frontier Sciences,
The University of Tokyo,
7-3-1 Hongo, Bunkyo-ku,
Tokyo 113-0033,
Japan

Takahisa Ando
SANYO Electric Co.,Ltd.
Technology R&D Headquarters
Digital Systems Development Center
1-1, Sanyo-cho, Daito City,
Osaka 574-8534,
Japan
t_andou@rd.sanyo.co.jp

Jun Arai
Science and Technical Research
Laboratories, NHK
1-10-11, Kinuta, Setagaya-ku,
Tokyo 157-8510,
Japan
arai.j-gy@nhk.or.jp

George Barbastathis
Mechanical Engineering Dept.,
Massachusetts Institute
of Technology,
Rm. 3-461c, 77 Massachusetts Ave.,
Cambridge, MA 02139,
USA
gbarb@mit.edu

David J. Brady
Director,
Fitzpatrick Center for Photonics
and Communication Systems,
Professor of Electrical
and Computer Engineering,
Box 90291,
Duke University,
Durham, NC 27708
USA
dbrady@duke.edu

Rachael Brady
Visualization Analysis Lab,
D212 Levine,
Science Research Center,
Box 90271,
Durham, NC 27708,
USA
rbrady@duke.edu

Evan Cull
Fitzpatrick Center for Photonics
and Communication Systems,
Department of Electrical
and Computer Engineering,
Box 90291,
Duke University,
Durham, NC 27708,
USA
evan.cull@duke.edu

Tomohiro Endo
Hongo 3D Research Center,
Telecommunications Advancement
Organization of Japan (TAO),
1-33-16-601, Hakusan, Bunkyo-ku,
Tokyo 113-0001,
Japan
yendo@3dpro.tao.go.jp

Jocelyn Faubert
École d'optométrie,
Université de Montréal,
3744 Jean-Brillant,
Montreal,
Canada H3T 1P1
jocelyn.faubert@UMontreal.CA

Steve Feller
Fitzpatrick Center for Photonics
and Communication Systems,
Department of Electrical
and Computer Engineering,
Box 90291,
Duke University,
Durham, NC 27708,
USA
sfeller@duke.edu

Lilian Fernandes
AIX Open Systems Communication,
IBM (Austin, TX),
USA
lilian_sf@hotmail.com

Yann Frauel
IIMAS - Universidad Nacional
Autónoma de México,
Apdo. Postal 20-726 Admon. No. 20,
Del. de Alvaro Obregón,
01000 México, D.F.
Mexico
yann@leibnitz.iimas.unam.mx

Toshiaki Fujii
Dept. of Information Electronics,
Graduate School of Engineering,
Nagoya University,
Furo-cho, Chikusa-ku,
Nagoya 464-8603,
Japan
fujii@nuee.nagoya-u.ac.jp

Tadahiko Hamaguchi
Hongo 3D Research Center,
Telecommunications Advancement
Organization of Japan (TAO),
1-33-16-601, Hakusan, Bunkyo-ku,
Tokyo 113-0001,
Japan
hamaguchi@3dpro.tao.go.jp

Hiroshi Harashima
Department of Information and
Communication Engineering,
The University of Tokyo,
7-3-1 Hongo, Bunkyo-ku,
Tokyo, 113-8656,
Japan
hiro@hc.t.u-tokyo.ac.jp

Toyohiko Hatada
Department of Photonic Information
and Media Engineering,
Tokyo Institute of Polytechnics,
Iiyama 1583, Atsugi,
Kanagawa 243-0297,
Japan
hatada@photo.t-kougei.ac.jp

Toshio Honda
Department of Information
and Image Sciences,
Chiba University,
1-33 Yayoicho, Inage-ku,
Chiba 263-8522,
Japan
honda@image.tp.chiba-u.ac.jp

Haruo Hoshino
Science and Technical Research Laboratories, NHK
1-10-11, Kinuta, Setagaya-ku,
Tokyo 157-8510,
Japan
hoshino.h-ii@nhk.or.jp

Tadashi Ichikawa
NTT Advanced Technology Corpration,
90-6, Kawakami-cho, Totsuka-ku,
Yokohama,
Kanagawa 244-0805,
Japan
ichi@hama.ntt-at.co.jp

Hiroyuki Imaizumi
Science and Technical Research Laboratories, NHK
1-10-11, Kinuta, Setagaya-ku,
Tokyo 157-8510,
Japan
imaizumi.h-oa@nhk.or.jp

Shoichiro Iwasawa
Telecommunication Advancement Organization of Japan,
401 Hortensia Hakusan Bldg.,
1-33-16 Hakusan, Bunkyo-ku,
Tokyo 113-0001,
Japan
shoichiro.iwasawa@acm.org

Ju-Seog Jang
Division of Electronics, Computers, and Telecommunications,
Pukyong National University,
599-1 Daeyun-Dong, Nam-Gu,
Pusan 608-737,
Korea
jsjang@pknu.ac.kr

Bahram Javidi
Electrical and Computer Engineering Department,
University of Connecticut,
260 Glenbrook Road,
U-2157 Storrs, CT 06269-2157,
USA
bahram@engr.uconn.edu

Yoshihiro Kajiki
Networking Research Laboratories,
NEC Corporation,
4-1-1, Miyazakidai, Miyamae-ku,
Kawasaki 216-8555,
Japan
y-kajiki@ah.jp.nec.com

David Kammeyer
702 W. Nevada St.,
Urbana, IL 61801,
USA
kammeyer@kammeyer.org

Toshifumi Kanamaru
Business Research Laboratory,
Corporate Research and Development Center,
Oki Electric Industry Co., Ltd.
550-5, Higashiasakawa-cho,
Hachioji-shi,
Tokyo 193-8550,
Japan

Jin Liu
Interactive Media-Human Factors,
Heinrich-Hertz-Institut,
Einsteinufer 37,
10587 Berlin,
Germany
liu@hhi.de

Ajay Luthra
Motorola,
6450 Sequence Dr.,
San Diego, CA-92121,
USA
ALuthra@gi.com

Osamu Matoba
Institute of Industrial Science,
University of Tokyo,
4-6-1, Komaba, Meguro-ku,
Tokyo 153-8505,
Japan
matoba@iis.u-tokyo.ac.jp

Akihiro Morinaga
Department of Information and
Image Sciences Chiba University
1-33 Yayoi-cho, Inage-ku,
Chiba 263-8522,
Japan
morinaga@vision.tp.chibau.ac.jp

Shigeo Morishima
Faculty of Engineering,
Seikei University,
3-3-1 Kichijyoji-Kitamachi,
Musashino
Tokyo 180-8633,
Japan

Takeshi Naemura
Computer Graphics Laboratories,
Stanford University,
Gates Computer Science Building
3B-390,
Stanford, CA 94305-9035,
USA
naemura@graphics.stanford.edu

Thomas J. Naughton
Department of Computer Science,
National University of Ireland,
Maynooth,
Ireland
tom.naughton@may.ie

Fumio Okano
Science and Technical Research
Laboratories, NHK
1-10-11, Kinuta-Setagaya-ku,
Tokyo 157-8510,
Japan
okano.f-la@nhk.or.jp

Makoto Okui
Science and Technical Research
Laboratories, NHK
1-10-11, Kinuta, Setagaya-ku,
Tokyo 157-8510,
Japan
okui.m-im@nhk.or.jp

Siegmund Pastoor
Interactive Media-Human Factors,
Heinrich-Hertz-Institut,
Einsteinufer 37,
10587 Berlin,
Germany
pastoor@hhi.de

Takahiro Saito
Department of Electrical, Electronics
and Information Engineering,
Kanagawa University,
3-27-1 Rokkakubashi, Kanagawa-ku,
Yokohama 221-8686,
Japan

Eiji Shimizu
Dept. of Electrical Engineering,
Osaka City University,
3-3-8, Sugimoto, Sumiyoshi-ku,
Osaka 558-0022,
Japan
shimizu@elec.eng.osaka-cu.ac.jp

Seung-Ho Shin
Department of Physics,
Kangwon National University,
Chunchon, 200-701,
Korea
shinsh@kangwon.ac.kr

Satoshi Shioiri
Department of Information
and Image Sciences,
Chiba University,
1-33 Yayoi-cho, Inage-ku,
Chiba 263-8522,
Japan
shioiri@image.tp.chiba-u.ac.jp

Jung-Young Son
Korea Institute
of Science and Technology,
39-1 Hawologog-dog, Sungbook-ku,
Seoul 136-791,
Korea
sjy@kist.re.kr

Ronald Stack
Distant Focus Corporation,
701 Devonshire Drive, C-17,
Champaign, IL 61820,
USA
rstack@distantfocus.com

Kenji Susami
Hongo 3D Research Center,
Telecommunications Advancement
Organization of Japan (TAO),
1-33-16-601, Hakusan, Bunkyo-ku,
Tokyo 113-0001,
Japan
susami@3dpro.tao.go.jp

Enrique Tajahuerce
Dept. de Ciències Experimentals,
Universitat Jaume I,
Av. 505 Bagnat, s/n,
12071 Castellón,
Spain
tajahuer@exp.uji.es

Hirohisa Yaguchi
Department of Information
and Image Sciences,
Chiba University,
1-33 Yayoi-cho, Inage-ku,
Chiba 263-8522,
Japan
yaguchi@image.tp.chiba-u.ac.jp

Kunio Yamada
Telecomunication Advancement
Organization of Japan,
401 Hortensia Hakusan Bldg.,
1-33-16 Hakusan, Bunkyo-ku,
Tokyo 113-0001,
Japan

Mitsuho Yamada
Science and Technical Research
Laboratories, NHK
1-10-11 Kinuta, Setagaya-Ku,
Tokyo 157-8510,
Japan
yamada.m-gy@nhk.or.jp

Ichiro Yuyama
Science and Technical Research
Laboratories, NHK
1-10-11 Kinuta Setagaya-ku,
Tokyo 157-8510,
Japan
yuyama.i-in@nhk.or.jp

Sumio Yano
Advanced Telecommunications,
Research Institute International,
ATR Human Information Science
Laboratories,
2-2-2 Hikaridai,
Seika-cho Souraku-gun,
Kyoto 619-0288,
Japan
yano@atr.co.jp

Part I

Capture, Display, and Transmission System

1 Stereoscopic HDTV

I. Yuyama and M. Okui

1.1 Introduction

Binocular parallax and convergence occur when the right and left eyes receive a horizontal displacement by an amount corresponding to the depth distance. Stereoscopic pictures based on this principle, in the form of handwritten pictures, were used even before the advent of photography. After photographs became available, several improvements were made on stereoscopes and binocular three-dimensional (3-D) photos also became popular during the latter half of the 19th century. The early 1900s already saw the invention of parallax barriers, integral photography, lenticular lenses, and other means of stereoscopic displays.

After the advent of movie technology, stereoscopic images were often shown at theaters and international expositions during the 1900s. The main means of stereoscopic display shifted from anaglyph to polarization, and 3-D screens of lenticular lenses were also used. This background led to the 3-D theaters found in many places today, where stereoscopic images are shown on a regular basis.

In Japan, television broadcasting started in 1953. Since then, television technology has progressed from black-and-white displays to color broadcasting and then to HDTV.

When NHK began studying future broadcasting services just after the Tokyo Olympic Games, two possible systems were considered. One was a high-definition television system, which creates a "sensation of reality" by increasing spatial resolution and widening the viewing angle. As a result of steady progress in the development of HDTV equipment and receivers, HDTV broadcasting has already begun in the USA and Japan.

The other system considered was 3-D television, which would bring a "sensation of depth" to the broadcasting service. Research has continued on developing a variety of 3-D display systems. As a result of a series of psychological tests, a wide screen or display was found to be necessary for 3-D images in order to produce adequate psychological effects. The stereoscopic HDTV was then developed as an experimental stereoscopic television system, aiming to deliver the majority of broadcasting services in the future.

Research on 3-D pictures, however, is nothing new and many outstanding studies have been published in the past [1, 2]. Most of these studies are related to 3-D movies, holography, and other photographic techniques, and very few concern television technologies. As television technology matures and HDTV produces pictures as clear as movies, 3-D TV is now capable of providing real-time stereoscopic pictures with high resolution. Thus, 3-D TV shows great promise as the next-generation imaging system.

This chapter outlines the studies that have been carried out on stereoscopic HDTV, focusing on such topics as human factors in relation to stereoscopic vision, and cameras, display systems as well as signal processing like bandwidth-reduction technology. We do not discuss research on holography and multiviewpoint systems such as IP.

1.2 Stereoscopic Human Factors

In many cases, the development of television systems began with research on visual science. By fully utilizing the results of such research, we can begin to construct a system that is efficient and valid. Similarly, research on stereoscopic vision is important in the development of stereoscopic television.

Three-dimensional information obtained by the visual system is processed by complex activities in the cerebral cortex. Some interesting observations have been made in this field [3]. For instance, researchers have confirmed the existence of binocular disparity cells that selectively react to specific parallax and other cells that react to the parallax of random dot stereogram (RDS) in the visual area I (V1) of the occipital lobe.

There are several clues to depth perception by the visual sense as shown in Fig. 1.1, and which clue will be dominant depends on depth distance [4].

Convergence as a depth clue is very effective when the distance is small, and is considered to be effective for up to 20 m. Convergence rapidly loses its effects as the distance increases (because the convergence angle gets smaller). Binocular parallax occurs when there are other objects in front of or behind the object on which the right-left eyes are focused. If the degree of this binocular parallax is sufficiently small, the images are fused and the observer has a clear sensation of depth before and after the object. Binocular parallax is effective and precise in recognizing the depth difference, i.e. the difference in the distance between several objects. It is an important factor in understanding the relative position of objects within 10 m.

Accommodation is only effective when the observation distance is no more than 2 to 3 m. Depth perception brought about by motion parallax also provides an important clue as to depth, and can be as effective as binocular parallax under certain conditions.

Some of those clues shown in Fig. 1.1 are especially important for the capabilities of 3-D display.

As effects such as linear perspective, light and shade, overlapping and aerial perspective can be easily achieved in actually shot 2-D images, they are not listed in Fig. 1.1. Primary research targets in 3-D image technology are binocular parallax, accommodation, convergence, and motion parallax (or

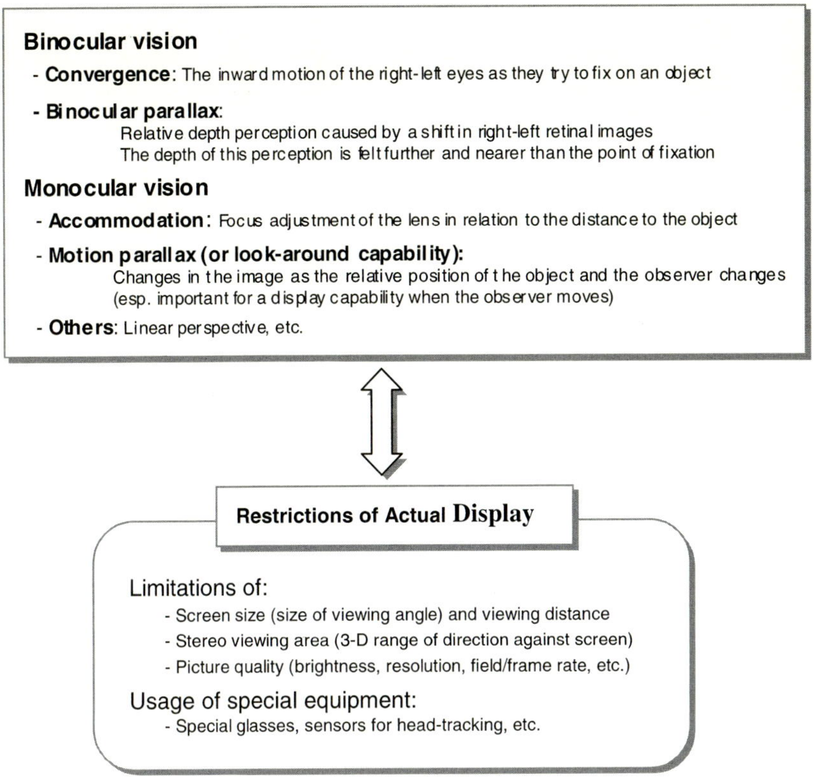

Fig. 1.1. Depth clues and display factors

look-around capability: changes of viewpoint accompanied with head movement). Figure 1.1 shows some of the properties that the stereoscopic display system should have. Also shown at the bottom of Fig. 1.1 are characteristics that may be traded-off frequently when displaying a stereoscopic image.

When observing a 3-D image on a display unit, such factors as the viewing angle and viewing area (the field where a stereoscopic image can be observed) in addition to the image's direct depth clues will be important in order to retain the perception of reality and a sense of immersion.

The average distance between the pupils of the human eyes is approximately 65 mm. Visual information received through one eye is not the same as that through the other eye, but we perceive these data as a single image. This ability, i.e. binocular vision, appears to be based on interaction of the right and left images, and this ability is an important function in the design of a 3-D television. Some results of experiments related to this function will be given in Sect. 1.4. The following section outlines some of the well-known phenomena [4].

Studies have been made on the differences between binocular and monocular visions in respect of the characteristics of spatial-frequency response. It has been reported that when an object is viewed with two eyes, sensitivity is 1.4 times greater than when it is viewed with just one eye. This effect is called "binocular summation".

The combinations have been examined where the observer feels that the right-left images have the same luminance level. The results show that when the level of luminance is not greatly different between the right and left eyes, the perception of brightness corresponds to the average of luminance levels given to the right-left eyes [5]. Studies have also been made on the level of contrast that can be perceived. Here, the effects of binocular summation are also felt, though not as much as those in the case of luminance [6].

"Binocular rivalry" occurs when there exist visual portions that do not match between right and left images. Studies have been made on the basic characteristics of binocular rivalry, though they are not explained here.

1.3 Psychological Effects on High-Definition and Stereoscopic Images

Stereoscopic images provide a strong sensation of reality because they have depth information, which most distinguishes them from 2-D images.

To clarify the properties of visual depth perception, the apparent depth and size of the image were measured by using random-dot stereograms and LEDs as a matching index. The result is shown in Fig. 1.2, where perceived

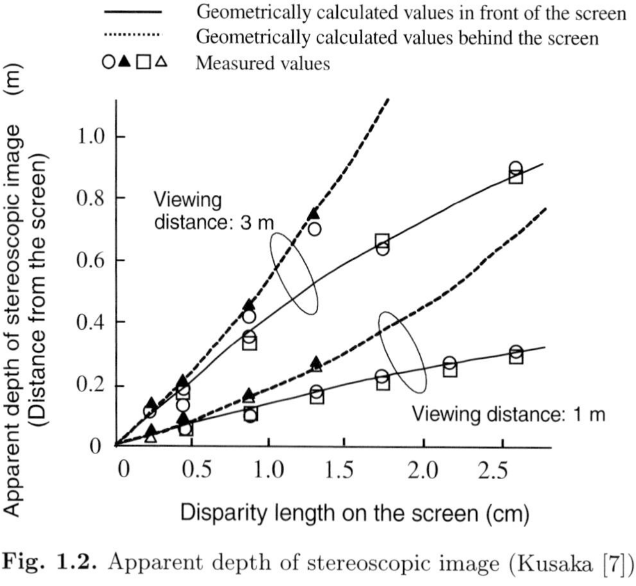

Fig. 1.2. Apparent depth of stereoscopic image (Kusaka [7])

depth agrees closely with the theoretical values [7]. As has been shown, the sensation of depth can be gained directly from binocular parallax clues. It is also clear that various psychophysical effects also influence the sensation of depth. To reproduce the sensation of reality efficiently and naturally, it is important to study the psychological effects of 3-D images. This section outlines the latest findings of research in this field and then discusses the basic features of 3-D images.

1.3.1 Psychological Effects of Stereoscopic HDTV – comparing 2-D and 3-D

Although it is a 2-D system, HDTV was designed as a medium to create a "sensation of depth" by widening the picture size and increasing spatial resolution.

Evaluation tests [8] have shown that still pictures are most acceptable when the aspect ratio of the screen is between 5:3 and 6:3, as shown in Fig. 1.3. Another psychological test [9] has shown that when the visual angle becomes greater than 30° an "inducing effect" appears, as illustrated in Fig. 1.4. Based on these factors, the initial specifications for HDTV in Japan were formulated.

How does the picture quality of binocular stereoscopic pictures compare to 2-D pictures? This was evaluated using a pair of slide projectors fitted with polarizing filters and using 8-inch by 10-inch transparencies [10]. The results are given in Figs. 1.5a and b. It was found that stereoscopic pictures have larger psychological effects than 2-D pictures. In order to achieve the same subjective values as stereoscopic pictures, the 2-D pictures need more than twice the visual angle of the stereoscopic pictures.

It was also shown that for stereoscopic pictures, those of less than 30° visual angle create inadequate psychological effects. From these results, it

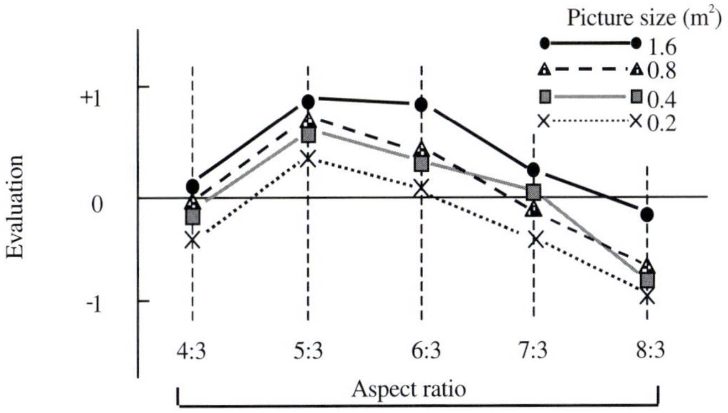

Fig. 1.3. Preference as a function of aspect ratio (2-D images at 2.5 m of viewing distance) (Mitsuhashi [8])

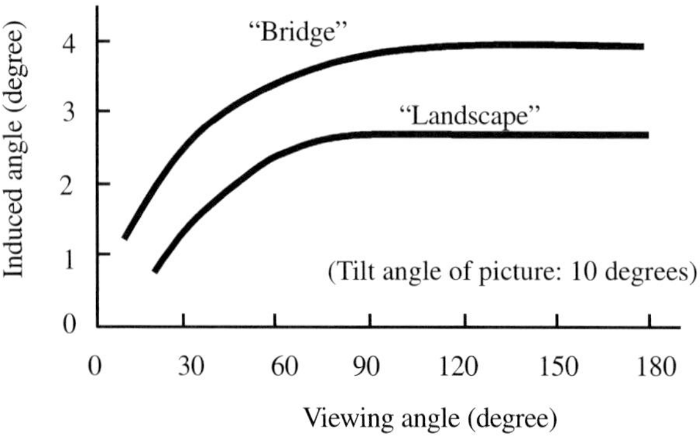

Fig. 1.4. "Inducing effect" as a function of image tilt (2-D images) (Hatada et al. [9])

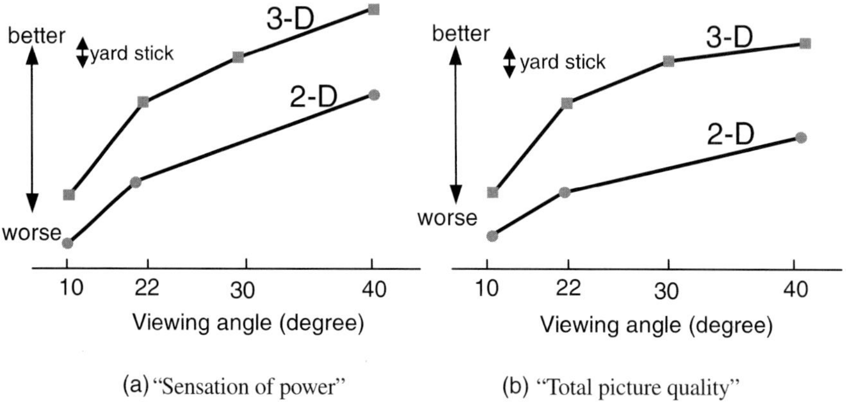

Fig. 1.5. Comparison of psychological effects between 2-D and 3-D images (Yano and Yuyama [10])

appears that at least the same or greater visual angle as in the HDTV systems is required for stereoscopic television.

The Heinrich-Hertz Institute in Germany conducted a similar experiment to evaluate the subjective effects of 2-D and 3-D pictures [11]. In this experiment, which used movie pictures, images were projected onto a $1\,\mathrm{m} \times 2.2\,\mathrm{m}$ screen. The experiment investigated such factors as overall image quality, the advantages and disadvantages of large-screen presentation for 2-D and 3-D picture, and concluded that 3-D pictures were preferable to 2-D images. They also performed a subjective evaluation of viewing distance, and concluded that a viewing distance of $3-4\,\mathrm{H}$ (H: screen height) was the most desirable.

These results indicate that the number of scanning lines and other standards of current HDTV can be used without large modification for a future

3-D TV system, and hence suggests a good possibility of coexistence between HDTV and a future 3-D TV system.

1.3.2 Study on Objective Evaluation of Psychological Effects

We have discussed some results of subjective evaluation experiments conducted to compare 2-D and 3-D pictures, but there have also been various attempts at objective comparisons. To objectively evaluate the psychological effects of 2-D and 3-D pictures, the viewer is asked to watch carefully a 2-D and a 3-D picture of an object that is rotating in a reciprocating manner. The body sway of the viewer is tracked, measured, and analyzed [12]. The analysis has revealed that the 3-D picture leaves a longer body sway trace than the 2-D picture, and that the frequency spectrum of the body sway for the 3-D picture has a distinctive peak. In other words, the viewer exhibits more pronounced reactions when watching a 3-D picture than a 2-D picture. These results mean that body-sway measurement can be used to objectively evaluate the psychological effects of 3-D pictures on the viewer.

The motion locus distance of the center of gravity was calculated in relation to the viewing angle, i.e. the total distance over which the center of gravity has traveled. The result is shown in Fig. 1.6. When the observer is shown the stimulus of a motion picture, this distance increases as the presented viewing angle increases, and more so in 3-D than in 2-D.

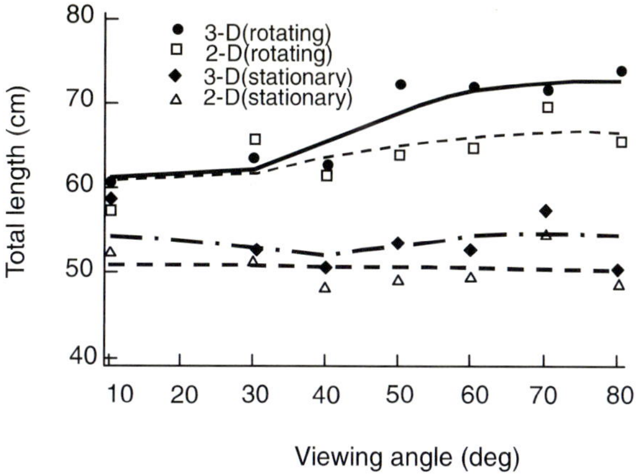

Fig. 1.6. Total length of body sway loci for various viewing angles (Shimizu et al. [12])

1.3.3 Accommodation Response to Binocular Stereoscopic Images

In the previous section, we discussed the advantages provided by 3-D pictures that are rarely provided with 2-D ones. However, many obstacles will have to be overcome before practical 3-D televisions can become a reality. Problems include eye fatigue that viewers of 3-D pictures may experience. One possible cause of eye fatigue is the inconsistency of eye movements when watching 3-D TV with binocular parallax, meaning the inconsistency between accommodation and convergence response. To better understand this point, the correlation between binocular parallax and accommodation response was measured [13].

Normally, the focal point, i.e. the point at which the eyes focus by accommodation when watching an object, and the point of convergence given through the rotating movement of the eyeballs, both fall on the point of fixation. When watching stereoscopic TV, a change in the amount of parallax causes the apparent image to move away from or come closer to the screen, and to make the viewer follow and converge on this movement. However, regardless of the amount by which the parallax is changed, the distance to the image on the screen remains the same, and so the eyes are forced to react in an exceptional manner. The inconsistency here is that binocular parallax causes a convergence response, subsequently altering accommodation. As a result, the focal point of the eyes may move away from the screen on which the image optically exists.

The difference between the average of the accommodation while the apparent position is on the screen and that while it is nearer or farther than the screen is plotted as a function of the L-R image distance in Fig. 1.7. From the experiment, it was found that the change in parallax while observing stereoscopic images causes the accommodation response, and that the amount of change in accommodation response is mainly influenced by the convergence angle created by parallax.

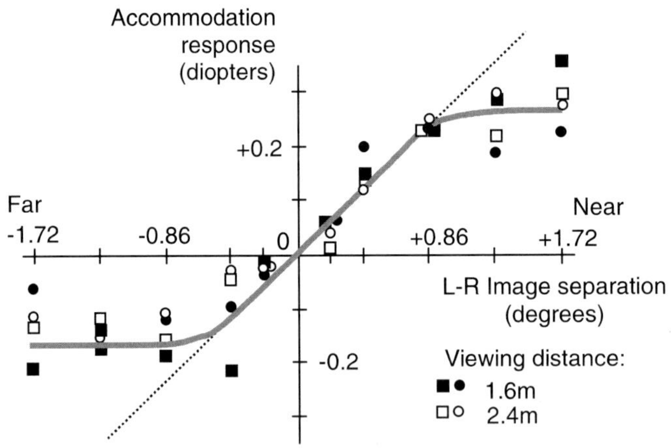

Fig. 1.7. Relationship between right-left image disparity and accommodation response (Hiruma and Fukuda [13])

These findings indicate that in order to increase the sensation of depth without overstraining the eye's accommodation system while watching 3-D TV, it is necessary to maintain an adequate distance between the viewer and a large display screen. Therefore, to develop a 3-D display that does not inflict much eyestrain, we need to pay closer attention to the focal depth of the eyes so as not to create a contradiction between accommodation and convergence. In other words, reproduced images should appear within the range of eye focal depth. One promising way is the development of a large, bright display screen which provides longer viewing distance and eye focal depth.

1.4 Requirements for Stereoscopic Television

There have been some attempts made to broadcast 3-D TV by applying the existing television system. However, the results of these experiments have not been satisfactory. With this background and recognizing the potential of 3-D TV broadcasting and its practical value, CCIR established in 1958 the study program IC-1/11. The organization unanimously agreed to carry out studies on the issue [14]. In response to these research topics, Report 312-4 [15] outlines a variety of 3-D display methods and band-reduction technologies, but satisfactory solutions has not achieved yet.

After these research activities, researchers generally believe that the following are required for the application of 3-D TV broadcasting: development of a display that reproduces natural 3-D pictures without causing viewer fatigue, compatibility with ordinary television signals as 2-D images and development of a system that does not require special glasses.

The areas indicated by the studies and essential items described above concern almost all the matters that are dealt with in this chapter. This section discusses some parameters that should be taken into consideration when designing a television system, including disturbance by noise, geometrical distortion, and differences between right and left images.

1.4.1 Bandwidth

In an experiment [16], the viewer compares 2-D and 3-D (stereoscopic) HDTV picture with limited bandwidths and determines which picture is sharper. Figure 1.8 shows the results of subjective evaluations when their frequency bandwidths are changed. The same object is reproduced in 2-D and 3-D. It is clear from Fig. 1.8 that, with the same bandwidth, the 3-D picture looks sharper than the 2-D picture. In other words, subjectively, a 3-D picture at 15 MHz looks about as sharp as a 2-D picture at 20 MHz.

We can say that 3-D pictures look sharper than 2-D pictures when their frequency bandwidth is the same, and this can be explained as follows. In 3-D pictures the distinction between the object and its background becomes

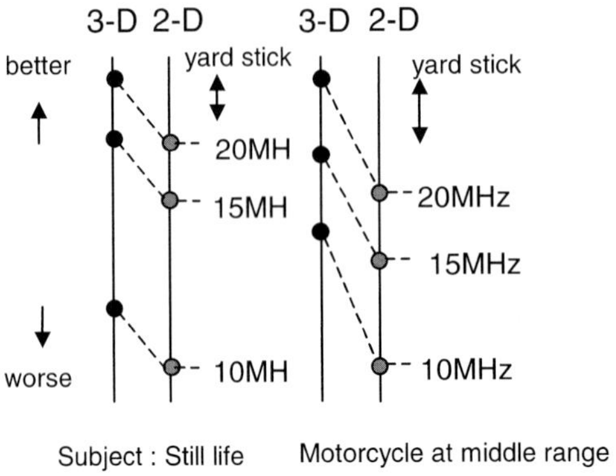

Fig. 1.8. Picture quality of stereoscopic pictures and 2-D pictures with various bandwidths (Yano [16])

clearer with the addition of binocular parallax, making the contours look sharper or enabling the viewer to see finer details.

The study of the required signal bandwidth of a stereoscopic image reveals that even if the bandwidth of one of the right-left images is limited, the signal will not deteriorate very much if most of the other is preserved. In an experiment with HDTV still pictures [16], as Fig. 1.9 shows, signal deterioration can be detected when the bandwidth-limiting frequency is about one third the frequency of the original image. The tolerance limit is about one fifth in this case.

1.4.2 Noise Characteristics

The Heinrich-Hertz Institute (HHI) conducted an experiment to evaluate the visibility of noise with correlated, uncorrelated and monocular patterns [17]. Random noise is less conspicuous in a stereoscopic image than in a 2-D picture

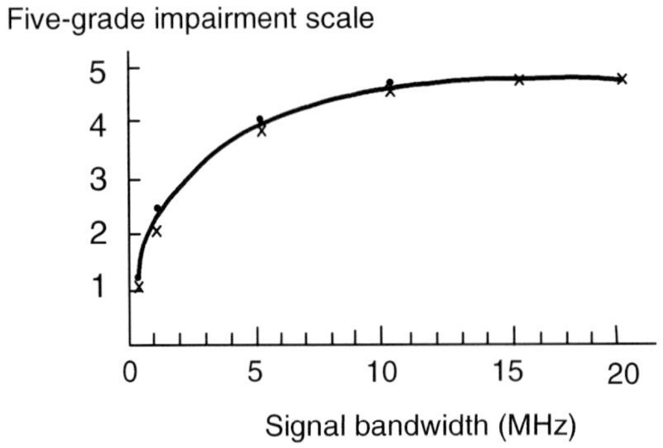

Fig. 1.9. Picture quality of one of a stereo-pair with reduced bandwidth (Yano [16])

– it is about 2 dB at the edges and 3 dB to 5 dB when there is no correlation between right and left noises. At CCIR impairment rating 4 (perceptible but not annoying), it is observed that the SNR of 3-D images with uncorrelated noise patterns is 3 dB lower than that in 2-D images.

1.4.3 Cross-Talk Level

Except for the head-mounted display, however, stereoscopic display devices are susceptible to crosstalk, i.e. leakage of right and left eye images to the opposite eye. This disturbance not only degrades the picture quality of stereoscopic images, but is also thought to cause eyestrain and fatigue for the viewer. To develop a stereoscopic display system for television broadcasting in the future, it is therefore important to know the relationship between the amount of crosstalk and the degree of disturbance.

The visibility threshold of crosstalk has been measured in the past at HHI [18], and we also examined the characteristics of crosstalk visibility. Our results [19] gave almost the same visibility threshold (HHI:0.3%, NHK:0.2% for a display contrast ratio of 100:1) and a similar dependence on the contrast and binocular parallax.

To consider the level at which crosstalk should be kept for practical purposes, we examined its disturbance using ordinary pictures in a subjective evaluation experiment. Three types of still picture from the 3-D standard chart were used. Figure 1.10 shows the evaluation results. The visibility thresholds, where evaluation value is equivalent to grade 4.5, were 1 to 2% in "football", where crosstalk was the most visible among the three pictures. However, just satisfying the allowable limits would not provide stereoscopic images of adequate quality; crosstalk of 1 to 2% is likely to serve as a criterion.

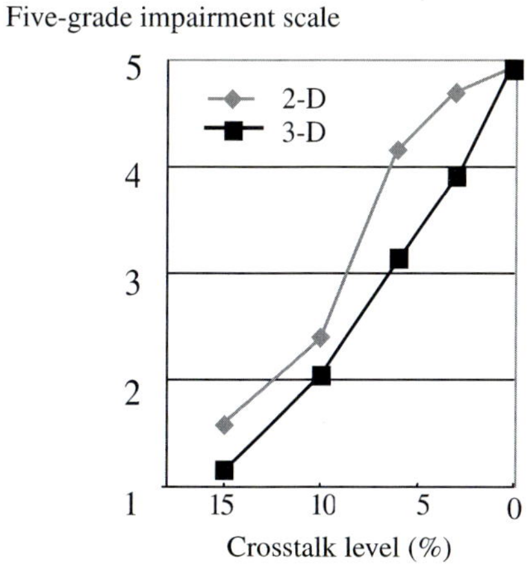

Fig. 1.10. An example of subjective evaluation results or crosstalk disturbance (3-D) and ghosting (simple double contour) disturbance (2-D) (Hanazato et al. [19])

Further, ghosting, a similar disturbance in the conventional 2-D television system, was compared with crosstalk. The results in Fig. 1.10 show that crosstalk affects stereoscopic images more than ghosting affects 2-D images.

Several approaches have been proposed to reduce such crosstalk visibility [20, 21]. The crosstalk can be cancelled to a certain extent by subtracting the left-eye and right-eye signals from each other with a coefficient corresponding to the crosstalk level.

1.4.4 Geometrical Distortions and Signal Differences

When shooting 3-D television programs, there can be geometrical distortions, such as vertical displacement, rotation error, and size inconsistency, between right and left images, caused by insufficient positioning of right and left cameras and/or different focal distances of these cameras.

The detection limits and tolerance limits of geometrical distortions have been studied by examining different images sampled from recent stereoscopic HDTV programs, and results [22] indicate no significant difference between ordinary images.

The detection limits have been reported as follows: right-left image rotation of 0.53°, size 1.21%, and vertical displacement 0.66%. As for the geometrical distortion of the right-left images, tolerance limits when viewing a PAL stereoscopic television image from a distance of 6 H were reported by CCETT as follows [23]: camera rotation error of 0.5° and lens focal distance of 1%.

The acceptable tolerance of contrast differences between the two channels of a stereoscopic camera was also studied by CCETT, and the following results [24] were reported as the maximum acceptable tolerance: signal-level differences of 1.5 dB (white level) and 0.1 dB (black level).

HHI also studied the visibility of luminance differences [25]. The visibility thresholds, where the evaluation value is equivalent to "perceptible but not annoying", were 3 to 6 dB when the luminance asymmetry pattern was stationary. When the pattern was moving the sensitivity increased by as much as 0.2 dB.

1.4.5 Depth Resolution and Other Related Issues

Measurements [26] of the spatial-frequency characteristics of stereosensitivity (depth information) show that spatial frequencies exhibit the characteristics of a bandpass type whose sensitivity peaks at 1 cycle/deg. For temporal frequencies, the sensitivity peak is located around 2 Hz. This is on the low side to a significant degree considering that the light/shadow pattern's sensitivity peaks at 10 – 20 Hz. These results point to the high possibility of depth-information compression.

For the purposes of a multiviewing point stereoscopic system, the image must change smoothly when the position of observation moves to the right or left. When the number of viewing points is small, the image may jump about, a phenomenon called "flipping," causing the picture quality to deteriorate. To avoid this problem, it has been reported that at least 60 viewing points are needed within the left-right eye distance (approx. 65 mm) in a stereoscopic film. If the image is not very large, however, nine viewing points are considered adequate [27].

1.4.6 Minimum Field Rate and Suitable Scanning Method for Field-Sequential Displaying

A single display device is used to create 3-D pictures by field-sequential displaying. One advantage of such a single-display system is that, since only one display device is in use, distortions or errors between right-left pictures are much less likely to occur. Further, such a system can be used both for 2-D and 3-D television.

An experiment has been conducted to investigate the detection limit for flicker in binocular parallax 3-D TV with field-sequential displays [28]. The critical fusion frequency (CFF) at which flicker becomes undetectable to the human eye (both eyes are alternately exposed to stimuli) is about 55 Hz for central fovea when the peak luminance through a liquid-crystal shutter is $180\,\mathrm{cd/m^2}$ and the horizontal field of view is 30°. It is desirable, from the viewpoint of eliminating flicker, that the field frequency of the time-division 3-D TV be set at about 110 Hz, double the CFF.

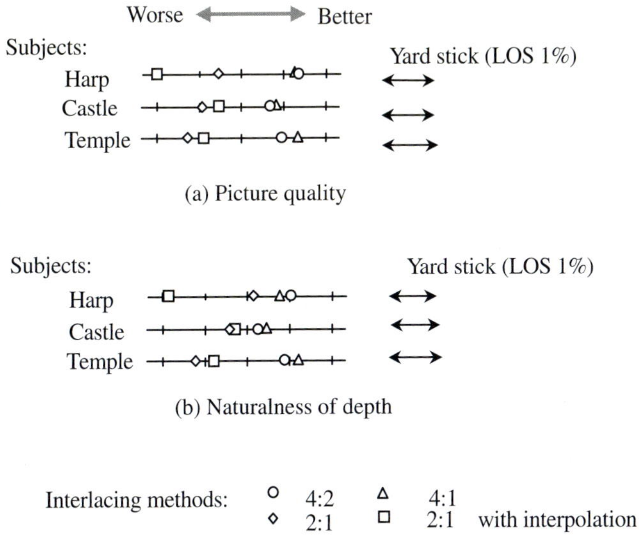

Fig. 1.11. Psychological effect of interlacing methods for field-sequential display (Okui et al. [29])

We examined the advantages and disadvantages of several scanning methods to identify the most suitable one for the field-sequential stereoscopic display technique [29]. An evaluation test using a system with 525 scanning lines and a 120 Hz field frequency was conducted. The results show that the picture quality of the 4:2 or 4:1 interlace method, which maintains high vertical resolution, is superior to that of the simple 2:1 interlace one (Fig. 1.11).

1.5 Shooting Method and Related Issues

Most of the stereoscopic image systems being developed, including stereoscopic HDTV, all reproduce 3-D images by providing both eyes of the viewer with images coming from two cameras representing human eyes. However, unlike real objects seen through human eyes, these 3-D images are greatly influenced by shooting and viewing conditions including camera separation, optical axis direction and focal length, as well as screen size and viewing distance. These conditions dominate such effects on the screen as sensations of massiveness, stereoscopy and reality or naturalness. Inappropriate shooting and/or viewing conditions could result in the reproduction of a 3-D image space that is quite different from reality and gives the viewer distorted sensations of massiveness and depth. Good examples here are the puppet-theater effect [30, 31] and the cardboard effect [32].

The distortion of reproduced 3-D space has been discussed by many researchers mainly from the standpoint of film [33–36]. Those studies are expected to play an important role, such as designing 3-D image systems. All these studies, however, were based on geometric calculation, and did not subjectively discuss the reproduction of 3-D image space. For broadcasting, it is necessary to discuss the reproduction of the 3-D image space of real objects from a subjective viewpoint as well as geometric analysis. The puppet-theater effect can be diminished by shooting an object in special ways, and some respective experiences were gathered in 3-D movie production. Still, little is known about this effect.

To understand it better, life-size images of a mannequin were projected onto a large 110-inch screen using a stereoscopic HDTV system and the size of images was subjectively evaluated [37]. The size of the mannequin images was maintained at life size and the viewing distance was kept at 4.4 m (3 times the screen height) throughout the experiment. The result shown in Fig. 1.12 indicates that the existence of a background significantly affects the subjective size of 3-D mannequin images, and a smaller subjective size is perceived as the depth sensation in the background is enhanced by the binocular disparity.

Another experiment with various focal length of the pickup camera lens showed that the perceived size of the 3-D image of a mannequin is smaller as the interaxial distance of the lenses widens and the focal length of the lens decreases [37].

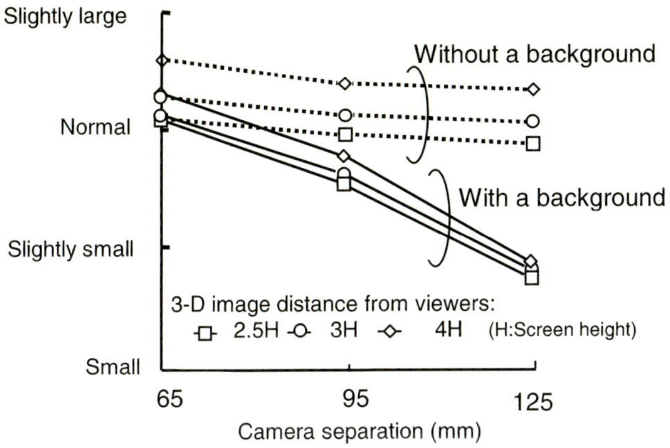

Fig. 1.12. Perceived size with various camera separations (Yamanoue [37])

Subjective evaluation test results indicate inconsistency between the depth information provided by the perspective of the lenses, and that provided by binocular parallax leads to the puppet theater effect. This effect is enhanced when the sensation of the depth of a background caused by binocular parallax dominates that caused by the perspective of the lenses used in shooting.

Through a geometrical-optics calculation the parallel-camera configuration does not cause such inconsistency under the specific shooting and viewing conditions called "the orthostereoscopic condition" (see Fig. 1.13). The orthostereoscopic conditions were examined by subjectively evaluating actually shot ordinary images. The result of this experiment [38] suggests that in order to satisfy the orthostereoscopic conditions subjectively, not only geometric conditions but also other factors such as screen size are involved.

In another experiment [39], we subjectively evaluated how the cardboard effect is influenced by such shooting variables as lighting conditions, lens selection, camera separation, and movements of shooting subjects during program production. The results show that binocular disparity represented by spatial thickness calculated from camera separation, lens selection, and convergence point is a dominant factor influencing the cardboard effect, and that the presence of movement and backgrounds are also factors.

1.6 Developments for Stereoscopic Displays in 1990s

Of 3-D technologies, 3-D display methods have long attracted much research attention. Here, some of the latest display technologies will be introduced. Although their current applications are limited to certain industrial applications (computer workstations, displays for communication terminals, interaction controlling, simulations), broadcasting and home entertainment are likely

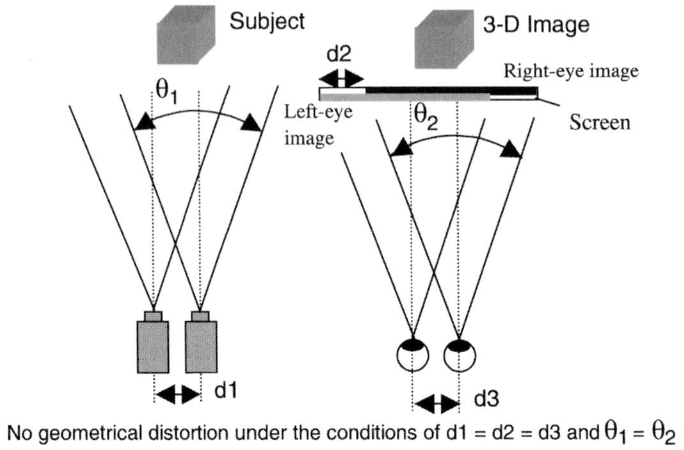

Fig. 1.13. Orthostereoscopic conditions

applications in the future. Most of them do not have HDTV picture quality, but some of them are likely to be applied to 3-D displays capable of showing full HDTV pictures in the near future.

Most of them are classified as autostereoscopic displays, which do not require special 3-D glasses, but nevertheless restrict the location of observers and the field of viewing. One of the methods proposed controls the viewing field in relation to the position of the observer's head. A multiviewpoint application is also possible by increasing the number of images spatially. With these multiviewpoints, the image will change as the observer's head moves, thus satisfying the conditions of motion parallax (look-around capability) to some degree.

On a flat panel display (FPD) precise positioning with the lenticular screen is possible as it has an accurate pixel arrangement. This is the case with a projection display using an FPD where images are projected onto a lenticular screen and also with a display where an FPD is directly attached to the lenticular screen. A series of experimental rear-projection stereoscopic displays using an LCD projector and a lenticular screen have been developed [40, 41]. One of them is an eight-viewing-point device with a 50-inch screen. All these displays use striped images produced by signal processing, instead of using the same number of projectors as viewing points.

When a lenticular lens is combined with a color LCD that has the RGB vertical stripe arrangement, chromatic separation will occur even if the observation point is correct. To solve this problem, a display using a mask with a special checkered pattern has been developed [42]. An input image of this display is a superimposed image with lateral stripes rather than one with vertical stripes. Pictures of this kind are compatible with video software used for the field-sequential method.

A display with a parallax barrier is also applied to stereoscopic television. However, it is more difficult with this method to secure necessary brightness than with the display using a lenticular lens array. Several improvements to resolve this problem have been made. In one example [43], a slip barrier is placed between the liquid crystal panel and the backlight, and an aluminum reflection film is then attached to the slit on the backlight side. It is reported that this arrangement increased the brightness on both sides by 1.4 times.

The development of an experimental stereoscopic LCD using this parallax barrier has been reported [44]. This system uses two LC panels put together, one serving as an image display plane and the other as a slit barrier. Since slit barrier patterns can be freely controlled with this structure, we can have an arbitrary number of viewing points from two eyes to multiple viewing points. It is also possible to make the barrier follow the movement of the head, and to display 2-D images at full resolution by eliminating the barrier altogether.

Ways to reduce moiré and crosstalk, including a double-barrier structure with an LC panel in between have also been reported. This type of 3-D display panel with a head-tracking system has been developed [45].

At Germany's HHI, a variety of autostereoscopic displays are being developed [46], including projection systems and flat panel displays using a lenticular screen, and others with LCD, DMD, or PDP. The chromatic separation mentioned earlier in this section, which occurs because of the RGB vertical stripe arrangement in an LCD, can be eliminated by making the pitch of the lenticular sheet correspond to each color subpixel of RGB, instead of each pixel. These stereoscopic displays use head-tracking methods according to their purpose. One head-tracking method precisely moves the lenticular screen itself back and forth and side to side, while another method controls the relative position between the observer and the display (the display is attached to the arm).

Other studies, such as image-shifting optics [47], have also investigated head tracking and several proposals have been made.

Another example is a stereoscopic display without viewing glasses that adopts a new head-tracking method [48]. Right and left images are displayed on each liquid crystal panel (LCD); viewing areas are formed for the right and left eyes through the use of a directional backlight for each LCD. This system is distinguished by the fact that as the observer's face image is used as the backlight pattern, viewing areas for the right and left eyes are automatically formed. The observer's face is detected by an image sensor, creating black-and-white images of the right-half face and left-half face. These images are then used as the backlights, which are enlarged by a convex lens (Fresnel lens) to form a viewing area. The right and left images are either combined by a half-mirror or time sequentially before they are displayed.

1.7 Coding and Signal-Processing Technology

Difficulties that often accompany the coding and signal processing of 3-D television signals are not a significant problem for the binocular television system. This is because existing television technologies can usually be applied to the two channels. Even so, demand is rising for more advanced image processing systems that utilize the 3-D structure of the object and scene. Such systems are attracting much research attention as they involve new technologies not used in 2-D image processing.

1.7.1 Coding Technologies

According to the Wang and Adelson classification [49], coding can be classified at three levels: low-level coding, mid-level coding, and high-level coding. Current technology is at the low-level. Ultimately, for ordinary pictures, we may be able to reach the high-level coding based on 3-D object recognition and other advanced technologies. Such a high level will take time to achieve, and it may be more realistic to concentrate on the mid-level coding, which offers good potential.

At the low-level coding, the element technology is a combination of motion compensation and waveform coding designed to reconstruct the original waveform. DCT, sub-band, and wavelets are the coding schemes of this class. At the mid-coding level, the element technology concerns processing of areas, layers, surfaces, depths, occlusions, and motions. At the high-level coding, the focus is on model-base coding by which the characteristics of 3-D objects are extracted by some means for coding.

The more advanced the method of coding, the greater the importance of technology in extracting and rebuilding information on the object's 3-D structure. It will be necessary to utilize not only video information, but also additional information called "metadata" including camera parameters and depth information. The next-generation coding method of this kind, which will not be discussed further here, would be essentially suitable for handling 3-D images or stereoscopic TV signals.

1.7.2 3-D Coding by MPEG

Simple simulcasting is a 3-D MPEG coding method that codes right and left images individually by MPEG. Right-left correlation is not employed by this method; rather, it simply uses two channels, one for the right image and the other for the left image.

Studies on the utilization of right-left correlation have been conducted in Japan, Europe, the USA, and Canada. This method has been added as a multiview profile by MPEG-2 (ISO/IEC 13818-2). Subjective tests using SDTV have been conducted, and the results of subjective evaluation tests [50] show that good ratings corresponding to the quality of "image impairment is perceptible, but not annoying" can be obtained for the multiview profile (left 6 Mbit/s, right 3 Mbit/s). Considering transmission channels of the same capacity (9 Mbit/s for right-left combined), subjective evaluation results of the multiview profile can be favorably compared with those of the simulcasting method (left 4.5 Mbit/s, right 4.5 Mbit/s). For the sake of compatibility, it is desirable to maintain the picture quality of one channel at a high level.

1.7.3 Disparity Detection in 3-D Pictures

Of the element technologies for 3-D coding, disparity detection is especially important. Generally, detection of correct disparity vectors is not guaranteed, and solutions cannot be obtained without some kind of restricting conditions. This is called the corresponding problem. Various approaches have been proposed for solving this problem, including the multiple-baseline stereo method using several cameras [51], a cooperative algorithm based on a global organization of local, interactive constraints [52], and the use of a four-layer neural-network model [53]. Other methods [54] detect disparity by enhancing robustness. Rubber matching [55] is a high-precision method for contour detection, where the contour and gradient of the contour are regarded as a rubber band; the amount of energy needed when this rubber band distorts is used in an evaluation function for the calculation. In this study, the matching error rates for each matching method are compared as shown in Fig. 1.14.

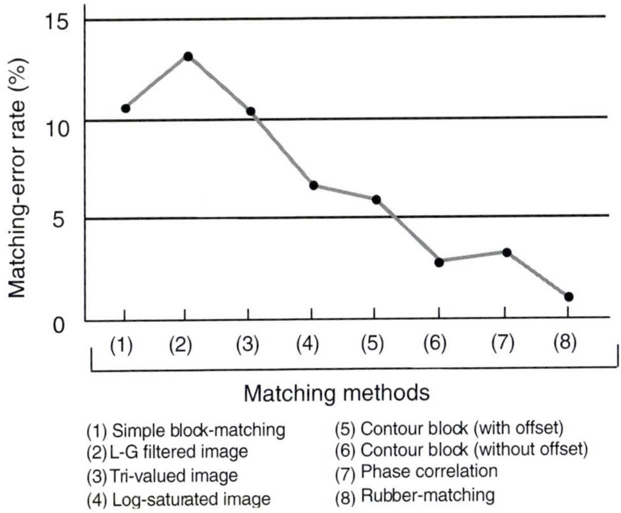

Fig. 1.14. Matching error rate for each matching method (Tadenuma and Yuyama [55])

1.7.4 2D-to-3D Conversion

Assuming that 3-D television broadcasting is fully operational, initially it may be difficult to provide enough stereoscopic programs for 3-D television on a regular basis. A 2D-to-3D technology for converting regular 2-D programs into 3-D programs is therefore required. The method described below has been studied periodically for 2D-to-3D conversion by the receiver. The earliest version of this technology was built inside wide-screen television models with field-sequential stereoscopic display functions.

With this method [56], still and moving pictures are generally processed differently. In the moving picture, motion vectors are detected. Next, a right-left eyes stereo-pair is created from the delay image controlled by the direction and size of the detected vectors. Processing is rather more complicated for the still image. Areas are identified by such information as contour, color, and contrast, giving rise to parallax. Stereoscopic images created from these two modes are then integrated by a special algorithm.

1.8 Equipment and Facilities for Stereoscopic HDTV System

The following discussion focuses on the present state of the stereoscopic HDTV system including its production and equipment.

1.8.1 Equipment for Program Production and Display

HDTV Systems – Cameras. At the initial stage of stereo HDTV development, tube-type cameras were employed. In order to create so-called pan focus pictures, the stereoscopic camera incorporated a supersensitive pickup tube called HARP, which is ten times more sensitive than normal saticon tubes (a recent "New Super Harp tube" is 640 times more sensitive than conventional ones). There were two types of pickup system with the HARP tube cameras [57]. In the first type, the right and left cameras are set side by side. In order to make the distance between the two lenses as close as possible to that of human pupils (approximately 6.5 cm), some improvements have been made to the cameras. The distance and convergence can be changed to some extent according to the camera lenses, scenes, and program production.

In order to shoot close-up objects, the second type of pickup system employs a half-mirror. The distance between the two lenses can be varied from 0 to 16 cm.

With this configuration, the scanning direction of the camera, which picks up the reflected pictures, should be the reverse of the normal direction. The transparency of the half-mirror is about 30%, however, the highly sensitive HARP tube recovers the loss of light and provides good focus both for far objects and near ones in the same frame. This is shown in Fig. 1.15a and b.

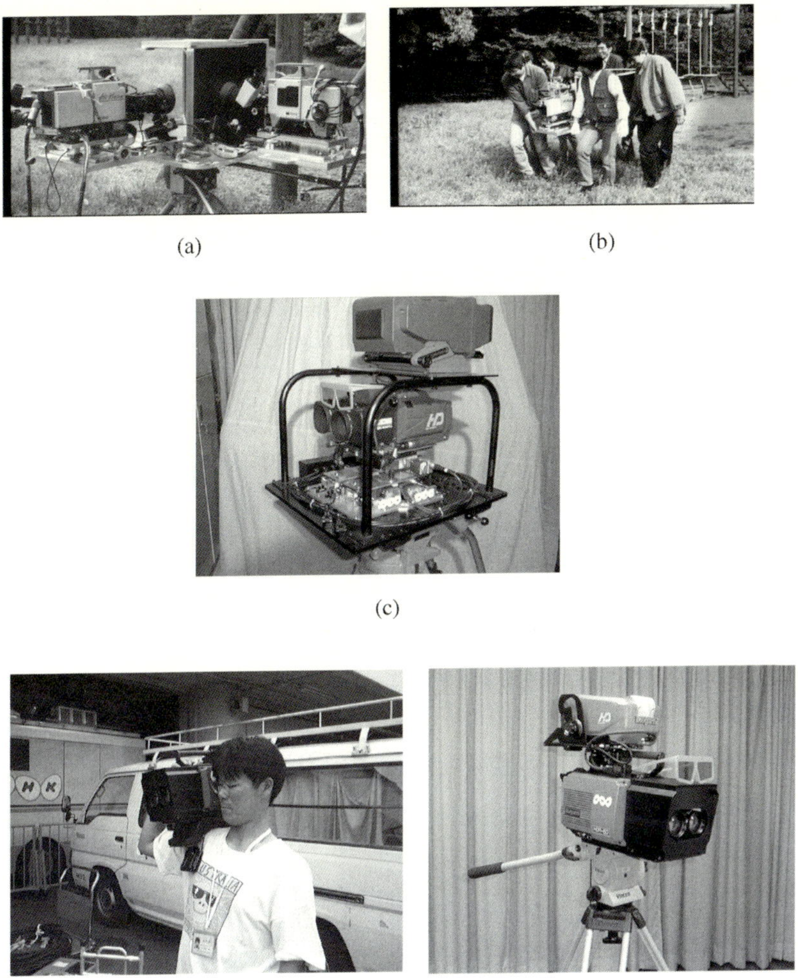

(a) (b) (c) (d) (e)

Fig. 1.15. Stereoscopic HDTV cameras: (**a**) and (**b**) First-generation stereoscopic HDTV camera with half-mirror (1988); (**c**) CCD type cameras equipped with 2/3 inch CCDs; (**d**) and (**e**) newly developed compact camera system

A stereoscopic HDTV camera setup shown in Fig. 1.15c is a conventional one each of which uses a 2/3-inch CCD. The distance between the lenses has been reduced to 65 mm. The use of 3-D cameras in actual shooting requires fine adjustment of the distance between the lenses and the convergence angle of two cameras. For such adjustment, these cameras can be mounted on a specially designed and fabricated base.

For monitoring purposes, these types of cameras are generally equipped with a 3-D monitoring system. The right and left signals switched on a field basis are displayed on a TV screen for the monitoring of 3-D effects by both directors and engineers. The field-sequential signals are also sent to the viewfinder of the cameras for convenience of camera operation. This instantaneous monitoring by video, repeatable recording by VCR, and easy processing by the electronic editing system are the most significant advantages over a conventional film-production system, particularly stereoscopic film production.

Recently, we have developed a new stereoscopic HDTV camera system capable of reproducing natural depth (see Fig. 1.15(d)) [58]. The lens axes of the two cameras are parallel and these two lines are kept apart by the distance between the average human pupils. The right and left cameras also zoom in and out with precision; a new zooming mechanism for 3-D shooting has been developed. This system is much smaller and lighter than conventional ones and significantly reduces the setup time for shooting.

Displays and Other Equipment. Usually, a stereoscopic HDTV display unit includes two HDTV projectors, such as those shown in Fig. 1.16; the right and left images are separated by the polarizing filters of the projectors and the polarizing glasses of the viewer. With a recent HDTV projector, it is possible to project pictures of up to 300-inch diagonal size or more. The projector comes in two types; one employs a CRT and the other a light valve such as an LCD. With the latter, new devices such as DMD are increasingly used as projector displays because of their very large light volume.

Fig. 1.16. Stereoscopic HDTV exhibition at NHK booth in NAB, 1991 (12-inch CRT projection units were used)

Since 1995, a stereoscopic 110-inch HDTV system without viewing glasses has been on ordinary display at NHK's studio park, which is open to the public, in Tokyo. This system features the practical application of a lenticular lens screen [40, 41]. The use of this screen contributes to less crosstalk and higher contrast, providing high-quality stereoscopic images. It is also easy to maintain.

Nagoya City has a hall with a giant 600-inch screen, on which an opera was displayed to the delight of the audience [59]. Polarization glasses are usually used at these facilities.

We developed an experimental field-sequential stereoscopic system for HDTV [29] with 4:2 interlacing with which it is possible to display high-quality stereoscopic HDTV pictures in full 1125-line vertical resolution without flicker. The display was utilized by the "Virtual Museum" system [60]. Virtual Museum is an interactive multimedia system for the appreciation of 3-D art objects, such as porcelain and sculptures. With the system, the viewer is able to see a representation of the object from various directions as if holding it in the hand.

We also developed an experimental stereoscopic display with 42-inch HDTV DC-PDP [21]. Grayscale representation on a PDP is achieved by subfields. In order to achieve a field-sequential stereoscopic display method, twice as many subfields as the usual 2-D display are required. However, this is difficult to implement because of the response speed of the PDP panels. At present, twelve subfields in a TV field are considered to be the limit of driving, in which case only six subfields are assigned to each of the right and left images. By using a newly developed stereoscopic light emission scheme, HDTV stereoscopic images were successfully displayed. The decay time of phosphor emission in PDPs is longer than that of CRTs, so the emission in the L period leaks to the R period and R to L.

To reduce the crosstalk, the signal processing shown in Fig. 1.17 has been employed. This method is suitable for signal processing on PDPs. Because the light-emission characteristics of PDPs is linear to signal level and the crosstalk can be canceled by subtracting the L and R signals from each other after inverse-gamma correction. It has been confirmed that the method improved the picture quality of displayed stereoscopic images though the minimum level of video signal must be limited to zero after subtraction.

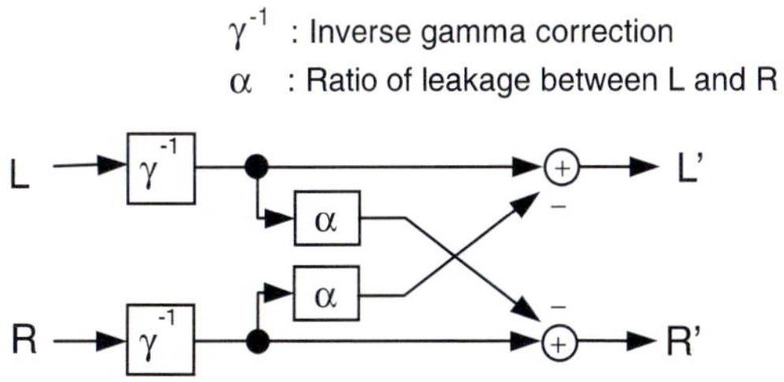

Fig. 1.17. Signal processing for canceling crosstalk on PDP display (Hamada et al. [21])

NHK have recently developed a new stereoscopic 50-inch PDP that has a larger screen, offers higher quality, and produces less crosstalk than the previous prototype. It was on display at NHK's exhibition in 2001 (see Fig. 1.18).

Fig. 1.18. A newly developed stereoscopic 50-inch PDP display (exhibited at NHK Broadcasting Center, Tokyo in May, 2001)

A compact bitrate-reduction processor for a stereoscopic HDTV VTR has been developed [61]. The processor can record stereoscopic HDTV signals (2 channel HDTV signals: right and left) on a conventional HDTV digital VTR using a bitrate-reduction technique of intrafield line-offset subsampling. The algorithm is shown in Fig. 1.19 and is easy to implement by a compact hardware and generates signals that are robust to the specific digital VTR circumstances such as the replacement of a damaged pixel. Furthermore, the processor enables both right and left eye channels to be edited simultaneously and facilitates playback using tricks such as slow motion and frozen images. Using this processor and a compact HDTV VCR (D6 VCR), program production is the same as for conventional HDTV.

1.8.2 Activities of the Stereoscopic HDTV Program

"Stereosaurus" CG images. "Stereosaurus" is a short stereoscopic HDTV computer graphics program produced in 1994 as a collaborative study by Media International Corporation (MICO), HD/cg NY and NHK Science and Technical Research Laboratories [62], and was first demonstrated at the NHK Open House. "Stereosaurus" depicts carnivorous dinosaurs chasing herbivorous dinosaurs, both of which lived on the North American Continent approximately 70 million years ago.

Computer graphics were an effective and efficient way to produce the images, with the visual points determined both freely and precisely. In "Stereo-

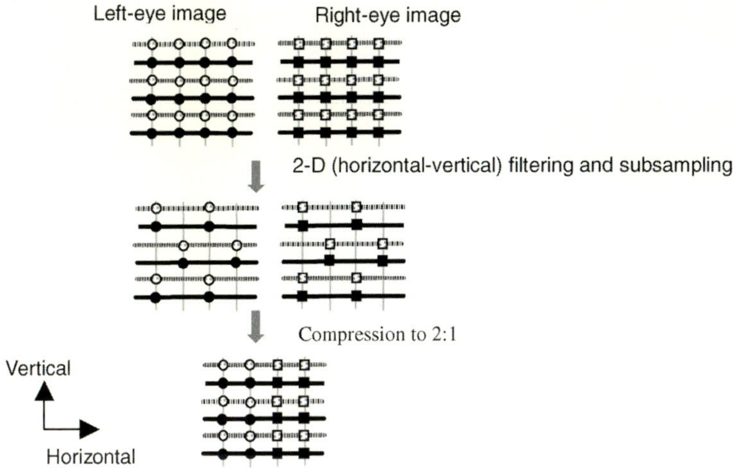

Fig. 1.19. Schematic algorithm of rate-reduction processor for VTR (Kuge et al. [62])

saurus", a localized 3-D sound-image corresponding to the images was also created using sound-image distance-control, in addition to the "bodysonic" sound system, which enables the listener to feel the vibrations. Furthermore, some sensation of fragrance and real airflows was authentically recreated in combination with the stereoscopic HDTV images and sound [63], thus making "Stereosaurus" a dramatic experience for viewers.

Stereoscopic Earth Images Converted from 2-D Images. To widen the scope of new visual services, it is also important to conduct more trials for expanding the possibilities of 3-D visual expressions and using new objects for 3-D shooting.

NHK's HDCAM was put aboard the space shuttle "Discovery" that went into orbit and shot realistic scenes of the earth in the fall of 1998. These pictures were then processed to produce 3-D images of the planet that looked very real and majestic [64].

A pair of stereoscopic images was created from the same continuous sequence of the earth's moving pictures; the one-eye images (left or right eye images) are derived from the time-shifted pictures while the other-eye images are unprocessed. This corresponds to a pair of viewing points tens of kilometers apart on the same orbit. Because the attitude of an orbiter craft is sufficiently stable, a pair of images could be considered to be shot with their optical axes in the same direction. From among these pictures of the earth, we chose ones that moved from left to right across the screen. For these 3-D images to look natural, it was necessary to adjust their horizontal and vertical positions.

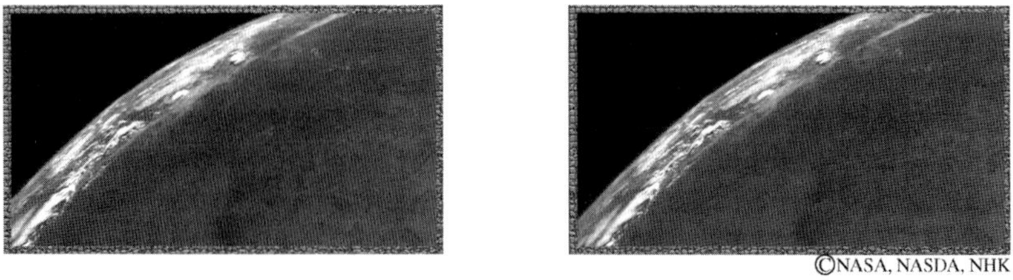

Left-eye image Right-eye image

Fig. 1.20. A pair of stereoscopic earth images converted from 2-D images

By taking the steps described above, we processed a series of earth pictures to produce stereoscopic effects (see Fig. 1.20). The two viewing points were about 32 km apart from each other, which would be equivalent to viewing a 26-m-diameter "earth" from a distance of 1.12 m with the naked eye. This picture was projected (front projection) onto a 120-inch screen. Fine geographical features of the earth such as land elevations and clouds that were drifting close to the surface are clearly seen.

This technique is widely used; it is similar to the one employed by Norling et al. [2] in an experiment using airplanes. The 3-D pictures created by this method, however, do not exist in nature and so will be very attractive to viewers.

Live from Nagano-Experimental Transmission of Olympic Games via CS. Live broadcasting is one of the strongest features of television, and this is even more true for stereoscopic HDTV. Live broadcasting by stereoscopic HDTV was tried during the Nagano Winter Games. In cooperation with other organizations, TAO showed stereoscopic HDTV images of the Games in Tokyo via a communications satellite. In this experiment in February 1998, right-eye and left-eye images were each sent at 45 Mbit/s via N-Star, a communications satellite, to NHK Broadcasting Center and Chiyoda Broadcasting Hall, both in Tokyo, where they were projected onto a large screen. These live 3-D images included those from hockey games and other events, strongly impressing the audience with their powerful sense of reality.

Medical Application. As described above, the stereoscopic HDTV system is being employed at theaters and amusement facilities across the country. It is also being experimentally used to support surgeries at hospitals and to help medical education (Fig. 1.21). For reduction of surgical intervention and surgical procedure, a new surgery system using stereoscopic HDTV and reference 3-D model such as CT images is being developed [65].

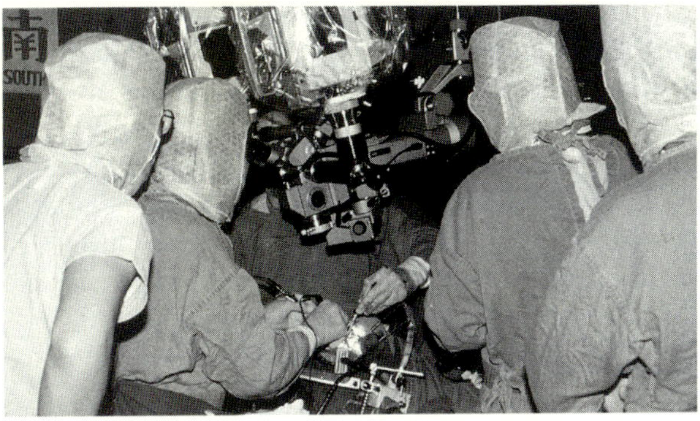

Fig. 1.21. Microscopic viewing of surgery using stereoscopic HDTV (at Shinshu University)

Stereoscopic Test Materials. In the area of research and application of 3-D images, it is often difficult to obtain appropriate video materials required by the researchers due to cost and facilities. This field of research would greatly benefit from broad access to a systematic set of 3-D pictures including high-quality natural images (people, scenery, etc.).

The Institute of Image Information and Television Engineers of Japan (ITE) is working to establish common conditions for assessing research results and has produced common test materials to promote research on stereoscopic images [66]. There are two types of these test materials: one for testing the stereoscopic visual functions of observers and the other for checking the quality of 3-D images.

Materials in the first group (those for testing the observer's stereopsis) are used to screen test subjects before the evaluation experiment starts. In the fine stereopsis test, as an example, the parallax is changed minutely to see if fine stereopsis is possible. As shown in Fig. 1.22, four circles are drawn at the four corners of each rhombus, and only one of these circles has binocular parallax. The observer chooses one circle that appears to pop out from the display surface and reports it.

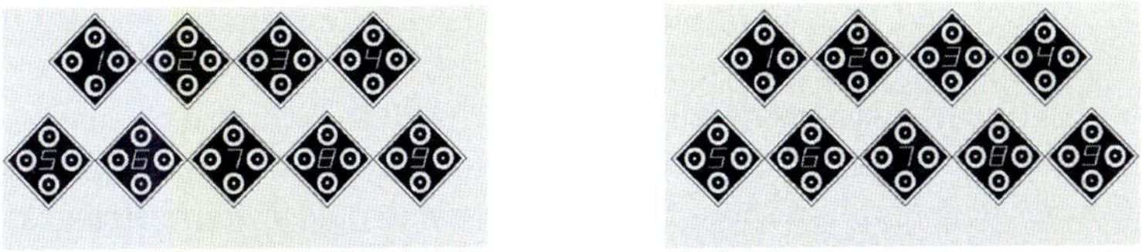

Left-eye image Right-eye image

Fig. 1.22. Pattern for fine stereopsis (Stereoscopic Test Materials, ITE [66])

Materials in the second group (those for assessing 3-D picture quality) were produced under controlled technical conditions, and can be used to test the performance of a variety of devices such as display units and coding algorithms. The second group contains natural images including still and moving sequences, which should be natural and easy to see. To satisfy this requirement, and also to avoid keystone distortion, orthostereoscopic conditions as described earlier in this chapter were employed. The main lens has a focal distance of 12 mm, which is most frequently used in program production; it has a field angle of 43° and is used for a 2/3-inch charge-coupled device (CCD) [67]. An example is shown in Fig. 1.23.

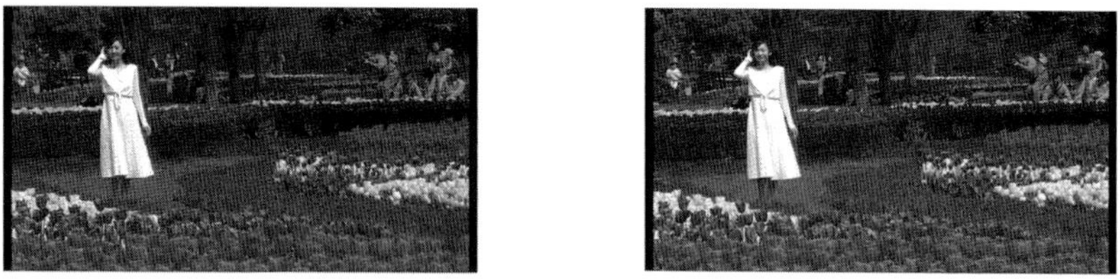

Left-eye image Right-eye image

Fig. 1.23. A sample test material "Tulip garden" (Stereoscopic Test Materials, ITE [66])

By setting the viewing angle also at 43°, which is the field angle of the main shooting lens, during observation (relative viewing distance of 2.2 H, HDTV aspect ratio, with H being the screen height), it is possible at least geometrically to recreate a distortion-free stereoscopic image space from the real space. For the observer to subjectively perceive this 3-D picture as a natural, distortion-free image, it has been reported that the display screen should be adequately large. In this study, the test materials were produced for a 120-inch screen (viewing distance: 3.3 m).

These materials are included in the new Recommendation ITU-R BT.1438 "Subjective Assessment of Stereoscopic Television Pictures". Although these pictures were originally produced for stereoscopic HDTV systems (1125/60/2:1 Rec.ITU-R BT.709), they can be converted and used for 3-D systems with SDTV quality and can be obtained from the ITE on digital HD VTR or CD-ROM in part.

In general, subjective assessment methods are used to evaluate psychological effects. That is, the methods described in Rec. ITU-R BT 500 could be used to evaluate the general picture quality of a stereoscopic system. However, when assessing 3-D picture quality, including the quality of depth reproducibility, viewing conditions must be set very carefully as they are influenced by such factors as shooting conditions, viewing distance, and screen size. To evaluate stereoscopic systems, the applicability of conventional methods for 2-D images has been examined [68]. Further studies will be necessary to develop a methodology that can be used to precisely evaluate the characteristic features of stereoscopic pictures.

1.9 Visual Fatigue Issue

The following are likely causes of eyestrain that an observer experiences from watching stereoscopic images.

a) Geometric differences between right and left images, and also luminance and chrominance differences
b) Parallax that is very large or that rapidly changes spatially or temporally
c) Inconsistency between accommodation and convergence

Factor a) is attributable to inadequate precision when setting the camera and display or incomplete uniformity of left and right signals in transmission and recording systems. To address these problems, it is necessary to set numerical tolerance limits as system requirements. At the moment, these tolerance limits are determined in relation to the degree of image impairment, but further studies on the effects of visual fatigue are needed in setting these limits.

Factor b) concerns the object position, shooting method, and viewing conditions. Scene changes and other means of expression in television programs are also relevant. The parallax of a stereoscopic image presented to the observer must be kept within a fusional area.

Factor c), accommodation-convergence contradiction, and factor b) are similar in that they both concern the functions of stereoscopic vision. This contradiction is a major phenomenon resulting from the difference between natural viewing and stereoscopic viewing; it is often cited as a cause of visual fatigue. Even so, there are not enough data to prove from experiments, and more studies are needed to identify the contribution of this contradiction to visual fatigue.

How visual fatigue can be measured objectively is an interesting research area. A variety of proposals (indices for measurement) have been made [69], and the growing trend in Japan now is to summarize the results of these studies [70, 71].

These are common topics not only for the stereoscopic system but for a number of others, except for a 3-D system with extremely numerous viewing points or holography. For stereoscopic images to play a greater role in broadcasting and for home entertainment, we need to continue studies in this area.

References

1. T. Okoshi: Proc. IEEE **68**, 548 (1980)
2. J.A. Norling: SMPTE J. **60**, 268 (1953)
3. T. Ito: *Foundations of the 3D Television*, (NHK Sci. & Tech. Res. Labs., Ohmsha, Ltd, Tokyo 1995) p. 93
4. M. Yasuda: *Foundations of the 3D Television*, (NHK Sci. & Tech. Res. Labs., Ohmsha, Ltd, Tokyo 1995) p. 9
5. W.J.M. Levelt: Br. J. Psychology **56**, 1 (1965)
6. G.E. Legge, G.S. Rubin: Perception and Psychophysics **30**, 49 (1981)
7. H. Kusaka: Proc. SPIE **1666**, 476 (1992)
8. T. Mitsuhashi: NHK Tech. Labs. Note, No. 256 (1980)
9. T. Hatada, H. Sakata, H. Kusaka: SMPTE J. **89**, 560 (1980)
10. S. Yano, I. Yuyama: SMPTE J. **100**, 14 (1991)
11. M. Wöpking: Subjective Assessments of 3D Versus 2D Presentation of Motion, Pictures for Large Screen Home Television, In *Proc. 1st Int. Symp. on 3-D Images*, (Paris, France 1991) p. 51–58
12. T. Shimizu, S. Yano, T. Mitsuhashi: Proc. SPIE **1666**, 457 (1992)
13. N. Hiruma, T. Fukuda: SMPTE J. **102**, 1137 (1993)
14. CCIR 1-1/11 SP1c-1/11, "Constitution of a system of stereoscopic television", 1958–1986
15. CCIR Rep. 312-4, "Constitution of a system of stereoscopic television", 1958-1966-1970-1978-1982
16. S. Yano: Displays **12**, 58 (1991)
17. S. Pastoor, I. Beldie: Proc. SID **30-3**, 211 (1989)
18. S. Pastoor: Human Factors of 3D Imaging: Results of Recent Research at Heinrich-Hertz-Institut Berlin IDW'95, *Proc. of 2th IDW*, Vol. 3, pp. 69–72, 1995
19. A. Hanazato, M. Okui, I. Yuyama: *Subjective Evaluation of Cross Talk Disturbance in Stereoscopic Displays*, The 20th International Display Research Conference, 2000, pp. 288–291
20. J. Konrad, B. Lacotte, E. Dubois: IEEE Trans. Image Process. **9**, 897 (2000)
21. K. Hamada, T. Yamamoto, T. Kurita, Y. Takano, I. Yuyama, A Field-Sequential Stereoscopic Display System with 42-in. HDTV DC-PDP, IDW98 *Proc. of 5th IDW*, PDP5-4, 1998, pp. 555–558
22. H. Yamanoue, M. Nagayama, M. Bitou, J. Tanada, T. Motoki, T. Mitsuhashi, M. Hatori, Systems and Computers in Japan, Vol. 29. No. 5, 1998
23. J. Fournier, T. Alpert: Admissible Rotation of One of the 2 Sensors of a Stereoscopic Camera, *Proc. 4-th European Workshop on Three-Dimension Television* 1993, pp. 41–48
24. J. Fournier, T. Alpert: Proc. SPIE **2177**, 45 (1994)
25. P. Beldie, B. Kost: SPIE **1457**, 242 (1991)
26. S. Nagata: (Title in Japanese), *Proc. 9th Joint Conference on Image Technology*, I-7, 1978, pp. 25–28
27. S. Pastoor, K. Schenke: Proc. SID **3**-3, 217 (1989)
28. H. Isono, M. Yasuda: SMPTE J. **99**, 136 (1990)
29. M. Okui, A. Hanazato, F. Okano, I. Yuyama: IEEE Trans. Circuit Systems Video Technol. **10**, 244 (2000)
30. R. Kutka: IEEE Trans. Commun. **42**, 29 (1994)

31. D.L. MacAdam: SMPTE J. **62**, 271 (1954)
32. S. Herman: SMPTE J. **80**, 539 (1971)
33. A.J. Hill: SMPTE J. **61**, 461 (1953)
34. E. Levonian: SMPTE J. **64**, 77 (1955)
35. R. Spottiswoode, N.L. Spottiswoode, C. Smith: SMPTE J. **59**, 249 (1952)
36. A. Woods, T. Docherty, R. Koch: Proc. SPIE **1915**, 36 (1993)
37. H. Yamanoue: SMPTE J. **106**, 225 (1997)
38. H. Yamanoue, M. Nagayama, M. Bitou, J. Tanada: Proc. SPIE **3295**, 111 (1998)
39. H. Yamanoue, M. Okui, I. Yuyama: IEEE Trans. Circuits Systems Video Technol. **10** (2000)
40. H. Isono, M. Yasuda, D. Takemori, H. Kanayama, C. Yamada, K. Chiba: Proc. of SPIE **1669**, 176 (1992)
41. H. Isono, M. Yasuda, D. Takemori, H. Kanayama, C. Yamada, K. Chiba: Euro Display '93 (SID), VIQ-P1, 1993
42. H. Morishima, H. Nose, N. Taniguchi, K. Inoguchi, S. Matsumura: Proc. SPIE **3295**, 193 (1998)
43. G. Hamagishi, M. Sakata, A. Yamashita, K. Mashitani, E. Nakayama, S. Kishumoto, K. Kanatani: *Proc. of the 15th International Display Research Conference on Asia Display '95*, 791 (1995)
44. H. Isono, M. Yasuda, H. Sasazawa: Japan Display '92, **S9-3**, 303 (1992)
45. G. Hamagishi, M. Sakata, A. Yamashita, K. Mashitani, M. Inoe: SID 00 Digest **54.3**, 1216 (2000)
46. R. Börner: Displays **20**, 57 (1999)
47. H. Imai, M. Imai, K. Kanazawa, K. Kubota: Proc. SPIE **2653**, 49 (1996)
48. T. Hattori: Proc. SPIE **1915**, 1 (1993)
49. J.Y.A. Wang, E.H. Adelson: IEEE Trans. Image Process. **3**, 625 (1994)
50. "Results of MPEG-2 Multi-View Profile Verification Test", ISO/IEC JTC1/SC29/WG11 N1373, 1996
51. M. Okumoto, T. Kanade: IEEE Trans. Pattern Anal. Mach. Intell., PAMI 15-4, 353 (1993)
52. D. Marr, T. Poggio: Science **194**, 283 (1976)
53. M. Fujii: Trans. IEICE **J73-D-II**, 1288 (1990)
54. W. Zheng, Y. Shishikui, Y. Tanaka, I. Yuyama: Trans. IEICE **J80-D-II**, 505 (1997)
55. M. Tadenuma, I. Yuyama: IWS 3DI'95, (1995)
56. T. Iinuma, H. Murata, S. Yamashita, K. Oyamada: SID 00 Digest 1212 (2000)
57. H. Yamanoue, M. Tadenuma, I. Yuyama: Proc. IBC'92, 533 (1992)
58. H. Yamanoue, M. Okui, F. Okano, I. Yuyama: IDW'99,3D2-3, 1071 (1999)
59. N. Iida, M. Mori, K. Suzuki, M. Nohara: ITE Tech. Report **20**, 19 (1996)
60. H. Mitsumine, H. Noguchi, K. Enami, Y. Ninomiya, H. Yamanoue, S. Yano, A. Hanazato, M. Okui: IEEE Trans. Broadcast. **42**, 200 (1996)
61. T. Kuge, S. Ogawa: S35-2, Asia Display '95 (1995)
62. H. Tamegaya, S. Sato, M. Sakakibara, K. Morihiro: IDW'95 **3**, 53 (1995)
63. H. Isono, S. Komiyama, H. Tamegaya: SID 96 Digest, 135 (1996)
64. M. Okui, H. Yamanoue, A. Hanazato, S. Ide, H. Hoshino, I. Yuyama: Proc. ITE Annual Convention, 2–7 (1999)
65. R. Mochizuki, K. Nannbu, H. Iseki, K. Takakura, T. Dohi: ITE Tech. Report **23**, 55 (1999)

66. ITE Test Chart Committee, *Stereoscopic Test Materials* (The Institute of Image Information and Television Engineers of Japan, 1998)
67. H. Yamanoue, M. Emoto, M. Okui, S. Yano, T. Yoshida: Proc. IDW'98, **3D3-2**, 815 (1998)
68. N. Narita, M. Kanazawa, I. Yuyama: Tech. Rep. IEICE, **EID98-174, IE98-165**, 63 (1999)
69. S. Saito: Proc. IDW'98, **3D3-1**, 811 (1998)
70. S. Nitta, T. Yambe: Proc. IDW'99, **3D1-1**, 1047 (1999)
71. K. Ukai: Proc. IDW'99, **3D2-1**, 1063 (1999)

2 Autostereoscopic Imaging Systems Based on Special Optical Plates

J.-Y. Son

2.1 Introduction

The basic principles of generating sense of depth in various three-dimensional (3-D) image-presentation methods can be categorized into the following: 1) drawing 3-D profiles on a scattering medium with a scanning laser beam [1], 2) presenting a time or a spatially multiplexed sequence of many different view images with a viewing zone forming optics [2, 3], 3) displaying interference fringes generated by the superposition of rays from each object point [4], 4) projecting or scanning layered images on a specially designed screen to create a volumetric image profile [5–7] and 5) inducing psychological effects with use of a super-large image-projection screen [8]. Each of these methods has some advantages and disadvantages. In Table 2.1, the performances of these methods are compared according to the conditions set by CCIR study program 1C-1/11: Constitution of a system of stereoscopic television (1958–86). The essence of the stipulation is that compatibility of 3-D image to the current 2-D image format and quality and comfortableness of viewing.

Among these methods, the method of presenting many different view images through the viewing zone, i.e. the multiview method, only can meet all conditions recommended by CCIR. Hence, most of the autostereoscopic imaging systems available today are based on the multiview image method [9–11] because its realization is relatively easy compared with the other methods due to its adaptability to the current plane image displaying technologies.

Since the term "autostereoscopic image" has the meaning of "presenting moving stereoscopic images without wearing special eyeglasses" [12], most of these systems employ optical plates, such as lenticular [13], parallax barrier [14], holographic screen [15] and Fresnel lens [16], as their main tools for forming (a) viewing zone(s) for inducing the depth sense without the eyeglasses. These plates are mostly for creating viewing zones with horizontal parallax only. The integral photography (IP) plate is also an optical plate for forming the viewing zone. With an IP plate, full parallax viewing zones can be created. Holographic grating array plates are also used to form the viewing zone [17, 18].

The viewing zone is a spatial location where viewers can see entire images displayed on the screen. It is formed such that rays from each view image in an image sequence that is aligned either under or on the optical plates, are getting together with and separated spatially from those from other view

Table 2.1. Comparisons between different 3-D methods

Method stipulation	Eye glasses stereo-scopic	Multiview	Holo-graphy	Volume-tric	Laser-beam scanning	Psycho-logical	Remark
Natural depth	✖ – ▲	▲ – ●	▲ – ●	▲	✖ – ▲	✖	
Viewer comfort	✖	▲ – ●	▲ – ●	▲	▲	✖	Within viewing zone
Group viewing	●	▲ – ●	●	●	●	●	
Compatibility :2D/3D	●	●	✖ – ▲	▲	✖	●	
No degrade picture	●	●	▲	●	✖	●	
Min. modification of video Standard	●	●	✖	●	✖ – ▲	●	
Moderate price	●	●	✖	▲	✖ – ▲	✖	

● : Possible ▲ : Some cases possible ✖ : Impossible

images in the sequence by the plates. The image sequence is composed of many different view images multiplexed either in time or spatially. The spatial separation of rays from different view images divides the viewing zone into many smaller sections corresponding to the number of different view images to be displayed. Each viewing zone section has the same horizontal size and is aligned side by side without any overlapping with its neighbor. The horizontal size of the viewing zone section should not be more than the distance between our eyes. Otherwise, viewers cannot locate their left and right eyes at two different viewing zone sections. Hence, no depth sense will be perceived. To make the viewer perceive depth sense, left and right eyes should be located to different viewing zone sections for acquiring the binocular parallax. The viewers can move their eyes within the viewing zone to see other view images. This moving provides motion parallax. It is obvious that the second method relies on both binocular and moving parallaxes of human eyes to provide the sense of depth.

The common problems related with the autostereoscopic systems based on the second method are that the size of the viewing zone is small for many viewers, no vertical parallax is provided and eye fatigue is probable due to the absence of real depth cues for accommodation and vergence of our eyes owing to providing only parallax information. The viewing zone size allows only a viewer to see the images. The lack of the vertical parallax is mainly caused by the enormous data amount contained in the multiview images in both horizontal and vertical directions. The consequences of lack of the vertical parallax are that the images seen at different vertical directions are distorted [19] and viewers need to fix their posture straight in the vertical direction. Despite all these problems, the method is still very attractive in realizing autostereoscopic imaging systems, because it is compatible with the current plane image systems and is relatively more developed than the other methods, and the optical plates for the systems are not difficult to find.

In this chapter, multiview image-processing methods, viewing zone forming principles of various optical plates and obtainable image depth, which are essential factors for designing autostereoscopic imaging systems, are described.

2.2 Distortions in Perceived Stereoscopic Images

The basic image unit in the autostereoscopic imaging systems is a stereo image pair that can be fused into a stereoscopic image to our eyes. This image pair provides binocular parallax to our eyes. However, with only the binocular parallax, the stereoscopic image perceived by the viewer will be distorted when the viewing conditions, i.e. position and stereo base are different from those of the photographic [19]. The distortions will be increased as the differences increase.

Distortions involved with the stereoscopic images are both introduced by psychological and geometrical reasons. Puppet-theater and cardboard effects are examples of psychologically generated distortions by the human-perception mechanism [20], however, they are also related with geometry, and keystone and nonlinearity are those of geometrical distortions in stereo image pairs introduced by the photographing and projection mechanisms [21]. The geometrical distortions are mostly caused by the stereo camera with the toed-in configuration. These distortions can be minimized by preparing stereo image pairs with a stereo camera having a parallel configuration with the stereo base equal to that of our eyes, and making photographing and displaying conditions such as image size, positions, and distances of both camera and projector from the image projection screen, equal [22]. However, in practice, difficulties of making photographing, displaying, and viewing conditions the same lead to the distortions.

2.2.1 Distortion Analysis

Distortions in the perceived stereoscopic images due to the differences in the conditions of photographing, displaying, and viewing can be quantified by

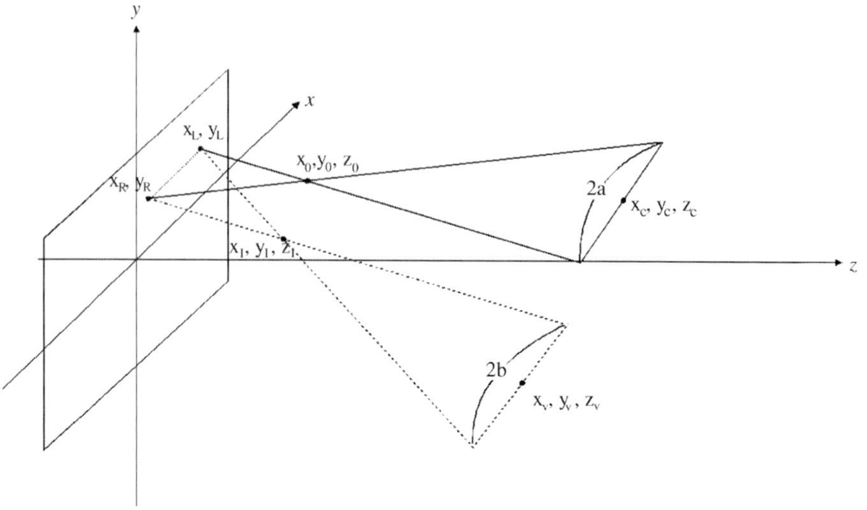

Fig. 2.1. Photographing and viewing geometry of a stereoscopic image pair: X_C, Y_C, Z_C and X_V, Y_V, Z_V represent camera and viewer positions, respectively

calculating the difference in the positions of the original object and the perceived image. Figure 2.1 shows the geometries of photographing and viewing. In Cartesian coordinates, it is assumed that a stereo camera, a viewer, an object and an image-projection screen are located at $(x_\mathrm{C}, y_\mathrm{C}, z_\mathrm{C})$, $(x_\mathrm{V}, y_\mathrm{V}, z_\mathrm{V})$, $(x_\mathrm{O}, y_\mathrm{O}, z_\mathrm{O})$, (0, 0, 0), respectively. The camera axis is normal to the screen surface and stereo bases of the camera and the viewer, i.e. intercamera and inter-eye distances are $2a$ and $2b$, respectively. It is also assumed that the viewer's two eyes and the camera are parallel to the x-axis. Since the stereo image pair recorded on a stereo camera can be considered as that projected on the screen surface with a stereoprojector having the same configuration and position as the camera, the image pair can be specified in the screen plane by $(x_\mathrm{R}, y_\mathrm{R})$ for right eye and $(x_\mathrm{L}, y_\mathrm{L})$ for left eye. In this case, $x_\mathrm{R}, x_\mathrm{L}, y_\mathrm{R}$ and y_L are represented as,

$$
\begin{aligned}
x_\mathrm{L} &= \frac{z_\mathrm{C} x_\mathrm{O} - z_\mathrm{O}(x_\mathrm{C} - a)}{z_\mathrm{C} - z_\mathrm{O}} , \\
x_\mathrm{R} &= \frac{x_\mathrm{C} x_\mathrm{O} - z_\mathrm{O}(x_\mathrm{C} + a)}{z_\mathrm{C} - z_\mathrm{O}} , \\
y_\mathrm{R} = y_\mathrm{L} &= \frac{z_\mathrm{C} y_\mathrm{O} - y_\mathrm{C} z_\mathrm{O}}{z_\mathrm{C} - z_\mathrm{O}} .
\end{aligned}
\tag{2.1}
$$

When the stereo image pair is viewed by a viewer at position $(x_\mathrm{V}, y_\mathrm{V}, z_\mathrm{V})$, the perceived image position $(x_\mathrm{I}, y_\mathrm{I}, z_\mathrm{I})$ is calculated as,

$$x_I = \frac{bx_O z_C + z_O(ax_V + bx_C)}{bz_C + z_O(a-b)},$$
$$y_I = \frac{by_O z_C + z_O(ay_V - by_C)}{bz_C + z_O(a-b)}, \tag{2.2}$$
$$z_I = \frac{az_O z_V}{bz_C + z_O(a-b)}.$$

From (2.1) and (2.2), the difference between positions in the perceived image and the original object is presented as,

$$x_O - x_I = \frac{(b-a)x_O + (ax_V + bx_C)}{(bz_C/z_O) + (a-b)},$$
$$y_O - y_I = \frac{(b-a)y_O + (ay_V - by_C)}{(bz_C/z_O) + (a-b)}, \tag{2.3}$$
$$z_O - z_I = \frac{(b-a)z_O + (az_V - bz_C)}{(bz_C/z_O) + (a-b)}.$$

Equation (2.3) shows that the difference in each coordinate direction has the same equation form, however, it depends on the object positions of the same coordinate direction and the z-axis. This means that the difference varies for different object positions. Hence, the perceived image will be distorted. For the case of $a = b$, i.e. the stereo bases of both camera and viewer's eye are equal, it can be seen that the difference in image size in each axis direction becomes proportional to the difference between the positions of the camera and the viewer, scaled by z_O/z_C. If the positions of the camera and the viewer are also the same, the difference will be zero, i.e. no distortion in the perceived image.

2.2.2 A Method of Minimizing the Distortion

The distortions in the stereoscopic images can be eliminated if a certain condition is met. In (2.1), when the viewer and the camera are positioned at z-axis, i.e. $x_C = y_C = x_V = y_V = 0$, (2.1) will be rewritten as,

$$x_R = \frac{x_O z_C - az_O}{z_C - z_O}, \quad x_L = \frac{x_O z_C + az_O}{z_C - z_O}, \quad \text{and} \quad y_R = y_L = \frac{y_O z_C}{z_C - z_O}. \tag{2.4}$$

If (2.2) is rewritten with x_R, x_L, and y_R

$$x_I = \frac{b(x_L + x_R)}{x_L - x_R + 2b}, \quad y_I = \frac{2by_R}{x_L - x_R + 2b}, \quad z_I = \frac{z_V(x_I - x_R)}{x_I - x_R + 2b}. \tag{2.5}$$

When projecting x_R and x_L, they are shifted horizontally to be opposite to each other by an amount Δ, x_R and x_L can be replaced by $x_R - \Delta$ and $x_L + \Delta$, respectively. Then, (2.5) will be rewritten as,

$$x_{\mathrm{I}} = \frac{b(x_{\mathrm{L}} + x_{\mathrm{R}})}{x_{\mathrm{L}} - x_{\mathrm{R}} + 2(b + \Delta)} ,$$
$$y_{\mathrm{I}} = \frac{2by_{\mathrm{R}}}{x_{\mathrm{L}} - x_{\mathrm{R}} + 2(b + \Delta)} , \tag{2.6}$$
$$z_{\mathrm{I}} = \frac{z_{\mathrm{V}}(x_{\mathrm{L}} - x_{\mathrm{R}} + 2\Delta)}{x_{\mathrm{L}} - x_{\mathrm{R}} + 2(b + \Delta)} .$$

By substituting (2.4) into (2.6), (2.6) is rewritten as,

$$x_{\mathrm{I}} = \frac{bx_{\mathrm{O}}z_{\mathrm{C}}}{z_{\mathrm{O}}(a - b - \Delta) + z_{\mathrm{C}}(b + \Delta)} ,$$
$$y_{\mathrm{I}} = \frac{by_{\mathrm{O}}Z_{\mathrm{C}}}{z_{\mathrm{O}}(a - b - \Delta) + z_{\mathrm{C}}(b + \Delta)} , \tag{2.7}$$
$$z_{\mathrm{I}} = \frac{z_{\mathrm{V}}\{z_{\mathrm{O}}(a - \Delta) + z_{\mathrm{C}}\Delta\}}{z_{\mathrm{O}}(a - b - \Delta) + z_{\mathrm{C}}(b + \Delta)} .$$

In (2.7), if $\Delta = a - b$, then

$$x_{\mathrm{I}} = \frac{b}{a}x_{\mathrm{O}} , \quad y_{\mathrm{I}} = \frac{b}{a}y_{\mathrm{O}} , \quad z_{\mathrm{I}} = \frac{a - b}{a}z_{\mathrm{V}} + \frac{bz_{\mathrm{V}}}{az_{\mathrm{C}}}z_{\mathrm{O}} . \tag{2.8}$$

Equation (2.8) shows that the perceived image size in the transverse direction will be magnified b/a times but in the depth direction the image will be shifted by the amount $(a - b)z_{\mathrm{V}}/a$ and magnified $bz_{\mathrm{V}}/az_{\mathrm{C}}$ times. If $z_{\mathrm{V}} = z_{\mathrm{C}}$, the perceived image will not be distorted, but it will be magnified equally b/a times in all three directions along with shifting in the depth direction by $(a - b)z_{\mathrm{V}}/a$.

When the projected image is magnified m times, if x_{R} and x_{L} in (2.5) are substituted by $mx_{\mathrm{R}} - \Delta$ and $mx_{\mathrm{L}} + \Delta$, respectively, the analysis will not be different from above. In this case, if $am - b - \Delta = 0$, x_{I} and y_{I} are the same as in (2.8) but z_{I} becomes,

$$z_{\mathrm{I}} = \frac{ma - b}{ma}z_{\mathrm{V}} + \frac{bz_{\mathrm{V}}}{maz_{\mathrm{C}}}z_{\mathrm{O}} . \tag{2.9}$$

In (2.9), if $b = ma$, $\Delta = 0$ and,

$$x_{\mathrm{I}} = mx_{\mathrm{O}} , \quad y_{\mathrm{I}} = my_{\mathrm{O}} , \quad z_{\mathrm{I}} = z_{\mathrm{O}}\frac{z_{\mathrm{V}}}{z_{\mathrm{C}}} . \tag{2.10}$$

If $z_{\mathrm{V}} = mz_{\mathrm{C}}$, then $z_{\mathrm{I}} = mz_{\mathrm{O}}$, i.e. the perceived image will be magnified m times in all three directions. Hence, there will be no distortions in the perceived image.

2.3 Methods of Presenting Images

The distortions in the perceived image can be eliminated in a few special cases as shown above, however, in most cases, it is impossible to eliminate

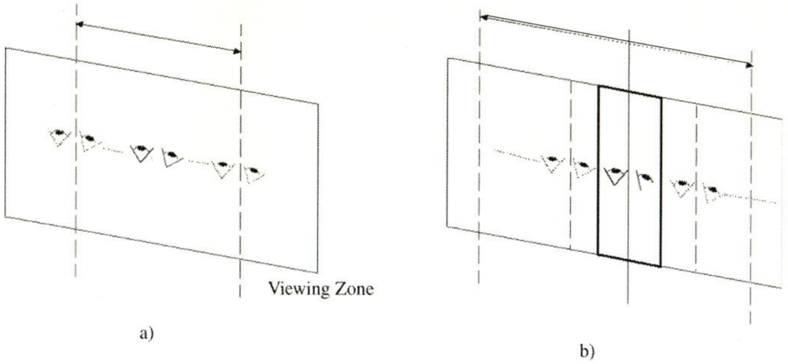

Fig. 2.2. Viewing zone changes with the viewer's head or eye movements: (**a**) Changing borderline in accordance with the viewer's head or eye movements; (**b**) shifting viewing zone in accordance with the viewer's head or eye movements

the distortions in the perceived stereoscopic images. A sure way of minimizing the distortions will be making the positions and the stereo bases of both the stereo camera and the viewer close to each other. Displaying different view images in accordance with viewer's viewpoint change, i.e. providing motion parallax can only do this.

To provide motion parallax, many different view images, i.e. multiview images are needed. These images can be presented in two different ways such as presenting a stereo image pair at a time to correspond to the viewer's viewing direction and presenting all view images through their corresponding viewing zone sections as shown in Fig. 2.2. There are two ways of presenting a different stereo image pair matched to the viewer's current viewpoint, such as dividing the viewing zone into two sections and then shifting the borderline between them [23], and shifting the entire viewing zone [24, 25] according to the viewer's head or eye movement.

To present all view images together, the viewing zone is divided into many equal width sections along the horizontal direction. The total number of the sections is equal to the total view number and hence each view can only be seen through its corresponding viewing zone section. Increasing the number of views will decrease either the amount of the borderline shift or the horizontal width of the section. As the number of view images increases, the perceived image can be changed more smoothly and continuously when the viewer shifts his/her head or position. If the width of each viewing zone section becomes smaller than our eye's pupil size, eye fatigue due to the absence of real depth cue can be reduced [26].

2.3.1 Multiview Image Sequences

To display multiview images, these images should be arranged as a specific image sequence for the real-time display. There are two known schemes of mixing multiview images as an image sequence to be displayed in real time on the autostereoscopic imaging systems; one is time-wise arrangement of the multiview images and the other is spatial arrangement.

For the time-wise arrangement, each view of the multiview images is sampled for a very short time period and then arranged with a specific order in a time sequence, and for the spatial arrangement, specific image columns of each view image are sampled and then arranged with a specific order in a spatial image sequence. The process of making each image sequence is called multiplexing.

Time Multiplexing. To display K different view images time sequentially, the display device should have the display speed of not less than $60K$ fields/s for the flickerless image at the usual TV brightness. The principle of arranging K different view images in a time sequence, i.e. the principle of time multiplexing is shown in Fig. 2.3.

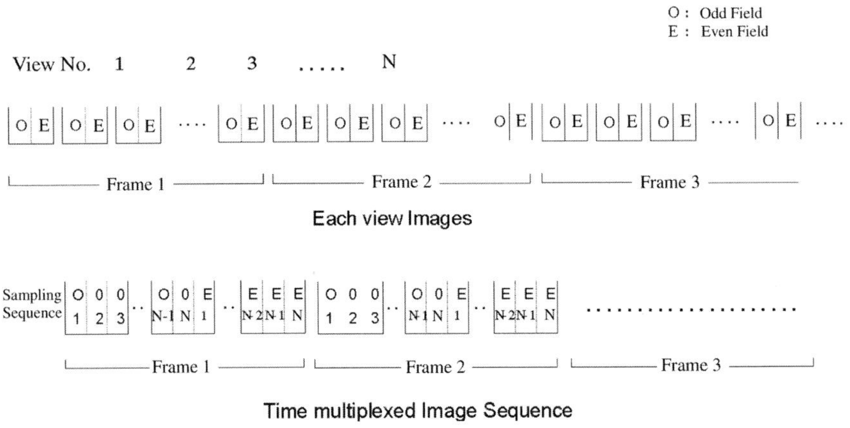

Fig. 2.3. Principle of time multiplexing

To make one frame of the time-multiplexed K different view images, each frame of the K different view images is divided into odd and even fields. Then the odd fields of each K different view images are sampled in time sequence in an order of odd field of view No. 1, odd field of view No. 2, – and odd field of view No. K followed by even field of view No. 1, even field of view No. 2, – and even field of view No. K, then repeats for the next frames. The sampling period for each field should be less than $1/(60K)$ s (for each second, K views $\times$ 30 frames /s$\times$ 2 fields/frame should be displayed). For example, to display 8 view images, a 480 fields/s sampling rate is required. This is the marginal speed of the commercially available high-speed CRTs for flickerless image. So each field should be sampled within 2.08 ms.

Spatial Multiplexing. Possible ways of displaying K different view images by spatial multiplexing are 1) projecting K different view images simultaneously with use of K equal projectors to the screen and 2) displaying multiview images on a single flat panel display device with a given frame speed.

The first case is applicable only for the projection-type autostereoscopic imaging systems. In this case, there will be no resolution loses in each view image but K projectors are required. For the second case, an example of the possible multiplexing schemes is depicted in Fig. 2.4 for the case when the horizontal resolution of both display device and each view image is the same, and equal to N times of K [27]. In this case, to make the 1st frame of the multiplexed K different view images, from the 1st frame of each view, $((n-1)K+1$, where $n = 1$ to $N)$th columns of view No. 1, $((n-1)K+2)$th columns of view No. 2, $((n-1)K+3)$th columns of view No. 3 and so forth up to $((n-1)K+K)$th columns, i.e. the last column of view No. K are sampled and then arranged in the same column order. The same sequence is repeated for the following frames. Other possible schemes will be arranging the same pixel lines for all different view images, i.e. $(n-1)K+1$ or a number between 2 to K, or half of the views $(n-1)K+1$ and the other half $(n-1)K+2$ or 3 to K, and so on. In this multiplexing scheme, the horizontal resolution of each view image will be reduced to 1/K times unless the display device has higher resolution than each view image. With this scheme, a compact system can be made because all view images can be displayed on a single display device.

Instead of the column-wise arrangement, rows of each view image can also be sampled and arranged in the same order as above. In this case, a special optical plate that can separate the vertically multiplexed multiview images into individual views and then direct them to their corresponding horizontally divided viewing zones. The optical plate for this scheme has the horizontal

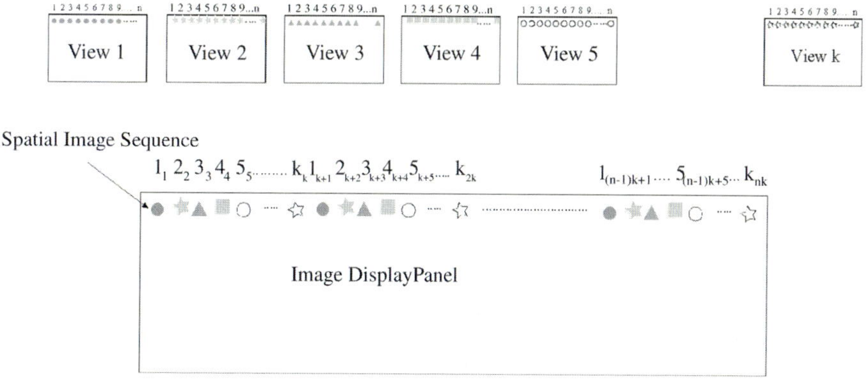

Fig. 2.4. Principle of spatial multiplexing

line grating patterns. Each line grating has its specific grating direction and period [18].

If the horizontal resolution of the display device equals K times that of each view image, K view images can be displayed without resolution loss.

2.3.2 Methods of Increasing View Numbers

Increasing view numbers will bring more comfortable and smooth 3-D images to the viewer. However, to increase view numbers it is necessary to increase the number of projectors or use either higher-resolution or higher-speed display devices. Increasing projector numbers will bring other problems of deteriorating image quality due to aberrations in optical plates and increasing bulkiness of the imaging system. Hence adopting higher-resolution or higher-speed display devices is a good solution, but currently available devices are limited in both resolution and speed to display multiview images.

However, there are ways of increasing number of different view images with the currently available display devices. Detecting viewer's viewing direction by head- or eye-tracking techniques and joining viewing zones of two multiview image channels together to make a single viewing zone are the possible ways. Displaying different view images in response to the viewer's viewing direction changes will allow as many view images as possible to be displayed. Many autostereoscopic systems are currently employing this technique [27–30].

The viewing zone joining method allows at least twice as many view images to be displayed with the currently available display devices. The method can be implemented by combining two multiview image channels where the images are in the time-multiplexed sequence. In this method, the viewing zone from each channel should be joined together without overlapping or a gap between them [31]. Figure 2.5 shows an example of an optical configuration that can combine two viewing zones together. Two 8-view CRT projection channels are combined by a folding mirror with a folding angle of 76° and then projected through a 16-strip LCD shutter by a projection lens, to the

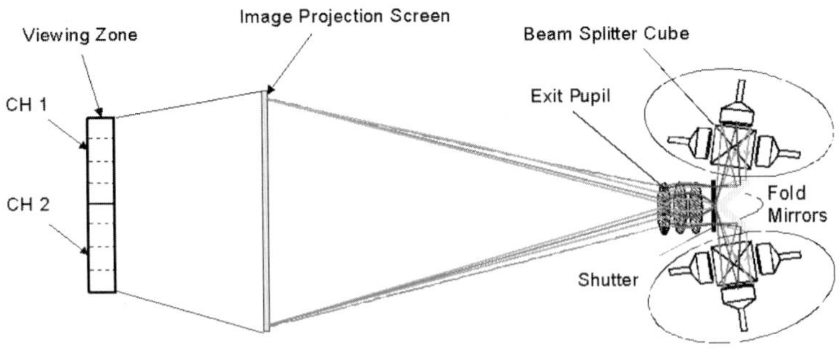

Fig. 2.5. Optical geometry of joining two different mutiview image channels to one

screen. The folding edge of the folding mirror is matched with the borderline between two center strips of the shutter that is located at the input pupil plane of the objective. Hence, the LCD shutter is divided into two 8-strip sections for the two CRT channels. This configuration allows 16 view images to be displayed.

2.3.3 Optical Plates for Viewing Zone Forming

In the autostereoscopic imaging system, a place where viewers can perceive stereoscopic images from the plane images displayed on the screen, i.e. a viewing zone, should be created. To do this, each imaging system needs an optical plate with a specific optical power. There are two types of optical plate: 1) forming viewing zones, and 2) forming viewing zones and working at the same time as the image-projection screen.

As for the optical plates of the 1st type, a lenticular plate, a parallax barrier plate, a grating plate and a microlens array plate like IP (integral photography) are well-known examples. These plates are usually laminated under flat panel display devices and separate different view images. For the case of a Fresnel lens, it is often laminated with LCD panels as a simple image forming optics for the viewing zone creation.

In this case, an extra optical plate that can separate different view images should be used together with the lens. A polarizing strip plate is an example of such a plate [32].

As for the 2nd type, there are a Fresnel lens, a spherical mirror, and holographic screens of both transmission and reflection types [33]. These plates are mostly for the projection-type systems. The Fresnel lens and the spherical mirror are classical optical elements for forming an image, focusing beams, and so on. The holographic screen is a holographic optical element having properties of a spherical mirror or a lens plus a diffuser. The lenticular and parallax barrier plates can also be used as the 2nd type. In this case, a diffuser plate and/or an extra lenticular plate should be used together with the lenticular and parallax barrier. The available size and characteristics of the plates are summarized in Table 2.2.

Figure 2.6 shows the actual samples of the plates. These plates reveal some undesirable characteristics. The lenticular plate is the most developed stereoscopic viewing plates at present, however, image clarity, and viewing-zone uniformity are limited. The parallax barrier plate is not much different from the lenticular, however, it has another problem of low image brightness. The microlens array plate has so far no standard form of manufacture and has a limited resolution. The Fresnel lens and the spherical mirror are well-known classical image-forming components, however, a Fresnel lens has a resolution problem and the mirror is too bulky and expensive, and the holographic screen has a limit in viewing zone size in the vertical direction.

Table 2.2. Characteristics of the Optical Plate

Type	**Variable size**	**Method of production**	**Image display surface**	
Lenticular	Several m^2	Mechanical optical	Direct projection	Discrete
Integral photography	Several 100 cm^2	Mechanical optical	Direct	Discrete
Parallax	Several 1000 cm^2	Film	Direct	Discrete
Holographic screen	Several m^2	Optical	Projection	Smooth (analog)
Fresnel lens	Several m^2	Mechanical	Projection	Discrete
Grating	Several 100 cm^2	Optical	Direct	Smooth

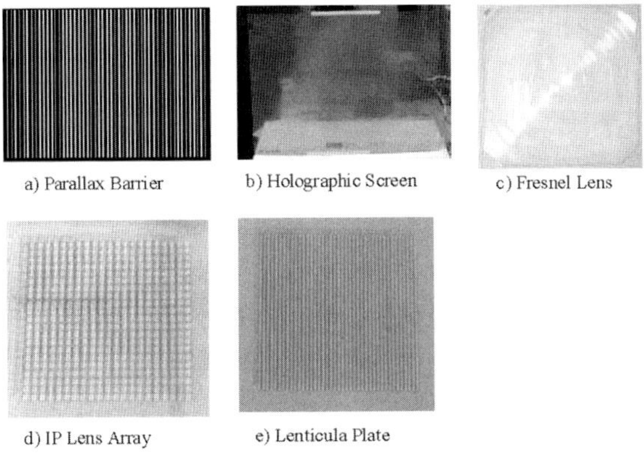

Fig. 2.6. Optical Plates

2.3.4 Methods of Forming Viewing Zone

Viewing zones are given as either images of pixels aligned under or above the optical plates, formed by the plates or the image of the output aperture of the projector's objective, formed by the optical plates. The former is for the cases of using lenticular, parallax barrier and IP, and the latter for other types of optical plates.

Viewing Zone as an Image of Pixels formed by an Optical Plates. In designing 3-D imaging systems with a lenticular or a parallax barrier plate, if the position of each viewing zone section is known, pixel locations of each view image can be defined for a given optical plate by backtracing the rays from the image because each optical element of the plate magnifies pixels aligned under or above it.

The relationship between multiview-image pixels and their corresponding viewing zones is shown in Fig. 2.7. Figure 2.7 depicts how the viewing zone

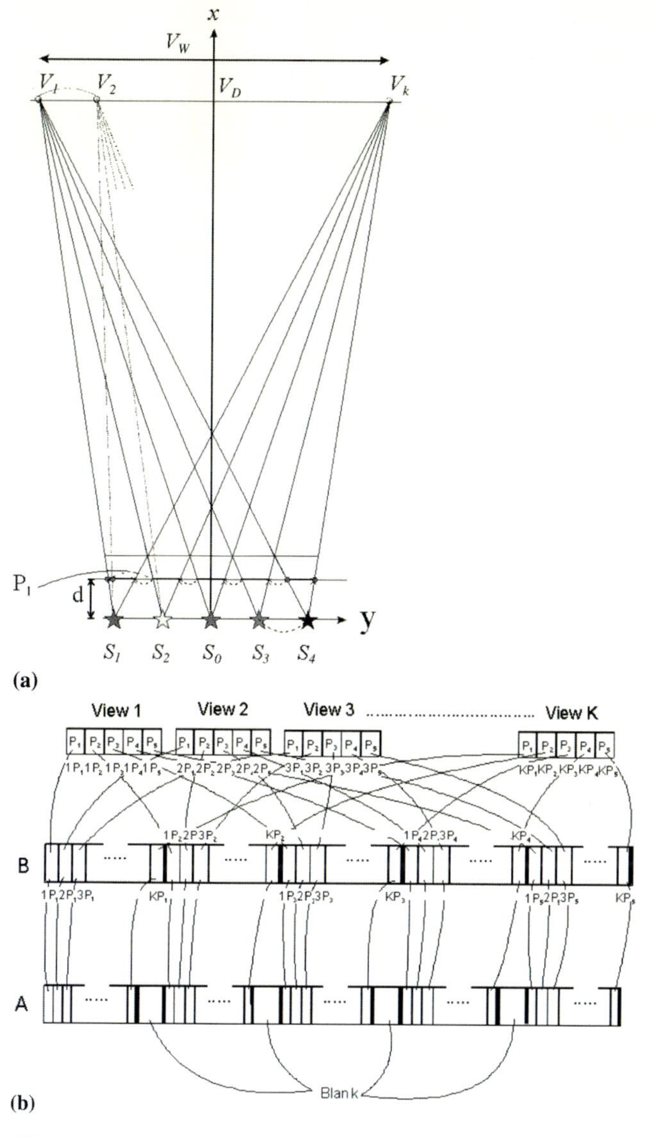

Fig. 2.7. Optical configuration of PLS array: **a**) viewing zone forming by the point light source array; **b**) the pixel arrangement in the image mask

is divided into K equal sections by K different view images numbered 1 to K from left to right. Each view image is assumed to consist of 5 pixels numbered 1 to 5, aligned in the horizontal direction. Figure 2.7a shows a point light source array with a one-dimensional image mask in front of it. The array consists of 5 equal point light sources specified by a star shape. Each source is identified by S_1, S_2, S_0, S_3, and S_4 and aligned along the y-axis with a distance between neighboring sources, p. In Fig. 2.7a, let us assume that S_0 is in the origin of the x and y-axes, and the viewing zone with a width V_W, is located at a distance in the x-axis direction from S_0 and is segmented into K equal size sections corresponding to K different view images. V_1 and V_K specify left and right edges of the viewing zone, respectively, and they are symmetric along the x-axis. V_1 and V_2 define the viewing-zone section VS_1 for the view image No. 1, V_2 and $V_3 VS_2$ for the view image No. 2, and V_K and $V_{K'} VS_K$ for the view image No. K. When lines are drawn from each PLS to V_1 and $V_{K'}$, these lines divide the horizontal line at position A into 5 equal sections. The position A is defined as that of the first intersection points between these lines. For the horizontal line at position B, it is segmented into 5 equal sections with a blank region between each of them. These blank regions have the same size and transmit no light from the PLS to the viewing zone. The position B can be any arbitrary location in-between positions A and PLS. When VS_1, VS_2 ... and VS_K have the same size, if lines are also drawn from each PLS to V_2, ... and V_K, these lines again divide each of the 5 equal sections into K equal subsections. The relative location of each subsection in each section is the same as that of the corresponding viewing zone section in the viewing zone, and the subsection size is proportional to its corresponding viewing-zone section size. Hence, the one-dimensional image mask at positions A and B should have the same structure as the viewing-zone line at V_D.

Figure 2.7b shows the pixel arrangement in the mask. In the mask, each section is working as a pixel cell. If each section and the K subsections are identified by numbers 1 to 5 and 1 to K from left to right, respectively, in the same way as the viewing-zone section, it can be said that the Nth ($N = 1$ to 5) pixel from the Lth ($L = 1$ to K) view image fills the Lth subsection of the Nth section. Each section is illuminated by its corresponding PLS. To make this kind of mask, a very high resolution display device is needed.

If the mask at position B is distance d apart from the PLS array, by the similar triangle relationship, the widths of each section, W_S and each blank region, W_B are calculated as,

$$W_S = \frac{dV_W}{V_D} \text{ and } W_B = P - \frac{d}{V_D}(P + V_W) . \tag{2.11}$$

In (2.11), $W_S + W_B$ represents the width of the mask corresponding to P, i.e. the distance between neighboring light sources. It is given as,

$$W_S + W_B = p\left(1 - \frac{d}{V_D}\right) . \tag{2.12}$$

When the mask is located at A, $W_B = 0$, then d and W_S are calculated as,

$$d = \frac{pV_D}{(V_W + p)} \text{ and } W_S = \frac{pV_W}{V_W + p} . \tag{2.13}$$

The size of each subsection in each section is given as W_S/K. The sizes of viewing-zone sections, $VS_1, VS_2, \ldots$, and VS_K can not be more than our eye distance to eliminate the chances of putting the viewer's two eyes in the same viewing-zone section. If the eye distance is assumed as 6.5 cm, the viewing-zone width V_W can not be more than $(6.5 \times K)$ cm. The viewing-zone size increases linearly in proportion to K.

The viewing zone specified by VZ_0 in Fig. 2.7 is not the only viewing zone for the configuration. It is just one of the viewing zones, located at the center. There will be many other viewing zones with a size V_W along the extended line of $\overline{V_1 V_{K'}}$ due to PLS illumination to other sections of the mask. For example, the illumination of the mask sections 2 and 1, and 4 and 5 by PLS S_0 will create viewing zones VZ_{-1} and VZ_{-2}, and VZ_1 and VZ_2, respectively as shown in Fig. 2.8.

The gap between each viewing zone has a size of $V_D W_B/d$. Hence, when $W_B = 0$, i.e. the image mask is at position A, no gap between neighboring viewing zones appears. In the border regions between neighboring viewing zones, there will be an abrupt view-image change from view No. 8 to No. 1. Hence, the depth of the perceived image will be reversed, i.e. the perceived image will be pseudoscopic. This pseudoscopic viewing regions can be diminished by increasing the gap between neighboring viewing zones more than our eye distance, i.e. $V_D W_D/d > 6.5$ cm.

Figure 2.9 is an extended view of Fig. 2.7a. If f is the distance between the PLS array and the line connecting crossover points at position A', it is expressed as,

$$f = \frac{pV_D}{(V_W - p)} \,. \tag{2.14}$$

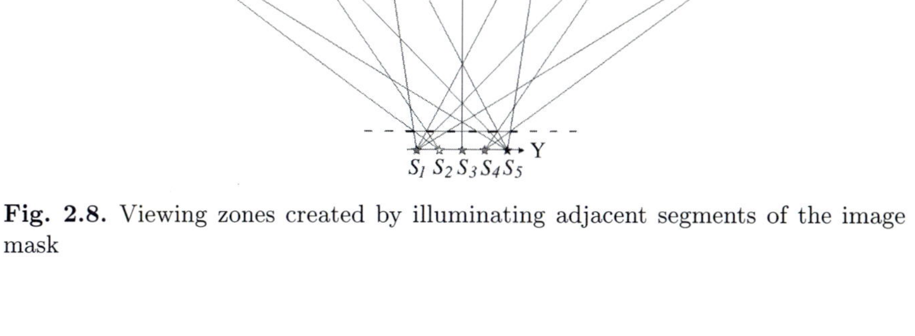

Fig. 2.8. Viewing zones created by illuminating adjacent segments of the image mask

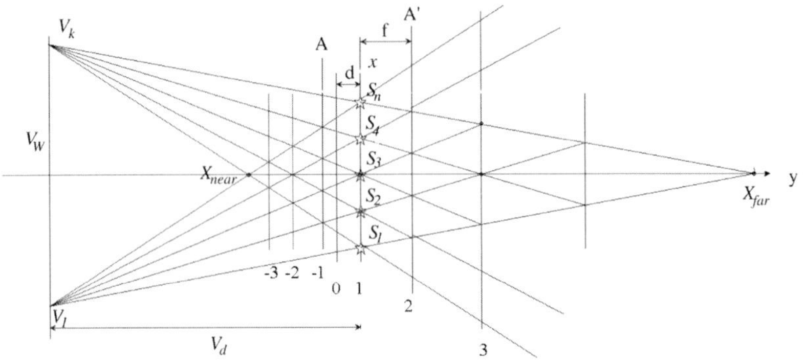

Fig. 2.9. Image depth obtainable with PLS array

In this case, the distance between neighboring crossover points, W'_S is calculated as,

$$W'_S = \frac{pV_W}{V_W - p} \, . \tag{2.15}$$

W'_S and W_S in (2.13) have the following relationship:

$$W'_S = W_S \frac{V_W}{V_W - p} \, . \tag{2.16}$$

Since triangles made of connecting each PLS and the line sections in positions A and A' are similar to each other, changing the line sections in positions A to A' will not make any difference in the geometrical point of view, if the sections in position A are magnified by $V_W/(V_W - p)$ times and rotated 180°. When this change is done with the image mask, each PLS can be replaced by a pinhole as shown in Fig. 2.10b without any change in geometry.

The basic performance of the configurations in both Fig. 2.10a and b will be the same. The replacement of the pinhole by a microlens with a diameter p and focal length f, as shown in Fig. 2.10c, will also make the basic performance unchanged. Figure 2.10c will make the image brighter than Fig. 2.10b because of a larger open area. The configurations in Fig. 2.10b and c will work both with or without illumination from the backside of the image mask. If the image mask has a 2-D shape with vertically aligned pixel lines of different view images, the pinhole array should be replaced by a slit array and the microlens array by an array of cylindrical lenses having a pitch p and focal length f. The slit array makes the parallax barrier plate and the cylindrical lens array, the lenticular plate. Since the basic performances of the PLS array, the parallax barrier plate and the lenticular plate are the same, all the derivations done for the PLS array will be equally applicable for the parallax barrier and the lenticular plates.

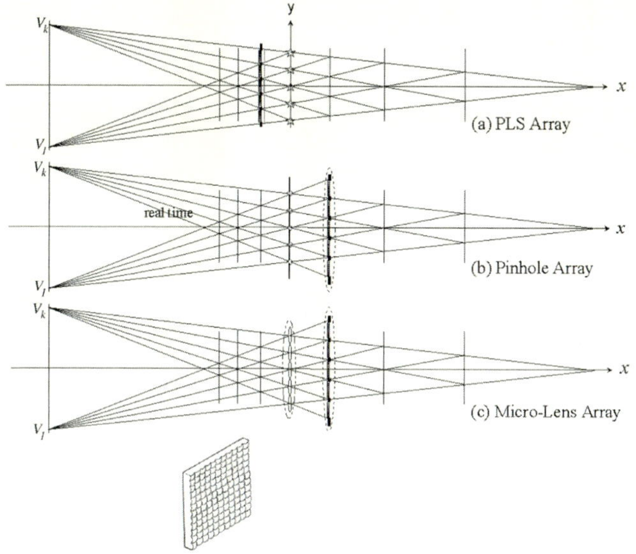

Fig. 2.10. Optical configurations of PLS array, pinhole array and microlens array: they perform the same roles and functions

With use of 2-D arrays of PLS, pinhole and microlens, along with the mask having 2-D arrangement of pixels, i.e. making the mask have a similar pixel structure in both vertical and horizontal directions, full parallax images can be displayed with the configurations in Fig. 2.10. Figure 2.11 shows the pixel structure of the image mask for the full parallax image consisting of

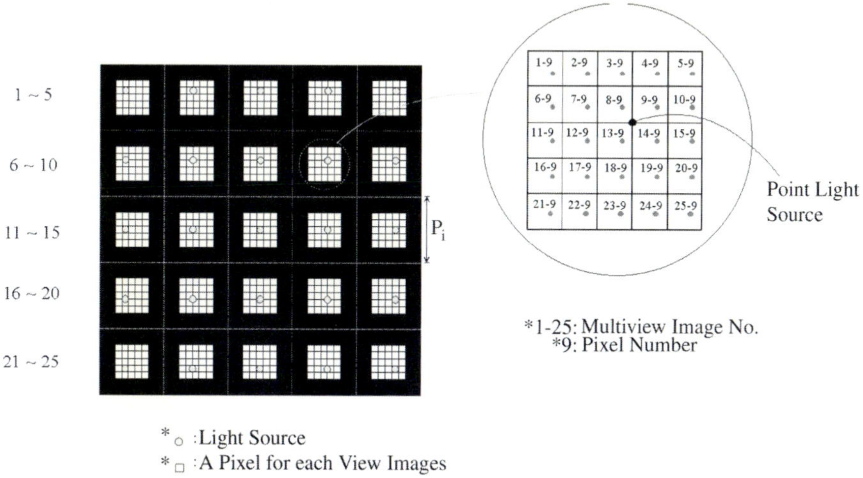

Fig. 2.11. Two-dimensional image mask structure for 55 view images

25 camera images. The figure in the circle is the magnified view of the ninth section of the image mask. The 2-D array of the microlens is just the IP plate. But in the conventional IP, in each section of the image mask, an image of the object, which is seen at each element lens position in the IP, is presented. The viewing zone is defined in the region where the field of view of each element lens in the IP plate is overlapping together. Hence, the viewing angle, i.e. the angle within which a viewer can move around is equal to the field of view of each element lens.

Viewing Zone as the Image of the Projector's Objective Exit Pupil. For the projection-type autostereoscopic imaging systems, an optical plate like a Fresnel lens or a holographic screen or a spherical mirror is used both as an image-projection screen and as a viewing-zone-forming optics. Among these plates, the Fresnel lens and the spherical mirror are simple imaging optics. Hence, the viewing zone is given as the image of the exit pupil of the projector's objective, formed by the plates. For the case of a holographic screen, the viewing zone is defined by the convolution of the exit pupil of the projector's objective and reconstructed object image. More explanations on the viewing-zone-forming principle of the holographic screen are given in the following section.

To display many view images, many projectors equal to the number of the view images to be displayed are required. These projectors are aligned behind the optical plate with an equal distance between them and then project images on the plate simultaneously. The optical plate images each projector's objective lenses in front of it into a neighboring space as a viewing zone. This is shown in Fig. 2.12. In this case, there should not be a gap or any overlapping between each viewing zone, and the viewing-zone size should not exceed our eye distance. This viewing-zone structure and the order of each

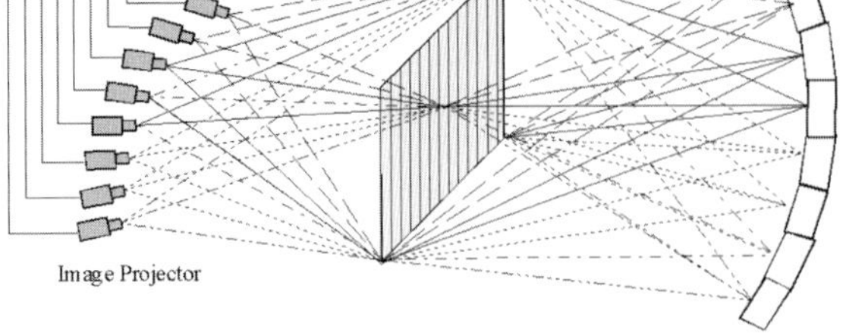

Fig. 2.12. Viewing zones for mutiprojectors

viewing-zone section will not be different from that in Fig. 2.7a. The viewing zone of each projector corresponds to each viewing-zone section.

For the case of using a very high speed display device like a CRT to display a time-multiplexed image sequence, a large-aperture objective lens has to be used to create a viewing zone with a reasonable size. In this case, the exit pupil of the objective is segmented into many vertical strips corresponding to the multiview image number. The segmentation is done by positioning a switching device such as an electro-optic shutter [34] *or* an optical scanner [35] either at the input pupil or at the exit pupil of the objective.

The electro-optic shutter consists of many vertical strip shutters aligned without gaps between them. The widths of the strip shutters are the same and they are parallel to each other. Each stripe shutter is imaged by the optical plate to form a viewing zone for a corresponding view image displayed on the display device. Hence, the total number of the strips in the shutter is the same as that of multiview images to be displayed by the system and each strip is ON only when its corresponding view image is appearing on the display device. Otherwise, the strip shutter is OFF. Since the viewing zone is just the magnified image of the electro-optic shutter, its structure is the same as that of the shutter. Hence, the appearance of each viewing-zone section will follow its corresponding strip shutter's ON/OFF sequence. This is depicted in Fig. 2.13. The viewing-zone sections are sequentially repeating ON and OFF actions.

The maximum number of multiview images being displayed by the system is solely determined by the response times of the display devices. The switching device should support the device speed. Since each view image is projected through a corresponding stripe shutter, if the width of this strip shutter image is smaller than our eye distance, the viewer's eyes can see different view images. Hence, 3-D images are perceived by the viewer. The

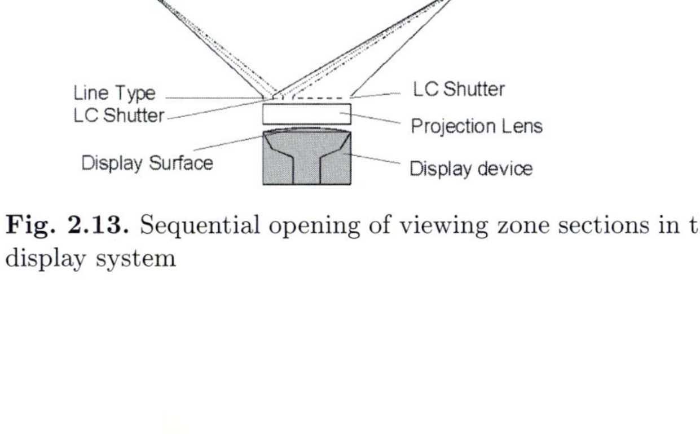

Fig. 2.13. Sequential opening of viewing zone sections in time-multiplexing image display system

optical scanner provides a fast mean of dividing the exit pupil virtually. The implementation of viewing-zone sectioning with a shutter device requires a large-aperture projection objective to view images with comfort. Larger viewing zones create more comfortable viewing conditions by providing the viewers more freedom of movement.

2.4 Holographic Screen

As an image-projection screen, the holographic screen (HS) will provide the highest resolution among 3-D image-projection screens available today because of its analog nature. The first HS was introduced in the year 1973 [36]. However, the full-color versions appeared in 1990. There are two holographic-screen types; reflection and transmission. The reflection type allows many separated viewing zones to be created with a single projection objective, for many viewers. The HS requires no mechanical process of manufacturing. A simple photographic process is enough to make it.

2.4.1 The Property of Holographic Screen

The holographic screen is a holographic optical element designed to work as a screen for the projection of 3-D images. It can be designed to have either the property of a spherical mirror or a diffuser plus a lens in the recording process. Hence, HS possesses a converging power. This converging power is additional to that by the HS's hologram nature. HS is a kind of imaging hologram.

The HS designed to have the property of a diffuser plus a lens is called a transmission type, a reflection type has the property of a spherical mirror. The transmission type is made on a usual holographic photoplate with a diffuser as an object. However, the high dispersive power of the holographic gratings recorded on the HS makes the full-color images projected on it disperse. Hence, the images on the transmission-type HS change colors as the viewing direction changes. To overcome this color dispersion, HS is made such that all spectral components of the reconstructed image overlap at a region where a viewing zone will be formed, by using a long narrow diffuser as an object [37].

The reflection-type HS is recorded on a holographic photoplate with emulsion layer thicknesses in the range of 5 to 100 μm, with a spherical mirror as an object. Its spectral selectivity, i.e. reflection bandwidth is too narrow to reflect full-color images.

To make a full-color image projection possible, either stacking three screens tuned to each of the three primary colors together [38] or forming a chirp-type fringe pattern in the active layer of a HS that was recorded with a diffused laser beam [39] can be done. Stacking three primary color HSs is an easy way of making a full-color HS because each primary color HS can be separately recorded and then stacked together. However, this screen has the problem of matching spatially different color beams reflected from each HOE at the viewing zone. The chirping process is to make the color separation in diffracted beams smaller and the recording process makes the HS have a diffuser property. Hence, the separated color beams are easily mixed together to form a full-color region. However, making a large size screen is difficult because of the problems related with recording the screen with a large-diameter spherical mirror of a required radius of curvature. The chirping process can currently be applied only to the DuPont's photopolymer film.

2.4.2 Forming the Viewing Zone

The full-color transmission-type HS is a hologram recorded with a long narrow diffuser as an object. However, for the full-color image display, the long axis of the diffuser is aligned on a straight line with the reference beam point and to make it near normal to the photoplate, as shown in Fig. 2.14 [37].

The viewing zone is defined as the image point where all spectral components of the reconstructed image of the diffuser are overlapped together. The image point is not actually a point because the projector's objective has a finite exit pupil size and also the diffuser has a finite width. As mentioned before, a hologram has an inherent converging power because it reconstructs images recorded on it. This converging power is represented by the following relationship [4].

$$\frac{\lambda}{\lambda_0}\left(\frac{1}{z_{\mathrm{O}}(\lambda)}-\frac{1}{z_{\mathrm{R}}}\right)=\frac{1}{z_{\mathrm{P}}}+\frac{1}{z_{\mathrm{I}}}, \tag{2.17}$$

where λ_0 and λ are wavelengths of recording and reconstruction, respectively, and $z_{\mathrm{O}}(\lambda)$, z_{R}, z_{I}, and z_{P} the distances of positions of the object beam, reference beam S_0, reconstructed image I_0 and projector's exit pupil P_0,

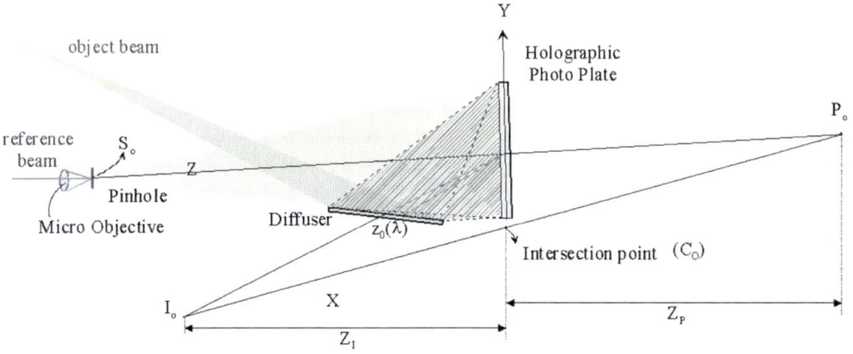

Fig. 2.14. Recording and reconstructions setup for full-color transmission-type HS

from the origin, as shown in Fig. 2.14. If the left side of (2.17) is replaced by $1/f_H$, it has just a typical form of a lens formula [40]. f_H is just the focal length of the hologram. From (2.17), for a given z_I value, f_H is found. From this f_H and a given value of z_R and λ_0, $z_O(\lambda)$ values for various reconstruction wavelengths are found. With these $z_O(\lambda)$ values, the required length of the diffuser for covering the spectral range of any projector can be calculated. Since f_H is independent of the projection and recording wavelengths, all object points specified by $z_O(\lambda)$ for different projection wavelengths will be imaged on the point I_0. This means that I_0 will be the full-color image point where all spectral components of the reconstructed image are overlapped together.

As shown in Fig. 2.14, since the long axis of the diffuser is near normal to the photograph, the fringe patterns recorded on the HS make a concentric pattern centered at C_0 as shown in Fig. 2.15. Hence, C_0 will be the center of the HS and make a straight line with I_0 and P_0 as shown in Fig. 2.14. Now, the screen is illuminated by a white-light projector, each spectral component of the white light will be diffracted to a specific direction by the screen and forms a real image of the diffuser along the line connecting I_0 and P_0. The reconstructed image of the diffuser for different spectral components appears at somewhat different locations along the line. These differences will be proportional to the relative position of the corresponding $z_O(\lambda)$ value on the diffuser. For example, the maximum values of $z_O(\lambda)$ are for the cases of $\lambda = 0.7\,\mu m$ (red) and $0.4\,\mu m$ (blue). The reconstructed images of the diffuser for both red and blue colors meet at point I_0; however, the red image extends toward the HS, but the blue image away from the HS.

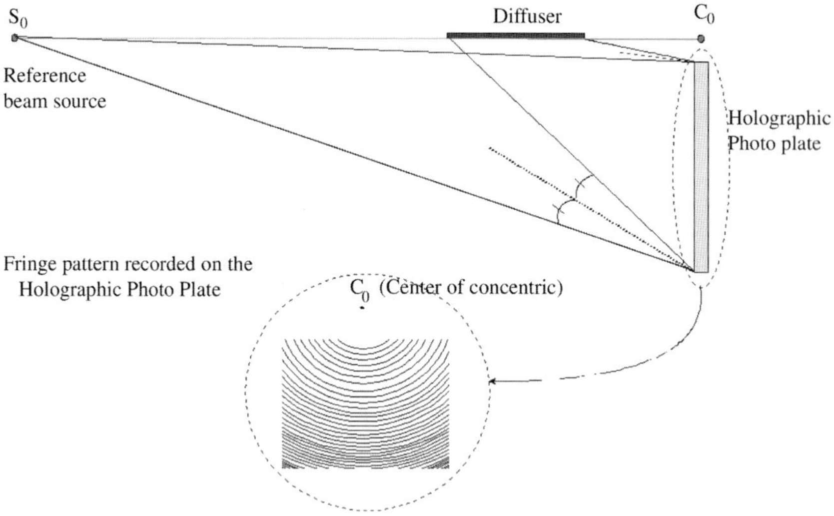

Fig. 2.15. Concentric fringe pattern recording

When the length and position of the diffuser are properly designed, all different spectral images of the diffuser are partially overlapped at I_0 and a full-color viewing zone is formed. Wherefrom, the viewer can see the full-color images displayed on the screen. Since two projectors are needed for a stereoscopic image, the viewing zone corresponding to each projector should be separated from each other by our eye distance.

2.4.3 Viewing Zone Size

Figure 2.16a, b and c show the effects of the diffuser length. Figure 2.16a shows no full-color viewing zone because the diffuser length is too short to cover all the spectral range. Figure 2.16b shows a full-color viewing zone when the length of the diffuser barely covers the entire visible spectral range. If it covers more than the spectral range, the overlapping area of different spectral images will be increased as shown in Fig. 2.16c. As a result, the viewer will have freedom to change the viewing distance from the HS within the overlapping region.

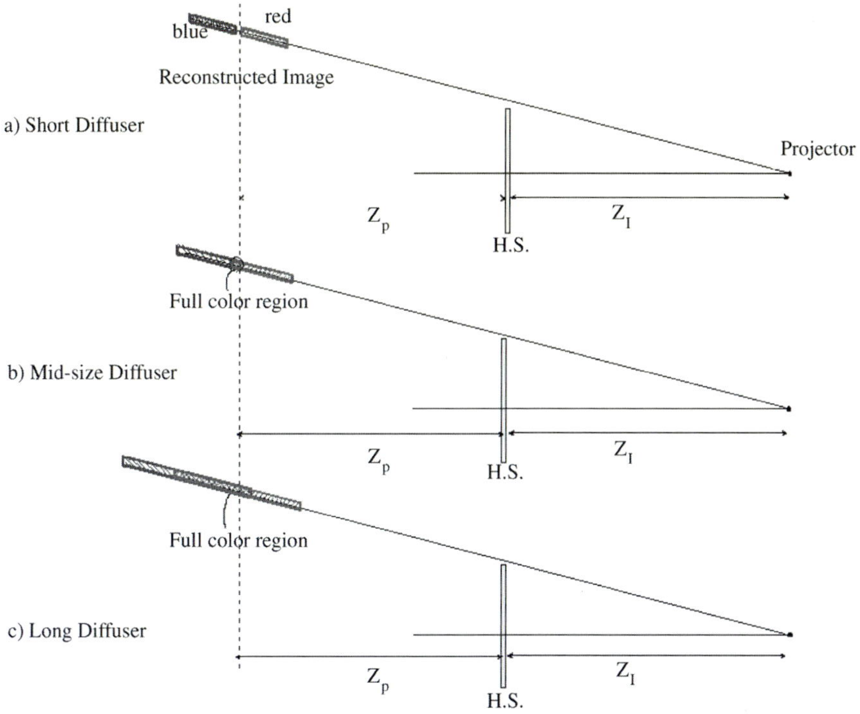

Fig. 2.16. Forming full-color viewing zone: the lengths of the full-color viewing zones depend on the diffuser length

The viewing-zone size of the transmission-type HS is proportional to the exit-pupil size of the projector objective in the vertical direction and to the size determined by the convolution between the reconstructed image of the diffuser and the image of the objective in horizontal direction. Hence, the vertical viewing-zone size cannot exceed 6.5 cm because the exit-pupil size of the projector objective cannot be larger than our eye distance. However, since the horizontal size of the full-color viewing zone is given as the convolution, the exit-pupil size of the projector objective should be less than our eye distance. This is the inherent factor for limiting the vertical viewing-zone size. The vertical size of the viewing zone is also proportional to the length of the overlapping region. If there is no aberration, the horizontal size can be controlled by adjusting the exit pupil size of the projector's objective and the diffuser width.

The light distribution in the viewing zone has a shape similar to that of the area in between two branches of a hyperbola because the 1st-order diffracted beams from the HS is converged at near I_0 and then diverged again. This is depicted in Fig. 2.17. The viewing-zone photograph in Fig. 2.17 shows that the viewing zone is bounded by curves similar to two branches of a hyperbola. The central bright region is the full-color viewing zone, and the upper part is bluish and lower part reddish.

The horizontal size of the viewing zone does not need to be more than viewer's eye distance, i.e. 6.5 cm. If it is too small, its shape will be unbalanced, very restricted movements are allowed for the viewer and the difficulty of finding the viewing zone increases. If the size is too large, viewing zones for left and right eyes can be fused together. The vertical size of the viewing zone is measured as near to 6 cm. This size is too small for the practical applications because viewers have different heights and often their eye positions in the vertical direction are not fixed by changing their posture.

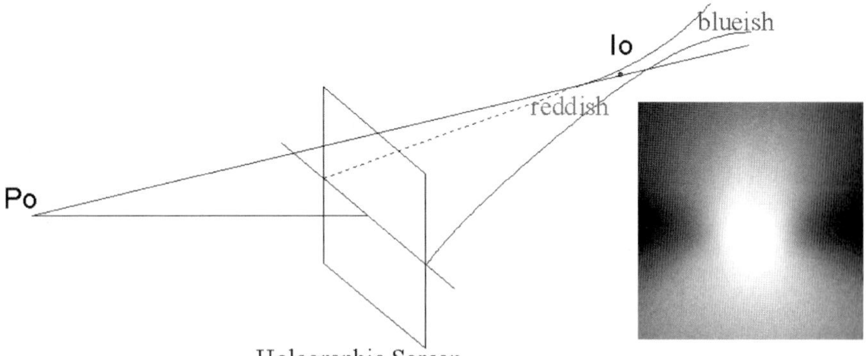

Fig. 2.17. Viewing zone shape

The viewing-zone size in the vertical direction can be increased by the following methods [41]: 1) Using a vertically extended projection objective of the projector, 2) changing the optical properties of the screen such that each point of the screen expands the light from the projected image vertically, 3) attaching a medium on the screen surface to scatter the light from the image displayed on the screen, vertically and 4) recording a phase grating that generate $\pm$1st-order diffracted beams that are apart by a viewing-zone diameter distance from each other.

The first method is not very promising, because the objective-size increase accompanies the objective's focal length increase. This focal-length increase causes increasing viewing and projection distances. A consequence of these increases is a projection-system size increase.

An example of the second method is the multiple-exposure method. When the screen is exposed several times by shifting the screen a predefined distance to C_0, if the distance is equal to one half of the exit pupil of the projector's objective, the viewing zone from each exposure will be joined together as a vertically extended viewing zone. It is possible to extend the vertical viewing-zone size up to 12 cm by exposing the screen twice. A problem involved with this method is that making a large-size screen is difficult because of difficulties in overcoming Bragg condition mismatch along with a diffraction efficiency decrease [42].

The third method is combining the holographic screen with an additional directional diffuser. The requirement for the diffuser is that it should not scatter light in the horizontal direction so as not to diffuse together the viewing zones for left and right eyes, and have high diffraction efficiency. This diffuser can be made holographically by recording the speckle pattern produced by illuminating a ground glass with a very thin laser beam strip. The diffuser allows the vertical size of the viewing zone to extend more than 10 cm. The problem related with this diffuser plate is that it causes image-quality degradation due to the visible speckle patterns on it. The visible speckle patterns appear because the spatial-frequency spectrum of the scattered light from the diffuser in the recording setup has the maximum intensity at near-zero frequencies. Due to this, the speckle pattern has quite a rough visible structure. This visible structure causes the displayed-image brightness to be modulated with it by diffracting some part of the image light to be outside of the viewing zone.

The fourth method is recording a phase grating on the screen directly with use of two reference beams. In this case, the vertical viewing-zone size will be increased without deteriorating the image quality because of the absence of low spatial frequency components in the beam illuminating the photoplate. In recording the grating, the grating period should be chosen such that it allows 1st-order diffracted beams from the grating to be located approximately a viewing-zone diameter distance from the zero-order beam.

In this case, the diffracted images of the projectors exit pupil can be joined with the zero-order images to form a vertically extended viewing.

The suitable grating periods for our holographic screens are in the range of 30 to 60 μm. Since our eyes cannot resolve the grating sizes, the image resolution will not be degraded by the grating. In joining the diffracted beams of different orders, care should be taken to make the intensity of each diffracted beam equal to that of the zero-order beam for a uniform-brightness viewing zone. The grating provides a 3 to 5 times increase in viewing-zone size. This makes the viewing-zone size in the vertical direction 10 to 15 cm for the 3 cm projection objective.

2.5 Obtainable Image Depth Range

In Fig. 2.7, the viewing-zone size V_{W} should be bigger than our eye distance to adopt many view images. However, since viewers can locate their eyes to any pair of viewing zone sections their eye distances apart, it is possible to consider that V_1 and $V_{K'}$ represent the pupil positions of a viewer's eyes. In this case, the lines connecting intersection points of lines from neighboring stars, lines from neighbor's neighboring stars, lines from stars from two star distance apart, and so on, will provide a means of determining the obtainable image depth for the autostereoscopic imaging systems.

Since the intersection points are virtual points, which can be seen simultaneously at both V_1 and $V_{K'}$, these points will work as volume picture elements of constructing stereoscopic images. In Fig. 2.9, for the generality of the analysis, it is assumed that there are n PLSs at position x_0, and the positions of lines connecting the intersection points that are in the same distance from the PLS array are specified by $x_{\pm j}(j = 0, 1, 2, 3, \dots, n-1)$. Then $x_{\pm j}$ is defined as,

$$x_{\pm j} = D_{\mathrm{D}} \cdot \frac{\pm jp}{V_{\mathrm{W}} \mp jp} . \tag{2.18}$$

In (2.18), V_{D} and V_{W} correspond to viewing distance and viewer's eye distance, respectively. Since the most protuding image point, i.e. the nearest image point to the viewers, x_{near} is when $j = -(n-1)$, x_{near} is given as,

$$x_{\mathrm{near}} = V_{\mathrm{D}} \cdot \frac{-(n-1) \cdot p}{V_{\mathrm{W}} + (n-1) \cdot p} . \tag{2.19}$$

Since $S = (n-1)p$ is the width of a display screen, the theoretically allowable maximum image protrusion is represented as $V_{\mathrm{D}} \times S/(V_{\mathrm{W}} + S)$. This relationship shows that the maximum image protrusion allowed for the autostereoscopic imaging systems is almost equal to the viewing distance. When, V_{D}, V_{W}, and S are assumed as 1 m, 6.5 cm, and 50 cm, respectively, the protrusion is calculated as 88 cm. This protrusion is too large to be watched comfortably by viewers. The protruded image depth for viewing the stereoscopic images comfortably is different for different people. However, empirically, it will be more comfortable for the viewers if the image protrusion does not exceed one third of the viewing distance. To make this value, S should equal $0.5V_{\mathrm{W}}$. In this sense, S will present the parallax between left and right eye images but not the screen size. Hence, it can be said that the parallax should not exceed one half of our eye distance for comfortable viewing of the protruded stereoscopic images.

From (2.19), the distance between two adjacent lines connecting the intersection points, d_{D} is calculated as,

$$d_{\mathrm{D}} = \frac{pV_{\mathrm{D}}V_{\mathrm{W}}}{V_{\mathrm{W}}^2 + (2n-3)pV_{\mathrm{W}} + (n^2 - 3n + 2)} . \tag{2.20}$$

If d_{D} is larger than our depth sensitivity at a given viewing distance, viewers will perceive discrete depth images. To make images with a smooth depth, the distance between each PLS, i.e. p and V_{D} should be smaller. In this case, the volume picture elements become finer.

The farthest image point, x_{far} is infinity when $V_{\mathrm{W}} = (N-1)p$, otherwise x_{far} equals to $x_{n-1}(j = n-1)$. That is,

$$x_{\mathrm{far}} = V_{\mathrm{D}} \cdot \frac{(n-1)p}{V_{\mathrm{W}} - (n-1)p} . \tag{2.21}$$

Equation (2.21) works only for the case of $V_{\mathrm{W}} > (n-1)p$ because when $V_{\mathrm{W}} < (n-1)p$, $x_{\mathrm{far}} > V_{\mathrm{D}}$. The possible minimum and maximum image depths behind the screen are given as f in (2.14), i.e. $n = 2$ case and infinity, respectively. Equations (2.19) and (2.21) show that $(n-1)p$ represents the parallax and the image depth is linearly proportional to the viewing distance V_{D}.

2.6 Methods of Increasing Number of Viewing Zones

For the projection-type stereoscopic or multiview 3-D imaging systems, the viewing-zone size is not enough to receive more than one viewer at a time, unless many projectors are used. It is possible to enlarge the viewing zone with use of a projection objective with a large aperture or, with high-magnification projection optics geometry. However, in this case, the number of multiview images to be projected should be increased and also a large-size projection objective should be used. Since increasing the multiview images and the size of the projection objective brings other difficult problems to solve, a more convenient method is to increase the viewing zone in number. The reflection-type HS [36], lenticular and the parallax barrier plates when used as an image-projection screen can multiply the viewing zone in number. It is shown that a beam-dividing prism plate can also effectively multiply the viewing zones

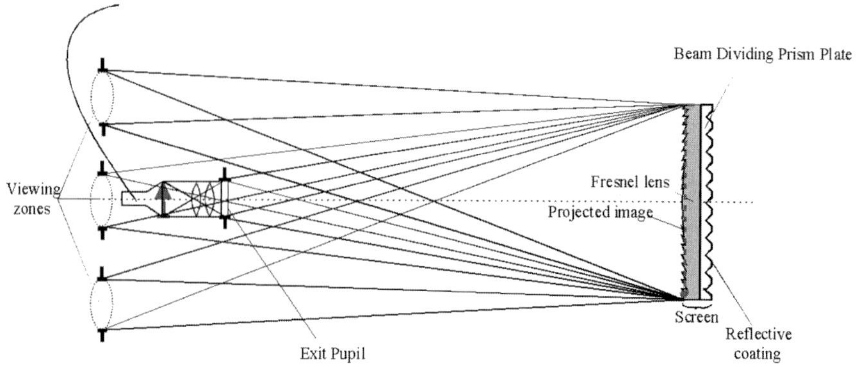

Fig. 2.18. Basic optical configuration of multiplying viewing zones by a beam-dividing prism

when it is used together with an image projection screen [43]. Figure 2.18 shows how a beam-dividing prism multiplies the viewing zone when it is combined together with the screen. In Fig. 2.18, the screen is assumed to be a Fresnel lens. The beam-dividing prism is located behind the screen with a small distance to eliminate the possible appearance of Moiré fringes. The surface of the prism plate is coated to reflect incoming light. Each prism in the plate has three facets of equal area. The rays passed through the Fresnel lens are reflected by the prism plate and divided into three different direction beams by the three facets of the prism. Hence, the Fresnel lens forms three images of the exit pupil of the projection objective, i.e. three viewing zones are formed. In Fig. 2.18, only a reflection-type beam-dividing prism plate is introduced. However, the prism plate will also work as a transmission type with no reflective coating. The area of each facet needs to be chosen to reflect the same amount of rays as other facets for equal brightness viewing zones. The prism size should not be more than a pixel size of the image projected on the screen for it not to deteriorate the image resolution. In Fig. 2.19a, the beam-reflection geometry of the beam-dividing prism with three facets is shown. The incident beam is divided into three different direction beams by the prism facets. In Fig. 2.19b, possible shapes of the beam-dividing prism for more than three viewing zones are shown. The shape Nos. 1 and 2 in Fig. 2.19b will create viewing zones aligned in a line, however, Nos. 3 to 6 will do so in different spatial locations.

The beam-dividing prism plate should be designed by considering the optical geometry and the desired number of viewing zones. An example of the beam-dividing prism plate having the surface profile as shown in Fig. 2.19a is shown in Fig. 2.20. The optical geometry of the projection system, which employed a Fresnel lens laminated with the beam-dividing prism as shown in Fig. 2.20, as an image-projection screen, is shown in Fig. 2.21. The distances between the screen and the projection optics, and between viewing zones

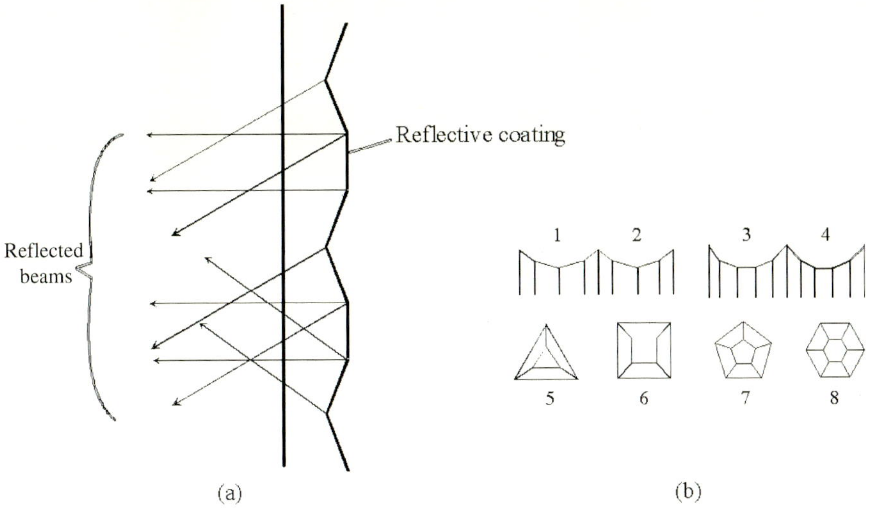

Fig. 2.19. Possible surface profiles of the beam dividing prism (**a**) beam reflecting geometry; (**b**) possible prism shapes

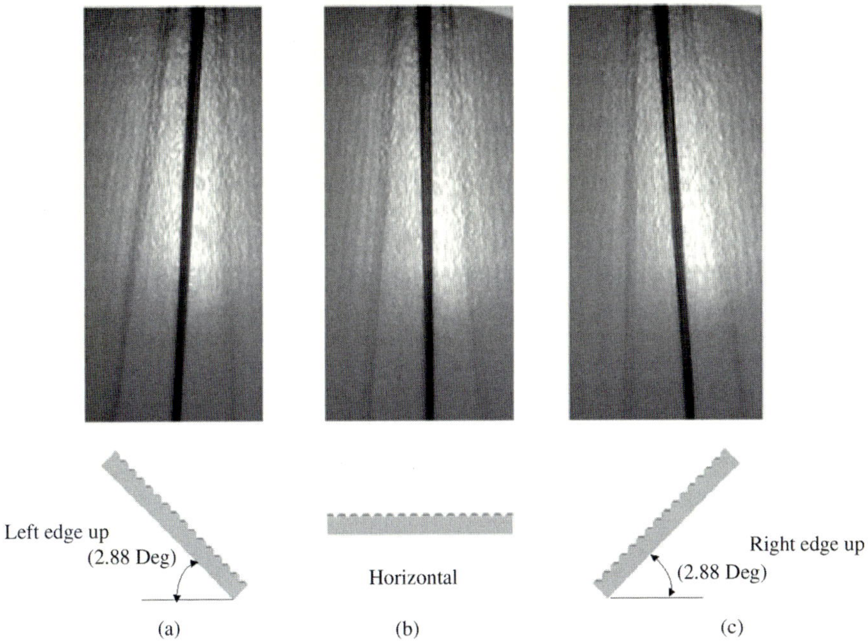

Fig. 2.20. Reflected images of a dark black strip fixed normal to the beam-dividing prism for three different slant angles: When the prism in a horizontal plane is slanted to $\pm 2.88^\circ$, it reflects the strip without distortion. This means that the prism is grooved to have three facets of corners angles 87.12° and 177.12°, respectively

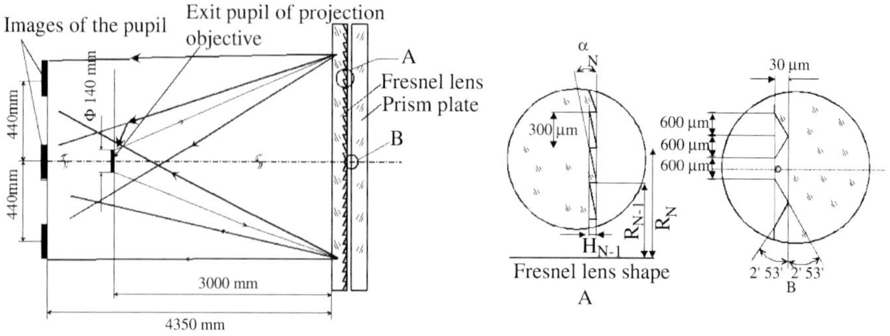

Fig. 2.21. The structures of beam-dividing prism plate and Fresnel lens beam-dividing prism

are 3 m and 440 mm, respectively, and the focal length of the Fresnel lens is chosen to give 1.45 times magnification, so the diameter of the viewing zones corresponding to that of exit pupil of the projection objective, 140 mm, is 203 mm. If it is assumed that the screen with the prism plate can resolve a 2 mm size object at 3 m away, i.e. a resolving power of 2.3 angular minute, the surface flatness of the prism should be comparable to 10 fringes/100 mm to achieve the resolving power. This flatness corresponds to approximately 1.1 mm lateral aberration in the viewing zone plane. The pitch of the prism is 1.8 mm. The groove slope of the prism corresponding to the given viewing-zone distance is 2.88°. The distance between the Fresnel lens and the prism plate should be enough to eliminate the appearance of Moiré fringes.

2.7 Acknowledgements

The author would like to acknowledge that this work would have been impossible without valuable contributions from Professor Victor I. Komar, Dr. Vladmir I. Bobrinev, Dr. Sergei A. Shestak, Dr. Vadim V. Smirnov, Dr. Vladmir V. Saveljev, Dr. Yu N. Gruts, Mr. Y-J Choi, Dr. H-S Lee, Mr. D-H Kang and Miss J-E Bahn.

References

1. E.J. Korevaar et al.: US Patent No. 4,881,068 (Nov. 14, 1989)
2. R. Börner: Fernseh und Kinotechnik **45**, 453 (1991)
3. A.R. Travis, S.R. Lang, J.R. Moore, N.A. Dodgson: J. SID **3**, 203 (1995)
4. R.J. Collier, C.B. Burckhardt, L.H. Lin: *Optical Holography* (Academic Press, New York 1971)
5. E.G. Rawson: Appl. Opt. **7**, 1505 (1968)

6. F. Budinger: Proc. SPIE **507**, 2 (1984)
7. J.-Y. Son, V.V. Smirnov, L.N. Asnis, J.H. Jeon, H.S. Lee, V.B. Volkonsky: Proc. SPIE **3639**, 137 (1999)
8. IMAX, OMNIMAX
9. S. Omori, J. Suzuki, S. Sakuma: J. SID **26**, 855 (1995)
10. G.J. Woodgate, J. Harrold, A.M.S. Jacobs, R.R. Moseley, D. Ezra: Proc. SPIE **3957**, 153 (2000)
11. A. Schwerdtner, H. Heidrich: Proc. SPIE **3295**, 203 (1998)
12. T. Okoshi: *3 Dimensional Imaging Technology* (Asakura Book Store, Tokyo, Japan 1991)
13. H. Isono, M. Yasudsa, D. Takemori, H. Kanayama, C. Yamada, K. Chiba: Proc. SPIE **1669**, 176 (1992)
14. J.B. Eichenlaub: Proc. SPIE **1256**, 153 (1990)
15. J.-Y. Son, V.I. Bobrinev, S.A. Shestak, H.-W. Jeon: Proc. SPIE **2951**, 168 (1996)
16. J.R. Moore, N.A. Dodgson, A.R.L. Travis, S.R. Lang: Proc. SPIE **2653**, 10 (1996)
17. K. Sakamoto, H. Takahashi, E. Shimizu, H. Ueda, K. Tanaka, M. Okamoto: Proc. SPIE **3011**, 36 (1997)
18. D.J. Trayner, E. Orr: Proc. SPIE **3639**, 167 (1999)
19. J.-Y. Son, Y. Gruts, J.-E. Bahn, Y.-J. Choi: In: *IDW'2000, Proc. 7th Int. Display Workshop, Kobe, Japan, Dec. 2000*
20. I. Takehiro (ed.): *Fundamentals of 3D Images* (NHK Broadcasting Technology Research Center, Ohm Sa, Tokyo, Japan 1995)
21. A. Woods, T. Docherty, R. Koch: Proc. SPIE **1915**, 36 (1993)
22. H. Yamanoue: SMPTE J. **106**, 225 (1997)
23. J.Y. Son, S. Shestak, S.S. Kim, Y.I. Choi: Proc. SPIE **4297**, 160 (2001)
24. R. Boerner: *Proc. 4th Int. Display Workshops IDW-97, Japan*, 835 (1997)
25. P.V. Hartman: Proc. SPIE **2653**, 56 (1995)
26. Y. Kajiki, H. Yoshikawa, T. Honda: IDW' 96 **2**, 489 (1996)
27. H. Heidrich, A. Schwerdtner, A. Glatte, H. Mix, Dresden Univ. of Technology: *Eye-position detection system*, in Proc. SPIE **3957**, 192 (2000)
28. H. Imai, S. Tsujikawa, M. Imai, NEC Corp.: in Proc. SPIE **3639**, 92 (1999)
29. N. Tetsutani, K. Omura, K. Kishino: Opt. Eng. **33**, 3690 (1994)
30. L. McMillan, G. Bishop: Proc. SPIE **2409**, 21 (1995)
31. J.-Y. Son, V.V. Smirnov, K.-T. Kim, Y.-S. Chun: Proc. SPIE **3957**, 184 (2000)
32. J.-Y. Son, V.V. Smirnov, Y.-S. Chun, K.-T. Kim, S.-S. Kim: Proc. SPIE **3639**, 132 (1999)
33. J.-Y. Son, V.I. Bobrinev: *Autostereoscopic Imaging Systems Based on Holographic Screen, Three Dimensional Video and Display: Devices and Systems*, ed. B. Javidi, F. Okano, Vol. CR76 (SPIE Press 2001)
34. J.-Y. Son, V.G. Komar, Y.-S. Chun, S. Sabo, V. Mayorov, L. Balasny, S. Belyaev, M. Semin, M. Krutik, H.-W. Jeon: Proc. SPIE **3295**, 218 (1998)
35. S. A. Shestak, J.-Y. Son, H.-W. Jeon, V.G. Komar: Proc. SPIE **3012**, 96 (1997)
36. V.G. Komar, O.B. Serov, *Art Holography and Holographic Cinematograthy*, (Iskusstvo, Moscow 1990)
37. S.H. Hwang, V.I. Bobrinev, J.Y. Son, S.A. Shestak, H.W. Jeon: Appl. Opt. **36**, 6605 (1997)

38. J.-Y. Son, V. G. Komar, Y.J. Choi: *Proc. of Workshop on 3D Display Technologies and Human Factor in The Third International Display Workshops* (IDW'96), Kobe, Japan, 509 (1996)
39. V.V. Smirnov, J.-Y. Son, H.-W. Jeon, H.-S. Lee, Y.-J. Choi: Proc. SPIE **3293**, 175 (1998)
40. F.A. Jenkins, H.E. White: *Fundamentals of Optics* (McGraw-Hill, New York 1976)
41. J.Y. Son, J.S. Kim, Y.J. Choi, V.I. Bobrinev: Proc. SPIE **4296**, 64 (2001)
42. J.-S. Kim, V.I. Bobrinev, J.-Y. Son, Y.-J. Choi, S.-H. Shin: Proc. SPIE **3957**, 148 (2000)
43. J.-Y. Son, V.V. Smirnov, J.-S. Kim, Y.-S. Chun, K.-T. Kim: *Proc. of 3D Image Conference '99, Japan*, 227 (1999)

3 A Head-Mounted Display Using a Holographic Optical Element

T. Ando and E. Shimizu

3.1 Introduction

Information media has steadily developed as civilization has progressed. Now, we are in the midst of a revolution related to various information media which has resulted in the realization of a multimedia information society, and further developments are anticipated. In the field of electrical information media such as CRTs and LCDs, a revolution in display technologies, in areas such as sense of presence and high resolution, is necessary. Image information media has evolved from non-electrical media such as printed still images to moving electronic pictures. This evolution is expected to continue with the next-generation display technology which is expected to offer stereoscopic display. The widely used two-dimensional display must evolve to have stereoscopic functionality. Therefore, various methods for realizing stereoscopic images have been proposed, and some stereoscopic displays have been released [1].

Holography is especially remarkable because it can provide us with complete three-dimensional images using the wave-front reconstruction method. However, holography can create only a still image, and more research and development is required to apply it for use with electronic moving pictures. Therefore, the mainstream method, binocular stereoscopy, is used for current stereoscopic displays in which the left and right parallax images are shown to the respective eyes. However, this method has the drawback of dissociation between accommodation and convergence.

Virtual Reality (VR) technology [2] has become a single technical field in computer and information media. In VR the display is a media through which bodily sensations are experienced related to the virtual world generated by the computer. It is not simply used as an information display device, but rather can provide us with the senses sound, touch, position, and sight. In addition, it can enable us to experience the interaction of those bodily sensations.

VR is a multimedia system that has three elements: sense of presence, autonomy, and interaction. While great developments have been made in VR, the computer and display that are necessary for VR technology have performance limitations. The present displays, which provide us with the visual information, cannot create a virtual world that is indistinguishable from the real world. Therefore, while VR research continues to advance, neither its practical use nor general markets for it have been highly developed.

Meanwhile, Mixed Reality (MR) technology [3] has recently been proposed as an approach for practical use of VR, since with the present calculation capabilities of computers it is not easy to express the real world through virtual images.

MR has two kinds of technology: Augmented Reality (AR) [4] and Augmented Virtuality (AV) [3]. AR is used to augment the human ability to process real-world information by adding computer-generated virtual information. AV expresses the virtual world from information that is captured from the real world. In the case of AV, it is possible to use a desktop monitor such as a CRT or LCD; however in the case of AR, it is necessary to use a see-through head-mounted display (HMD) in which both the real world and virtual world can be seen at the same time.

When two-dimensional information like a document is displayed on a monocular HMD, observers can understand it although it has no relation to real-world objects. However, the information on a see-through HMD must be expressed as three-dimensional images because the real world is a three-dimensional world. What an observer sees needs to be augmented by virtual information (three-dimensional images) in accordance with the real objects. Consequently, a see-through HMD for AR needs to be able to combine both the transparence of the real world and reflection for the virtual images.

A conventional HMD uses a half-mirror as a combiner which prevents both real-world and virtual images from being seen simultaneously at high intensity because the half-mirror has a tradeoff between transparence and reflection. Moreover, the wide field of view of a conventional HMD with a half-mirror is limited because a half-mirror must be focused on each eye separately and be tilted at 45° in front of the face. In addition, dissociation occurs between accommodation and convergence when stereoscopic images are seen through a conventional HMD because it has a fixed focal plane.

In this research, we propose a new HMD using a Holographic Optical Element (HOE) instead of a half-mirror as a combiner in order to resolve the above-mentioned defects of a conventional HMD. The HOE is a diffraction grating made using the holography technique and it can diffract a certain incident ray to a certain direction because of its optical characteristics, such as angular selectivity, wavelength selectivity, multiple exposures, and wavefront transformation. If a HOE is applied to a see-through HMD, a single HOE can produce the binocular parallax images for each eye, and have sufficient transparency for the outside world and a wide field of view for both real-world and a virtual images. Further, a HOE can be equipped with the Maxwellian View [5] function and produce images with extremely long focal depth. If a binocular stereoscopic image is displayed with an extremely long focal depth, the conventional problem of the binocular stereoscopic display, the dissociation between accommodation and convergence, may be solved.

In this paper, we will introduce our recent research to produce a prototype of a see-through HMD using a HOE. Specifically, we propose two types of

HOEs for application to see-through HMDs, one is the diverging HOE and the other is the converging HOE for the Retinal Projection Display.

3.2 Characteristics of HMD

This chapter will outline the structure and features of a conventional HMD. Various stereoscopic displays have recently been developed and the see-through HMD is very popular in VR. However, there are some drawbacks with the conventional see-through HMD, such as using a half-mirror as a combiner and the problem of dissociation between accommodation and convergence.

3.2.1 Structure of the See-Through HMD

The fundamental role of a HMD is to produce a distinguishable image of the Spatial Light Modulator (SLM) and enable the outside world to be seen through the see-through HMD. This requires both a see-through optical system and a magnifying optical system.

There are two kinds of magnifying optical systems: the ocular type and the relay type. The former is shown in Fig. 3.1a and it magnifies the images on the SLM (for instance, LCD) by only the ocular (for instance, concave mirror). The latter is shown in Fig. 3.1b and it consists of the ocular and a relay optical system. The image on the SLM is focused on the first imaginary plane, and then it is magnified by the ocular optics. The half-mirror is needed in both types to fuse the real and virtual scene. And through the half-mirror, the rays on the SLM are reflected and the rays of the outside world are penetrated. However, it cannot effectively produce both rays with high intensity because of the tradeoff between transmissivity and reflectance. The ocular type, in particular, reduces the intensity because the rays are transmitted many times through a half-mirror, and while it is suitable for miniaturization, it is difficult to attain a wide field of view. On the other hand, with the relay type a wide field of view can be attained, but it is difficult to miniaturize.

3.2.2 Drawbacks of HMD

A conventional HMD is equipped with a half-mirror so both the real and virtual world cannot be seen simultaneously at high intensity because of the half-mirror tradeoff between transmissivity and reflectance. In addition, a conventional HMD has a fixed focal point for the virtual images on the SLM, while the eye dynamically accommodates to the changing fixation point of real objects. Therefore, both real and virtual objects cannot be steadily observed because an observer can only accommodate to one of them. This phenomenon is shown in Fig. 3.2.

For a stereoscopic conventional HMD, the parallax images for the left and right eye are displayed onto each eye separately, and the ocular optical

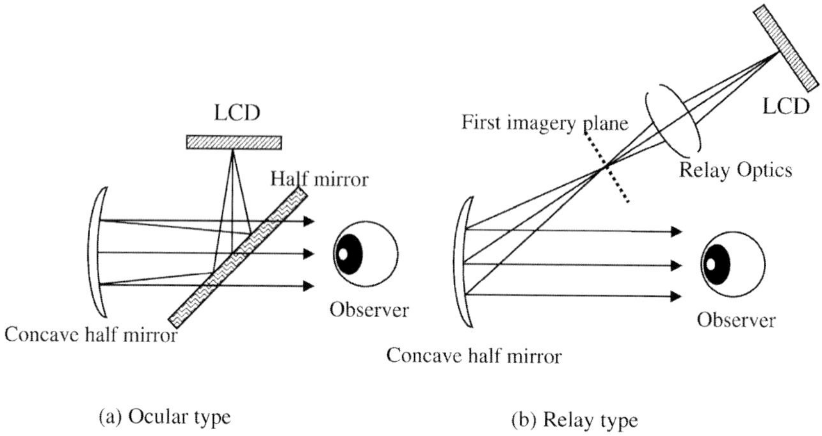

Fig. 3.1. Optical system of see-through HMD

systems are isolated from each other. The field of view to the real world for each eye is limited, as shown in Fig. 3.3, and the dissociation between accommodation and convergence, which is the fundamental drawback of binocular stereoscopic displays, occurs. Accommodation will correspond to the screen with the eye bringing the image which is displayed on the screen into focus. At the same time the convergence of the virtual object which is displayed as a binocular parallax image will be in front of or behind the screen and consequently the accommodation and convergence do not correspond. The dissociation between accommodation and convergence is shown in Fig. 3.4, where the convergence is on the facade of the screen. The distance from an observer's eye to the screen, shown as D_a, is the depth an observer perceives as the focal point to the screen, while the converging point, shown as D_v, is the distance of convergence for binocular stereoscopic images. D_a (perceptible distance by accommodation) and D_v (perceptible distance by convergence) cannot be made to correspond in the case of binocular parallax images, and eyestrain is induced.

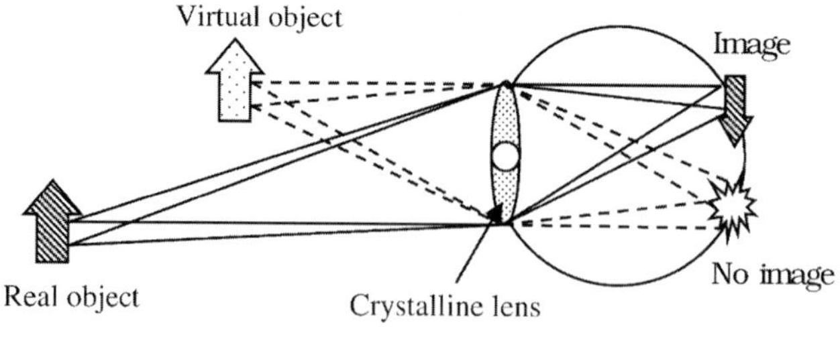

Fig. 3.2. Dissociation between real and virtual objects

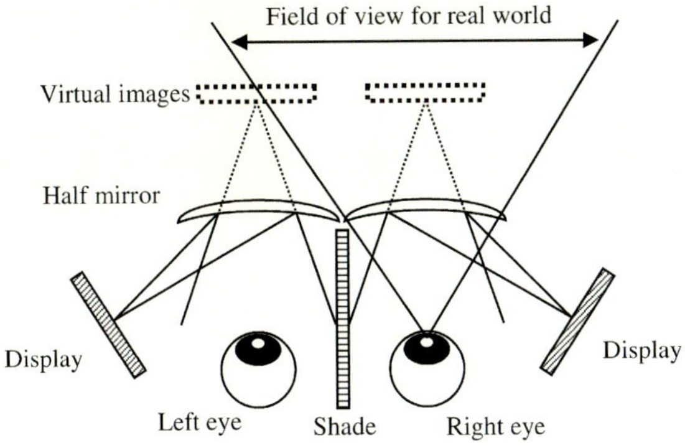

Fig. 3.3. Field of view of HMD

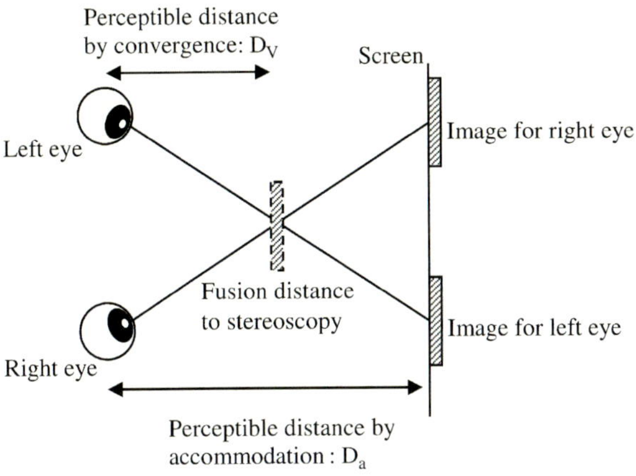

Fig. 3.4. Dissociation between accommodation and convergence in binocular stereoscopy

3.3 Principle of a Holographic Optical Element for a See-Through HMD

In holography an image of an object is regenerated by wave-front reconstruction after recording the interference fringe. The holographic optical element (HOE) is used as the lens or optical element to transform the wave front. The HOE has superior optical characteristics such as angular selectivity, wavelength selectivity, multiple exposure, and wave-front transformation. In this chapter, we will give the outline of the main optical characteristics of a HOE, and the features of a HMD using a HOE. At the end of this chapter, we will propose two kinds of HOEs: "diverging HOE" and "converging HOE", which can be applied to a see-through HMD.

3.3.1 Optical Characteristics of HOE

In this section, angular selectivity and wavelength selectivity are described.

Angular Selectivity of HOE A HOE is a diffraction grating and has the characteristic of angular selectivity and wavelength selectivity. Angular selectivity refers to the fact that the diffracted intensity of the HOE in reconstruction depends on the incident angle of the illuminating beam, which is equivalent to the reference beam used in recording. The diffracted light cannot be seen if the viewpoint is off slightly. A HOE can be exposed to multiple beams and diffract many illuminating lights in specific directions. Figure 3.5 shows the basic principle of angular selectivity of a HOE in relation to transmission type and reflection (Lippmann) type. In both cases, ray 1 can be diffracted when it is illuminated by a certain incident angle to the HOE; however, ray 2 cannot be diffracted when it is illuminated by an angle which is different from ray 2's incident angle to the HOE.

Wavelength Selectivity of HOE The wavelength selectivity of a HOE refers to the fact that the diffracted intensity of the HOE in reconstruction depends on the wavelength of the illuminating beam when the incident angle of the illuminating beam is equivalent to that of the reference beam used in recording. In other words, the HOE can diffract the slight width of the wavelength in the illuminating beam. The characteristic of the reflection type of a HOE is very remarkable. It is possible to transmit visible rays and to reflect the narrow band of a specific wavelength of rays (for example, red). Therefore, the holographic combiner enables both the outside world and the specific wavelength image shown on the display to be seen. Figure 3.6 shows the principle of the holographic combiner.

Ray 1, which includes all wavelengths (λ_{-n}–λ_{+n}) for the outside world, will penetrate through the HOE, and only the specific wavelength ray 2 (λ_0)

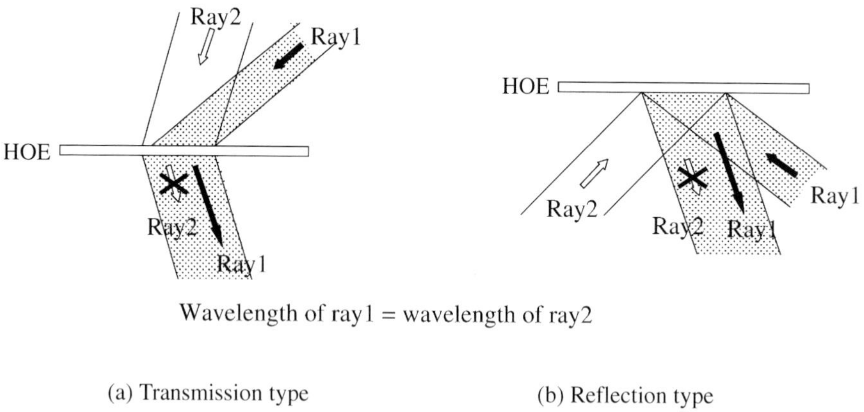

Fig. 3.5. Basic principle of angular selectivity of HOE

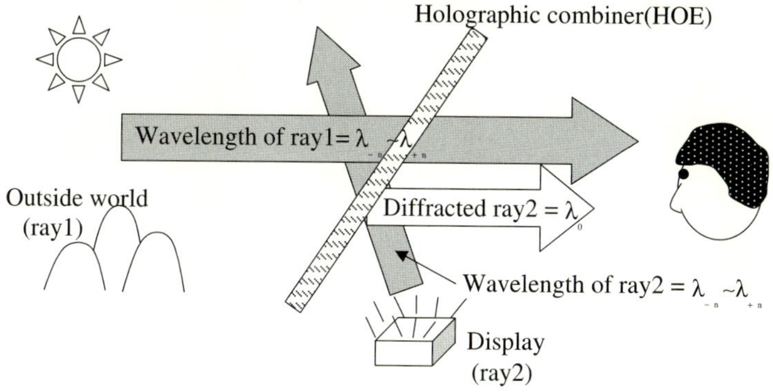

Fig. 3.6. Principle of holographic combiner

will be reflected on the HOE, although ray 2 has all wavelengths (λ_{-n}–λ_{+n}) on the display. The principle of this is the same as for a Head-Up Display (HUD). In the case of a HUD, the stereoscopic images on a display like a LCD cannot be seen. Normally, the viewing distance for the HUD is long (focal distance of 60 cm–2 m) and the screen size is relatively small (about 10 cm^2), therefore for a HUD small images have to be displayed on both eyes. Furthermore, if the head is even moved slightly, the diffracted images cannot be seen because the exit angle of a HUD is very narrow.

3.3.2 Principle of HOE for See-Through HMD

In this section, the method of binocular stereoscopy using the angular selectivity and wavelength selectivity of HOE and the features of a HMD using HOE are described.

Binocular Stereoscopy of a HOE Due to angular selectivity and wavelength selectivity, a single HOE exposed to multiple recording beams can display binocular images, which are diffracted from two illuminating beams to the left and right eye separately. Then a single HOE can produce binocular parallax images. This function is the same as a parallax barrier display or lenticular display without special glasses. We propose that binocular stereoscopic images can be seen through a HOE like with a HUD if it is located close to the eyes and two rays with the parallax information are illuminated from a specific direction. Figure 3.7 shows the principle of the binocular parallax display of the HOE.

Feature of HMD using a HOE We propose a HMD using a HOE as the combiner by taking advantage of the optical characteristics of a HOE, as shown in Fig. 3.8. A HOE is most suitable to realize a see-through HMD for MR because it has the function of being both a combiner and an optical lens. Figure 3.8a shows the optical configuration of a conventional HMD

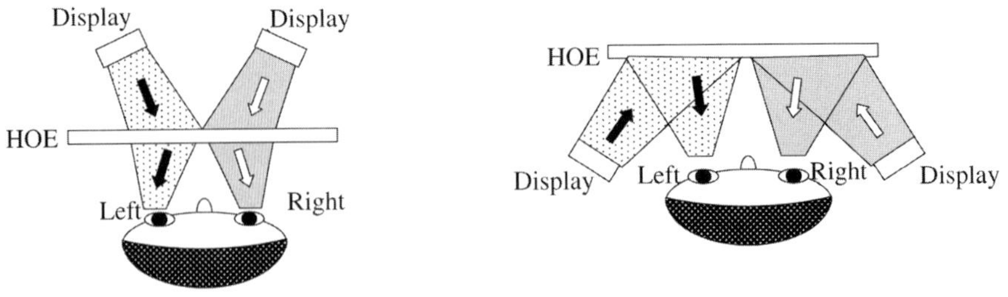

(a) Transmission type (b) Reflection type

Fig. 3.7. Principle of binocular parallax display of a HOE

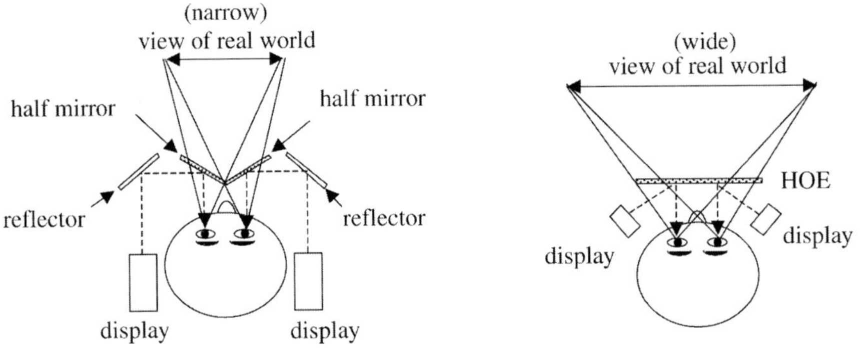

(a) Conventional HMD (b) New HMD using HOE

Fig. 3.8. Concept of optical configuration of HMD

and Fig. 3.8b shows the optical configuration of a HMD using a HOE. The HMD using a HOE is more useful than the HMD using a half-mirror as the combiner, especially for a MR environment. A conventional HMD has the optical system close to the eyes and this limits the real-world field of view. Figure 3.8a shows the real-world field of view for a conventional HMD. The field of view of the virtual image on the display is very narrow because the two half-mirrors are set on each eye separately. Meanwhile the HMD using a HOE has just one grating in front of the face and its real-world field of view is very wide. Figure 3.8b shows the real-world field of view for a HMD using a HOE. Furthermore, all wavelength rays of the real world can be transparent and specific wavelength rays illuminated by a display can be diffracted in a specific direction. Therefore, this HOE can provide us with binocular stereoscopic images.

3.3.3 Proposal of a HOE for HMD

We propose two kinds of HOEs in this research: "diverging HOE" and "converging HOE". The outline for them is given in this section.

The Diverging HOE This HOE has the optical characteristics required to produce a virtual image. Figure 3.9 shows the principle for recording and reconstruction. For recording, shown in Fig. 3.9a, the interference fringe is recorded on the HOE by two-beam interference of the object rays O and the reference rays S that are diverging from the laser point source. For reconstruction, shown in Fig. 3.9b, if the point source E is set in the position which is equivalent to reference point source S in recording, the interference fringe on the HOE reconstructs the wave front as if its rays were diverging from position E' which is equivalent to the object point source O in recording.

In this case, the HOE emits the diverging rays and an observer sees this virtual image diverging from the interference fringe on the HOE by using the function of imagery on the human eye (accommodation). In addition, it can be seen through the HOE in the area where the reconstructed rays are diverging. In other words, the point source E' which is reconstructed as the

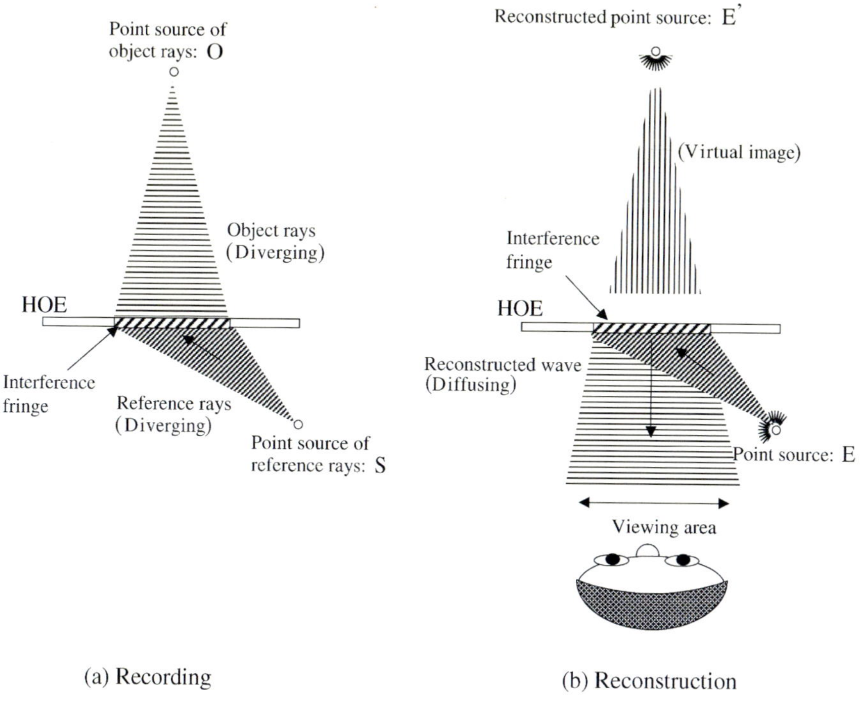

Fig. 3.9. Optical characteristics of the diverging HOE

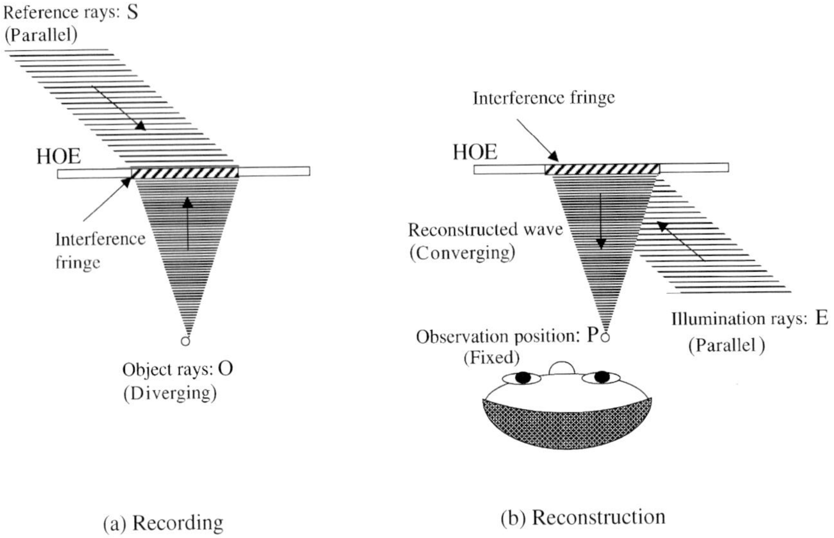

Fig. 3.10. Optical characteristics of the converging HOE

virtual image by the HOE can be observed in the viewing area only, as shown in Fig. 3.9b.

The Converging HOE This HOE has the optical characteristics required to produce a real image. Figure 3.10 shows the principle for recording and reconstruction. For recording, shown in Fig. 3.10a, the interference fringe is recorded on the HOE by two-laser beam interference of the object rays O and the reference rays S that are parallel rays. For reconstruction, shown in Fig. 3.10b, if the parallel illumination rays E are set from the direction which is conjugate to the reference rays S in recording, the interference fringe on the HOE reconstructs the wave front converging at an observer's eye position P which is equivalent to the position of the object point source O in recording. In this case, the HOE emits the converging rays and an observer sees this real image converging from the interference fringe on the HOE without experiencing the function of imagery of the human eye (accommodation). In addition, it can be seen through the HOE in the area where the reconstructed rays are diverging. If an observer's eye does not correspond with the converging point P, the rays cannot be seen. By using this converging HOE, the Retinal Projection Display we mention later can be realized.

3.4 See-Through HMD Using a Diverging HOE

As was described in Sect. 3.3.3, we have proposed two kinds of HOEs. In this chapter, we will give a detailed explanation of the see-through HMD using a diverging HOE [6].

3.4.1 Principle of Binocular Separation

In the previous section we explained how a diverging HOE produces a virtual image, and that the viewing area in which an observer can see the virtual image is the region where the wave front diverges from the interference fringe on the HOE. This condition is shown in Fig. 3.9b. If this viewing area is narrow and the viewing distance from an observer to the HOE is short, the virtual image cannot be seen by both eyes. Therefore, if two interference fringes were recorded on one HOE separately, as shown in Fig. 3.11a, the LCD_L image could be seen by the left eye and the LCD_R image could be seen by the right eye. In short, one HOE can provide an observer with both binocular images separately, as shown in Fig. 3.11b.

3.4.2 Optical Specifications of the Diverging HOE

A prototype of this HOE is 4 in. long and 5 in. wide. The distance between the eyes and the HOE is 4 in. The distance between the pupils is 65 mm. Figure 3.12 shows the optical characteristic of the diverging HOE. In recording, an object beam and a reference beam are diverging rays; the angle between them is 120°: the incident angle of the object rays to the HOE is 90° and that of the reference rays is 30°. The incident angle of the reference rays can prevent the reflected rays on the HOE from being incident upon the human eyes when the small LCD displays on both sides of the head are put at the position of the reference rays in reconstruction. The focal length is 50 cm, which means that the images which are reconstructed by the interference fringe on the HOE can be seen on the LCD as if they were 50 cm away.

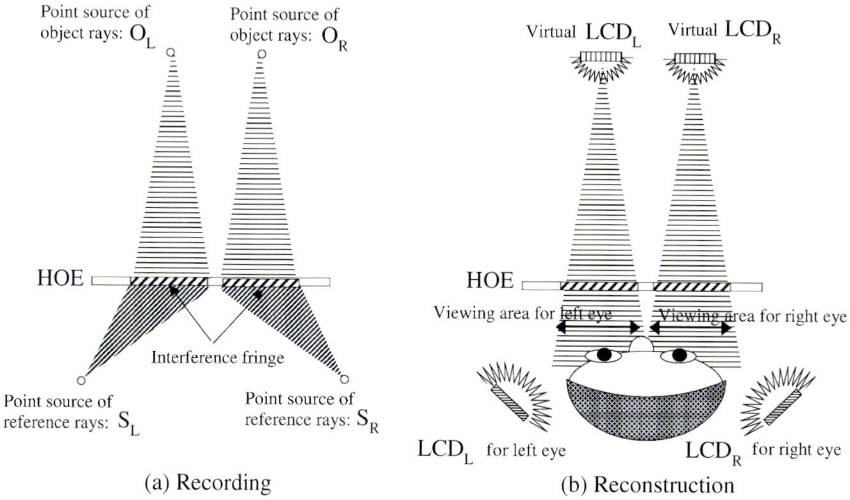

Fig. 3.11. Principle of binocular separation

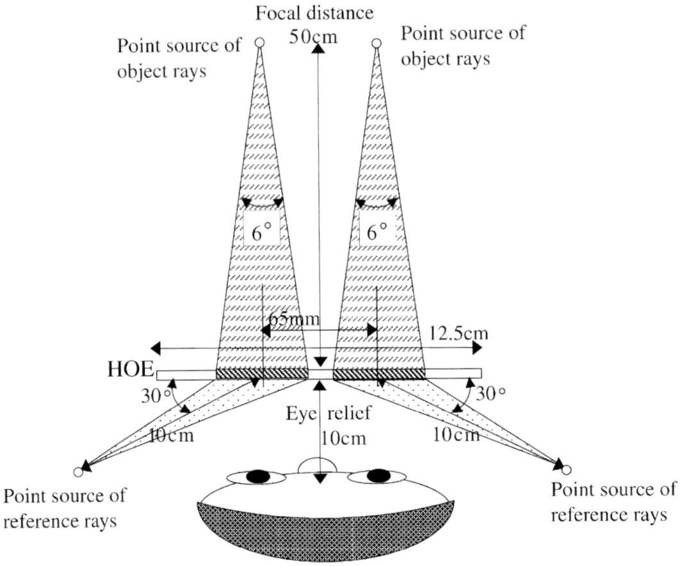

Fig. 3.12. Optical specification of HOE

The binocular images modulated by illuminating light for reconstruction can be seen because two interference fringes are recorded on a HOE. And each image for the left and right eye can be seen in the narrow viewing area, and at a distance, which is called eye relief, from the eye to the HOE. The rays from the left are diffracted by the HOE and can be seen on the left eye, and the rays from the right are diffracted and can be seen on the right eye. In this way a single HOE can separate the stereoscopic images to the left and right eyes. In addition, both the rays for the real world and the diffracted images can be seen simultaneously because the HOE can diffract specific wavelength rays without reducing the intensity of the visible rays.

3.4.3 Reconstruction Optical System

We made a prototype of a diverging HOE as shown in Fig. 3.12 by way of the two-beam interference method and we made an experimental optical system for reconstruction by using a semiconductor laser as the illuminating light. Figure 3.13 shows the experimental optical system for reconstruction. It uses a He–Ne laser (wavelength: 632.5 nm) in recording and a semiconductor laser (wavelength: 635 nm) in reconstructing. The reconstructed image is red color only.

This optical system uses a 1.3-in. LCD as the spatial light modulation device. The resolution of the LCD is VGA. Diffusing laser illumination is used as the backlight of the LCD. The left-side LCD displays the image for the left eye and the right-side LCD displays the image for the right eye. They

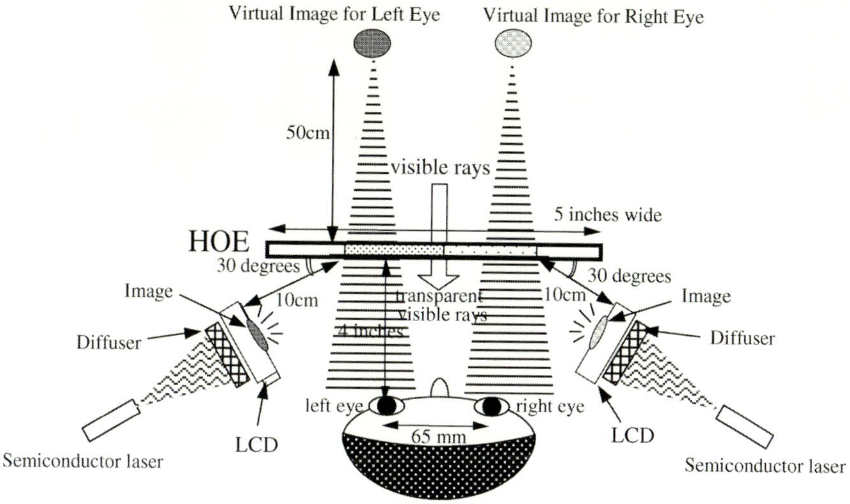

Fig. 3.13. Optical system for reconstruction

are binocular parallax images. When an observer places this optical system close to his/her eyes the binocular separated images can be seen due to the above-mentioned principle. An observer can see the real outside scene through the HOE with high transmissivity and at the same time see the reconstructed virtual images on the LCD.

Figure 3.14 shows the appearance of the HOE mounted on a helmet. Figure 3.15 shows the diffracted images displayed by this optical system. Figure 3.15a shows the diffracted image for the left eye and Fig. 3.15b shows the diffracted image for the right eye. Both the diffracted images and the outside world through the HOE can be seen in an ordinary office room in which the illumination is about 900 lux. Figure 3.15 shows the diffracted images. The left and right parallax about a triangle and an ellipse can be

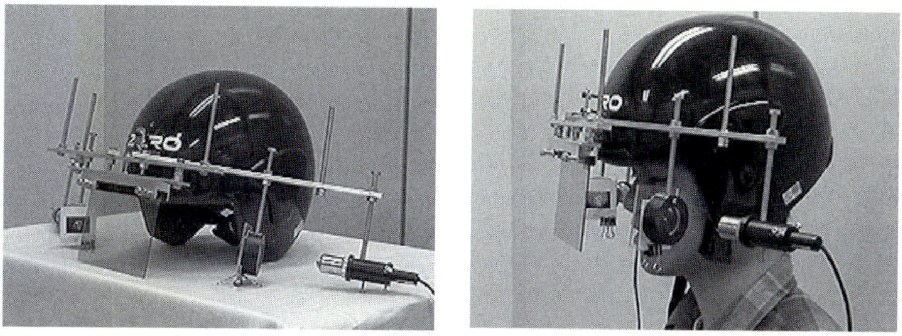

Fig. 3.14. Appearance of the HMD using the diverging HOE

(a) Image for left eye

(b) Image for right eye

Fig. 3.15. Diffracted images for binocular parallax recognition

seen. In this case, the parallax refers to the overlap between the triangle and the ellipse.

3.4.4 Interactive MR System

We have worked to produce a prototype in which one can experience the MR environment by using our new HMD with the diverging HOE. As part of this we propose a new media expression in the field of the multimedia: "Floating Image Media". We will describe this concept in this section.

Structure of MR System We have developed an interactive MR system in order to evaluate the validity of the HMD using a diverging HOE. In this MR system, virtual stereoscopic images can be observed among the real world and virtual objects floating in the real environment can be felt through the HMD because the view of the real world behind the virtual hologram images is very wide. Figure 3.16a shows the structure of this MR system. It uses a graphic workstation for generating stereoscopic images and a magnetic sensor for head tracking. The graphic workstation creates a pair of VGA-resolution images for the left and right parallax, and those images are transmitted to the LCD attached to the helmet. The images on the LCD are diffracted by the HOE; holographic images are reconstructed by the HOE; and virtual images can be seen through the HOE. When an observer looks around at the real world by moving his/her head, the appropriate virtual objects are observed according to the head position which is determined with a magnetic head tracking sensor. In order to create high-definition images for interaction, like walking or grasping objects, SGI Indigo2 Maximum IMPACT is used as the hardware, and Division dVISE, which is a virtual-reality application package software, is used.

Concept of the "Floating Image Media" We propose "Floating Image Media" as a new expression related to using our new HMD. "Floating Image

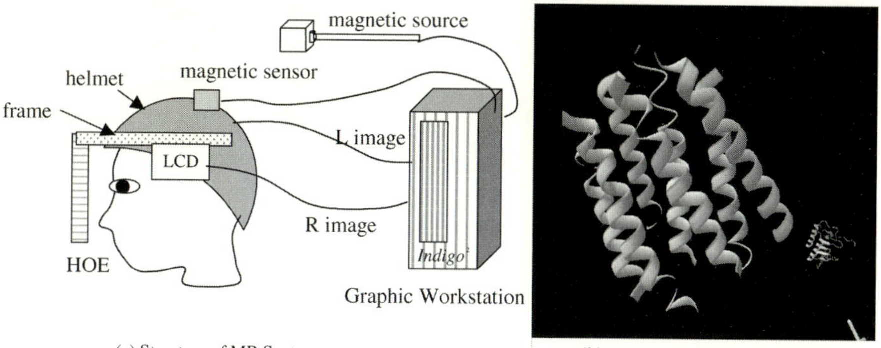

(a) Structure of MR System (b) Example of Floating Images

Fig. 3.16. Interactive MR system

Media" means to express the virtual images within the real environment and is a new media expression related to MR technology. Our ultimate goal for MR is to display electroholography among the real world to the point where real world images can be seen naturally and also the natural electronic stereoscopic images can be seen with full parallax.

But it is very difficult to achieve such a high standard of electroholography because there is currently no device which can display such a large number of interference fringes. An optical see-through HMD is now used for this purpose, but the current HMD does not have a high level of capability to combine both real-world and virtual images. Therefore, we have developed a new HMD using a diverging HOE, which is very effective for viewing the natural real environment.

By using our new HMD, the "Floating Image Media system" has been realized without using electroholography. The software developed for the first prototype of the "Floating Image Media" was for a molecular simulation, although it is possible to express any shape and motion. Figure 3.16b shows an example of a molecular simulation using the "Floating Image Media". The various molecular models can be viewed from different perspectives, depending on the position of the observer's head. We demonstrated this HMD on an interactive MR system at SIGGRAPH99. A total of 656 subjects tried it, and all except three subjects, who were stereo-blind to begin with, could observe the binocular stereoscopic images among the real world through the HOE.

3.4.5 Results

We have completed a prototype of a HMD using a diverging HOE for use in an MR environment. We proved that a single HOE can separate the binocular parallax images onto the left and right eye. Furthermore, we showed that a HOE as a combiner could improve upon the drawback of using a half-mirror, the tradeoff between transmissivity and reflectance. This current prototype is a little heavy and was not miniaturized because it was handmade. In addition, the field of view of the virtual image is a little narrow because the angle of view is about 20° horizontally, and the virtual image is monochrome red because only a red semiconductor laser was used. In future, we have to improve these elements.

3.5 Characteristics of the Converging HOE

A diverging HOE was discussed in the previous section. In this section, the optical characteristics of the converging HOE shown in Sect. 3.3.3 are described. In particular, a detailed explanation is given of its specific features related to the Maxwellian View [5] in which an image is projected onto a retina directly, and we explain how the principle of the Maxwellian View and the specification of the converging HOE have resulted in the Retinal Projection Display. We propose in this paper that a Retinal Projection Display enables production of electronic images with long focal depth by using both the principle of the Maxwellian View and an electronic Spatial light Modulator (SLM).

3.5.1 Principle of the Maxwellian View

In the Maxwellian View, parallel rays are converged directly at the center of the pupil, and projected onto the retina directly. This is used to measure the sensitivity of the human vision system or in experiments of psychological visual perception [7]. Figure 3.17 shows the principle of the Maxwellian View. The diverging rays from the point source are collimated by the lens L1 and irradiate the transparent object M. Object M is located on the front focus plane of lens L2, and the pupil of the human eye is located on the back focus plane of lens L2. In this case, Object M and the retina are conjugate, and the light stimulus, which has an extremely long focal depth, can be observed. The function of this optical system is to have the real image from the small aperture (pinhole) be seen on the pupil [8]. Using this principle, we propose a Retinal Projection Display that can provide an extremely long focal depth image without causing ocular accommodation.

3.5.2 HOE for Retinal Projection Display

The optical elements to converge the parallel rays are needed to realize the Maxwellian View, and we have adopted a diffraction grating as our HOE. This HOE can be used as the combiner for the see-through display because it can bend the optical axis and it is transparent. In addition, as was described in Sect. 3.3, through angular selectivity and wavelength selectivity, this HOE as the combiner for a see-through HMD is efficient for both transparency and

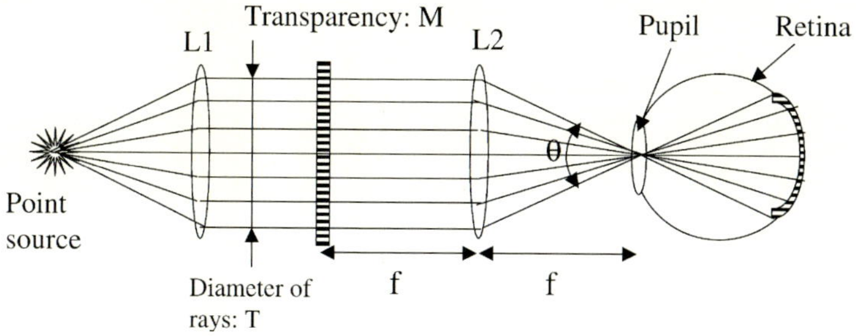

Fig. 3.17. Principle of the Maxwellian View

diffraction. Figure 3.18 shows the characteristics of our HOE made using the two-beam interference method. The rays from the left are diffracted by the HOE and converged at the left eye. The rays from the right are diffracted and converged at the right eye. A single HOE can separate the stereoscopic images to the left and right eye, and both real-world and diffracted images with high-intensity rays can be seen simultaneously because the HOE can diffract specific wavelength rays without reducing the intensity of the visible rays. Figure 3.19 shows the experimental optical system for reconstruction, in which the SLM is transparent film (OHP) on a slide mount [9].

A Retinal Projection Display, which can produce electronic dynamic images, is designed by using an LCD or DMD (digital micromirror device) as the SLM instead of still image film (OHP on the slide mount), as shown in Fig. 3.20. In this system, a semiconductor laser is used for illumination, and the color of this display is monochrome red only. In the future a full-

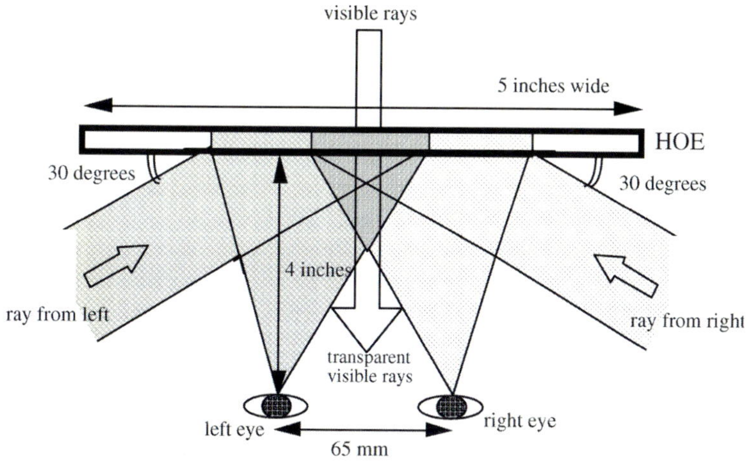

Fig. 3.18. Specification of the HOE

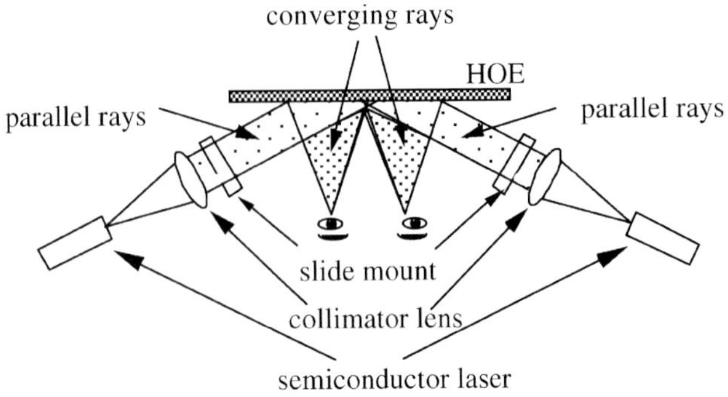

Fig. 3.19. Optical system for reconstruction

color display will be designed by using a 3-laser source (red, green, blue). We have proved the validity of a converging HOE that can produce binocular parallax images with long focal depth by using the basic experimental optical system shown in Fig. 3.19 [10]. As the first step of our research concerning the converging HOE, we made a basic optical system using both a HOE and a still image on a slide mount. However, we had to realize a prototype that can produce an electronic dynamic image based on the principle of the Maxwellian View, therefore we will describe a Retinal Projection Display that uses both the converging HOE and electronic SLM (DMD and LCD) in the next section.

3.6 Monocular Retinal Projection Display

A HOE for Retinal Projection and the verification of its capability by using a still slide image such as the SLM were introduced in the previous section. However, when put into practical use, an electronic display SLM like an LCD should be used. In this chapter, we will describe an electronic Retinal Projection Display using both converging HOE and an LCD such as the SLM.

3.6.1 Structure of Retinal Projection Display

The structure of the Retinal Projection Display with the Maxwellian View is shown in Fig. 3.20. It converges the parallel rays at the center of the pupil and projects the rays directly onto the retina. If parallel rays are emitted from the aperture of the SLM, for example an LCD or a DMD, one straight ray is emitted from each pixel and all the straight rays from the SLM are converged at one point, which in this case is the center of the pupil. These are then projected onto the retina directly. One point source of each pixel on an SLM stimulates a certain point on the retina and each straight ray of each aperture of an SLM stimulates a different point on the retina. In this way the

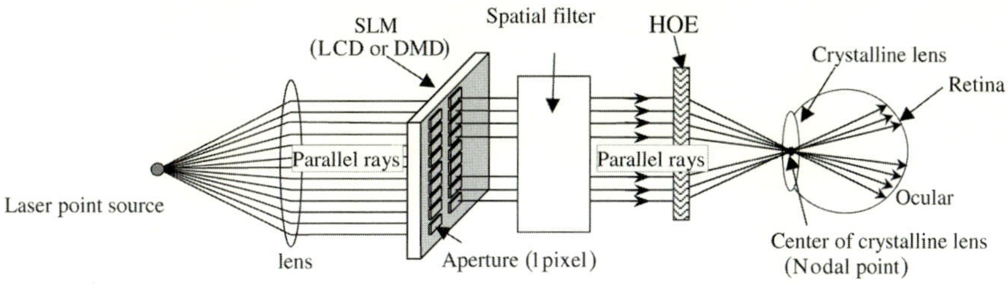

Fig. 3.20. Structure of the Retinal Projection Display

image pattern projected onto the retina directly can be recognized. The image using the Maxwellian View can be recognized without ocular accommodation because the rays from each pixel pass through the center of a crystalline lens (nodal point). In this optical system, an electronic SLM with a small aperture is used. If the SLM is illuminated by coherent laser rays, the rays passing through the SLM are diffracted by the small aperture of the SLM. They contain high-order diffracted rays. This condition is different from that in which a still image is used as the SLM, as shown in the previous section (Fig. 3.17). Therefore, to prevent this phenomenon, a spatial filter, which will be discussed later, is installed in this optical system. A prototype of the Retinal Projection Display was fabricated by using the following components: the converging HOE, the electronic SLM, the Spatial Filter, and the laser point source.

3.6.2 DMD as the Spatial Light Modulator (SLM)

We can use an LCD or a DMD as the SLM instead of transparent film, as shown in Fig. 3.20. We made a prototype using a DMD for this monocular experimental Retinal Projection Display [11]. The DMD is capable of reflecting rays with a 16 μm^2 microscopic mirror. We remodeled a market model DLP projector and extracted a part of the DMD optical engine to irradiate the laser light onto the surface of the DMD chip. Figure 3.21 shows the appearance of the DMD optical engine extracted from the product. Figure 3.22 shows the schematic of the optical system of the Retinal Projection Display using the DMD. The DMD chip (0.67 in.) itself could not be used because a part of the DMD optical engine was extracted from the DLP product and it could not be broken down to a smaller size. Therefore, this prototype is a desktop type rather than a head-mounted type because of the limitation of miniaturization. In addition, it is a monocular system. A stereoscopic system will be the next generation of our work, and will be described in the next section. The resolution of the DMD is SVGA, and the grayscale level is 256, because the backlight source is only a red semiconductor laser. In this way we completed the first prototype of the Retinal Projection Display using both a HOE and a DMD, which can produce an electronic dynamic picture. Figure 3.23

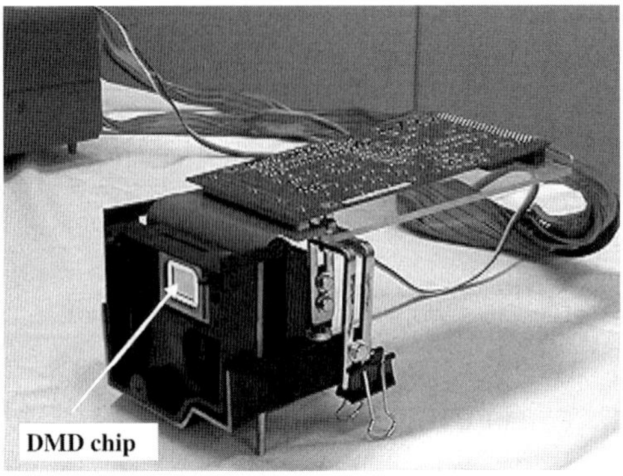

Fig. 3.21. DMD optical engine unit

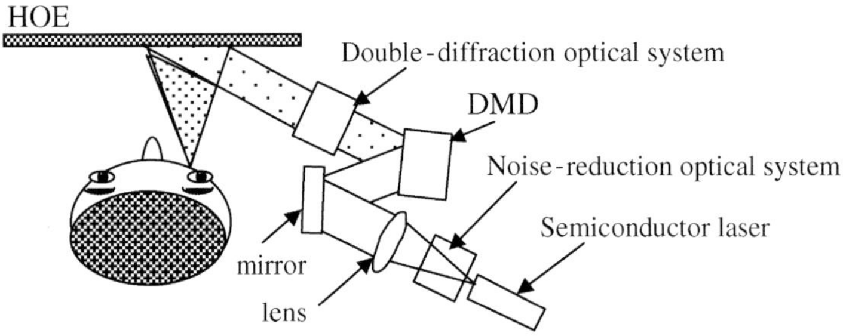

Fig. 3.22. Schematic of the display using the DMD

shows an experiment in which an observer is viewing Maxwellian View images through the HOE with a Retinal Projection Display using a DMD.

Coherent noises are observed in this optical system because a laser light source is used as the illumination on the DMD, so it is equipped with a noise-reduction optical system. The laser beam is waved slightly by the diffuser to reduce the noise [7]. The diffuser is attached in front of the magnified lens which converts the beam to parallel rays, and the smoothing to reduce the noise is achieved by rotating the diffuser.

3.6.3 Spatial Filtering

In this optical system, diffraction by the mirror of the DMD occurs because laser light is used as the illumination source. In order to avoid this phenomenon, it is equipped with an optical system to eliminate high-order diffracted rays. If the parallel rays are incident on the HOE just as they are, due to high-order diffracted rays multiple images will be seen. In order to get the parallel rays to be incident on the HOE without the high-order diffracted rays, a double-diffraction optical system has been designed to extract the zero-order diffracted rays by spatial filtering.

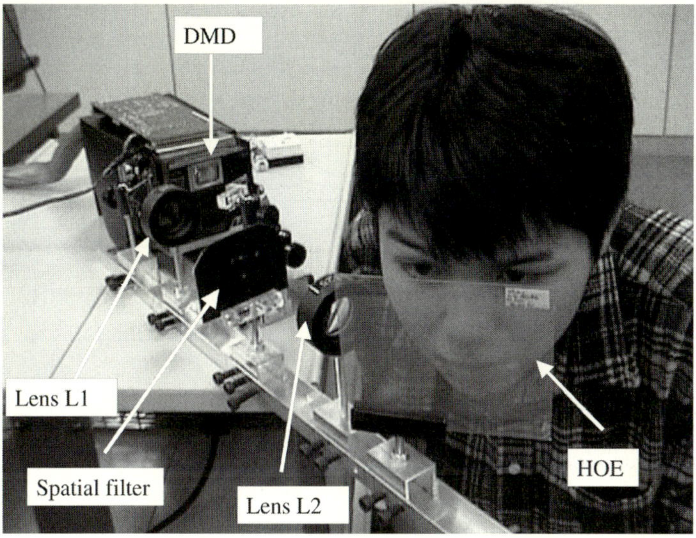

Fig. 3.23. The observer watching the images through the converging HOE

Figure 3.24 shows the concept of the double-diffraction optical system for filtering to extract only zero-order/diffracted rays. The rays, which are emitted from the SLM as parallel rays, are Fourier transformed by optical lens L1. The SLM, that is the DMD, is put at the first focal point (focal length: f_1) of lens L1, then the Fourier-transformed plane is achieved on the second focal point (focal length: f_1) of lens L1. The Airy patterns that include both the zero- and higher-order diffracted rays can be seen on the Fourier plane of lens L1. Next, the spatial filter, which is a pinhole, is installed on the Fourier plane. Then, lens L2 is used, it is installed at the second focal point of lens L1, which is the same point as the first focal point (focal length: f_2) of lens L2. Consequently, the rays emitted from lens L2 are parallel rays without high-order diffracted rays. These rays are the image with only zero-order diffracted rays on the surface of the Fourier plane. If the rays after lens L2 are incident on the HOE and the pupil is fitted at the converging point of the HOE, images on the SLM are seen by the Maxwellian View.

3.6.4 Results

The results we achieved with the experimental optical system, as shown in Fig. 3.22, allowed us to see an outside scene through the HOE as well as the dynamic images displayed on the computer by the Maxwellian View. In the case of the Maxwellian View, ocular accommodation is unnecessary, and we proved that even a person with myopia could observe the Maxwellian View images without glasses.

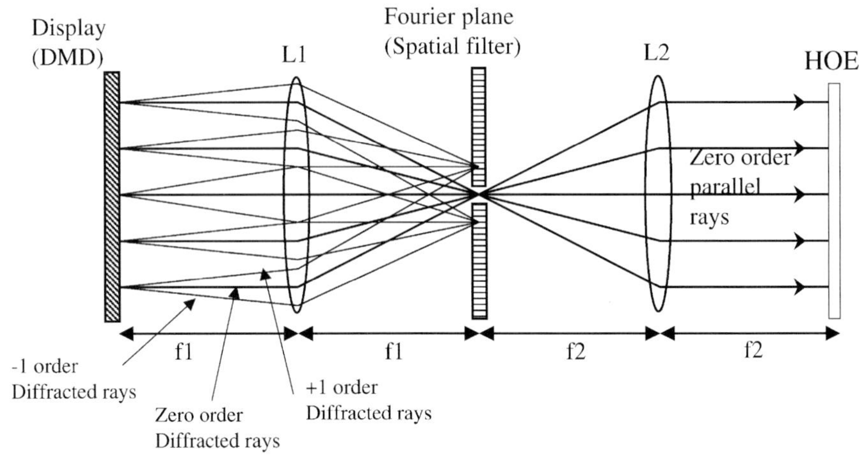

Fig. 3.24. Double-diffraction optical system

Experimental results photographed using a CCD camera are shown in Fig. 3.25 for when the focus to the outside world through the HOE was changed. Figure 3.25 a shows the case of focusing on real objects correctly, and the virtual images produced by the Maxwellian View can be observed clearly. Meanwhile, Fig. 3.25b shows the case of defocusing of the CCD camera. Even though the focus to the outside world is dim, the virtual image produced by the Maxwellian View is clear. These results show that the virtual images are recognized clearly even when the focus of the camera changes. These virtual images are dynamic pictures displayed on the computer. We proved that the Retinal Projection Display produced by the Maxwellian View could be

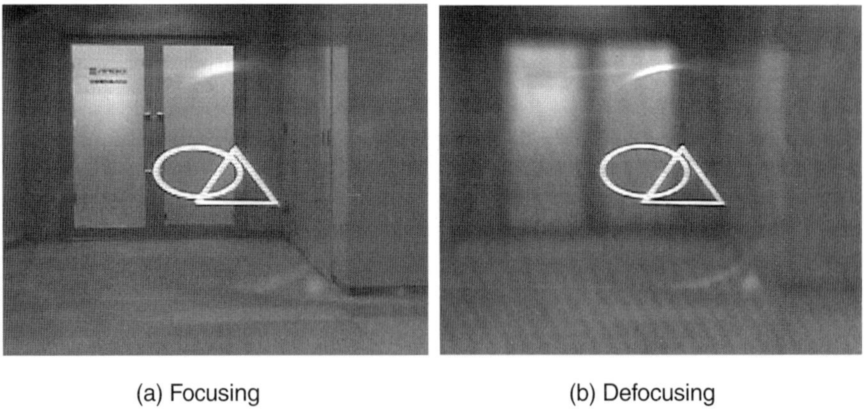

Fig. 3.25. Images through HOE

observed clearly any time the ocular accommodation was changed because its image had an extremely long focal depth.

3.7 Binocular Stereoscopic Retinal Projection Display

The Retinal Projection Display can produce an electronic image that has a long focal depth. In the previous section, we introduced monocular Retinal Projection Display using a DMD as the SLM. In this section we will introduce binocular stereoscopic Retinal Projection Display using an LCD.

In order to realize a binocular stereoscopic display, a monocular display should be set up for each eye. If the monocular display is large, it cannot be set up for a single eye; therefore, a Retinal Projection Display needs to be small-sized. The monocular display using a DMD discussed in the previous chapter is too large for just a single eye. Furthermore, a binocular display uses an LCD instead of a DMD as the SLM. In this chapter, the structure of the binocular display and the method of adjusting to the pupil distance are described.

3.7.1 Structure of Binocular Stereoscopic Retinal Projection Display

Figure 3.26 shows the structure of the binocular stereoscopic Retinal Projection Display. This display system uses a HOE that has the same specifications as those detailed in Sect. 3.5, and an LCD with 270,000 pixels over 0.7 in. as the SLM, instead of the DMD described in the previous section. A laser light source is used as a backlight for the LCD. The double-diffraction optical system and the noise-reduction optical system are similar to those of the monocular display. For the monocular optical system, an observer sees an image by head movement and setting the pupil at the converging point of the diffracted rays. For the binocular optical system, the pupil distance can be adjusted for each eye so that binocular stereoscopic images are displayed on both eyes separately. A jaw and forehead rest fixes the position of the observer's pupils, and the adjustment of the pupil distance is made by moving the optical system, which is mounted on an optical bench.

The incident angle of the parallel laser rays is 30°. If this angle is changed slightly, the converging point will move slightly too. In this way, the two converging points of the diffracted rays on the HOE can be adjusted for each pupil. Figure 3.27 shows how this optical system looks. The elements of the optical system, such as the semiconductor laser, the collimator lens, and the double-diffraction optical system, are mounted on an optical bench, as shown in Fig. 3.27. The optical bench can be adjusted by hand to adjust the incident angle of the parallel laser rays modulated by the LCD. Figure 3.28 shows an observer looking at an image through this optical system.

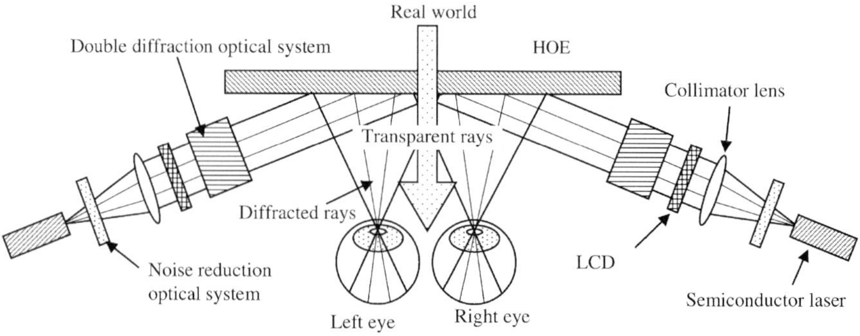

Fig. 3.26. Structure of binocular stereoscopic retinal projection display

3.7.2 Adjustment for the Pupil Distance

To see stereoscopic images through this optical system the center of the pupil has to be precisely set at the center of the converging point. The pupil distance is different for each pupil, so an adjustment mechanism for the two converging points of rays is necessary. When the incident angle to the HOE of the parallel rays is changed, the converging point moves according to the equation of holography imagery. If the incident angle is changed $\pm 3°$ from the designed standard angle of $30°$, the converging point when the distance from the HOE is 100 mm moves 3 mm right or left, as shown in Fig. 3.29. The pupil distance can be changed from 59 to 71 mm by adjusting each angle of the parallel rays.

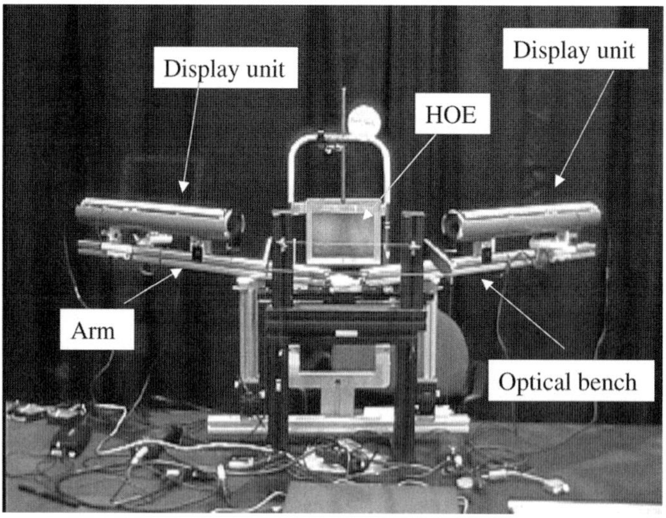

Fig. 3.27. Appearance of optical system

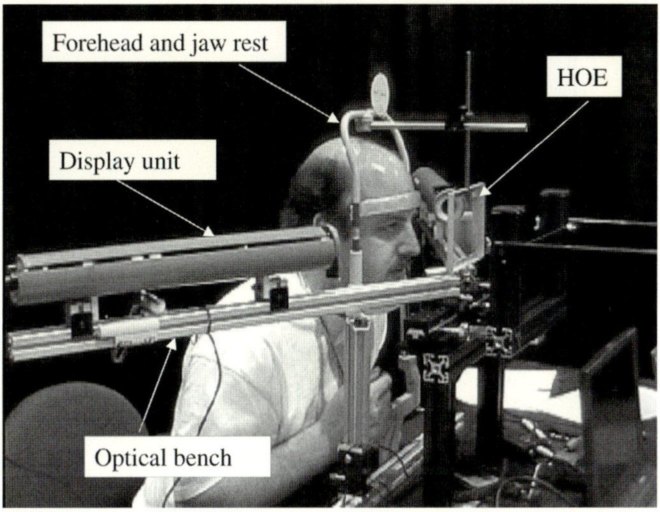

Fig. 3.28. Observer using the optical system

However, the diffraction efficiency decreases, due to the angular selectivity of the HOE, if the incident angle is changed from the designed standard.

Therefore, the relationship between a change in the incident angle and the diffracted efficiency is discussed in this section. The angular width $\Delta\theta$ is expressed below (n, average refractive index of hologram; x, interference fringe interval in the thickness direction; λ_0, wavelength used in recording; s, thickness of hologram; and θ_r, angle of incident ray)

$$\Delta\theta = n\frac{x}{s} = n\frac{\lambda_0}{s}\sin\theta_r .$$

Substitute in the above $s = 6\,\mu\text{m}$, $n = 1.63$, $\theta_r = 60°$, $\lambda_0 = 632.8\,\text{nm}$, and the result is $\Delta\theta = 11°$. This value means that the diffraction efficiency will be zero when the incident angle changes to 11° from the designed standard incident angle of 30°. A displacement within 3° can compensate for the individual differences in human pupil distance in this optical system. Since 3° is less than 11°, this reduction in the diffraction efficiency is within allowable parameters. Concerning the actual diffraction efficiency, it is 35.9% for the standard incident angle of 30°; when the displacement is +3°, it is 34.8%; and when the displacement is −3°, it is the same, 34.8%.

3.7.3 Results

Figure 3.30 shows diffracted images that can be seen as electronic images on this optical system. Figure 3.30a shows the left diffracted image, and Fig. 3.30b shows the right diffracted image. In these images we can recognize

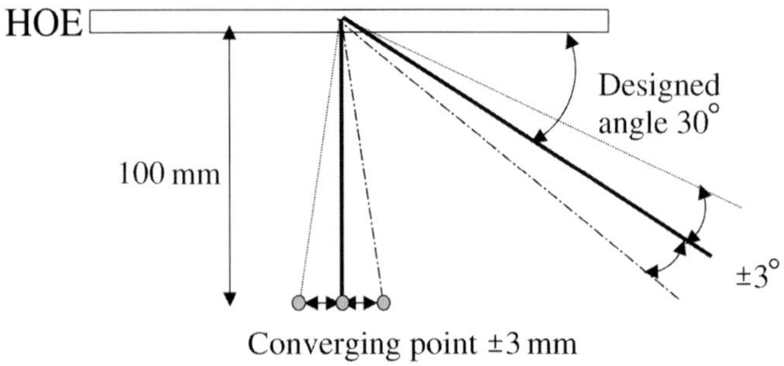

Fig. 3.29. Relationship between the incident angle and the converging point

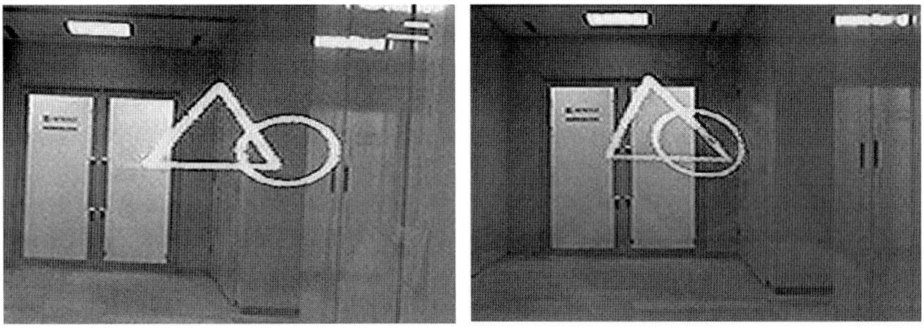

(a) Image for left eye (b) Image for right eye

Fig. 3.30. Stereoscopic images on the binocular retinal projection display

the left and right parallax between a triangle and an ellipse. The resolution and intensity of an image on this system using an LCD are inferior to those of a monocular system using a DMD; however, the field of view of a reconstructed image and the transparent efficiency for the outside world are the same as for the monocular system.

We developed a prototype in which an electronic binocular image can be seen by Maxwellian View that has an extremely long focal depth by using a converging HOE and an LCD. We developed a binocular Retinal Projection Display using both the HOE and the LCD. In addition, we proved that virtual images that had the extremely long focal depth could be seen by using the Maxwellian optical system. In our prototype, stereoscopic dynamic images on the computer could be seen by an SLM that is an LCD. In the future, we have to achieve miniaturization and produce a head-mounted optical system.

3.8 Discussion

We have introduced our work on a see-through HMD using a diverging HOE and a Retinal Projection Display using a converging HOE. In particular, we showed that the Retinal Projection Display has the unique feature of producing a long focal depth by Maxwellian View. In this section, we will consider the viewing area of this Retinal Projection Display, which in the future will be a Head-Mounted Optical system.

3.8.1 Viewing Area of Maxwellian View

When viewing an image by Maxwellian View, the viewing point of the display is limited to being a pinpoint because the pupil center has to be the converging point of the rays. This correlates with the back focus plane of lens L2, as shown in Fig. 3.17. Therefore this optical system requires a calibration capability as to the position of the pupil center. In addition, a change in eye relief or the position of the pupil caused by any head movement has to be considered. In this section, the viewing area of the Maxwellian View will be discussed.

Figure 3.17 shown in Sect. 3.5.1 shows a typical case in which the center of the pupil corresponds with the converging point of the rays. However, a certain amount of deviation between the pupil center and the converging point is permitted. The permissible range of deviation between the pupil center and the converging point is within the diameter of an observer's pupil when his eye moves in the vertical direction. In this case the converging point of the rays can be located within the diameter of the pupil, as shown in Fig. 3.31a. Supposing that the diameter of the pupil is D and the deviation of the centerline of the pupil to the optical axis is $\mathrm{d}y$, the range of viewing area $\mathrm{d}y$ in the case of the vertical direction to the optical axis is expressed by

$$-\frac{D}{2} < \mathrm{d}y < \frac{D}{2} . \tag{3.1}$$

In the case of the horizontal direction to the optical axis, the pupil can be put within the range in which all rays diverging from the focus point are incident on the diameter of the pupil, as shown in Fig. 3.31b. Suppose that this deviation is $\mathrm{d}x$, and the range of the viewing area $\mathrm{d}x$. In this case the horizontal direction to the optical axis is expressed by

$$\mathrm{d}x = \frac{D}{2} \cdot \cot\left(\frac{\theta}{2}\right) . \tag{3.2}$$

Consequently, the range for which all rays are incident within the eyeball is found to be within a diamond-shaped field, as shown in Fig. 3.31c. Except for within this field, the rays cannot be entirely seen or only partial rays can be seen because of the shading by a pupil's edge. Now, supposing that D is 2–8 mm and the field-of-view angle θ of the image is 30°, $\mathrm{d}y$ is equal to 1–4 mm and $\mathrm{d}x$ is equal to 4–15 mm. Therefore, we consider that if the precision of the calibration is kept within a 1 mm extent then the center of

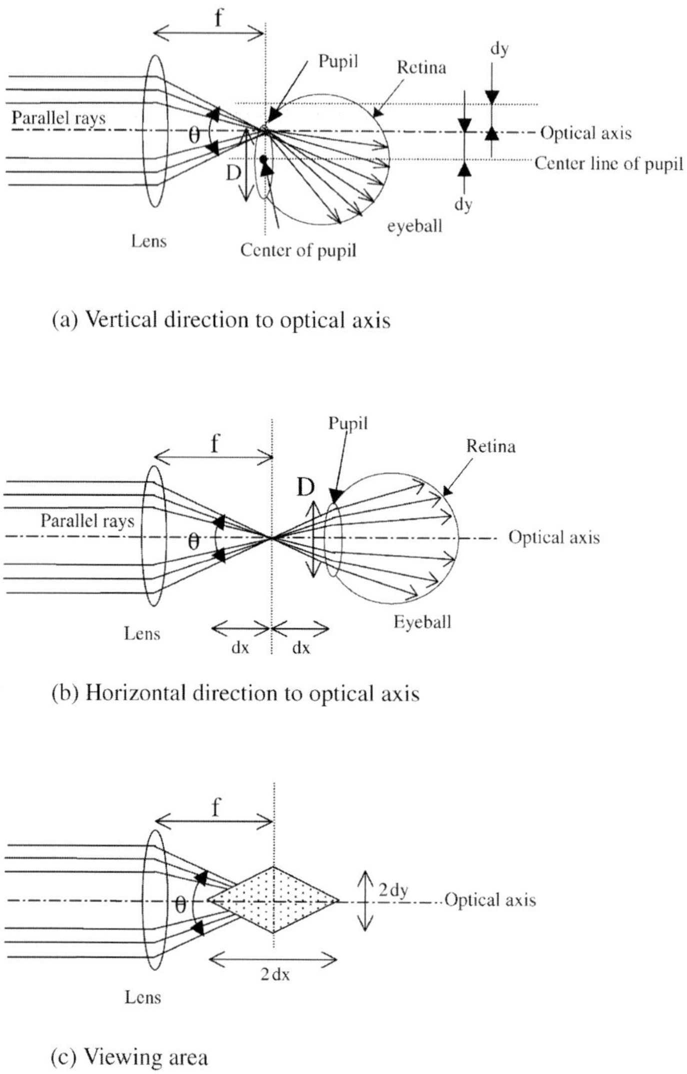

Fig. 3.31. Viewing Area of Maxwellian View

the pupil can be kept within the diamond-shaped field, as shown in Fig. 3.31c by supporting the headset although the eye relief, the center of pupil, or the optical axis is changed from its original position. We believe it is possible to have a Maxwellian View optical system on a Head-Mounted Optical system.

3.8.2 Viewing Area in Monocular Movement

In the previous section we described how it is possible to produce a Maxwellian View optical system on a Head-Mounted Display although the viewing area in the Maxwellian View is very narrow. In this section, we will consider whether or not the viewing area can be maintained when there is ocular movement, especially adduction in the monocular condition.

Supposing that for the initial calibration the optical axis corresponds with the line of sight in which the direction "OC" is viewing an infinitely far point, as shown in Fig. 3.32a. From this condition, the line of sight "OP" can be adducted (O, the center of rotation; R, the distance between the center of rotation and the center of the pupil; f, the focal length of the lens; D, the diameter of the pupil).

Then, supposing that the critical angle of adduction in which the rays of the Maxwellian View can be seen is ϕ and ϕ is the condition in which the line of sight "OP" will correspond with the edge of the pupil, as shown in Fig. 3.32b. ϕ is expressed as

$$\phi = \tan^{-1}\left(\frac{D}{2R}\right) . \tag{3.3}$$

Meanwhile, supposing that the critical angle of adduction to the image which has a 30° field of view angle is δ, and δ is the condition in which the line of sight "OP" will correspond with the edge of the field of view of the displayed image, as shown in Fig. 3.32c. δ is expressed as

$$\delta = \tan^{-1}\left[\frac{f\tan(\theta/2)}{(R+f)}\right] . \tag{3.4}$$

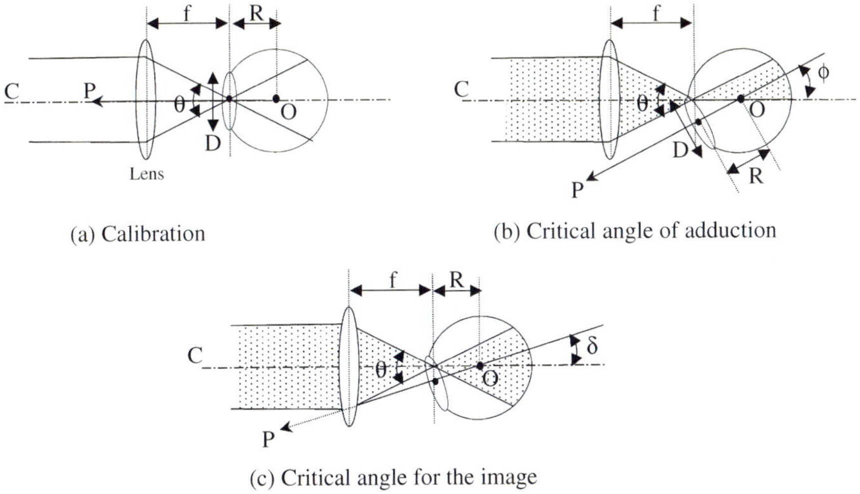

Fig. 3.32. Viewing area for adduction

Supposing that R is half of the standard 24 mm diameter of an eyeball, in the case of a minimum diameter $D = 2\,\text{mm}$, $\phi = 5°$, and in the case of a maximum diameter $D = 8\,\text{mm}$, ϕ is equal to 18°. Then, supposing that the field of view θ is 30°, in the case of $f = 30\,\text{mm}$, δ is equal to 11°.

The present HMD has about a 30° field-of-view angle. If the Maxwellian View image is almost equal to this angle, eye movement is permitted within the whole image field ($2\phi = 36° > 2\delta = 22°$) with a maximum pupil diameter. However, the range of eye movement is narrow with a minimum pupil diameter ($2\phi = 10° < 2\delta = 22°$). Therefore, it is desirable that a HMD with a Maxwellian View should be equipped with eye tracking in the future. However, we believe that a Retinal Projection Display can be useful even without eye tracking because an image can be seen if the field-of-view angle is from 15° to 20°.

3.8.3 Viewing Area in Binocular Movement

In this section, we will consider whether or not the viewing area can be maintained when there is ocular movement, especially convergence in the binocular movement.

The range of eye movement for the Maxwellian View depends on the critical angle ϕ mentioned in a previous section, and we consider the range of convergence in binocular movement by using the critical angle ϕ. The distance of the near point ds is expressed in (3.5) according to Fig. 3.33 by using ϕ and "Pd" which is the pupil distance.

$$\mathrm{d}s = (\mathrm{Pd}/2)/\tan\phi = \mathrm{Pd}\cdot\frac{R}{D}\,. \tag{3.5}$$

In this condition, the near point by convergence ds in binocular stereoscopy can be changed from 390 mm to infinity for a minimum pupil diameter (= 2 mm), and from 98 mm to infinity for a maximum pupil diameter (= 8 mm) ("Pd" is 65 mm).

In conventional binocular stereoscopic display, there is the problem of dissociation between accommodation and convergence. The most suitable

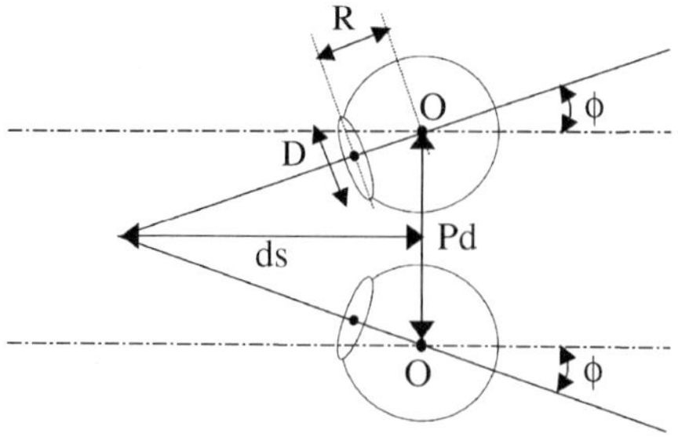

Fig. 3.33. Near point of convergence in the Maxwellian View

condition, which can be seen as natural stereoscopy, has been proposed so far. This is that the convergence should be set from 18 cm to 2 m when the screen position is 50 cm. Taking account of this condition, we consider that the range of convergence in the Maxwellian View is almost the same or superior to conventional stereoscopic display. Therefore, it can be concluded that the Retinal Projection Display has enough range for convergence and focal depth without ocular accommodation.

3.9 Application

We produced a binocular Retinal Projection Display as a desktop optical system. As previously stated, we believe that a retinal projection display should be equipped with an eye movement tracking function because the position of the pupil center when seeing images through this display was restricted. Even though this was the case, stereoscopic images could still be seen through it without the tracking. In the future, we anticipate that the binocular Retinal Projection Display developed so far will be miniaturized and used on a head mount. While our final goal is a binocular retinal projection HMD, at the same time we are investigating other applications. In this section, we will introduce the Retinal Projection Display for visually handicapped persons which is a recent significant application.

A visually handicapped person can perceive an image using the retinal projection display because it has a long focal depth and it can project the images on the retina directly through a spot like an artificial pupil. Therefore, we have developed this kind of experimental display for clinical examination in order to prove that it would be effective for a visually handicapped person. Our main purpose in this prototype for clinical examination is to prove the effectiveness of a visually handicapped person using this type of display, and not producing a head-mounted and binocular optical system. The prototype for clinical examination is a desktop optical system. However, if the optical length is shortened and the exterior package is lightened, it will be possible to make it a head-mounted optical system.

The monocular retinal projection display for clinical examination is shown in Fig. 3.34. It uses a 0.6 in. LCD and a semiconductor laser, and its field of view is 30°. The ocular element of this optical system is not a HOE, but is a combination of a concave mirror and a half-mirror. The focal length of the ocular element is 20 mm in order to realize a field of view of 30° with the 0.6 in. LCD. The concave mirror is used as the short-focus lens without color aberration, and the half-mirror is used as the see-through combiner.

Visually handicapped persons suffer from various symptoms and it is difficult for them to observe the images on a display when the visual axis and optical axis are different. This optical system is equipped with an XYZ stage and a jaw rest. An observer will see the images on the display with their face fixed on the forehead and jaw rest.

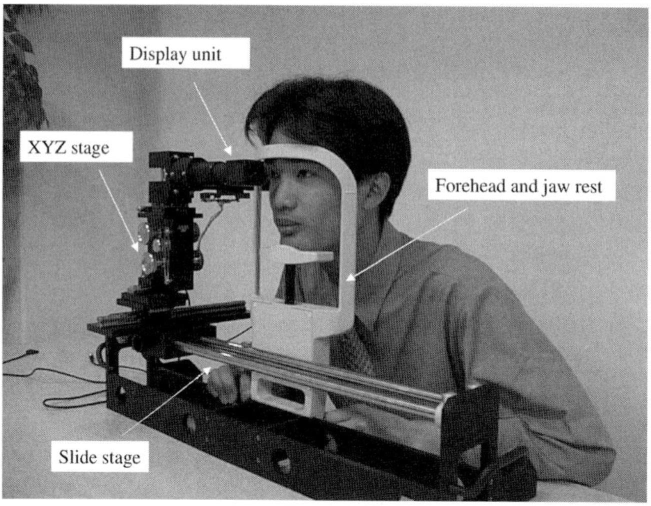

Fig. 3.34. Monocular retinal projection display for clinical examination

For a visually handicapped person whose retinal cells near the fovea are damaged, the images must be projected around the fovea. This optical system is equipped with rotating mechanics that can intersect the visual axis and optical axis in the center of the pupil in order to project the images onto an optional area of the retina. When applied to persons whose vision is ametropic, clear images, in particular, can be seen without the influence of an astigmatic lens. In the future we intend to experiment with the application of our display to persons with low vision or retinitis pigmentosa. Furthermore, we would like to produce a new display as a vision aid to persons with low vision. If it were a binocular stereoscopic display, visually handicapped persons who had no stereopsis so far would be able to enjoy stereoscopic entertainment.

3.10 Conclusions and Outlook

We proposed two kinds of displays using a holographic optical element (HOE). First, we produced a head-mounted display (HMD) using a diverging HOE. Second, we produced a Retinal Projection Display using a converging HOE.

The HMD using a diverging HOE was realized as a see-through HMD. We demonstrated the superiority of this new HMD which has a wide field of view to the outside world and excellent see-through capability by using a HOE instead of the half-mirror used for a conventional HMD. In addition, we proved the fundamental validity that a HOE could be used as the combiner of the see-through HMD. We realized an interactive mixed reality (MR) system with this HMD, calling this new media concept "Floating Image Media". In this new world, virtual objects can be seen as if they are floating in the air, and the outside world can be seen clearly, with both a wide field of view and high intensity, through our new HMD. However, in order to produce an HMD using a HOE for practical use, miniaturization and full color for the optical system are required. In the future, the durability and precision of the HOE also need to be dealt with, because our HOE was made by hand in a laboratory. Furthermore, to improve the system we have to consider not only the hardware but also the software. For example, it is necessary to determine what information is important for industrial applications.

The Retinal Projection Display prototype proved the basic validity of the converging HOE for application in a see-through HMD. We produced an optical system as the Retinal Projection Display which could produce electronic moving pictures with long focal depths. Images can be seen without ocular accommodation. In other words, our new optical system can produce images with a pan-focus capability. The Retinal Projection Display consists of a converging HOE, laser illumination, an LCD, and some optical lenses. By using laser rays as the backlight of the LCD, high-order diffracted rays appear in the rays which are emitted from the aperture in the LCD, and they produce multiple images. In order to eliminate them, we proposed a double-diffraction optical system and we proved its validity by making a prototype. It has not yet been completed as an optical system for a HMD yet. However, we were able to propose a new concept for an optical system with images that have extremely long focal depths. We produced a prototype in which electronic stereoscopic images could be seen as long focal depth images through the Retinal Projection Display we proposed in this paper. This may solve the drawback found in binocular stereoscopic display related to the dissociation between accommodation and convergence. In addition, with the Retinal Projection Display using the HOE, we have developed a new image media that can provide us with a new world which we have not seen so far. We have to consider whether this new media will be accepted in our image information society or not.

The display is a nucleus device for multimedia. Our information process depends on visual information, and, as long as the saying, "Seeing is believing" holds true, the display, whether it is two-dimensional or three-dimensional, will be necessary in our daily lives. We think that the ultimate stereoscopic display technology may be electroholography; however, more extensive research and some breakthroughs are needed in order to apply it practically. It is our judgment that we have to use binocular parallax stereoscopy until electroholography has been developed enough to be put to practical use.

References

1. G. Hamagishi, M. Sakata, A. Yamashita, K. Mashitani, E. Nakayama, S. Kishimoto, K. Kanatani: 'Stereoscopic LC display without Special Glasses'. SID Digest of Applications Papers, **XXVI**, 75 (1995)
2. G. Burdea, P. Coiffet: *Virtual Reality Technology* (Wiley 1994)
3. H. Tamura, Y. Ohta: *Mixed Reality* (Ohmsha 1999)
4. R.T. Azuma: Presence: Teleoperators and Virtual Environments **6**, 355 (1997)
5. G. Westheimer: Vision Res. **6**, 669 (1966)
6. T. Ando, H. Takahashi, E. Shimizu: 'Head mounted display for mixed reality using holographic optical element'. Mem. Fac. Eng., Osaka City University, **40**, 1 (1999)
7. A. Felipe, J.M. Artigas, A.M. Pons: J. Opt. Soc. Am. A **14**, 972 (1997)
8. M. Inami,N. Kawakami, T. Maeda, Y. Yanagida, S. Tachi: 'A stereoscopic display with large field of view using maxwellian optics'. *ICAT'97 Proc*, pp. 71–76 (1997)
9. T. Ando, K. Yamasaki, M. Okamoto, E. Shimizu: SPIE Proc. **3293**, 183 (1998)
10. T. Ando, K. Yamasaki, M. Okamoto, T. Matsumoto, E. Shimizu: SPIE Proc. **3637**, 110 (1999)
11. T. Ando, K. Yamasaki, M. Okamoto, T. Matsumoto, E. Shimizu: SPIE Proc. **3956**, 211 (2000)

4 Three-Dimensional Television System Based on Integral Photography

F. Okano, H. Hoshino, J. Arai, M. Yamada, and I. Yuyama

4.1 Introduction

The integral photography (IP) method was first proposed by Lippmann in 1908 based on a photographic technique [1, 2]. The IP method is classified as an image reproduction type the same as holography, because it reproduces three-dimensional (3-D) images identical to the objects by integrating the light rays from the elemental images behind each aperture or lens. The method creates autostereoscopic 3-D images that do not require special viewing glasses. It also facilitates the continuous viewpoint of horizontal and vertical direction without flipping. These advantages make it one of the favored methods for producing 3-D images [3–5].

In this chapter, we discuss four problems that have been identified: a) simple lens arrangements produce pseudoscopic images; b) optical cross talk between elemental image; c) 3-D images reproduced far from the lens array have lower resolution than those around the lens array under some conditions; d) difficulty of realizing the real-time pickup and display system. We propose some means to overcome these problems, and describe an experimental system that provides reproduced images in real time. The results show that the system provides full-color, full-parallax, and autostereoscopic images with motion parallax using natural light during pickup and display [6]. Resolution of the reproduced image is low at present, however, this system has potential for a 3-D television system.

4.2 Principle of Integral Photography

Figure 4.1 shows the principle of the basic IP method proposed by Lippmann. An array of microlenses is placed in front of a film. Numerous small images of the object, equal in number to that of the microlenses, are projected onto this film. These so-called elemental images are inverted when the object is placed in front of the lens array (on the opposite side of the lens array to the film). An object is shot from many different angles that are slightly different from each other; so that each elemental image differs slightly from the neighboring one. These differences include 3-D information. This film is developed into a transparent photograph, which is then placed where the film was when the object was captured. Next, the photograph is irradiated by white diffused

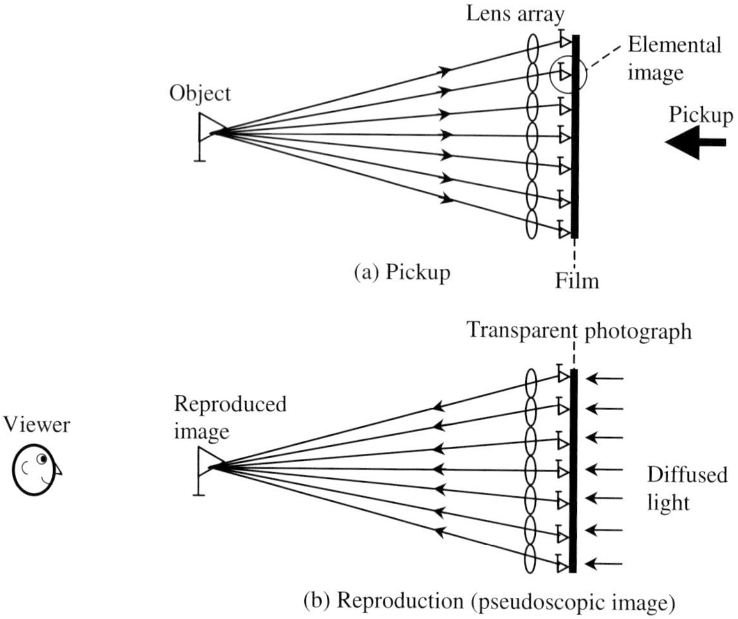

Fig. 4.1. Principle of IP

light from behind. The light rays from the photograph will travel along the same path as the light did when the object was captured, thus forming a 3-D image at the point where the object was initially located. As the light rays of the original object are reproduced, the 3-D image appears as if the object itself is present. Therefore, no special viewing glasses are necessary to observe this 3-D image. The viewer changes his position and still can see the image corresponding to his position. The IP method is therefore regarded as one of the ideal 3-D system.

If W_{L} represents the width of the lens array, P_{e} the lens array pitch, N_{m} the number of microlenses, and N_{e} the number of pixels in an elemental image, P_{c} pixel pitch of pickup device, then we obtain horizontally or vertically:

$$W_{\mathrm{L}} = N_{\mathrm{m}} P_{\mathrm{e}} = N_{\mathrm{m}} N_{\mathrm{e}} P_{\mathrm{c}} \,. \tag{4.1}$$

From this equation, the total number of pixels that the film requires is the product of the number of microlenses and the number of pixels constituting the elemental images. The number of microlenses determines the upper limit of resolution, and should be the same as that of pixels of the conventional television in an ideal IP system. The number of pixels in the elemental image effects the resolution of the 3-D image away from the lens array. The film requires N_{e} times the number of pixels compared with the conventional television, horizontally and vertically, therefore N_{e}^2 times the numbers as a whole. The same is true for the number of pixels for the display. Thus, extremely high resolution is required for both the film and photograph.

4.3 Pseudoscopic Image and Optical Crosstalk

According to the principle of the IP method as shown in Fig. 4.1, the direction of shooting an object is opposite to that of observation. A 3-D image, which is supposed to form behind the photograph, is positioned in front of the photograph. Further, convex shapes will appear concave when displayed, hence depth is inverted. Several methods have been proposed to solve them. One of the methods was given by Ives in 1931, which as shown in Fig. 4.2, is to capture the depth-inverted image once more [7, 8]. Depth inversion occurs twice, and the result is an image with correct depth direction. Hamasaki and others have proposed an alternative to these approaches called orthoscopic-pseudoscopic image-conversion optics (ICO [9]) using a prism array.

Although these methods take different approaches to solving depth inversion, the characteristics of the elemental image remain the same. As shown in Fig. 4.3, the viewer's eyes must be placed in the direction of the light rays, and the film must be placed on the opposite side of the lens array. To preserve the light ray direction, each elemental image needs to be erected when shooting an

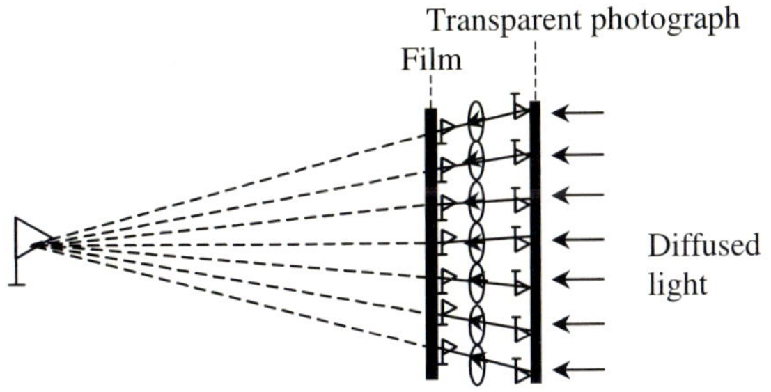

Fig. 4.2. Second step pickup

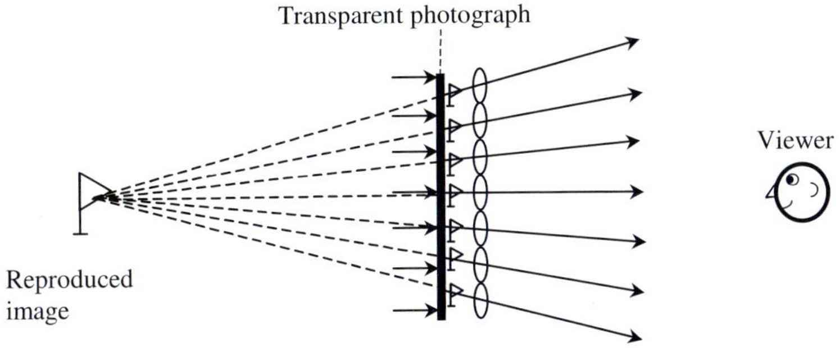

Fig. 4.3. Orthoscopic reproduced image

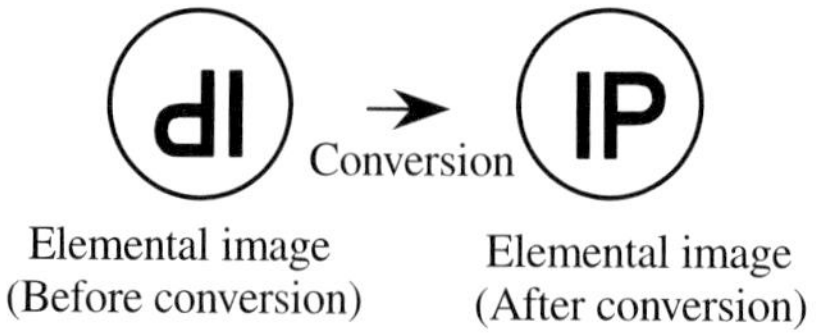

Fig. 4.4. Orthoscopic-pseudoscopic conversion

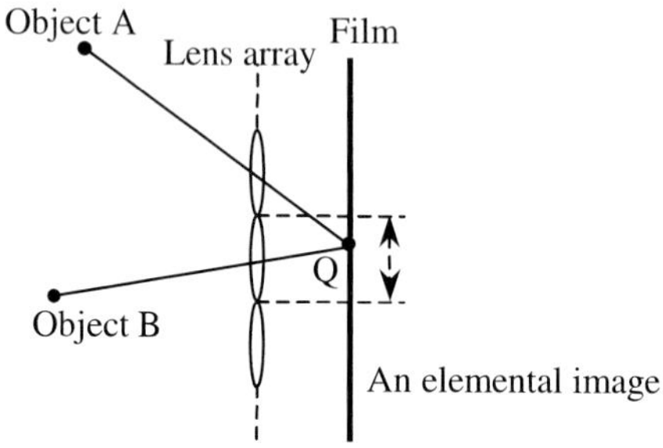

Fig. 4.5. Optical crosstalk

object placed in front of the microlenses array. The elemental image becomes inverted if conventional convex lenses are used as elemental lenses for pickup. When displayed, the elemental image must be erected (Fig. 4.4) which is done by turning it by 180° around its center. This conversion to a noninverted image can also be done by other optical or electronic means.

Another problem is optical crosstalk between elemental images. Figure 4.5 shows a schematic of an enlarged portion of the lens array. Light rays from two different objects (A, B) converge at the same point Q, causing optical crosstalk. But they cannot be properly separated during image reproduction. Consequently, a double-reproduced image results. To obtain clear images free from double or multiple images, it is essential to prevent such optical crosstalk between elemental images during pickup. Optical barriers have to be set between neighboring elemental lenses.

4.4 Gradient-Index Lens Array

A gradient-index (GRIN) lens, well known as a multimode fiber, was first proposed by Marcuse and Miller in 1964 [10]. In the following discussion, we show that this lens can be used to avoid pseudoscopic images with depth inversion [11]. In the paraxial approximation case, a gradient-index lens should have a square-law index dependence, which can be represented by:

$$n(r) = n_0 \left(1 - \frac{1}{2} A_c r^2\right) , \tag{4.2}$$

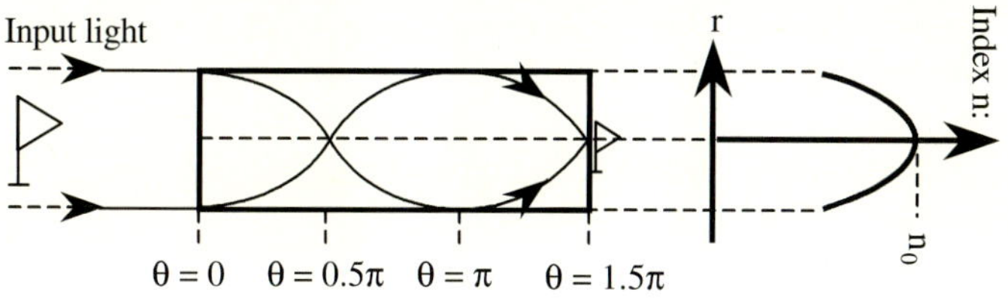

Fig. 4.6. GRIN lens

where, r is the distance from the optical axis, $n(r)$ is the refractive index in the direction of r, n_0 is the refractive index on the optical axis, and A_c is the gradient constant.

When the distance between an object and the input plane of the GRIN lens is large enough, as shown in Fig. 4.6, the optical path curves periodically because the index changes continuously along the radius within the optical lens, that is, the optical path and the ray trajectory within the GRIN lens are periodic. Whether the lens provides an inverted or erected image [12] depends on the length of the GRIN lens. The following equations represent focal length f_p and the relation between the position of an object and that of the focused image.

$$f_p = \frac{1}{n_0\sqrt{A_c}\sin\theta}\,, \tag{4.3}$$

$$L_2 = \frac{n_0 L_1\sqrt{A_c}\ \cos\theta + \sin\theta}{n_0\sqrt{A_c}\ \left(n_0 L_1\sqrt{A_c}\ \sin\theta - \cos\theta\right)}\,, \tag{4.4}$$

$$M = \frac{-1}{n_0\sqrt{A_c}\ \sin\theta\left(L_1 - \frac{\cot\theta}{n_0\sqrt{A_c}}\right)}\,, \tag{4.5}$$

$$\theta = \sqrt{A_c}\ G\,, \tag{4.6}$$

where, G is the length of the gradient-index lens, L_1 is the distance between the incident plane and the object, and L_2 is the distance between the output plane and the image. The normalized length θ is given by (4.6). Its unit is radians. Lateral magnification M in (4.5) shows the ratio between the size of the object and that of its image. If M is a positive number, the image is erect in relation to the object. By calculating $(2k-1)\pi\theta < 2k\pi$ (k is a natural number), we obtain the condition for a positive M, meaning that the image is erect, when L_1 is set properly. When $k = 1$, in particular, $\pi < \theta < 2\pi$. As mentioned above, because the IP method requires erect elemental images on the display plate to avoid pseudoscopic images, in this condition, the pickup device is set on the exit plane of the GRIN lens.

The elemental images are produced on the pickup device. The images are isolated from each other. Optical crosstalk between them is eliminated. This condition is achieved by $L_2 = 0$ in (4.4), giving

$$\tan\theta = -n_0 L_1 \sqrt{A_c} \,. \tag{4.7}$$

In particular, when the distance between the object and the GRIN lens, L_1 is large enough, $\theta = 1.5\pi$. This length is applied to the elemental lens. Consequently, the two problems, of reversed depth and optical crosstalk are solved at the same time.

4.5 Description of MTF

The authors have published detailed reports [13] on the resolution of the display system. Their results can also be applied to the image pickup system. Based on the results, this chapter discusses the overall resolution of the display and image pickup systems. Coordinates are set as shown in Figs. 4.7 and 4.8 in order to describe the pickup and display systems. The zero point of the Z axis is where the lens array is located. The object's frequency is u_s (cycle/mm) and the distance is Z_s during pickup, then the spatial frequency α_s (normalized by the distance, cpr: cycle per radian) when the object is observed from the lens array is expressed by the following:

$$\alpha_s = \arctan\left(|z_s| \cdot u_s\right) \cong |z_s| \cdot u_s = s_p \cdot g_p \,, \tag{4.8}$$

where g_p is the distance between the microlens array and the pickup plane. The α_s can also be expressed using the frequency s_p (cycle/mm) on the pickup plane.

The optical image of the object that is formed through the microlens is then picked up as an elemental image by a pickup device or film. The resolution characteristics of this elemental image are expressed by the product of the resolution characteristics of the lens and those of the image-pickup device. When the device is equipped with an optical low-pass filter, the resolution characteristics include those of the optical filter. In the display system, the captured elemental image is shown on the display plane. Light rays from this plane travel back through the microlens to form a reproduced image. The resolution characteristics of the display system are obtained by the product of the resolution characteristics of the display device and those of the microlens. The overall resolution characteristics are therefore influenced by the resolution characteristics during pickup and display shown as Fig. 4.9, and are expressed by the product of four modulation transfer functions MTFs:

$$\mathrm{MTF}(\alpha_s) = \mathrm{MTF}_{\mathrm{pl}}(\alpha_s)\mathrm{MTF}_{\mathrm{pd}}(\alpha_s)\mathrm{MTF}_{\mathrm{rl}}(\alpha_s)\mathrm{MTF}_{\mathrm{rd}}(\alpha_s)\,, \tag{4.9}$$

where, $\mathrm{MTF}_{\mathrm{pl}}$, $\mathrm{MTF}_{\mathrm{pd}}$, $\mathrm{MTF}_{\mathrm{rl}}$, and $\mathrm{MTF}_{\mathrm{rd}}$ are for the pickup microlens, pickup device, display microlens, and display device, respectively. If the display system and the pickup system are of different size, the reproduced image and the object will then differ from each other in their position and size. If these two systems are of the same size, the reproduced image will then appear at the same place and with the same size as the object. For simplicity of description, this discussion assumes that these two systems are of the same size. The u_s (cycle/mm) and α_s (cpr), which are defined for the pickup system, can also be regarded as the frequency of the reproduced image and frequency component of the light ray from the microlens during display, respectively.

As for $\mathrm{MTF}_{\mathrm{pl}}$ of the microlens in the pickup system, assuming that the microlens is an ideal lens, its characteristics are given by the following well-known equation, which is the product of the diffraction, first part, and off-focusing, second part, characteristics:

$$\mathrm{MTF}_{\mathrm{pl}}(\alpha_s) = \left(1 - \frac{\lambda\alpha_s}{w_a}\right) \sin c\left(w_a\alpha_s e_p\left(1 - \frac{\lambda\alpha_s}{w_a}\right)\right)\,, \tag{4.10}$$

$$e_p = \left|\frac{1}{z_s} - \frac{1}{g_p} + \frac{1}{f_p}\right| = \left|\frac{1}{z_s} - \frac{1}{z_{fp}}\right|\,, \tag{4.11}$$

$$\frac{1}{z_{fp}} = \frac{1}{g_p} - \frac{1}{f_p}\,, \tag{4.12}$$

where, λ is the wavelength of the light, g_p the distance between the lens array and the pickup plane, f_p the focal length of the microlens, w_a the width

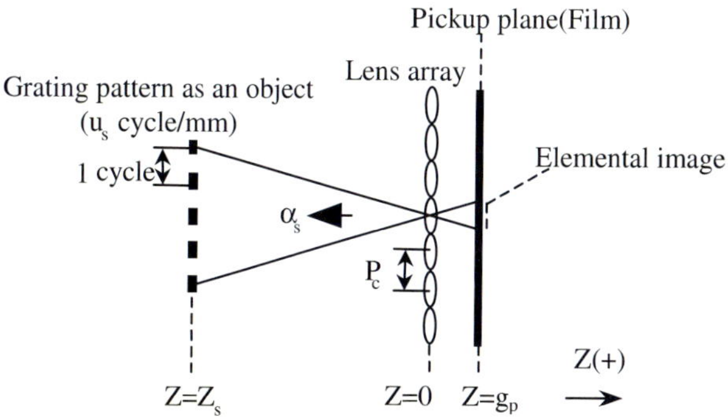

Fig. 4.7. Spatial frequency α_s

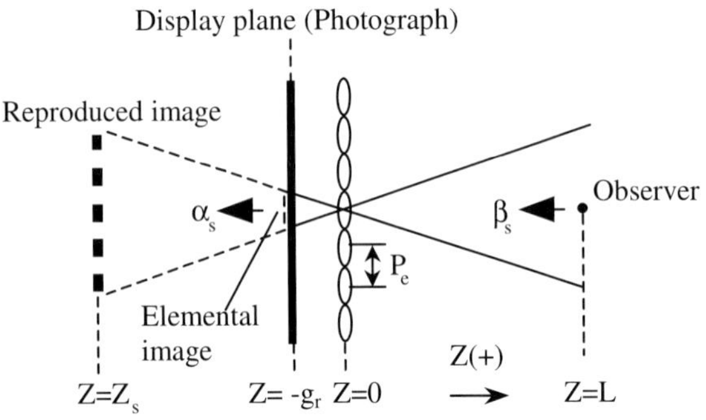

Fig. 4.8. Spatial frequency β_s

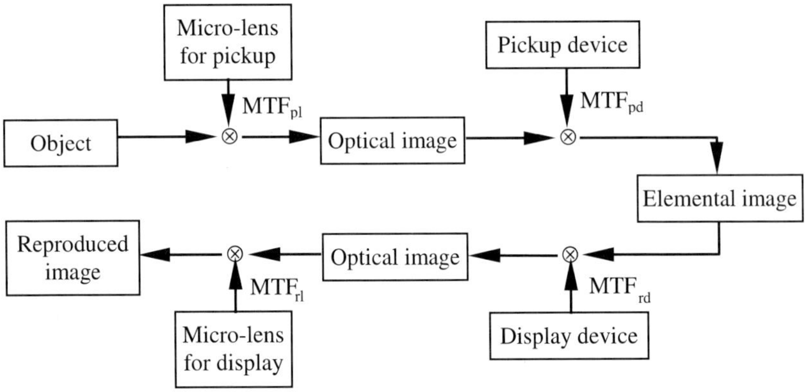

Fig. 4.9. Flowchart of pickup and display

(diameter) of the microlens, e_p the off-focusing component, Z_{fp} the conjugate point of which microlens is focused on the pickup plane. By adjusting the position of the pickup plane such that the object is located at the conjugate point, the off-focusing component given by (4.11) becomes zero. When there is more than one object, as often happens in practice, these objects have different depth directions and it is not possible for all of them to be just focused. Figure 4.10 shows the resolution characteristics of the lens using two examples where distances to the object are different. In both these cases, g_p is the same as the focal length f_p, meaning that the focus is at infinity from the lens array. The resolution characteristics in this case are expressed by an MTF that is influenced by the off-focusing. The closer the object positions to the lens array, the greater the effects of the off-focusing component and the lower the resolution. This means that resolution of elemental image degrades, but does not mean that the reconstructed image degrades, as described later.

Assuming that the microlens of the reproduction system is ideal, like the microlens of the pickup system, then MTF_{r1} of the reproduction system is given by the following equations, which closely resemble (4.10)–(4.12):

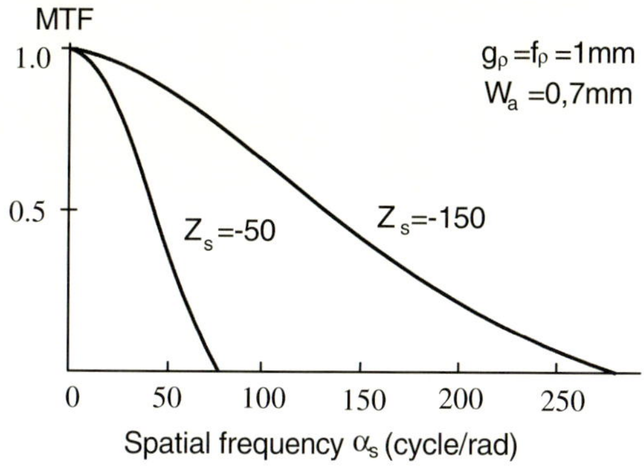

Fig. 4.10. Examples for resolution characteristics of an elemental image

$$\mathrm{MTF}_{\mathrm{rl}}(\alpha_{\mathrm{s}}) = \left(1 - \frac{\lambda\alpha_{\mathrm{s}}}{w_{\mathrm{a}}}\right) \sin c\left(w_{\mathrm{a}}\alpha_{\mathrm{s}}e_{\mathrm{r}}\left(1 - \frac{\lambda\alpha_{\mathrm{s}}}{w_{\mathrm{a}}}\right)\right), \tag{4.13}$$

$$e_{\mathrm{r}} = \left|\frac{1}{g_{\mathrm{r}}} + \frac{1}{z_{\mathrm{s}}} - \frac{1}{f_{\mathrm{r}}}\right| = \left|\frac{1}{z_{\mathrm{s}}} - \frac{1}{z_{\mathrm{fr}}}\right|, \tag{4.14}$$

$$\frac{1}{z_{\mathrm{fr}}} = \frac{1}{f_{\mathrm{r}}} - \frac{1}{g_{\mathrm{r}}}, \tag{4.15}$$

where, g_{r} is the distance between the lens array and the display plane, f_{r} the focal length of the microlens, w_{a} the width (diameter) of the microlens, e_{r} the off-focusing component of the reproduced image, and Z_{fr} the image point on which microlens is focused. The z_{s} is the position of the reproduced image, which corresponds to the position of the object when it was shot, as explained earlier. The focusing during the display can be adjusted by changing the distance g_{r} between the lens array and the display plane. This adjustment is independent of the pickup. As reproduced images also have different depth positions, it is not possible for all of them to be focused. Depending on their positions, they are therefore influenced by off-focusing (4.14). The characteristics expressed by (4.10) and (4.13) each have different cutoff frequencies, of which the lower one becomes dominant as the overall MTF contains their product. It is therefore important, for efficient optimal designing, that these two cutoff frequencies are the same. Equations (4.10) and (4.13) become the same when $g_{\mathrm{p}} = g_{\mathrm{r}} = g$ and $Z_{\mathrm{fp}} = Z_{\mathrm{fr}} = Z_{\mathrm{f}}$ by adjusting the focal length of the microlens f_{p} and f_{r}:

$$\begin{aligned}&\mathrm{MTF}_{\mathrm{pl}}(\alpha_{\mathrm{s}}) = \mathrm{MTF}_{\mathrm{rl}}(\alpha_{\mathrm{s}}) \\ &= \left(1 - \tfrac{\lambda\alpha_{\mathrm{s}}}{w_{\mathrm{a}}}\right) \sin c\left(w_{\mathrm{a}}\alpha_{\mathrm{s}}e_{\mathrm{pr}}\left(1 - \tfrac{\lambda\alpha_{\mathrm{s}}}{w_{\mathrm{a}}}\right)\right),\end{aligned} \tag{4.16}$$

$$e_{\mathrm{pr}} = e_{\mathrm{p}} = e_{\mathrm{r}} = \left| \frac{1}{z_{\mathrm{s}}} - \frac{1}{z_{\mathrm{f}}} \right| , \tag{4.17}$$

where,

$$f_{\mathrm{p}} \cong g + \frac{g^2}{z_{\mathrm{f}}} , \tag{4.18}$$

$$f_{\mathrm{r}} \cong g - \frac{g^2}{z_{\mathrm{f}}} . \tag{4.19}$$

Therefore, (4.9) becomes the following:

$$\begin{aligned} \mathrm{MTF}(\alpha_{\mathrm{s}}) = & \left\{ \left(1 - \frac{\lambda \alpha_{\mathrm{s}}}{w_{\mathrm{a}}}\right) \sin c \left(w_{\mathrm{a}} \alpha_{\mathrm{s}} e_{\mathrm{pr}} \left(1 - \frac{\lambda \alpha_{\mathrm{s}}}{w_{\mathrm{a}}}\right)\right)\right\}^2 \\ & \times \mathrm{MTF}_{\mathrm{pd}}(\alpha_{\mathrm{s}}) \, \mathrm{MTF}_{\mathrm{rd}}(\alpha_{\mathrm{s}}) . \end{aligned} \tag{4.20}$$

Here, the null point α_{cdif} of diffraction and the first null point α_{cerr} of the off-focusing component can be expressed by:

$$\alpha_{\mathrm{cdif}} = w_{\mathrm{a}} / \lambda , \tag{4.21}$$

$$\begin{aligned} \alpha_{\mathrm{cerr}}(z_{\mathrm{s}}) = & \frac{2}{e_{\mathrm{pr}}(z_{\mathrm{s}})} \cdot \frac{1}{w_{\mathrm{a}} + [\, w_{\mathrm{a}}^2 - 4\lambda / e_{\mathrm{pr}}(z_{\mathrm{s}}) \,]^{1/2}} \\ \geq & \frac{1}{e_{\mathrm{pr}}(z_{\mathrm{s}}) \, w_{\mathrm{a}}} . \end{aligned} \tag{4.22}$$

Therefore, the overall cutoff frequency $\alpha_{\mathrm{s\,max}}$ of the microlens is the lowest of α_{cdif}, α_{cerr}, α_{cp}, α_{cr}. It is given by:

$$\alpha_{\mathrm{s\,max}} = \min \left[\alpha_{\mathrm{cdif}} , \alpha_{\mathrm{cerr}}(z_{\mathrm{s}}) , \alpha_{\mathrm{cp}} , \alpha_{\mathrm{cr}} \right] , \tag{4.23}$$

where, α_{cp} and α_{cr} are the cutoff frequencies of the pickup device and the display device, respectively. This frequency is also regarded as the maximum frequency of the ray emitted from the microlens during image reproduction.

4.6 Sampling on Image Reproduction

The spatial frequency β_{s} when the reproduced image is viewed by the observer, and α_{s}, when the object is viewed from the lens array, have the following relation:

$$\beta_{\mathrm{s}} = \arctan \{ u_{\mathrm{s}} \cdot (L - z_{\mathrm{s}}) \} \cong u_{\mathrm{s}} \cdot (L - z_{\mathrm{s}}) \cong \alpha_{\mathrm{s}} \frac{L - z_{\mathrm{s}}}{|z_{\mathrm{s}}|} \quad (z_{\mathrm{s}} \leq L) . \tag{4.24}$$

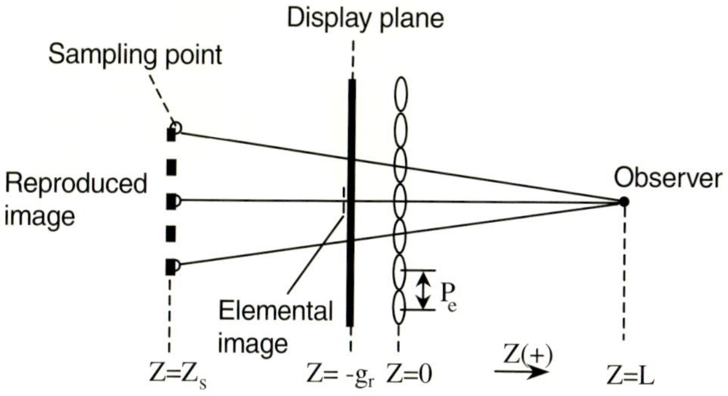

Fig. 4.11. Sampling on image reproduction

The β_s shows a large value near the lens array. $\beta_{s\,max}$ corresponding to the $\alpha_{s\,max}$ is shown as follows:

$$\beta_{s\,max} = \alpha_{s\,max}(z_s)\,\frac{L - z_s}{|z_s|}\,. \tag{4.25}$$

The $\beta_{s\,max}$ changes in relation to the distance z_s. As shown in Fig. 4.11, not all the light rays that constitute the reproduced ray will be seen by the observer at once. Of the light rays from the elemental images, the viewer observes only those that pass through the microlens array and move toward him. The reproduced image is sampled by the light ray, which connects the microlens and the observer. If p_e and L represent the microlens pitch and the viewing distance, respectively, then the sampling interval is given by p_e/L rad. The spatial Nyquist frequency β_{nyq} (cpr) is therefore given by:

$$\beta_{nyq} = \arctan\left(\frac{L}{2p_e}\right) \cong \frac{L}{2p_e}\,. \tag{4.26}$$

If the reproduced image contains frequencies larger than the Nyquist frequency β_{nyq}, meaning the following relation, aliasing will occur. In this case, the spatial frequency, like ordinary aliasing, can be expressed by the following equation as a component modulated by higher-order carrier frequencies, namely, this frequency β_i is converted into a lower one, shown in Fig. 4.12,

$$\tilde{\beta}_i = \pm\beta_i + 2N\beta_{nyq} \tag{4.27}$$

where N is an integer.

The maximum frequency β_{max} of the reproduced image, viewed by the observer, is provided by either the spatial Nyquist frequency β_{nyq} or the maximum frequency $\beta_{s\,max}$ contained in the reproduced image, whichever is the smaller, and is expressed as follows. Here, we call β_{max} the limiting resolution.

$$\beta_{max} = \min(\beta_{s\,max}, \beta_{nyq}) = \min\left(\alpha_{s\,max}(z_s)\cdot\frac{L - z_s}{|z_s|}, \beta_{nyq}\right)\,. \tag{4.28}$$

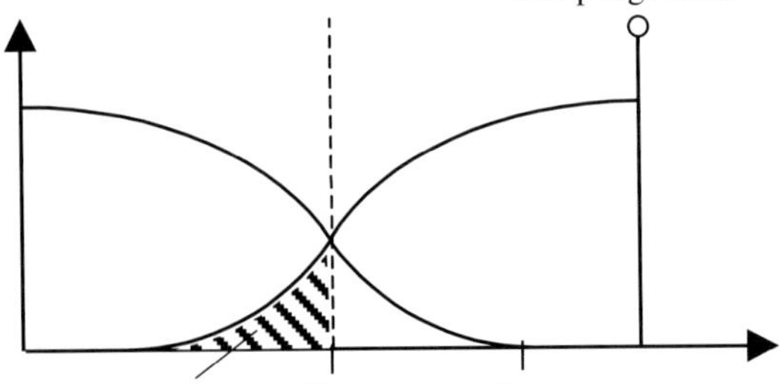

Fig. 4.12. Aliasing component by sampling

4.7 Limiting Resolution

The limiting resolution depends on the distance between object and elemental lens, and other factors such as the Nyquist frequency, number of pixels of pickup device and display device, effects of off-focusing and diffraction. The general characteristics of resolution are described here.

When $\alpha_{\mathrm{s\,max}}(\mathrm{far}) < \beta_{\mathrm{nyq}}$, the limiting resolution calculated by (4.23) and (4.28) is shown in Fig. 4.13a. Here, $\alpha_{\mathrm{s\,max}}(\mathrm{far})$ is the value of the $\alpha_{\mathrm{s\,max}}(z_{\mathrm{s}})$ shown by (4.23) in the $z_{\mathrm{s}} = -\infty$. It is possible to produce images around the lens array without resolution degradation, because β_{nyq} is less than $\beta_{\mathrm{s\,max}}$, which is determined by the off-focusing component. However, the limiting resolution of the images produced far from the lens array approaches the maximum projectable frequency $\alpha_{\mathrm{s\,max}}(\mathrm{far})$ through the elemental lens, and resolution degradation occurs. Also, the limiting resolution of the images produced near the observer decreases markedly.

When, $\alpha_{\mathrm{s\,max}}(\mathrm{far}) \geq \beta_{\mathrm{nyq}}$, the limiting resolution is calculated as shown in Fig. 4.13b. Images can be produced from far behind the lens array without resolution degradation. The limiting resolution of the images produced near the observer is degraded but is still higher than when $\alpha_{\mathrm{s\,max}}(\mathrm{far}) < \beta_{\mathrm{nyq}}$. Consequently, to produce images distant from behind the lens array without resolution degradation, we consider $\alpha_{\mathrm{s\,max}}(\mathrm{far}) = \beta_{\mathrm{nyq}}$ to be sufficient. Therefore, this condition is recommendable as the system.

Next, the effects of focusing are described under the condition corresponding to the Fig. 4.13b. It is assumed that the number of pixels of the pickup device and the display device is enough. Focusing both of the pickup and the display is able to be set freely. The first case is the *infinite distance focusing type*. For pickup, the focusing is set for an object at infinite distance from the microlens. Figure 4.14a and b show two examples of the limiting resolution, when the lens width is relatively small or large. The viewing distance is 2 m in both examples. When the lens width is small (0.694 mm), the cutoff frequency determined by the focusing error at any depth position z_{s} is higher than either the Nyquist frequency β_{nyq} or the limiting resolution determined

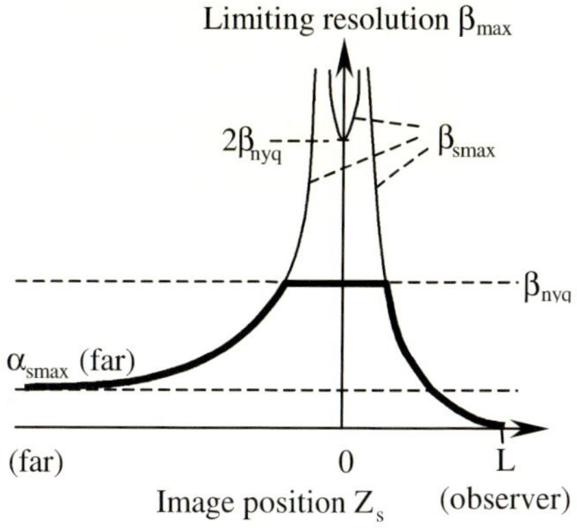

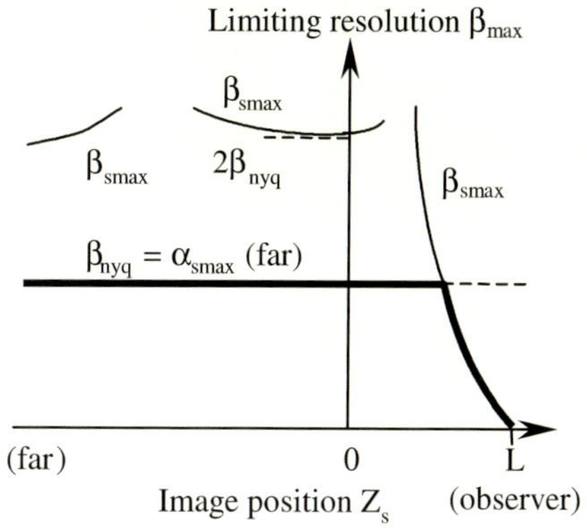

Fig. 4.13. General characteristics of limiting resolution. The *thick line* indicates limiting resolution

by the diffraction limit. In this case, the focusing error has no effect on the overall resolution. However, when the lens width is large (2.78 mm), the limiting resolution of the image reproduced near the observer is lower, and determined by the focusing error.

The second case is the *observer's position focusing type.* For pickup, the focusing is set for the object whose position is the same as that of the viewer during display. For display, the elemental image is also focused at the viewer's position. When the lens width is 0.694 mm, the limiting resolution is just the same as the infinite distance focusing type, shown in Fig. 4.14a. Figure 4.14c shows an example of the limiting resolution, when the lens width is 2.78 mm. In this cases, the limiting resolution determined by the focusing error is higher than the Nyquist frequency β_{nyq}; therefore, the focusing error has no effect on the overall limiting resolution. The limiting resolution near the viewer is improved, compared with the infinite distance focusing type, shown in Fig. 4.14b.

4.8 Viewing-Zone Angle

Here, we calculate the width of the viewing zone when the viewing distance and the limiting resolution requirement are given. The viewing zone is expressed as the angle measured at the center of the display surface (the plane of the lens array) and indicates how far an observer can move laterally. Out of the viewing zone, the viewer can still observe the reproduced image, however, the image does not correspond to the viewer's position. As shown in Fig. 4.15, the viewing zone becomes widest when all of the rays from the center of the elemental images crossover at a single point Q_{v}. When the width w_{a} of the

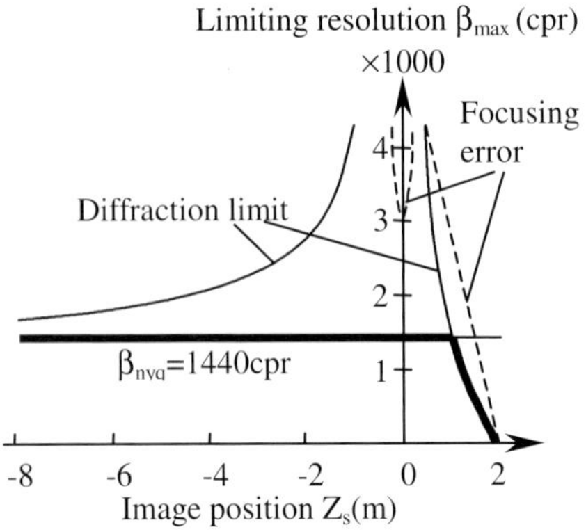

(a) The lens width and pitch = 0.694 mm
Infinite focusing type/Observer's position focusing type

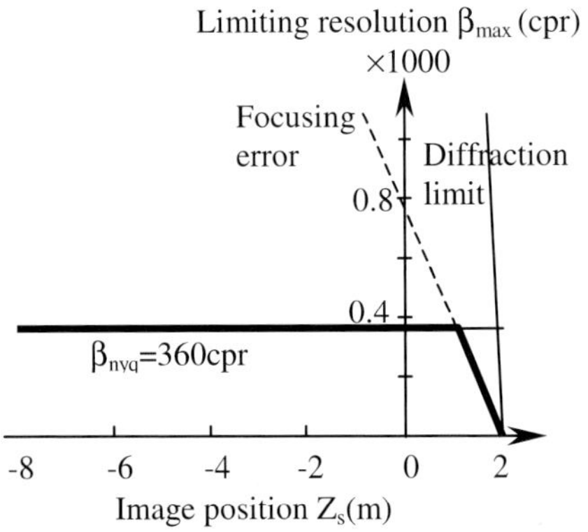

(b) The lens width and pitch = 2.78 mm
Infinite focusing type

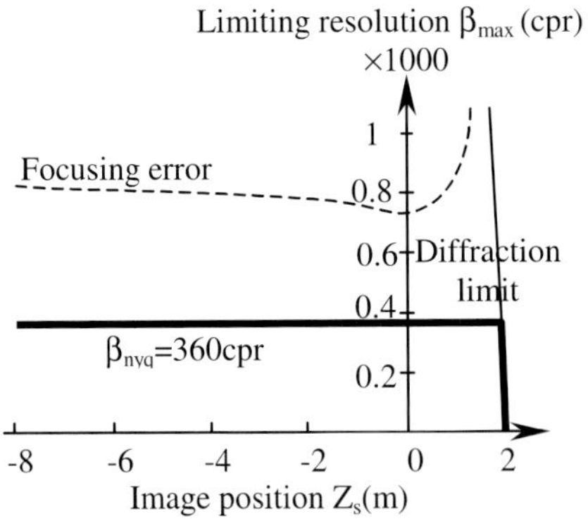

(c) The lens width and pitch = 2.78 mm
Observer's position focusing type

Fig. 4.14. Characteristics of limiting resolution. The *thick line* indicates limiting resolution. Viewing distance = 2 m

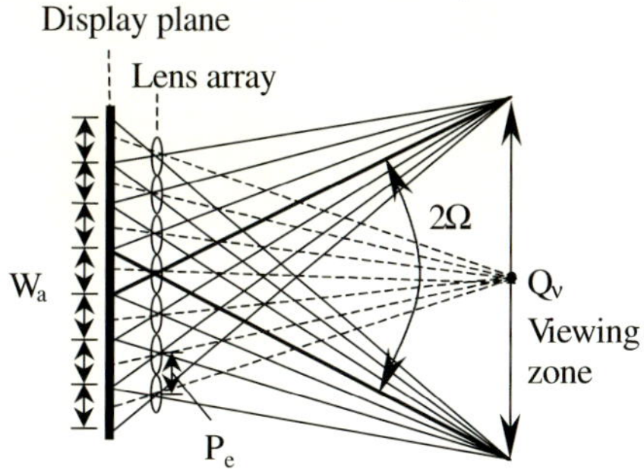

Fig. 4.15. Viewing zone

elemental image and the gap g between the elemental lens and the elemental image are used, the width of the viewing zone angle 2Ω is

$$2\Omega = 2\arctan\left(\frac{w_a}{2g}\right). \tag{4.29}$$

The elemental images should not overlap each other. Therefore, the width w_a of the elemental image can be expressed by the pitch p_e of the exit pupil as follows:

$$w_a = \frac{p_e\,(L+g)}{L} \cong p_e\,. \tag{4.30}$$

From (4.26) and (4.29), (4.30) is rewritten as follows:

$$2\Omega = 2\arctan\left(\frac{L}{4\beta_{nyq}g}\right). \tag{4.31}$$

Figure 4.16 shows the relation between the viewing-zone angle and the Nyquist frequency β_{nyq} calculated by (4.31). This figure shows that the

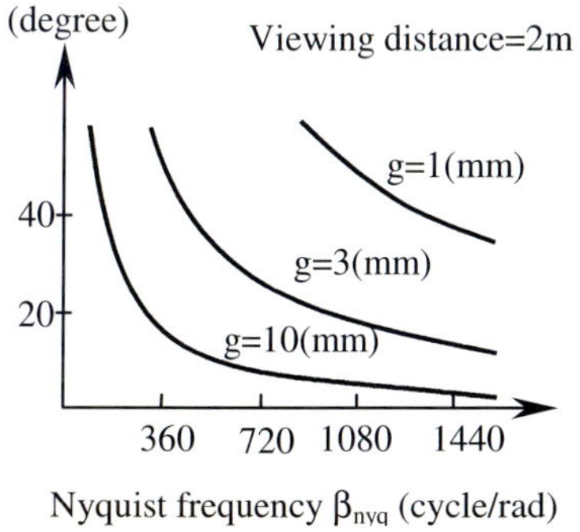

Fig. 4.16. Viewing-zone angle versus Nyquist frequency β_{nyq}

smaller g produces the wider viewing zone. The smaller g increases the frequency of the elemental image on the pickup or display plane shown as (4.8). In the case that the cutoff frequency of the pickup system or display system is not enough as follows, the smaller g degrades the resolution of the reproduced image.

$$\alpha_{\mathrm{cdif}}\,,\ \alpha_{\mathrm{cerr}}\left(z_{\mathrm{s}}\right) \geq \alpha_{\mathrm{cp}}\,,\alpha_{\mathrm{cr}}\,. \tag{4.32}$$

4.9 Depth Control

3-D images reproduced far away from the lens array for reproduction have a lower resolution than those around the lens array under the condition $\alpha_{\mathrm{s\,max}}(\mathrm{far}) < \beta_{\mathrm{nyq}}$. This problem is remarkable when the number of pixels of pickup or display is not enough. Furthermore, as explained also in Sect. 4.2, after the orthoscopic-pseudoscopic image conversion is performed, orthoscopic-reproduced images are positioned just the same as the objects had been, and so they exist only behind a lens array for reproduction. Therefore, it is preferable to control the position of the 3-D images around the lens array for reproduction, so that the real images of objects are produced around the GRIN lens array for pickup. A depth-control lens, which is a large-aperture convex lens, is employed to solve these problems. As shown in Fig. 4.17, this condition can be performed such that the depth-control lens makes real images of objects around the GRIN lens array. It is similar to image holography [14, 15] in which real images of objects are positioned around the hologram by a convex lens. As is well known, the position of the real image produced by the convex lens is expressed as follows.

$$L_{\mathrm{b}} = \frac{L_{\mathrm{a}} f_{\mathrm{d}}}{L_{\mathrm{a}} - f_{\mathrm{d}}}\,, \tag{4.33}$$

where, L_{a} is the distance between the object and the depth-control lens, L_{b} is the distance between the depth-control lens and the real image, and f_{d} is the focal length of the depth-control lens. L_{a} is larger than f_{d}. If the distance between the depth-control lens and the object is set adequately, the real image

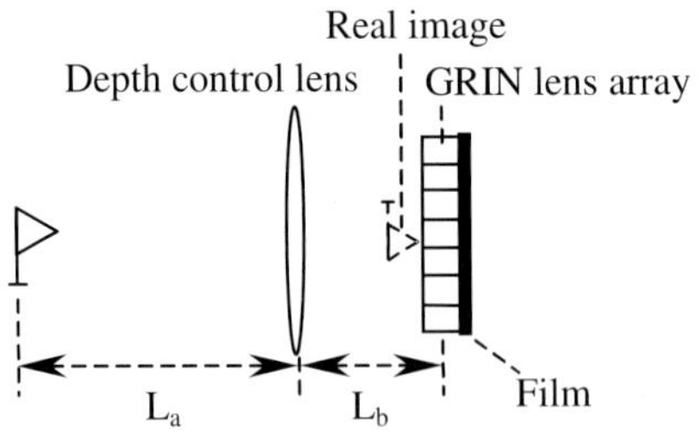

Fig. 4.17. Depth-control lens

is positioned around the lens array. Whether the real image is positioned in front of or behind the lens array depends on L_a. The real image of the object around the lens array is captured by the lens array as each elemental image. D_r the ratio of lateral magnification M_L to longitudinal magnification M_Z shows distortion:

$$D_r = \left| \frac{M_Z}{M_L} \right| = \left| \frac{f_d}{L_a - f_d} \right| , \tag{4.34}$$

$$M_L = -\frac{L_b}{L_a} = -\frac{f_d}{L_a - f_d} , \tag{4.35}$$

$$M_Z = M_L^2 = \left(\frac{f_d}{L_a - f_d} \right)^2 . \tag{4.36}$$

The D_r depends on L_a. When the L_a is larger than twice f_d, the M_Z is less than 1, namely, depth is compressed. Therefore, the real image produced by the depth-control lens around the GRIN lens array is distorted in the depth direction. In order to avoid the distortion, the depth-control optics with $D_r = 1$ is employed as the depth-control lens.

4.10 An Experimental Real-Time System

The basic IP method using photographic film is not suited to taking pictures of moving objects in real time, and therefore cannot be applied to television without modifications. One way to obtain moving pictures is to use a pickup device, such as a CCD, in place of the photographic film. These pictures can then be taken by a pickup device placed just behind the lens array. For reproduction, a display panel, such as an LCD, is placed behind the lens array rather than the transparent photograph. Video signals are transmitted from the pickup device to the display device.

Generally speaking, the size of the lens array for pickup needs to be larger than the average distance between human pupils (about 65 mm) in order to obtain suitable parallax. On the other hand, the size of commercially available pickup devices for moving pictures is about 10 mm, smaller than the lens array. In the actual system, a television camera using a pickup lens is set in order to pick up all elemental images formed by microlenses just behind the lens array. In this case, the camera lens is focused on the output plane of the GRIN lens array. In the future, it is preferable that a pickup device of the same size as the lens array be developed.

Figure 4.18 shows a schematic of an experimental real-time IP system, and Fig. 4.19 is the overview of the pickup system. A depth-control lens, a GRIN lens array, a converging lens, and an HDTV(high-definition television) camera

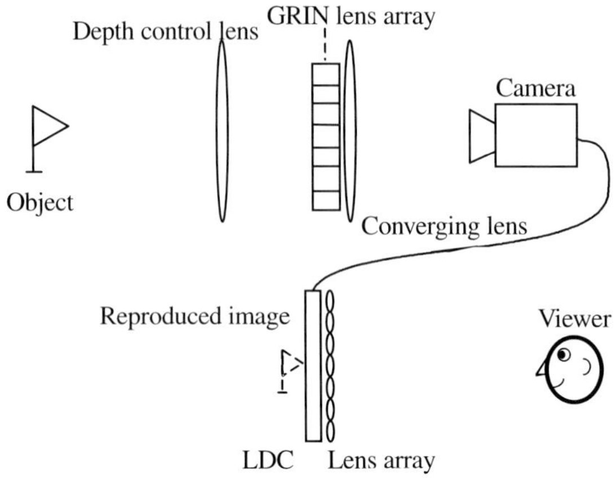

Fig. 4.18. An experimental real-time IP system

Table 4.1. Equipment specifications for pickup

CCD camera	Number of pixels	1920(H)×1035(V)
	Size of pickup device	14.0(H)×7.9(V) mm^2
	Focal length of lens	90 mm
	Color system	3 CCDs (R, G, B)
GRIN lens array	Number of lenses	54(H)×63(V)
	Pitch	1.085 mm
	Lens length	20.25 mm
	Focal length	– 2.65 mm*
	Size	60(H)×60(V) mm^2
Elemental image	Pitch	20 pixels
Depth-control lens	Focal length	250 mm
Converging lens	Focal length	1000 mm

* The focal length of the GRIN lens is a negative value, because it is specified as $\theta = 1.5\pi$, meaning 3/4 of 1 period of the optical path.

are introduced for pickup. Table 4.1 shows the experimental specifications of the pickup setup. Figure 4.20 shows the two-dimensional arrangement of the GRIN lens array employed in the experiment. The arrangement has offset line structure, which is efficient compared with a grid structure. The pitch between the elemental images corresponds to 20 pixels of HDTV system. The converging lens, set just behind the GRIN lens array, brings together light from the lens array to a HDTV camera. The HDTV camera, set on the focal plane of the converging lens, is focused on the output plane of the GRIN lens, and captures a group of elemental images. If optical loss is acceptable for the system use of the converging lens is not mandatory.

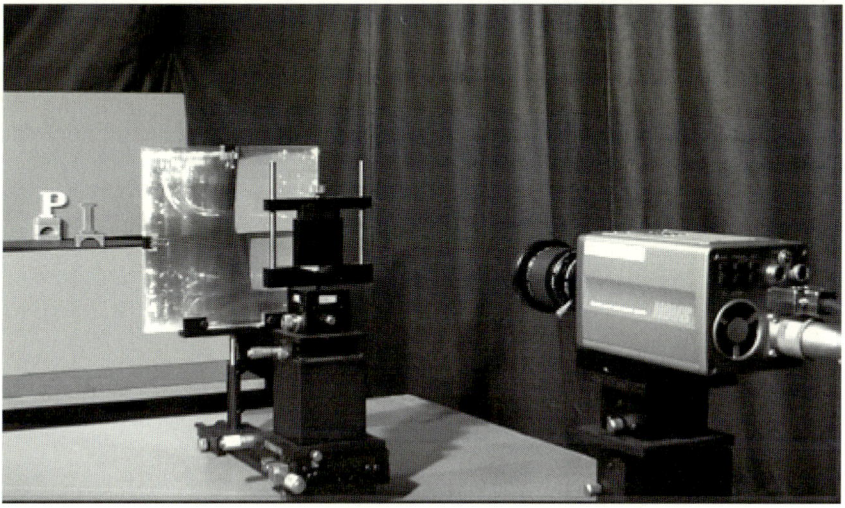

Fig. 4.19. Overview of the experimental system for pickup

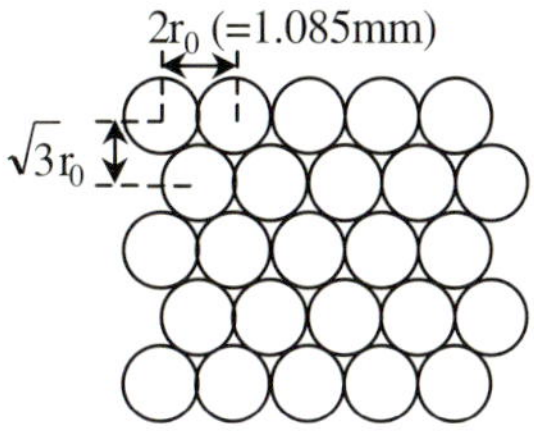

Fig. 4.20. Arrangement of the GRIN lens array

The video signal of a group of elemental images is directly introduced to a display device. Table 4.2 shows the experimental specifications of the display setup. The display device combines a color LCD panel and an array of convex microlenses. The microlens array has the same offset line structure as the GRIN lens array. The top and bottom of the LCD panel are inverted

Table 4.2. Equipment specifications for reproduction

LCD	Number of pixels	1280(H)×1024(V)
	Size	13 inches (diagonal)
	Active display area	1080(H)×972(V)
	Color system	R/G/B stripe filter
Lens array	Number of element-lens	62(H)×55(V)*
	Pich of element-lens	4.05 mm
	Size of element-lens	3.75 mmϕ
	Focal length	10.09 mm

* The active number of microlenses is 54(H)×55(V)

to obtain erected 3-D images, because real images produced by the depth control lens are inverted. The focal length of each elemental microlens should be the same as that of the GRIN lens, in principle. But, in this system, the pitch of the array of convex microlenses for reproduction is about 4 times that of the GRIN lens array, and the focal length of the elemental microlens

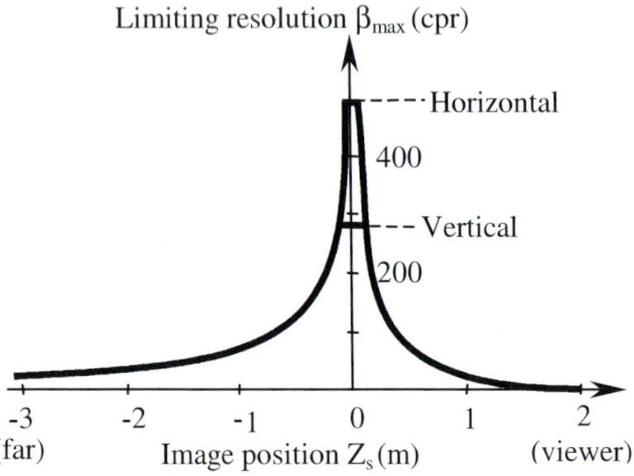

Fig. 4.21. Limiting resolution of the experimental system

Fig. 4.22. Overlooking objects

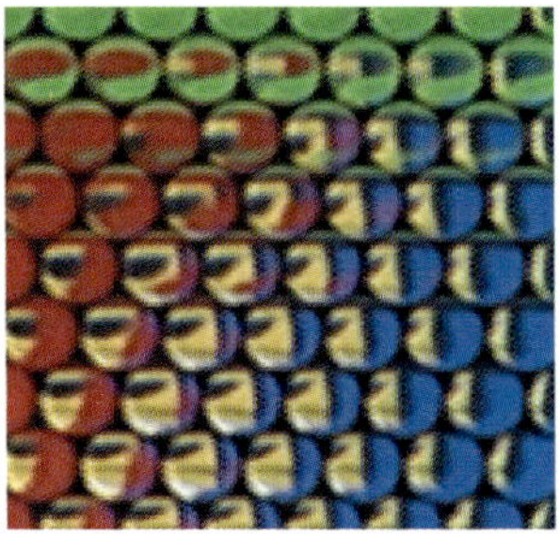

Fig. 4.23. Enlarged portion of elemental images

is also about 4 times that of the GRIN lens. Therefore, the size and depth of reproduced images are about 4 times those of the images provided by the depth-control lens. We can modify the focal length of the convex microlens, which is different from that of the GRIN lens, in order to control the viewing zone and the depth of the reproduced 3-D images according to the system design [16, 17].

Figure 4.21 shows the characteristic of the limiting resolution of this experimental system. It is degraded far from the lens array. The upper limit of resolution, the spatial Nyquist frequency of the horizontal direction is different from that of the vertical direction. The reason is that the active pitches of the lens array, whose arrangement is an offset line structure, are different horizontally and vertically. As the upper limit of resolution depends on the pitch of elemental images, the pitch should be fine to match with the resolution of conventional television (for example, 480(V)×720(H) pixels) in the future.

Figure 4.22 is a picture of overlooking objects. Figure 4.23 shows an enlarged portion of the elemental images, which are slightly different from neighboring elemental images. These elemental images are isolated from each other. The number of active elemental images is 54(H)×55(V), which is quite insufficient for television. The viewing-zone angle of this system is about 20° horizontally and vertically, which is also not sufficient. Figure 4.24a,b,c and d show a reproduced 3-D image taken from the upper, the lower, the left and the right viewpoints, respectively. These indicate that different images are obtained according to the viewpoint. Double or multiple images were not observed, indicating that interference between neighboring elemental images did not occur. The reproduced 3-D images are orthoscopic, not pseudoscopic, with horizontal and vertical parallax in real time.

4.11 Conclusions

We described a 3-D television system based on integral photography, which provides autostereoscopic 3-D images. In this system, real erect images of an object are formed by a GRIN lens array as elemental images, and are directly captured by a television camera. The use of a GRIN lens array for pickup solves two problems simultaneously: the appearance of pseudoscopic images and optical crosstalk between elemental images. The resolution of reproduced images far from the lens array is degraded under the same conditions. A depth-control lens is introduced during pickup to avoid this problem. Video signals of elemental images picked up by a television camera are introduced to a display device composed of an LCD panel and an array of convex microlenses without any signal processing.

We observed full-color, autostereoscopic 3-D images with horizontal and vertical parallax, and these images were free from these problems. As a result, we confirmed that this system could be applied to a 3-D television system.

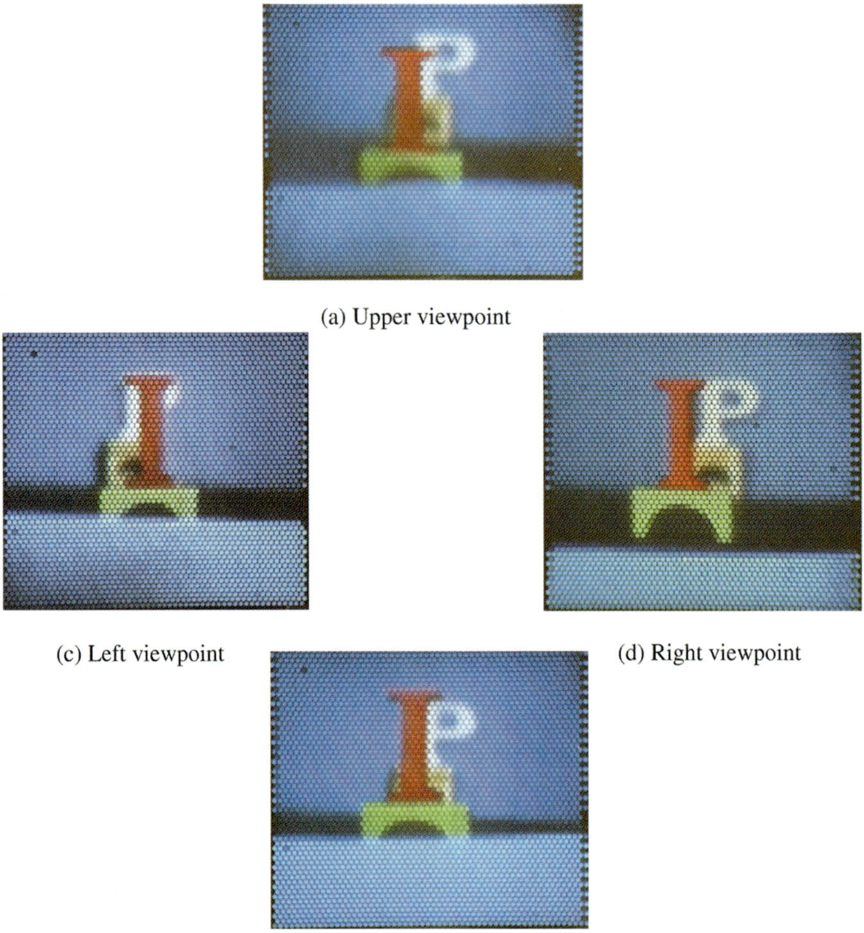

(a) Upper viewpoint

(c) Left viewpoint

(d) Right viewpoint

(b) Lower viewpoint

Fig. 4.24. Reproduced image at different viewpoints

However, the number of elemental images and the number of pixels per elemental image need to be increased to reproduce high-quality 3-D images. Because elemental images produced by the GRIN lens array contain 3-D information, this approach can also be applied to many other applications such as conversion to holography and 3-D measurement.

References

1. G. Lippmann: Comptes-Rendus **146**, 446 (1908)
2. T. Okoshi: *Three Dimensional Imaging Techniques* (Academic, New York 1971)

3. A.P. Sokolov: *Autostereoscopy and Integral Photography by Professor Lippmann's Method* (Izd. MGU, Moscow State University Press 1911)
4. H. Higuchi, J. Hamasaki: Appl. Opt. **17**(24), 3895 (1978)
5. M. McCormick: *Proc. of the 2nd Int. Display Workshop*, ed. by T. Hatada, 3-D-9, pp. 77–80 (ITE, Hamamatsu 1995)
6. F. Okano, J. Arai, H. Hoshino, I. Yuyama: Opt. Eng. **38**(6), 1072 (1999)
7. H.E. Ives: J. Opt. Soc. Am. **21**, 171 (1931)
8. R.L. DeMontebello: *Proc. 1977 SPIE Annu. Tech. Conf., San Diego*, Seminar 10, No. 120-08, Tech. Digest 73–91 (1977)
9. J. Hamasaki, M. Okada, S. Utsunomiya: Monthly J. Inst. Industrial Sci., University of Tokyo, **40**(3), 127 (1988)
10. D. Marcuse, S.E. Miller: The Bell System Tech. J. **43**, 1759 (1964)
11. J. Arai, F. Okano, H. Hoshino, I. Yuyama: Appl. Opt. **37**(11), 2034 (1998)
12. M. Sakamoto, H. Ueshima, M. Furukawa, M. Toyama, T. Yamada: Electrophotography **12**, 12 (1973)
13. H. Hoshino, F. Okano, H. Isono, I. Yuyama: J. Opt. Soc. Am. A **15**(8), 2059 (1998)
14. G.W. Stroke: Phys. Lett. **23**(5), 325 (1966)
15. L. Rosen: Appl. Phys. Lett. **9**(9), 337 (1966)
16. D. Gabor, UK Patent 541753 (1940)
17. D. Daly, M.C. Hutley, R.A. Hunt, K. Khand, R.F. Stevens, R. Wilson: The Image Science Journal **45**, 45 (1997)

5 Improved-Quality Three-Dimensional Integral Imaging and Projection Using Nonstationary Optical Components

J.-S. Jang and B. Javidi

5.1 Introduction

Integral imaging (II) is a method to obtain a three-dimensional (3-D) image by use of a set of two-dimensional (2-D) elemental images with different perspectives. It was first proposed by Lippmann in 1908 and is being actively studied currently for 3-D TV and visualization [1–12].

Figure 5.1 shows a conventional II. The multiple-elemental images that have different perspectives for a given 3-D object are usually generated by micro-optics array of lenslets, and then they are recorded in a 2-D light-sensitive device such as a charge-coupled device (CCD). A 3-D image of the object is reconstructed by displaying the recorded 2-D elemental images in front of another lenslet array. For a given observation point, we see a group of subimages sampled from their own elemental images with different perspectives through the corresponding display lenslets. Thus, it is possible to detect, transmit, and manipulate 3-D images in their 2-D forms.

For II to be practical, however, many important issues should be solved. In this chapter we address three of them: viewing resolution, field of view (or viewing angle) limited by interference of the elemental images, and pseudoscopic nature of reconstructed 3-D images. As a possible solution to the three problems we investigated a nonstationary lenslet array technique [11].

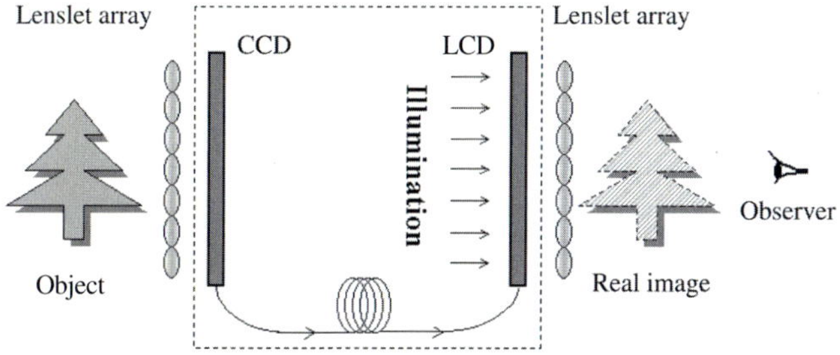

Fig. 5.1. Direct pickup and display in II

5.2 Nonstationary Lenslet Array Technique to Overcome the Viewing Resolution Limited by the Nyquist Sampling Theorem

The resolution of the reconstructed 3-D image is determined by many system parameters, for example, the size and the pitch of lenslets, the resolution of the CCD and the display device, etc. [5, 9, 13, 14]. It is obvious that the number of pixels in the pickup and display devices is important in determining the 3-D image resolution, because all the elemental images should be detected and displayed with sufficient resolution. In II, however, one of the most fundamental factors that limits the viewing resolution of the reconstructed 3-D image is the pitch of the lenslet arrays, which determines the sampling rate of the elemental subimages in the spatial dimension [5]. From the Nyquist sampling theorem, the upper limit of the viewing resolution in a lateral dimension is given by [5]

$$\beta_{\mathrm{nyq}} \approx \frac{L}{2p}, \tag{5.1}$$

in cycles per radian (cpr), where p is the pitch of the lenslet array in the lateral dimension and L is the distance between the observation point and the display lenslet array. However, if we reduce p (i.e. the size of the lenslet element is too small), diffraction of the lenslets limits the resolution even though the resolution of the detector and the display is infinite. Many authors have studied an optimal lenslet size, which is around 1 or 2 mm in reasonable viewing circumstances [9, 13]. For example, suppose that $p = 1\,\mathrm{mm}$ and $L = 1\,\mathrm{m}$ (i.e. $\beta_{\mathrm{nyq}} = 500\,\mathrm{cpr}$). This means that we cannot discern the images whose size is less than about 2 mm from a distance of 1 m. Such low resolution limits the usefulness of II in practical applications.

We propose the use of nonstationary micro-optics such as moving lenslet arrays for both pickup and display in II to overcome the upper resolution limit imposed by (5.1). Because our eyes have the effect of after-images, we can generate as many elemental images of slightly different perspectives as possible by changing the position of the lenslets for pickup and display processes within the time constant of the eyes' response time. Of course, the lenslets' movement should be synchronized so that the relative position of the pickup and display lenslet arrays can be maintained. Otherwise, the 3-D image cannot be reconstructed correctly, because the elemental images detected by the moving lenslet arrays move out of phase.

Figure 5.2 shows our system setup to study the effect of moving lenslet arrays. To illustrate the effects of nonstationary lenslets, let us assume that the resolutions of both the detector and the display devices in Fig. 5.1 (see the dashed box) are infinite. The first lenslet array is used to obtain multiple elemental images and the second lenslet array is used to reconstruct a 3-D image again. The gap between the two lenslet arrays was adjusted to minimize the focusing error [5].

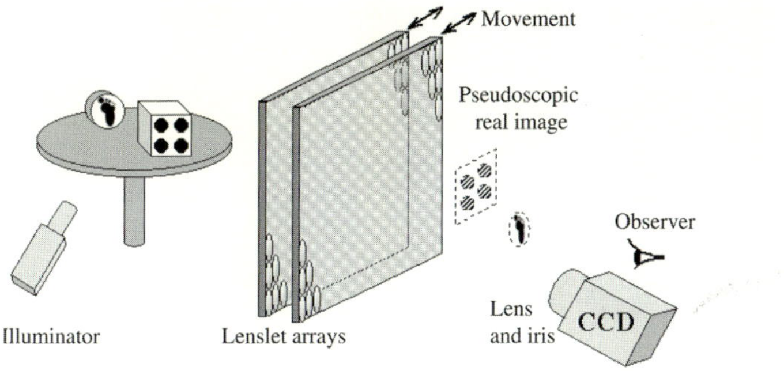

Fig. 5.2. Optical setup to study the effect of moving lenslet arrays

The object used for 3-D imaging is composed of a die and a plastic button with a footprint figure as shown in Fig. 5.3. The side length of the die is 1.5 cm and the diameter of the button is 1 cm. The distance between the die surface with four dots and the button surface with the footprint figure is about 3 cm. The lenslet arrays we used have 53 × 53 lenslets. Each lenslet element is square-shaped and has a uniform base size of 1.09 mm × 1.09 mm, with less than 7.6 μm separating the lenslet elements. The focal length of the lenslets is 5.2 mm.

The distance between the pickup lenslet array and the front surface of the die is approximately 7 cm. The real pseudoscopic image is detected by a CCD with a standard camera lens and a built-in iris. The focal length and the maximal f number of the lens are 50 mm and 1.4, respectively. The distance between the real image and the display lenslet array is about 10 cm, and that between the real image and the camera lens is about 50 cm. The image is also examined with the naked eye.

When we view the reconstructed real image with the naked eye, we can visualize the 3-D nature of the image, although it is pseudoscopic. The footprint figure looks closer to us than the die surface does. However, the image

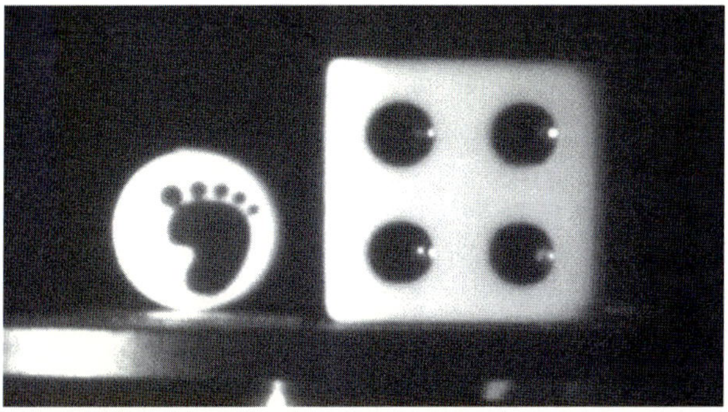

Fig. 5.3. 3-D objects used in the experiments

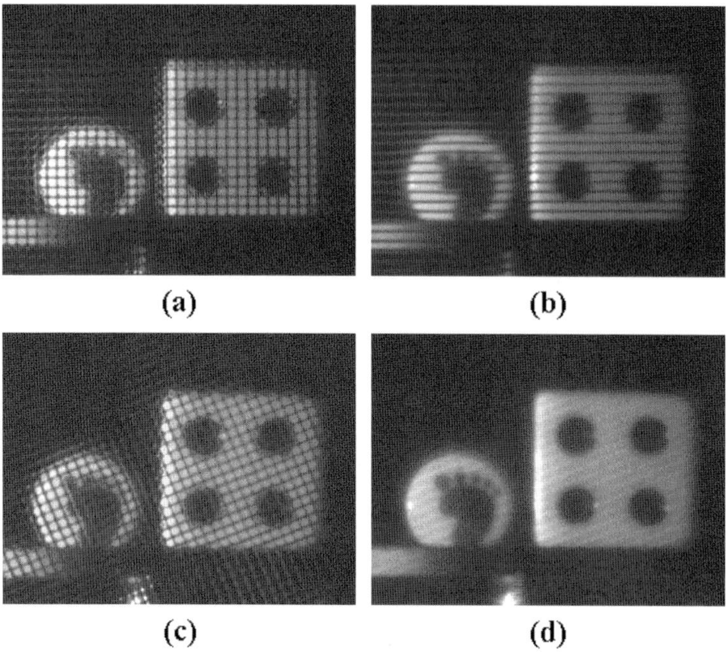

Fig. 5.4. Experimental results for the reconstructed 3-D images. When the lenslet arrays are (**a**) fixed; (**b**) moving in the horizontal direction; (**c**) tilted and fixed; and (**d**) tilted and moving in the horizontal direction

quality is really poor, because the pitch of the lenslet arrays is approximately 1 mm. The image is shown in Fig. 5.4a. Note that the pupil diameter of the camera iris was set to its minimum, approximately 2.3 mm, in order to make the condition of the camera lens and the CCD similar to the human eye, whose pupil diameter is about 3 mm in an ordinary environment. The die dots that are larger than the size of each lenslet element are distinguishable. However, the toes in the footprint figure, which are similar to or slightly smaller than the size of each lenslet element, are not clear.

When we set in motion the two lenslet arrays for both pickup and display synchronously in the horizontal direction, the resolution in that direction is improved and the toes are clear, as shown in Fig. 5.4b. To improve the resolution in both the vertical and the horizontal directions, the sampling rate should be increased in both directions. To illustrate this effect, we tilt the two lenslet arrays by an angle of 20°, and then move them horizontally. The 3-D image for the still lenslet arrays is given by Fig. 5.4c and that for the moving lenslet arrays is given by Fig. 5.4d. This result shows that the resolution limit imposed by the pitch of the lenslet arrays is relaxed. The improvement of the viewing resolution by the moving lenslet arrays was also confirmed with our naked eye.

The movement range of the lenslet arrays need not be large: The range of one pitch in each direction will be enough, in principle, because the lenslets are periodic. For still objects, the velocity components of the lenslet arrays along the horizontal (x) and the vertical (y) axes, V_x and V_y, respectively, should satisfy the following conditions: $V_x > p_x S$ and $V_y > p_y S$ where p_x (p_y) is the pitch of the lenslets in the x (y) direction, and S is the inverse of the response time constant of the human eye (or the electronic shutter speed of the CCD). When the objects to be imaged move with a maximal velocity of $V_{\max}$ in either direction, the velocity of the lenslets should be increased by $V_{\max}$ in that direction. If we apply our technique to the 3-D TV, the frame rate (scanning speed) should be increased by $\max(V_x,\ V_y)$, which is the price we have to pay for overcoming the resolution limit given by (5.1).

In fact, the movement of the lenslet arrays need not be linear. For practical applications, we propose a circular motion of the lenslets as depicted in Fig. 5.5. In this case, the radius of the circular motion should be larger than the half pitch of the lenslets. To maintain strict synchronization between the two lenslets for pickup and display in 3-D TV systems, a clock signal will be necessary in controlling their circular motions.

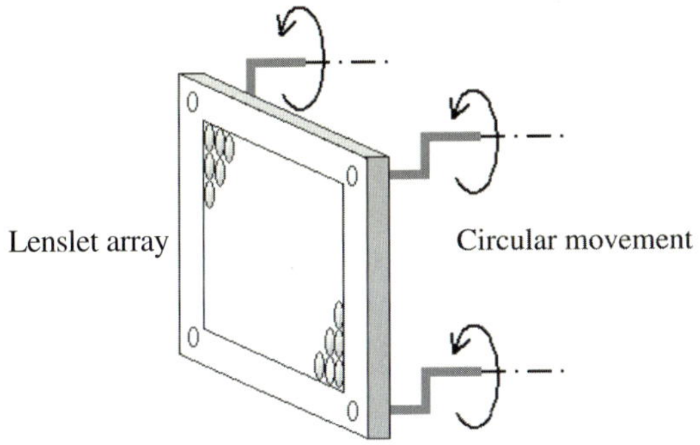

Fig. 5.5. Circular movement of the lenslet arrays

It is very instructive to consider our moving lenslet array technique in a more intuitive way. Suppose that the lenslet arrays are now fixed and that the pupil size of the camera lens in front of the CCD is kept to a minimum. If we put a screen (a transmissive diffusion plate) at the image plane of the footprint figure, a clean footprint image is obtained whose resolution is similar to that obtained by our moving lenslet array technique, as shown in Fig. 5.6a. Similarly, we can also see a clean die image if we move the screen to the image plane of the die surface, as shown in Fig. 5.6b. In fact, when the screen is close to the lenslet array to see the die surface, multiple die images were captured. This is because the elemental images are also displayed through their neighboring lenslets and these multiply reconstructed images

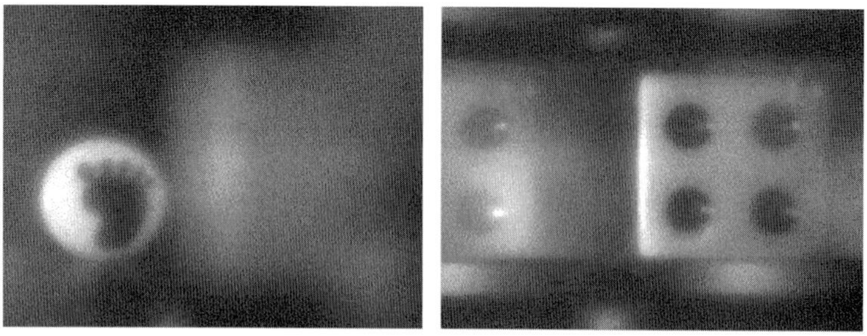

Fig. 5.6. Detection of the reconstructed 3-D image with a screen. The lenslet arrays are fixed. When the screen is positioned at the image plane of (**a**) the footprint; and (**b**) the die surface

are not sufficiently separated. This is closely related with the limitation of the viewing angle and will be further discussed in Sect. 5.4.

We can see that the resolution of the reconstructed 3-D image itself is as good as what we obtain by our technique. The deterioration of the resolution of the reconstructed 3-D image is caused by the grid structure of the lenslet arrays. The reason why we see the grid structure and the 3-D image together is that our eyes have a small pupil diameter and thus have a long depth of focus. This fact is easily confirmed by capturing the real image using the CCD with its iris opened maximally, as shown in Fig. 5.7. In this case, the grid structure is not well detected because of a short depth of focus of the camera lens. When we move the lenslet arrays synchronously, only the grid structure moves but the reconstructed 3-D image does not. The eyes average out only the image of the grid structure because of the after-image effect, and thus the resolution improves.

From this intuitive argument, we can see that the viewing resolution depends on the pupil size of our eyes as well. The upper resolution limit given in (5.1) does not take the pupil size of our eyes into consideration. Strictly speaking, the limit is correct when the sizes of the lenslet and our eye's pupil are infinitely small. When the reconstructed 3-D image is viewed by us, according to the ray diagram shown in Fig. 5.8, the size of the subimage sampled by each lenslet (i.e. the size of the subimage that we see through each lenslet), denoted by w, becomes

$$w = \frac{gL_2}{L_1 + L_2}\left(\frac{a}{L_1} + \frac{b}{L_2}\right). \tag{5.2}$$

Equation (5.2) holds regardless of the lenslet position within the array. Here, g is the gap between the elemental image and the display lenslet array, L_1 is the distance between the lenslet array and the image plane, L_2 is that between the image plane and the eye's lens, a is the lenslet size, and b is the pupil

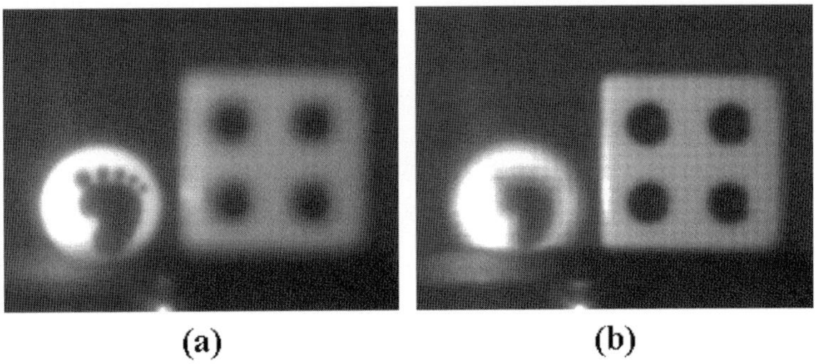

Fig. 5.7. Detection of the reconstructed 3-D image with the iris maximally opened. (**a**) When the footprint is focused. (**b**) When the die surface is focused

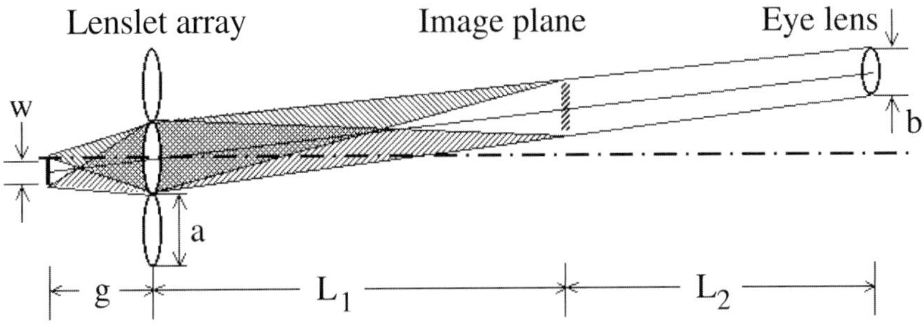

Fig. 5.8. Calculation of the sampling width (the size of each subimage we can see through each lenslet)

size of the eye. When $w \ll p$ (i.e. delta-function sampling), the limit of the viewing resolution is imposed by (5.1). As either a or b or both increase, w for each elemental image (the sampling rate is fixed by $1/p$) also increases. In this case, the viewing resolution can be greater than that obtained by delta-function sampling. The improvement of the viewing resolution by increasing the pupil size of the camera lens proves this, if we compare the images shown in Fig. 5.4a ($w/p \approx 0.07$) with those in Fig. 5.7 ($w/p \approx 0.3$).

Unfortunately, the maximal pupil size of the human eye is limited. So, one may think of improving the viewing resolution by increasing the lenslet size a. In this case, however, the 3-D aspect of the reconstructed image that we feel reduces. In other words, the resolution in the depth direction reduces. This is because the depth of focus of each elemental image is shallow so that integration of the elemental images occurs in a limited depth (z) direction. Thus, the binocular parallax, by which we feel the 3-D aspect, decreases for the reconstructed image, as we view the 3-D image with both eyes. This is similar to using a single large lens for both pickup and display in an extreme case. The moving lenslet array technique, we believe, is the correct solution to improve the viewing resolution without reducing the 3-D aspect of the reconstructed image.

5.3 All-Optical Three-Dimensional Integral Imaging Projector Using an Optically Addressed Spatial Light Modulator

Another factor that deteriorates the viewing resolution of reconstructed 3-D images, in addition to aberration and misalignment of micro-optics, is the resolution of the image pickup and display devices [10]. This is because the 2-D elemental images, which are the components of the 3-D image to be reconstructed, should have sufficient resolution.

We use an optically addressed spatial light modulator (SLM) (liquid crystal light valve (LCLV)) as an image pickup and display device. The optically addressed SLM can potentially provide higher image resolution than commercially available liquid crystal displays do [15]. The LCLV has about 20 lines/mm over an approximately 5 cm × 5 cm area. In addition, the use of the LCLV makes all-optical parallel pickup and display possible, in which converting optical signals to video signals and vice versa is not necessary. We also demonstrate that a nonstationary lenslet array technique can be applied to our system to reconstruct 3-D images with improved viewing resolution.

Figure 5.9 presents our 3-D integral imaging projector. The detector and display devices depicted in the dashed box of Fig. 5.1 are replaced with the LCLV. The pickup lenslet array denoted by LA1 generates multiple 2-D elemental images of an object. These elemental images are transmitted, through relay optics such as a fiber bundle, to the writing surface of the LCLV. The written elemental images are read out with a polarizing beam splitter (PBS), an analyzer or polarizer (P1), and a polarized incoherent beam that is obtained by passing incoherent light through another polarizer (P2). The

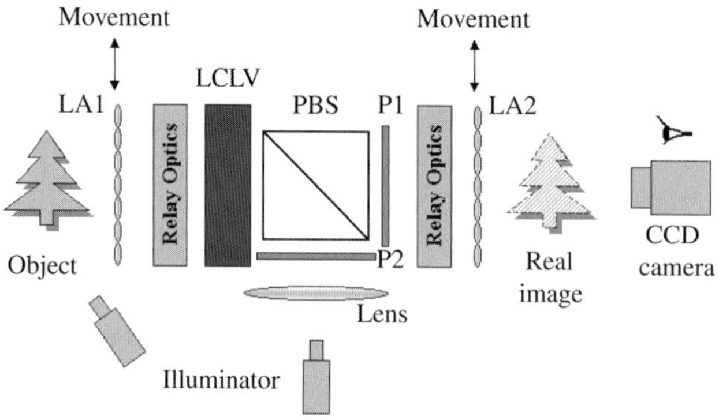

Fig. 5.9. All-optical 3-D integral imaging projector

read-out elemental images can be transferred through additional relay optics to the display lenslet array (LA2). Finally, the reconstructed 3-D image is captured by a CCD camera, and is also examined with our naked eyes.

As described in Sect. 5.1, the object used for 3-D imaging is composed of a die and a plastic button with a footprint figure, as shown in Fig. 5.10a. The distance between the button surface with the footprint figure and the die surface with four dots is about 3 cm. The lenslet arrays (LA1 and LA2) are both tilted around the optic axis (z-axis) by an angle of about 30° from the horizontal direction of the objects (x-axis). The lenslets are mounted on translational stages, and can be moved together simultaneously. The distance between LA1 and the surface of the button is approximately 4 cm. We used a LCLV developed by Hughes Research Laboratories (Model H-4060). The diameter of the LCLV aperture is 50 mm. For simplicity, we used a camera lens whose focal length is 50 mm and f number is 1.4 as the relay optics. The CCD camera that is used to capture the reconstructed output image has a standard camera lens with a built-in iris. The focal length and the maximal f number of this camera lens are 50 mm and 1.2, respectively.

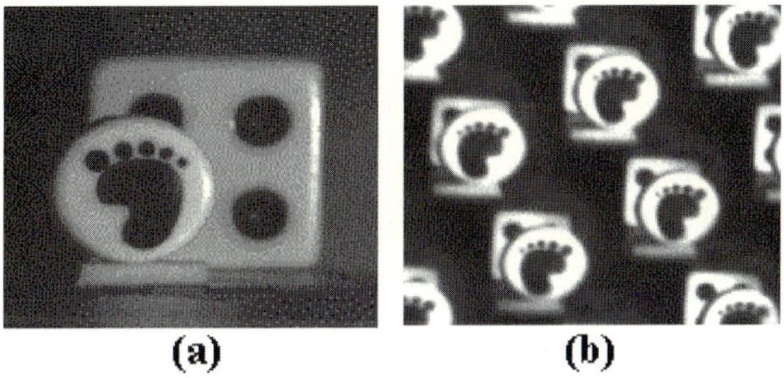

Fig. 5.10. (**a**) 3-D objects used in the experiments; (**b**) elemental images read out from the LCLV

The size of each elemental image exposed to the writing surface is approximately 2 mm. The diameter of the smallest toe image is approximately 80 μm in this case. A few read-out elemental images from the LCLV are shown in Fig. 5.10b. We can see that the LCLV has enough resolution to represent the toe images clearly. When we view the reconstructed real image with our naked eyes, we can visualize the 3-D nature of the image, although it is pseudoscopic. The die surface looks closer to us than the footprint figure does.

The reconstructed 3-D image captured by the camera is shown in Fig. 5.11a. Here, the pupil diameter of the camera lens is approximately 10 mm or less. Note that the pupil diameter of the human eye is about 3 mm in an ordinary environment. Thus, when we observe the reconstructed images, the toes in the footprint figure are not clear, although the dots in the die are

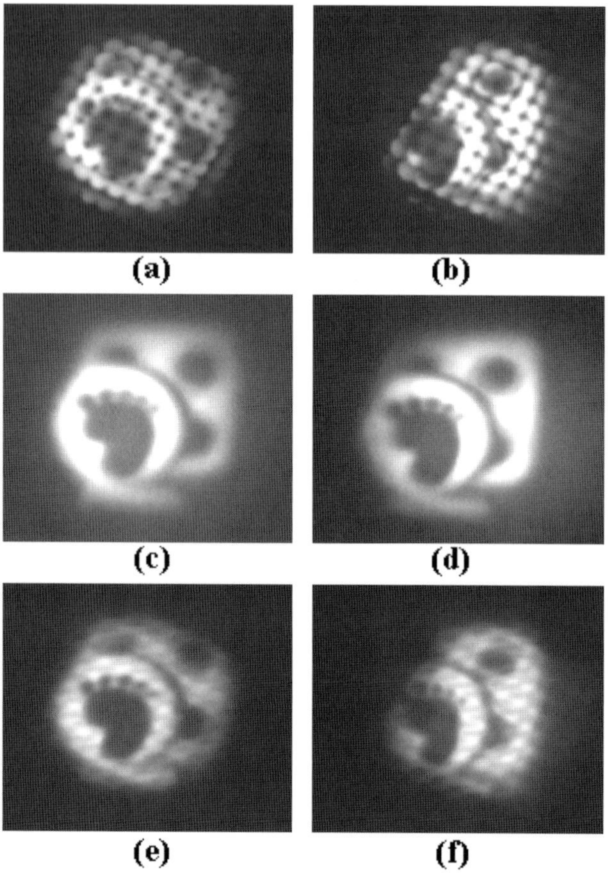

Fig. 5.11. Experimental results for 3-D image reconstruction. (**a**), (**c**), and (**e**) are the images captured when the CCD camera is on the optical axis. (**b**), (**d**), and (**f**) are 3-D images captured when the camera is off the axis. (**a**) and (**b**) 3-D images when the iris pupil diameter of the CCD camera lens was reduced to 10 mm. (**c**) and (**d**) 3-D images when the iris was maximally opened. In this case, its pupil diameter was about 42 mm. (**e**) and (**f**) 3-D images when the CCD pupil diameter was reduced to 10 mm and lenslet arrays are set in motion

clear. This is because the toes are similar to or smaller than the lenslet pitch. To show that the reconstructed image has the 3-D nature, we changed the position (and accordingly the viewing angle) of the CCD camera and captured the reconstructed 3-D image again. This image is shown in Fig. 5.11b. The reconstructed images in Fig. 5.11a and b indicate that even if the resolution of both the pickup and display device is high enough, the image is poor quality because of the array structure of lenslets.

The resolution of the reconstructed image captured by the CCD camera is dependent on the pupil size of the CCD camera lens [11]. For a given lenslet-array pitch, it is possible to capture a reconstructed image of improved resolution if the pupil diameter of the CCD camera is increased. For comparison, Fig. 5.11c and d illustrate the captured images when the iris of the CCD camera is opened maximally. In this case, the toe images become clear. Unfortunately, the pupil diameter of our eyes is small so we cannot observe the images with equal resolution.

To improve our viewing resolution, we adopt the nonstationary micro-optics (lenslet array) technique described in Sect. 5.1. The temporal response time of the LCLV used in our experiment is approximately 100 ms. Thus, we cannot directly apply the nonstationary lenslet-array technique because the temporal response time of our eyes is less than 100 ms. To illustrate the effect of the nonstationary lenslet-array technique in the 3-D imaging projector, we used an indirect method. First, we reduced the pupil diameter of the camera lens as used when the images in Fig. 5.11a and b were obtained. Then, we captured a sequence of 34 reconstructed images by simultaneously setting in motion both lenslet arrays (LA1 and LA2) horizontally and with 50 μm steps. The movement range of the lenslet arrays does not need to be large: The range of one lenslet pitch in each direction will be enough, in principle, because the lenslets are periodic. Because both lenslet arrays are already tilted by an angle of 30°, shifting the lenslet arrays in the horizontal (x) direction increases the spatial sampling rates in both x and y directions. Finally, the 34 images are averaged in a personal computer to mimic the time-averaging effect of the eyes. The resulting images for the on-axis and off-axis camera positions are given in Fig. 5.11e and f, respectively. Although our nonstationary lenslet-array technique is applied in an indirect and discrete manner, a noticeable improvement is evident because the toe images have become clear.

The pitches of LA1 and LA2 may not be equal. In this case, to keep one-to-one correspondence between the lenslets and the elemental images in both pickup and display, the elemenal images shoud be properly magnified (or demagnified) by the relay optics depicted in Fig. 5.9. Thus, the synchronous movement of the lenslet arrays does not necessarily mean physical synchronization. It should be image-based synchronization, because synchronization should be measured as a reference to the elemental images picked up and displayed.

It is possible, in principle, to reduce the response time of the LCLV by increasing the writing-beam intensity [15]. In addition, optically addressed SLMs, such as ferroelectric devices, with faster response time are available. If a fast SLM is used, the nonstationary lenslet-array technique can be applied in a continuous manner, and thus direct observation of the reconstructed 3-D image with improved resolution will be possible.

5.4 Improvement of Both Field of View and Viewing Angle Using Nonstationary Micro-Optics

When the fill factor of lenslet arrays is high, interference between neighboring elemental images in the pickup process may occur. In other words, neighboring elemental images can overlap. One solution to avoid the interference problem is to isolate each lenslet with optical barriers [6]. This solution results in reduced fields of view for the pickup lenslets and thus for the reconstructed image.

In the same manner, an opposite problem exists in the display process. The elemental images can also be displayed through their neighboring lenslets. Therefore, multiple 3-D images are reconstructed as shown in Fig. 5.6b. To isolate each lenslet with optical barriers is also a solution to the problem of reconstructing only one image in the display process [6]. In this case, the viewing angle of the reconstructed image is limited.

If two identical lenslet arrays are used for pickup and display, we can use one angle parameter ψ to describe both the field of view and viewing angle. Simply, let us call ψ the viewing angle, and it is given by [6]

$$\psi = 2 \arctan\left(\frac{p}{2g}\right), \tag{5.3}$$

where p is the pitch of the display lenslet array and g is the gap between the display plane and the lenslet.

Our approach to improve ψ is to separate lenslet elements enough [12]. In other words, for a given size of lenslet element, the pitch of the lenslet arrays is increased as (5.3) implies. In this case, the fill factor of the lenslet arrays reduces, and thus the viewing resolution of the reconstructed image decreases as expected from (5.1). However, by adopting the nonstationary lenslet-array technique, the resolution deterioration caused by the use of low fill factor (high-pitch) lenslet arrays can be easily relaxed.

5.5 Orthoscopic Three-Dimensional Real Images Using Two-Step Integral Imaging

The reconstructed 3-D real image in the direct pickup and display process (i.e. one-step II) is pseudoscopic. Obviously, two-step II, which repeats the pickup and display process twice, results in an orthoscopic image [16]. It is generally believed, however, that two-step II is impractical because of its low image resolution [3, 4, 17]. Therefore, one-step II that gives orthoscopic images was investigated by inverting every elemental image at their own center optic axis [4, 6]. In this case, the reconstructed images are orthoscoptic but virtual images. Previous studies on two-step II is also limited on formation of orthoscopic virtual images [16, 17]. For 3-D TV or movies, formation of real images is preferable, because the 3-D images can be reconstructed close to viewers even if the display lenslet array is far from the viewers.

In this section, we investigate two-step II that gives orthoscopic real 3-D images. To overcome the viewing resolution limit imposed by the lenslet pitch, we adopt our nonstationary lenslet-array technique in this two-step II system. Experimental results show the potential usefulness of our two-step II.

Figure 5.12 illustrates the structure of our two-step 3-D integral imaging system. Here, we use a liquid crystal light valve (LCLV) [15] as an image pickup and display device, because the LCLV provides higher image resolution. The pickup lenslet array denoted by LA1 generates multiple 2-D elemental images of an object. These elemental images are transmitted, through relay optics such as a fiber bundle, to the writing surface of the LCLV. The written elemental images are read out with a polarizing beam splitter (PBS), an analyzer, namely a polarizer (P1), and a polarized incoherent beam that is obtained by passing incoherent light through another polarizer (P2). The read-out elemental images can be transferred through additional relay optics to the display lenslet array (LA2). The real image reconstructed by LA2 is pseudoscopic. To obtain an orthoscopic real image, we put an additional pair of lenslet arrays (LA3 and LA4) in front of the viewer or the CCD camera, which means two-step II. The role of LA3 is to generate the elemental images of the pseudoscopic image formed by LA2, and the role of LA4 is to reconstruct an orthoscopic image from the elemental images generated by LA3. Thus, one can regard the additional lenslet pair, LA3 and LA4, as a simple pseudoscopic-to-orthoscopic image converter. The final reconstructed 3-D image can be examined with our naked eyes, and is recorded by a CCD camera.

Note that the pair, LA3 and LA4, may be positioned in front of LA1 to pickup a pseudoscopic image. In this case, the pair, LA3 and LA4, becomes an orthoscopic-to-pseudoscopic image converter. Then, the real image reconstructed by LA2 becomes orthoscopic. In our experiment, we used the pair, LA3 and LA4, as a pseudoscopic-to-orthoscopic image converter in order to

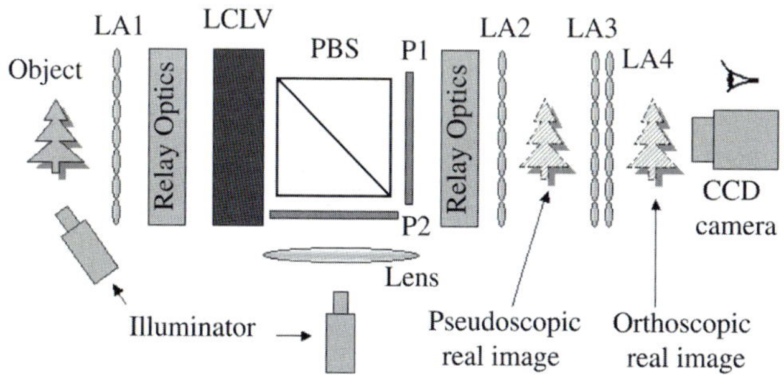

Fig. 5.12. Two-step 3-D integral imaging system

avoid reducing the light intensity of the elemental images to be written onto the LCLV.

The object used for 3-D imaging is composed of a die and a plastic button with a footprint figure, as shown in Fig. 5.10a. LA1 and LA2 are both tilted around the optic axis (z-axis) by an angle of about 30° from the horizontal direction of the objects (x-axis), while LA3 and LA4 are both rotated by an angle of about 60°. The distance between LA1 and the surface of the button is approximately 4 cm.

When we view the reconstructed real image with our naked eyes in one-step II, we can visualize the 3-D nature of the image, although it is pseudoscopic. The die surface looks closer to us than the footprint figure does. After two-step II, we confirmed the pseudoscopic image is correctly converted to its orthoscopic version.

The resolution of the reconstructed image captured by the CCD camera is dependent on the pupil size of the CCD camera lens. The pseudoscopic image after one-step II and the orthoscopic image after two-step II are shown in Fig. 5.13a and b, respectively, when the pupil diameter of the camera lens is approximately 10 mm or less. Note that the pupil diameter of the human eye is about 3 mm in an ordinary environment. Thus, when we observe the reconstructed images for both one-step II and two-step II, the dots in the die are clearly viewed, but the toes in the footprint figure are not. This is because the size of the toes is similar to or smaller than the lenslet pitch. As shown in Fig. 5.13a and b, the orthoscopic image quality obtained in two-step II is about the same as the pseudoscopic image obtained in one-step II. Both images have poor-quality reconstruction.

For comparison, one-step II and two-step II images are captured when we maximally opened the iris of the CCD camera as shown in Fig. 5.13c and d, respectively. In this case, the toe images become clear for both one-step II and two-step II. Unfortunately, the pupil diameter of our eyes is so small that we cannot observe the 3-D images with high resolution.

To improve the viewing resolution observed by the naked eyes, we also use the indirect nonstationary lenslet-array technique in two-step II. First, we reduced the pupil diameter of the camera lens as used in obtaining the images in Fig. 5.13a and b. Then, we captured a sequence of 34 reconstructed images by moving the four lenslet arrays (LA1 - LA4) simultaneously in the horizontal direction in 50 μm steps. Because the lenslet arrays are already tilted by an angle of either 30° or 60°, shifting the lenslet arrays along the horizontal (x) direction increases the spatial sampling rates in both x and y directions. Finally, the sequence of the images are averaged in a personal computer to mimic the time-averaging effect in the eyes. The resulting images for both one-step II and two-step II are given in Fig. 5.13e and f, respectively. Although our nonstationary lenslet-array technique is applied in an approximate and discrete manner, a noticeable improvement is evident because the

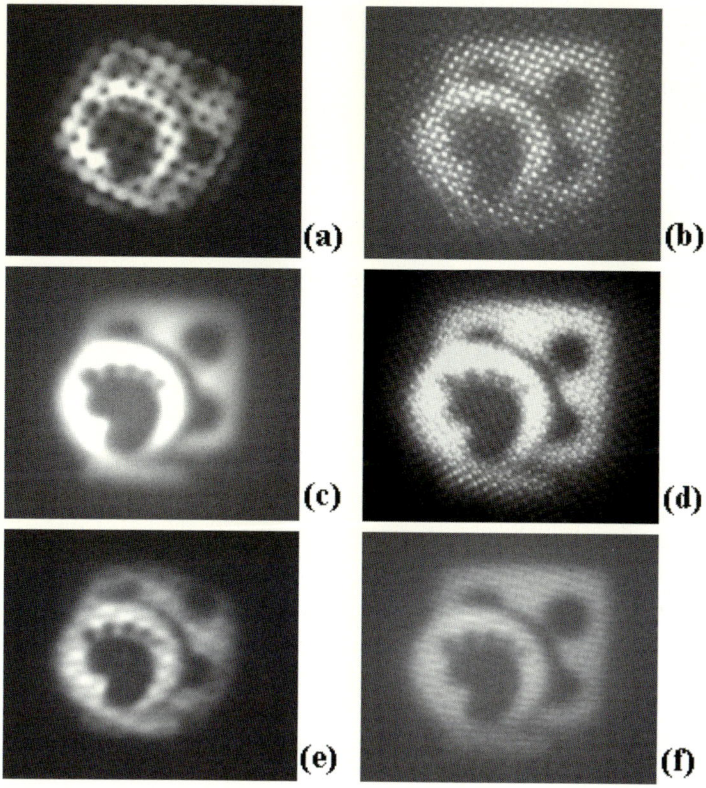

Fig. 5.13. Experimental results for the 3-D image reconstruction. (**a**), (**c**), and (**e**) are the images obtained by one-step II. (**b**), (**d**), and (**f**) are the images obtained by two-step II. These images were captured with a CCD camera. In (**a**) and (**b**) the iris pupil diameter of the CCD camera lens was reduced to 10 mm. In (**c**) and (**d**) the iris was opened maximally to a pupil diameter of about 42 mm. In (**e**) and (**f**) the CCD pupil diameter was reduced to 10 mm and the nonstationary lenslet-array technique was applied

toe images have become clear. The toes whose diameter is about 0.8 mm can be reconstructed.

To show the 3-D nature of the reconstructed image, we changed the position (and accordingly the viewing angle) of the CCD camera and repeated the experiments. The reconstructed 3-D images captured by the CCD camera are shown in Fig. 5.14. We can see that if the viewing position is changed, one dot of the die is occluded by the button. Note that the camera positions for the pseudoscopic image and the orthoscopic image are mutually opposite from the center optic axis to get the images of a similar perspective. When the images are viewed in an off-axis direction, the astigmatic aberration of the lenslets is increased. And thus the toe images of two-step II are somewhat

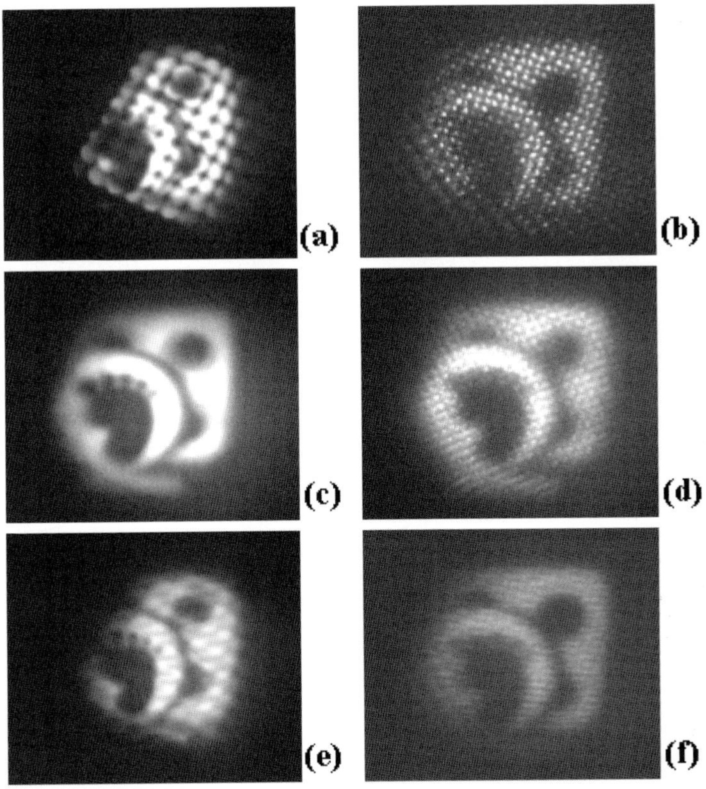

Fig. 5.14. Experimental results for the 3-D image reconstruction. The position and thus the direction of the CCD camera was changed to show the 3-D nature of the reconstructed images. In (**a**)–(**f**), the CCD pupil diameter was reduced to 10 mm and the nonstationary lenslet-array technique was applied

smeared, as shown in Fig. 5.14d and f, while those of one-step II are still clear as shown in Fig. 5.14c and e.

In conclusion, two-step II is a simple solution to produce real orthoscopic 3-D images. From the experimental results, we can conclude that the use of the moving lenslet array technique in our two-step II system improves the viewing resolution efficiently. The image resolution of one-step II is slightly better than that of two-step II. However, two-step II can be used if our moving lenslet array technique is employed.

5.6 Summary

In this chapter, we have dealt with three problems in II: viewing resolution, viewing angle or field of view limited by interference of the elemental images, and the pseudoscopic nature of reconstructed 3-D images. First, we have shown that the nonstationary lenslet-array technique can be a powerful tool because the Nyquist resolution limit is relaxed easily with the technique. Second, we have presented an all-optical II projector in which an optically addressed SLM is used and the nonstationary micro-optics technique is adopted. The optically addressed SLM provides potentially better image resolution than the conventional CCD and liquid crystal display pair. Third, we have discussed how both the viewing angle and the field of view can be improved by use of lenslet arrays with low fill factors. In general, the use of low fill factor lenslet arrays results in lower viewing resolution compared with the use of high fill factor lenslet arrays, if other lenslet parameters are the same. Even in this case, however, the viewing resolution can be spared when the nonstationary lenslet-array technique is used. Finally, a two-step integral imaging system to obtain 3-D orthoscopic real images is also investigated. We have shown experimentally that the potential usefulness of two-step integral imaging when the nonstationary micro-optics technique is adopted.

References

1. G. Lippmann: Comptes-Rendus Academie des Sciences **146**, 446 (1908)
2. N. Davies, M. McCormick, M. Brewin: Opt. Eng. **33**(11), 3624 (1994)
3. F. Okano, H. Hoshino, J. Arai, I. Yuma: Appl. Opt. **36**(7), 1598 (1997)
4. J. Arai, F. Okano, H. Hoshino, I. Yuma: Appl. Opt. **37**(11), 2034 (1998)
5. H. Hoshino, F. Okano, H. Isono, I. Yuyama: J. Opt. Soc. Am. A **15**(8), 2059 (1998)
6. F. Okano, H. Hoshino, J. Arai, I. Yuma: Opt. Eng. **38**(6), 1072 (1999)
7. B. Javidi, F. Okano, eds.: Proc. SPIE, **CR 76**, Boston, MA, Nov. 2000
8. H. Arimoto, B. Javidi: Opt. Lett. **26**(3), 157 (2001)
9. J.-H. Park, S.-W. Min, S. Jung, B. Lee: Appl. Opt. **40**(29), 5217 (2001)
10. L. Erdmann, K.J. Gabrial: Appl. Opt. **40**(31), 5592 (2001)
11. J.-S. Jang, B. Javidi: Opt. Lett. **27**(3), 324 (2002)
12. J.-S. Jang, B. Javidi: *SPIE Conf. Electronic Imaging 2002: Science and Technology, 19–25 January 2002, San Jose, California USA*
13. C.B. Burckhardt: J. Opt. Soc. Am. **58**(1), 71 (1968)
14. T. Okoshi: Appl. Opt. **10**(10), 2284 (1971)
15. W.P. Bleha, L.T. Lipton, E. Wiener-Avnear, J. Grinberg, P.G. Reif, D. Casasent, H.B. Brown, B.V. Markevitch: Opt. Eng. **17**(4), 371 (1978)
16. H.E. Ives: J. Opt. Soc. Am. **21**, 171 (1931)
17. T. Okoshi: Proc. IEEE **68**(5), 548 (1980)

6 Real-Time Three-Dimensional Integral Imaging Display and Object Recognition Using Volume Optical Storage

S.-H. Shin and B. Javidi

6.1 Introduction

Display and recognition systems for two-dimensional (2-D) objects have been extensively studied using optical, digital, and hybrid methods. The optical method, although less flexible than the digital method, remains attractive due to faster processing. In recent years, there has been much activity in the area of three-dimensional (3-D) information-processing systems because of their potential applications in the field of computer vision, graphics and displays. Therefore, many conceptual developments of 3-D systems have been reported, including 3-D object recognition and display [1–6]. In contrast to conventional 2-D processing, however, all-optical 3-D information-processing systems for display and recognition are not extensively developed. Recently, the integral imaging method has received renewed interest as one of the best techniques for practical application of 3-D display, because it provides continuous viewpoints and does not require special viewing devices. Since integral imaging was first proposed by Lippmann [7], it has taken a long time to beimproved by use of the many pickup and display devices currently under development. The resolution of the pickup and display device is very important, because the 3-D object should be decomposed into small-size elemental images in the integral imaging system.

In this chapter, we propose to use the photorefractive volume holographic memory for 3-D integral imaging with a massive optical data storage system. It can be implemented as a database system for 3-D real objects. The photorefractive volume holographic memory system is able to provide high resolution, unprecedented capacity, fast data transfer rate, and fast random access performances for diverse applications. In the volume holographic storage mechanism, the holograms are recorded in photosensitive media by interferring light beams. The hologram can be recorded with a page of information (i.e. a 2-D spatial information) and parallel processed page by page. There are also many multiplexing methods to record multiple holograms superimposing in volume media. In view of integral imaging, which retains 3-D information by use of many different perspectives of 2-D images, volume holographic storage is an attractive method for storing 3-D images.

All optical 3-D object display and recognition systems by use of the integral imaging and volume holographic methods are described in this chapter. Sect. 6.2 discusses 3-D image display, including a 3-D video system using integral imaging and volume holography. Sect. 6.3 discusses 3-D image recognition using volume holography. In Sect. 6.4, we present 3-D object recognition using integral imaging and volume holography. Experimental results are presented in each section.

6.2 Speckle Reduced Three-Dimensional Display System Using Volume Holography and Integral Imaging

For three-dimensional (3-D) display, holographic methods can produce a 3-D image of the original object with complete parallax. However, the use of holography has drawbacks, such as speckle noise and narrow viewing angle caused by Bragg mismatching [4, 8]. In holographic display, a speckled image is not only unpleasant but also a cause of information loss. Although several methods to reduce speckle noise were suggested for coherent 2-D imaging [9, 10], thesecannot be applied to 3-D object display. To remove speckle noise, random speckle noise contained in the images should be averaged out incoherently, but this is not possible with a conventional holographic 3-D display system. In this section, we present a method for coherent 3-D display by use of volume holographic memory based on integral imaging to reduce the speckle noise.

The integral imaging method has been studied for many applications [11–13]. In the integral imaging method, the resolution of the pickup and display device is very important, because small-sized elemental images with different perspectives should be captured and displayed with sufficient resolution. For 3-D image formation, as many as hundreds of times more pixels than lens elements are required for the pickup and display devices [14]. However, until recently, there has been no proper pickup and display device able to provide an elemental image in high fidelity. In our approach, we use a photorefractive crystal, which has very high resolution and unprecedented storage capacity [15, 16]. Therefore, the proposed system not only reduces speckle noise but also provides pickup, storage, and display of small elemental images with sufficient resolution. We discuss the optical properties of the proposed system, such as field of view and resolution in Sect. 6.2.1. The speckle reduction is discussed in Sect. 6.2.2 in detail. Experimental results for the 3-D display system are presented in Sect. 6.2.3. In Sect. 6.2.4, we present a 3-D video system using integral imaging and volume holographic storage.

6.2.1 Integral Imaging by Use of Volume Holography

Optical System for Integral Imaging. In integral imaging, 3-D images can be created by real elemental images that are formed by a microlens array as shown in Fig. 6.1. The figure presents three different pickup methods. In the film pickup method shown in Fig. 6.1a, a photographic film located at the imaging plane of the lens array is used to record elemental images. This method cannot be applied to real-time display because of the developing process needed for reproduction. Therefore, a direct pickup method (Fig. 6.1b), in which a camera and various types of spatial light modulators [17, 18] are used to record and display 3-D images in real time, was suggested [14]. In the direct pickup method, the pickup device is smaller than the size of the lens array, in general, so additional optical systems for adjusting the size of the elemental image formed by the microlens array are required, as shown in Fig. 6.1b. Therefore, the display device, which should have as many pixels as that of the pickup device, is needed. Figure 6.1c shows the system we propose, in which a volume holographic crystal is used instead of the camera for pickup. In the volume holographic pickup method, a single crystal can perform the functions of pickup, storage, and display for the elemental images.

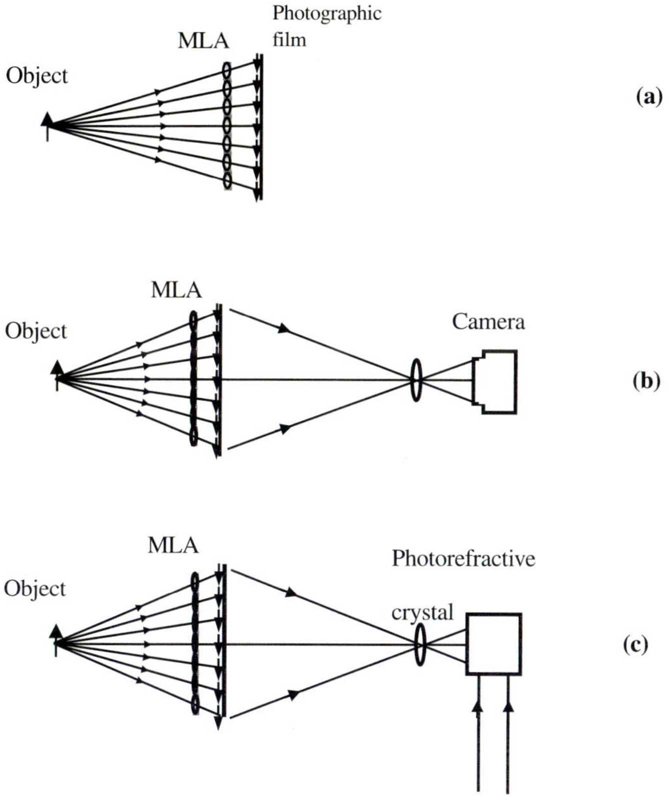

Fig. 6.1. Various pickup methods for integral imaging; (**a**) film pickup suggested by Lippmann; (**b**) direct pickup using camera; (**c**) direct pickup using volume holographic storage

We will show that use of integral imaging allows formation of speckle-reduced holographic 3-D images.

Unlike the film pickup method, the elemental images do not record at the imaging plane of microlens array in the volume holographic pickup method. Therefore, when a small-aperture lens is employed for elemental image recording into the crystal, as shown in Fig. 6.1c, optical losses due to vignetting effects are increased, especially from the microlens located far from the center of the microlens array. This introduces the position error for the principal rays and makes the reconstructed image different from the 3-D object that is being recorded. To solve this problem, the directions of the rays that pass through the center of each microlens need to be parallel to each other. We meet this condition by using a large-aperture convex lens, which is located close to the focal plane of the microlens array [14].

Figure 6.2 shows a schematic diagram for the direct pickup method, as shown in Fig. 6.1b and c. In this geometry, the width of the pickup device W_S is given by

$$W_S \cong \frac{f_c}{Z_{MC}} W_L \,, \tag{6.1}$$

$$W_L = W_M N \,, \tag{6.2}$$

where W_L is the width of the lens array, f_c is the focal length of the imaging lens, Z_{MC} is the distance between the lens array and the imaging lens, W_M is the pitch of the microlens, and N is the number of microlenses. The focal length, f_F, of the large-aperture convex lens should be equal to the distance, Z_F, between the large-aperture convex lens and the imaging lens to achieve parallel projection:

$$Z_F = f_F \,. \tag{6.3}$$

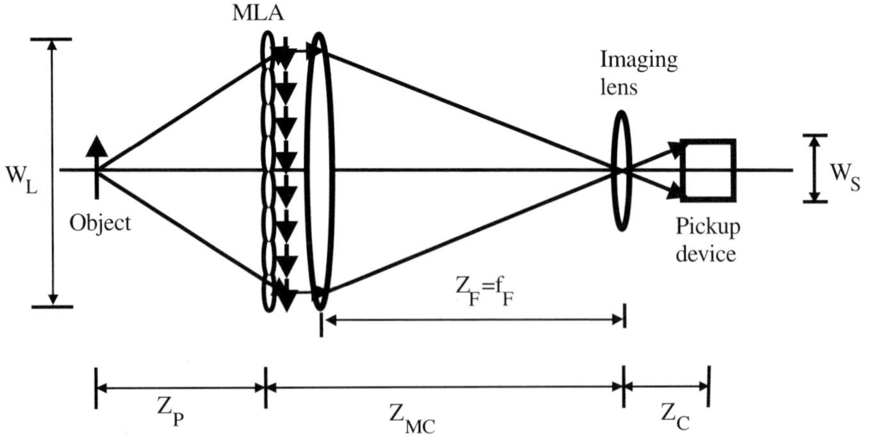

Fig. 6.2. Schematic of the photorefractive pickup integral imaging system

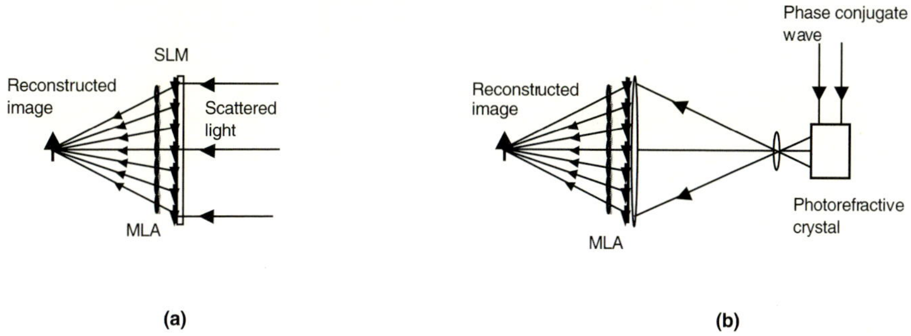

Fig. 6.3. Various reconstruction methods; (**a**) reconstruction with photographic film or spatial light modulator; (**b**) reconstruction with phase-conjugate wave: PRC is photorefractive crystal; MLA is microlens array; SLM is spatial light modulator

Therefore, using (6.3), (6.1) is simplified as follows:

$$\frac{W_{\mathrm{L}}}{f_{\mathrm{F}}} = \frac{W_{\mathrm{S}}}{Z_{\mathrm{C}}} \cong \frac{W_{\mathrm{S}}}{f_{\mathrm{c}}} \, . \tag{6.4}$$

Equation (6.4) shows that the object beams that propagate along the direction parallel to the optic axis can be collected by the imaging lens, as shown in Fig. 6.2. Thus the position error can be reduced in direct image pickup.

In a conventional system, the processed elemental images, which are either recorded on the photographic film or represented on the LCD panel, are exposed to the scattered light as shown in Fig. 6.3a. In our system, unlike the conventional integral imaging system, the 3-D image can be reconstructed by use of a phase-conjugate reading beam as shown in Fig. 6.3b. In this case, the single lens-array optical system can be used for the recording and display. The phase-conjugate reading beam provides several advantages in terms of aberration and alignment of optical system because the aberrations or misalignments of the imaging systems are compensated.

Resolution and Field of View. Similar to holographic 3-D imaging systems [8], weneed to consider several important optical parameters for the proposed display system. These include resolution, viewing angle, and speckle noise. Regardless of reconstruction methods, there are several major factors that limit total resolution of the integral imaging system. The most fundamental factor is the resolution of the elemental image, i.e. 2-D resolution, because the reconstructed image is composed by integration of many different sections of the elemental images. The sampling frequency and cutoff frequency due to the diffraction are also important factors that limit resolution. For the resolution analysis, we apply the angular spatial frequency method suggested by Hoshino et al. [11] to our system. Assuming the spatial frequency that can provide the pickup-and-display device is α_{i}, viewing spatial frequency at the viewpoint as shown in Fig. 6.4 is

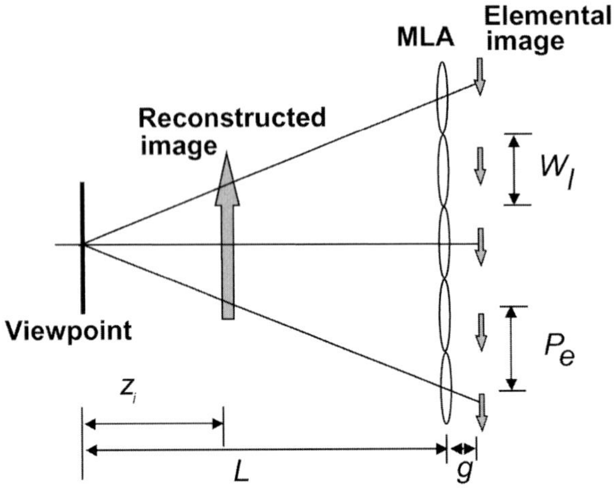

Fig. 6.4. Schematic diagram of 3-D image formation

$$\beta_{\mathrm{ele}} = \alpha_{\mathrm{i}} \frac{z_{\mathrm{i}}}{|L - z_{\mathrm{i}}|} \,. \tag{6.5}$$

The Nyquist frequency of the sampling, which has the period P_{e}/L, is given by

$$\beta_{\mathrm{nyq}} = \frac{L}{2P_{\mathrm{e}}} \,, \tag{6.6}$$

where L is the distance between the viewpoint and lens array, and P_{e} is the period of the lens array, as shown in Fig. 6.4. In addition, the spatial frequency to be cutoff by diffraction of the microlens is

$$\beta_{\mathrm{dif}} = \frac{w_{\mathrm{l}}}{\lambda} \frac{z_{\mathrm{i}}}{|L - z_{\mathrm{i}}|} \,, \tag{6.7}$$

where w_{l} is the lens width. Although there is a focusing error, it has a limited effect on the overall resolution of the proposed system due to the phase-conjugated reading beam. Therefore, the resolution of the reconstructed image that is observed at viewpoint as shown in Fig. 6.4 is limited by the minimum of these three factors:

$$\beta_{\mathrm{max}} = \min(\beta_{\mathrm{ele}}, \beta_{\mathrm{nyq}}, \beta_{\mathrm{dif}}) \,. \tag{6.8}$$

In the case of integral imaging by using a CCD camera and LCD panel, the resolution of the reconstructing 3-D image is limited by the number of lenses as well as the number of pixels for an elemental image in the pickup and display devices. For acceptable image reconstruction, the resolution of the camera and the display for integral imaging must be much higher than that of conventional TV displays [19]. Figure 6.5a shows the plots of three resolution

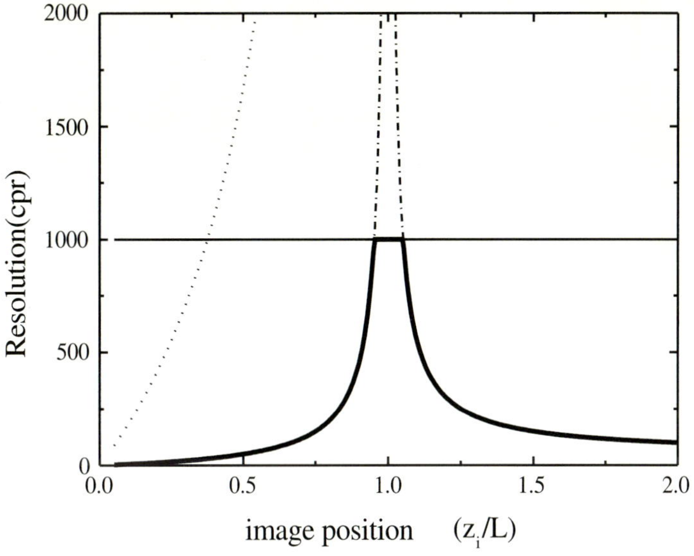

(a)

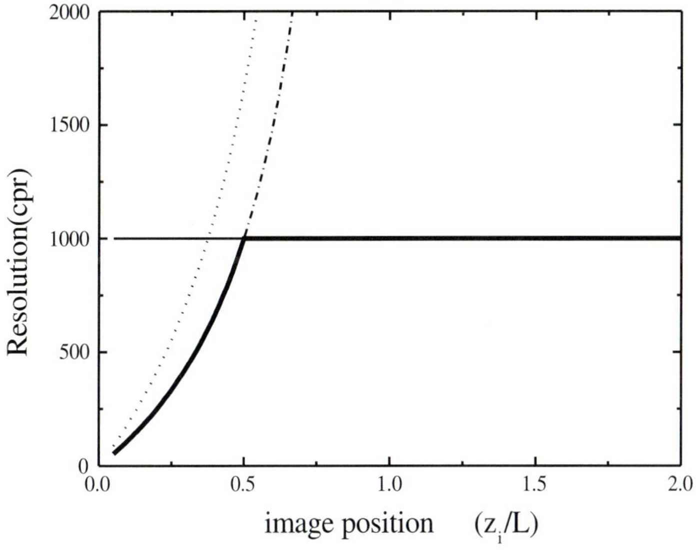

(b)

Fig. 6.5. Overall resolution of the integral imaging system obtained from (6.8) using (**a**) LCD; (**b**) photorefractive crystal. The resolution due to the viewing spatial frequency of 2-D elemental image is illustrated by the *dash-dotted line*. The resolution due to Nyquist frequency is illustrated by the *solid line*. The cutoff spatial frequency caused by diffraction is shown by the *dotted line*. The overall resolution is illustrated by the *thick solid line*. The parameters for resolution calculation are as follows: $L = 2\,\mathrm{m}$, $P_\mathrm{e} = W_\mathrm{l} = 1.01\,\mathrm{mm}$, $\alpha_\mathrm{i} = 50\,\mathrm{mm}^{-1}$ (LCD), $\alpha_\mathrm{i} = 1000\,\mathrm{mm}^{-1}$ (photorefractive crystal) factors, which are calculated using (6.8), when a LCD panel is employed for display. The overall resolution of the integral imaging system is fundamentally limited by the resolution of the LCD panel, which is the same resolution as the elemental image. According to (6.5) and (6.7), if the position of the 3-D image is close to the viewpoint, i.e. $z_i \approx L$, the viewing spatial frequency and cutoff frequency are increased sharply, so that the overall resolution is maximized and limited only by the Nyquist frequency. However, this area is very narrow and too close to see the reconstructed image. In our case, the photorefractive crystal has very high spatial resolution compared with a CCD camera or LCD panel. Figure 6.5b shows the overall resolution of the proposed system using a photorefractive crystal for pickup and display. In this case, 2-D resolution is increased significantly, so that the overall resolution of the reconstructed 3-D image is improved up to the Nyquist frequency.

The other important optical parameter is the field of view, which is the angle measured at the center of the display surface. In a volume holographic display system, the field of view is limited by the Bragg-matching condition of the reconstructed beam. Although the proposed system is used as a volume holographic storage, however, the field of view "θ" is given by a function of the pitch P_e and the gap g between the lens array and the elemental image formed at the imaging plane:

$$\theta = 2 \arctan\left(\frac{P_e}{2g}\right). \tag{6.9}$$

Since the resolution of the lens-array integral imaging system is independent of the gap g between the lens and the elemental image, the field of view depends on the f-number of the microlens. Therefore, a wide field of view can be achieved in spite of the diffraction angle, which satisfies a fixed Bragg-matching condition.

6.2.2 Speckle-Reduced Reconstruction

As with many coherent imaging systems, the speckle-noise problem significantly corrupts the performance of the imaging system [4, 20]. Speckle noise reduces the spatial resolution and degrades the signal-to-noise ratio of any coherent imaging process. Although the speckle problem has been extensively analyzed, it has not been solved in the case of a 3-D projection system. In our system, we employ a method to reduce speckle noise, which is based on the time integration of multiple coherent images having independent speckle patterns at the image plane of the microlens array. The diffuser, which is employed during the reconstruction procedure and rotates with the appropriate frequency, plays a key role in speckle-noise reduction. The time averaging of the elemental images by use of a rotating diffuser is performed at the imaging plane of the lens array [21]. If the time constant τ of the pickup device, which responds to the intensity, is large compared with the vibrating period of the speckle pattern, the random speckle noises are averaged out. Therefore, the signal-to-noise ratio in the image plane is improved by $1/\sqrt{N}$ [9], where N is the number of random-phase images participating in the time averaging. In the proposed system, the diffuser, which is located in the image plane of the lens array, is rotating at a constant speed for averaging elemental images reconstructed from the photorefractive crystal. When the diffuser is rotating continuously, the speckle pattern tends to be averaged to a uniform background. The minimum rotating frequency depends on the response time of the imaging device.

6.2.3 Experimental Results

The experimental setup for the proposed system is shown in Fig. 6.6. An Ar^+ laser ($\lambda = 514.5\,\text{nm}$) is used for elemental image recording and readout in the photorefractive crystal. The expanded laser beam is split into a plane-wave reference beam and an object beam illuminating the 3-D object (two dice whose dimensions are $5 \times 5 \times 5\,\text{mm}^3$).

We used a 0.03 mol% iron-doped $LiNbO_3$ crystal for storage. The 90-degree recording geometry was used to take advantage of a minimum detuning Bragg angle for multiple-image recording. The microlens array used in this experiment has a base size of $1.01\,\text{mm} \times 1.01\,\text{mm}$ for each lens, $f = 3.5\,\text{mm}$, and total numbers of 53×53 elements. Figure 6.6a and b show the recording and reconstructing scheme, respectively. During the reconstruction process, the phase-conjugate beam and the rotating diffuser in the imaging plane of microlens array are employed as shown in Fig. 6.6b. Figure 6.7a shows a reconstructed 3-D image of the die with speckle noise. Figure 6.7b shows a speckle-reduced 3-D image using a rotating diffuser, which is a light-shaping diffuser having a $10°$ circular diffusion angle. The result shows that the speckle noise is significantly reduced and the details of the 3-D image are clarified. In the case of reconstruction without a diffuser, there is no fixed plane for imaging by the microlens array. So the defocusing error can be neglected. When we employ a rotating diffuser, however, the elemental images are confined on the surface of the diffuser and the focusing error should be considered. The focusing error can be reduced by using focal-length adjustment as proposed by Hoshino et al. [11] without significant impact on the overall resolution of the proposed integral imaging system.

The proposed system can increase the viewing angle. Figure 6.8 shows a series of 3-D images of a pair of dice viewed from different perspectives. A different number of black dots are displayed when viewed from different perspectives along the horizontal plane. The maximum viewing angle of this system is larger than that of the conventional volume holographic display. It is measured at $\sim 13°$, which is comparable to the calculated results using (6.9). In the conventional volume holographic display system, the perspective of the image is severely restricted by the Bragg-matching condition (less than

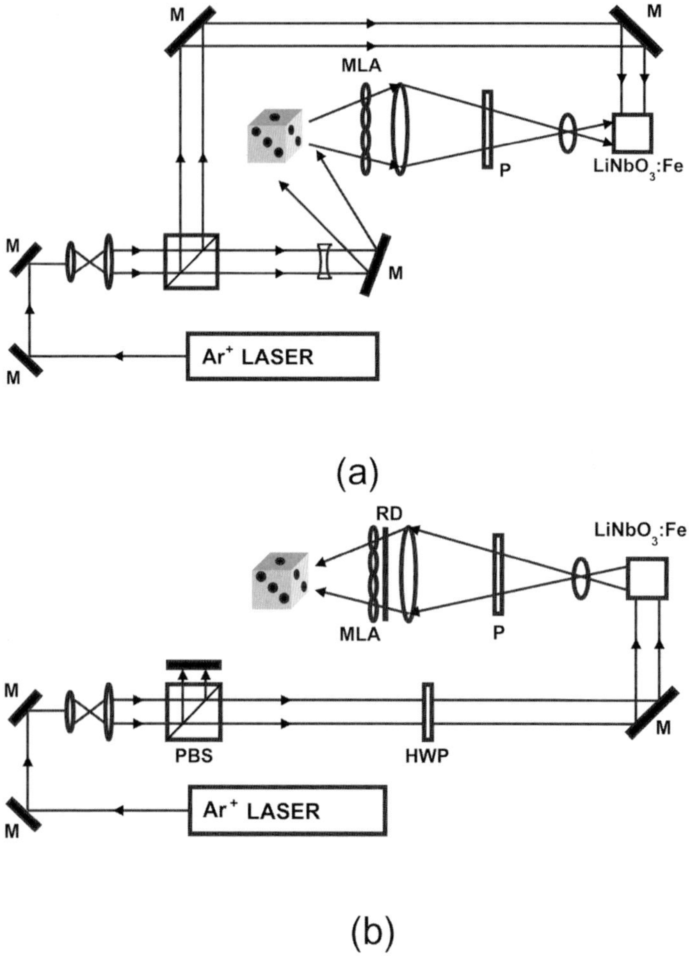

Fig. 6.6. Experimental setup for the speckle-reduced 3-D display system: (**a**) recording; (**b**) reconstruction. M is mirror; P is polarizer; HWP is half-wave plate; PBS is polarizing beam splitter; MLA is microlens array; RD is rotating diffuser

one degree). In the double phase-conjugate mirror (DPCM) system, which offers the largest range of perspective, the maximum viewing angle was given by $\sim 5^\circ$ [4].

6.2.4 3-D Video System Using Integral Imaging and Volume Holography

In contrast to conventional storage methods, volume holographic storage uses a parallel page-by-page processing technique. The signal arm is spatially

(a) (b)

Fig. 6.7. Reconstructed images by using phase-conjugate reading beam (**a**) without rotating diffuser; (**b**) with rotating diffuser

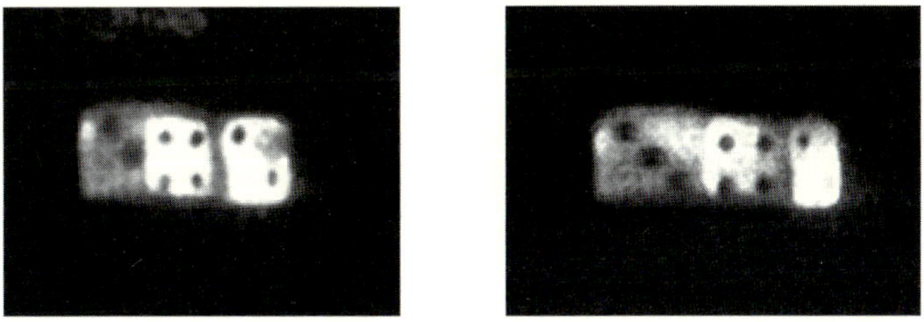

(a) (b)

Fig. 6.8. Speckle-noise reduced images reconstructed at two different perspective angles

modulated by a page of information either after the beam passes through a spatial light modulator or reflects from a 3-D real object. In this case, the reference arm provides a selective recording and reconstructing by changing the incident angle or wavelength. It thereby provides massive and fast processing abilities. In the practical realization of photorefractive holographic memories, the volatality of the stored data is a serious obstacle. Recently, photorefractive holograms have been fixed up to 10,000 pages in $LiNbO_3$:Fe by using a thermal fixing method [22]. In the integral imaging display, 3-D information for a real object is converted to many 2-D element images, which provide 3-D perspectives of the real object. We combine the integral imaging method and volume holographic storage to take advantage of each method and provide a 3-D video system. Figure 6.9 shows a diagram for a 3-D object display using integral imaging and volume holography. The object, which is moving, is recorded by interferring with the reference beam diffracted from the acousto-optic deflector.

The recording is accomplished based on angular multiplexing using Bragg-selectivity properties of volume holograms [23]. There are many different multiplexing methods for multiple-page recording in a single crystal, including

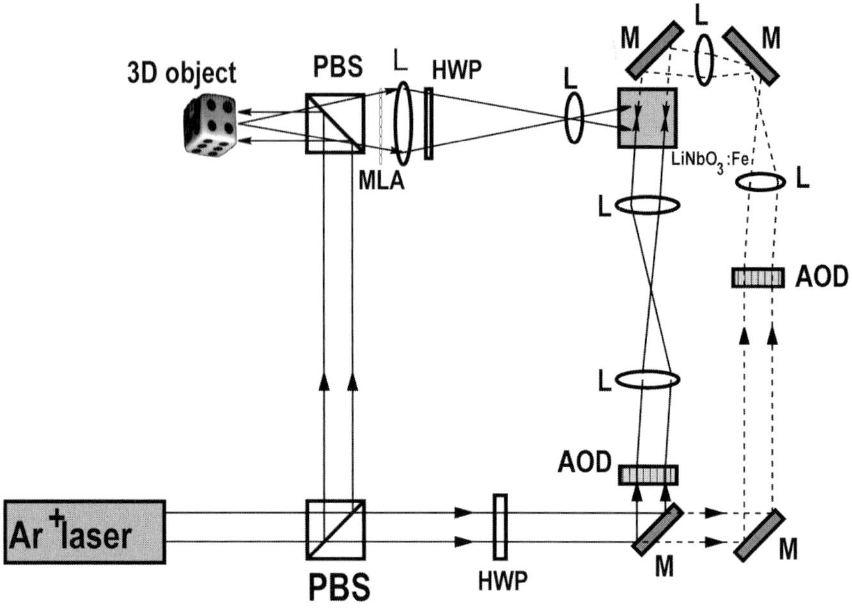

Fig. 6.9. Schematic diagram for 3-D video system using integral imaging and volume holographic storage: M is mirror, L is lens, PBS is polarizing beam splitter, HWP is half-wave plate, AOD is acousto-optic deflector, and MLA is microlens array

wavelength multiplexing, peristrophic multiplexing, and phase-coded multiplexing [24–26]. In Fig. 6.9, dots indicate the reference beam for recording and solid lines indicate the read-out beam, which is a phase-conjugate beam of the recording beam.

Two half-wave plates are employed to keep the ordinary polarization in the photorefractive crystal. After recording, the gratings can be fixed by thermal means to eliminate their sensitivity to the light of the recording wavelength as they are being read out. The speckle-reducing technique mentioned above can be applied during reconstruction processing. To show fast 3-D video real-object projection, we can control the scanning speed of the acousto-optic deflector. This system has some advantages, such as distortion elimination by using phase-conjugate read out, high resolution, speckle reduction, and real 3-D observation over a wide perspective range.

In this section, we have proposed and demonstrated a speckle-reduced volume holographic 3-D display system based on integral imaging. The small multiple elemental images formed by the microlens array are stored in a photorefractive crystal and reconstructed by using a phase-conjugate reading beam. We have also analyzed this 3-D display system in terms of resolution, field of view, and speckle reduction. The system is able to display a 3-D image with relatively large field of view compared with a conventional volume holographic display system. We have achieved speckle reduction by using a rotating diffuser. This system can be extended to a massive database for 3-D object display including 3-D video.

6.3 Three-Dimensional Object Recognition by Using Photorefractive Volume Holographic Processor

6.3.1 Introduction

The demands for image recognition and display of 3-D objects have rapidly increased in many areas of robotic vision, medical imaging, and encryption [27–29]. Several methods for performing 3-D optical correlation have been proposed by researchers. These include successive mapping of 2-D images, 3-D joint transform correlator, and 2-D filtering [30–33]. However, most of these suggestions are based on the 3-D to 2-D mapping, which needs sequential recording of several different 2-D perspectives. Other approaches for 3-D object recognition are digital holography using a single hologram combined with a digital image-processing technique [34, 35]. Because of the intrinsic 3-D nature of the hologram, a volume holographic correlator should be an attractive candidate to recognize 3-D objects. For the past few decades, volume holographic systems have received considerable interest for use in optical memory, which provides high-capacity data storage and high-rate data transfer using massive parallelism [36, 37]. These advantages of holographic memory systems can be transferred to 3-D object-recognition applications directly.

In this section, we present an application of volume holographic recordings to 3-D object recognition [38]. The advantage of this method is that it does not require recording of multiple 2-D images and is implemented in the Fresnel diffraction region. In general, an optical correlator operates in the Fourier plane to take advantage of shift invariance and spatial filtering [8, 39]. However, the Fourier plane is no longer precisely defined for 3-D real objects. The correlator is implemented in the Fresnel diffraction region to avoid the nonlinear response of the recording material due to a series of high-intensity peaks generated from the Fourier transform of regular or periodic components of the object.

6.3.2 3-D Object Recognition by Using Volume Holography

To describe the 3-D object recognition suggested in this section, let us consider the coordinates for the 3-D object and recording region. Compared to a planar recording device, such as a plane hologram or a CCD camera, which has a single imaging plane, we need to introduce the z coordinate in the recording region for the volume holographic medium. As shown in Fig. 6.10, the object illuminated by coherent light is assumed to lie in (x', y', z') coordinates and the coordinates for recording media are denoted (x, y, z), where x, x', and y, y' are transverse coordinates and z, z' are the longitudinal coordinates along the optical axis. So, the resulting expression for the field at (x, y, z) becomes [20];

$$H(x,y,z) = \frac{\exp[ikz_0]}{j\lambda z_0} \iiint U(x',y',z') \exp\left[\frac{jk}{2}(z-z')\right] \times \exp\left[\frac{jk}{2z_0}\{(x-x')^2 + (Y-y')^2 + (z-z')^2\}\right] \mathrm{d}x' \mathrm{d}y' \mathrm{d}z', \quad (6.10)$$

where z_0 is the distance between the origins of the 3-D object and recording-media coordinates, as shown in Fig. 6.10, and k is the wave number.

The object beam $H(x, y, z)$ propagating in the Fresnel diffraction region can be recorded in the photorefractive crystal by interferring with an off-axis plane wave reference beam $R(x, y, z)$. Neglecting the proportionality constants and omitting the coordinates, the complex irradiance I recorded in the crystal is given by

$$I = |H + R|^2 = |H|^2 + |R|^2 + R^*H + RH^*, \quad (6.11)$$

where the superscript * denotes the complex conjugate. As is well known, the last two terms constitute the diffraction. In the photorefractive crystal, a holographic grating can be recorded proportional to the intensity of interference fringes and the diffraction efficiency is given by [37];

$$\eta = \left(\frac{\varepsilon_1 kL}{2\varepsilon_0 \cos\theta}\right)^2 \mathrm{sinc}^2\left(\frac{2L(\Delta\theta)\sin\theta}{\lambda}\right), \quad (6.12)$$

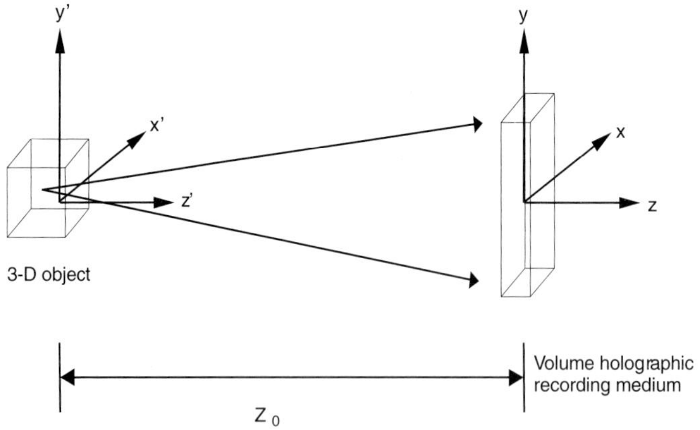

Fig. 6.10. Coordinate system for the 3-D object and volume holographic correlator

where ε_1 is a constant expressing the hologram strength, L is the thickness, θ is the angle between the signal and reference beam, and $\Delta\theta$ is the angular displacement.

When this hologram is read by another object beam H', generated by a 3-D object to be compared with the original 3-D object, the amplitude of the light diffracted by the recorded template and propagating in the direction of the original reference wave is given by

$$(RH^*)H'. \tag{6.13}$$

The diffracted beam is collected by a Fourier transform lens, so the output arriving at the detector plane is given by the Fourier transform of (6.13) as follows:

$$\delta * h \otimes h', \tag{6.14}$$

where * denotes convolution, $\otimes$ denotes correlation, and δ is the Fourier transform of the plane-wave reference beam R. In (6.14), h and h' are the Fourier transform of H and H', respectively. As shown in (6.8), H and H' are the Fresnel transforms of the reference 3-D object U and the 3-D input U', respectively. The degree of similarity between two inputs can be measured by $h \otimes h' = h_0 \otimes h_0'$ [40], where h_0 and h_0' are the Fourier transforms of the two inputs U and U', respectively.

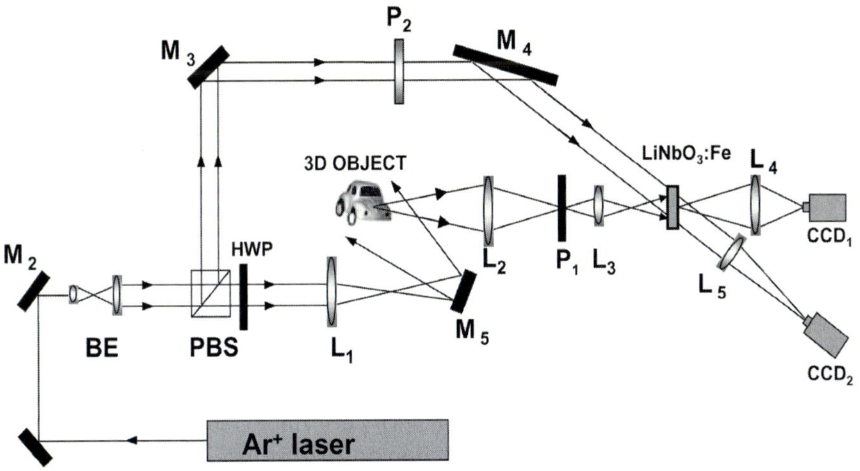

Fig. 6.11. Experimental setup of the volume holographic processor for 3-D object recognition: M_1–M_5: mirrors, P_1, P_2: polarizers, L_1–L_5: lenses, BE: beam expander, PBS: polarizing beam splitter, HWP: half-wave plate

6.3.3 Experiment

The experimental setup for the photorefractive volume holographic 3-D object recognition system is shown in Fig. 6.11. An Ar^+ laser ($\lambda = 514.5\,\text{nm}$) is employed as the recording and reading light source. The laser beam, expanded and spatially filtered to provide a Gaussian beam, is split into an object beam enlarged by lens L_1 illuminating the 3-D object, and a plane-wave reference beam. The reference beam is polarized in the vertical direction to minimize the beam-fanning effect inside the crystal and the vertical polarization component of the scattered light from the 3-D object is selected by polarizer P_1.

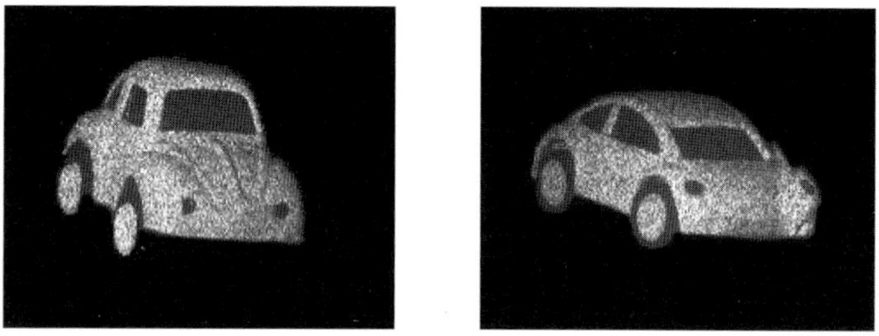

(a) (b)

Fig. 6.12. (**a**) Reference 3-D object recorded in the photorefractive crystal as the template; (**b**) different 3-D object used to test the discrimination capability of the system

In the correlation experiments, we employed two small cars with sizes of about 25 mm × 25 mm × 45 mm as the 3-D template and 3-D object. The template is recorded in the 0.1 mol% Fe-doped $LiNbO_3$ crystal with a size of 15 mm × 10 mm × 2 mm with an expanded Ar^+ laser ($\lambda = 514.5\,\text{nm}$) beam. During the template-recording procedure, the signal and reference beam ratio is kept at 1:3 to avoid nonlinear signal amplification and to produce maximum diffraction efficiency. To image a real 3-D object in the proximity of the crystal and to adjust to a 3-D object of varying size, we use a standard 4f-imaging setup with two different focal length lenses, L_2 ($f = 250\,\text{mm}$) and L_3 ($f = 50\,\text{cm}$), which provide 1/5 demagnification at the imaging plane. In the experiments, the real 3-D object is located at a distance, $d = 40\,\text{mm}$, from the photorefractive crystal. Diffraction efficiency is monitored using a CCD camera (CCD_1) placed at the second imaging plane behind lens L_4. First, we recorded the template with the 3-D object as shown in Fig. 6.12a. In general, the holographic recording time constant in a photorefractive crystal strongly depends on the total beam intensity and that is the weak point of Fourier-transform processors. However, in the Fresnel region, the intensity distribution is properly uniform to record a hologram in the linear dynamic range of the crystal. In the experiment, we expose for a sufficient time to record the weak components of the object under the condition of lower beam ratio without nonlinear beam amplification, as discussed earlier. Figure 6.12b shows the 3-D object to be correlated with the recorded template.

Figure 6.13 illustrates the experimental result obtained using the photorefractive-based processor. In Fig. 6.13a, the autocorrelation peak is shown, which is taken at the Fourier-transform plane of the reference beam using a CCD camera (CCD_2). The 3-D input object matches the template stored in the optical memory. It can be seen from the figure that the correlation peak is clearly visible with a good signal-to-noise floor ratio. This peak is normalized with a neutral-density filter located in front of the CCD camera. We use a different input 3-D object to perform a cross-correlation experiment, as shown in Fig. 6.13b. There is no correlation peak on the noise floor, as shown in Fig. 6.13b. We also investigated spatial tolerance of this processor by moving the 3-D object along the z-axis. Figure 6.14 shows the relative height of the correlation peak as a function of object displacement along the z-axis. As shown in the figure, this processor is more sensitive than the standard 2-D correlator working in the Fourier-transform plane. One of the possible explanations for this sensitivity is based on an intrinsic property of volume hologram: the so-called Bragg mismatch. In volume holography, the small changes in angle of the readout beam in the horizontal plane result in the Bragg mismatch, which eliminates the convolution term of the hologram and acts as a matched spatial filter [41].

From (6.12), the first sinc null of the diffraction efficiency is given by $\Delta\theta_{\mathrm{B}} = \lambda/2L\sin\theta$, which is calculated to be about 0.02° for our experimental setup. So, the corresponding longitudinal tolerance is about 0.17 mm. Transverse tolerance due to the Bragg mismatch is ten times more sensitive than

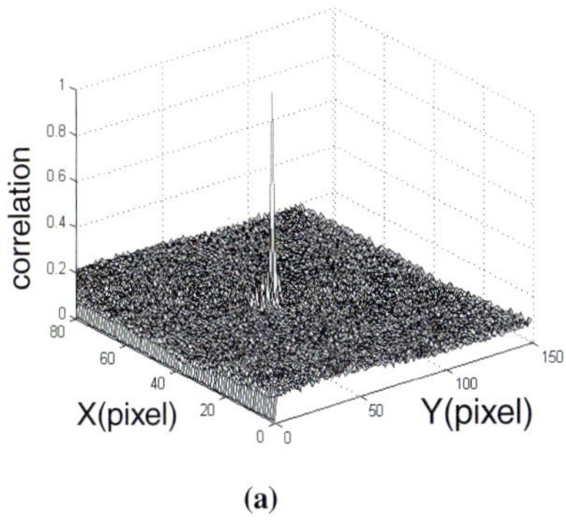

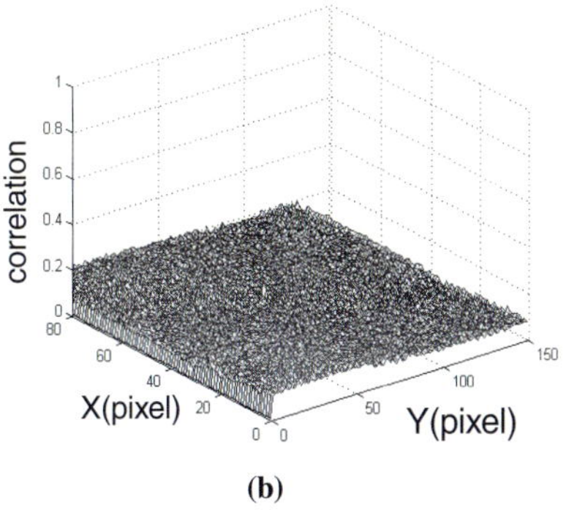

Fig. 6.13. (**a**) Autocorrelation; (**b**) cross-correlation for the 3-D objects shown in Fig. 6.12

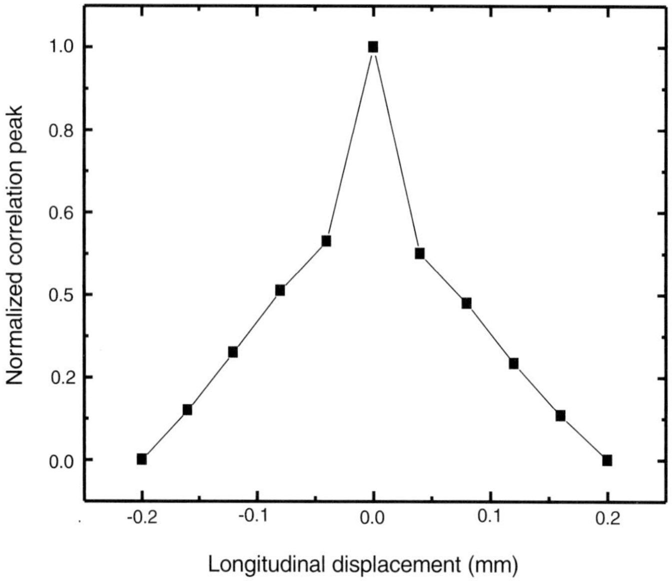

Fig. 6.14. Normalized correlation peak as a function of longitudinal displacement of the 3-D object

the longitudinal one. For comparison, we performed correlation experiments using conventional 2-D input images of the 3-D object, as shown in Fig. 6.13a and b, instead of the 3-D real object. Figure 6.15 shows the experimental results of autocorrelation (Fig. 6.15a) and cross-correlation (Fig. 6.15b) using 2-D inputs. For 2-D input images, there is no significant difference between autocorrelation and cross-correlation peaks. The discrimination capability of the 3-D recognition system is substantially better than its 2-D counterpart. Distortion and rotational invariance for 3-D object recognition can be implemented by recording multiple templates using the proper version of the 3-D object.

6.4 3-D Object Recognition Using Integral Imaging and Volume Holography

As mentioned above, the volume holographic method is able to recognize a 3-D object with very high sensitivity. Another approach for recognizing 3-D objects is with the integral imaging technique, using multiple perspective parallel processing [42]. In this case, a whole set of 2-D element images is supposed to recognize a 3-D object. In this subsection, we describe an all-optical technique for 3-D object recognition by use of integral imaging and volume holography. Figure 6.16 shows an experimental setup for the 3-D object recognition system. In addition to the optical processor described in

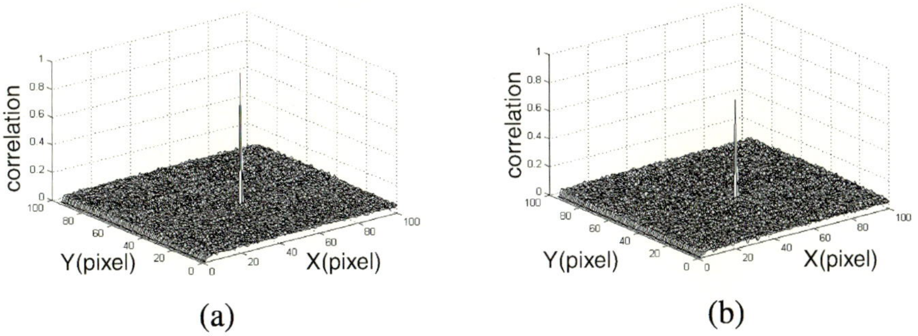

Fig. 6.15. Conventional correlation of the 2-D images of 3-D object shown in Fig. 6.13, (**a**) autocorrelation; (**b**) cross-correlation

the previous section, a single lens array, which consists of total numbers of 53×53 microlenses, is used to generate multiple perspective element images. As for the method described in the previous section, a set of element images is recorded as a template in the photorefractive volume holographic crystal.

To minimize the effect of illumination-angle change for the rotation application, we employ a polarizing beam splitter, which is placed between the 3-D real object and the microlens array, as shown in Fig. 6.16. In the experiments we employ dice with a size of approximately $15\,\text{mm} \times 15\,\text{mm} \times 15\,\text{mm}$ as input and reference 3-D objects. Figures 6.17a,b and c illustrate the experimental results obtained with the real-time integral imaging volume holo-

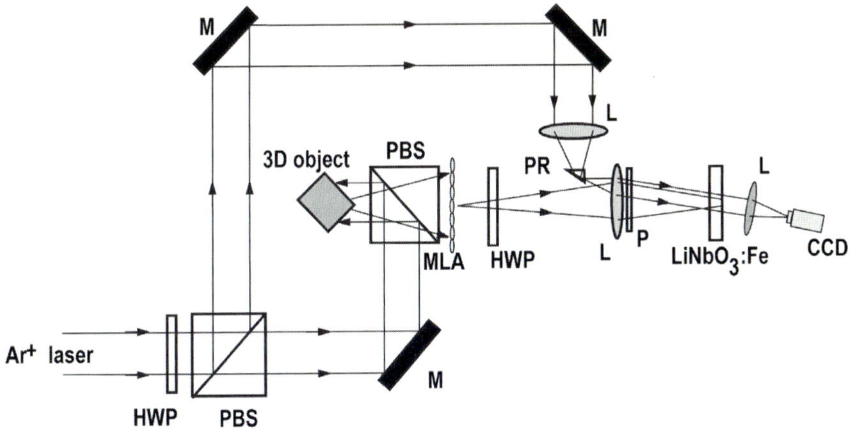

Fig. 6.16. Experimental setup for 3-D object recognition by using integral imaging and volume holographic processor; M is mirror, P is polarizer, L is lens, PBS is polarizing beam splitter, MLA is microlens array, PR is prism, and HWP is half-wave plate

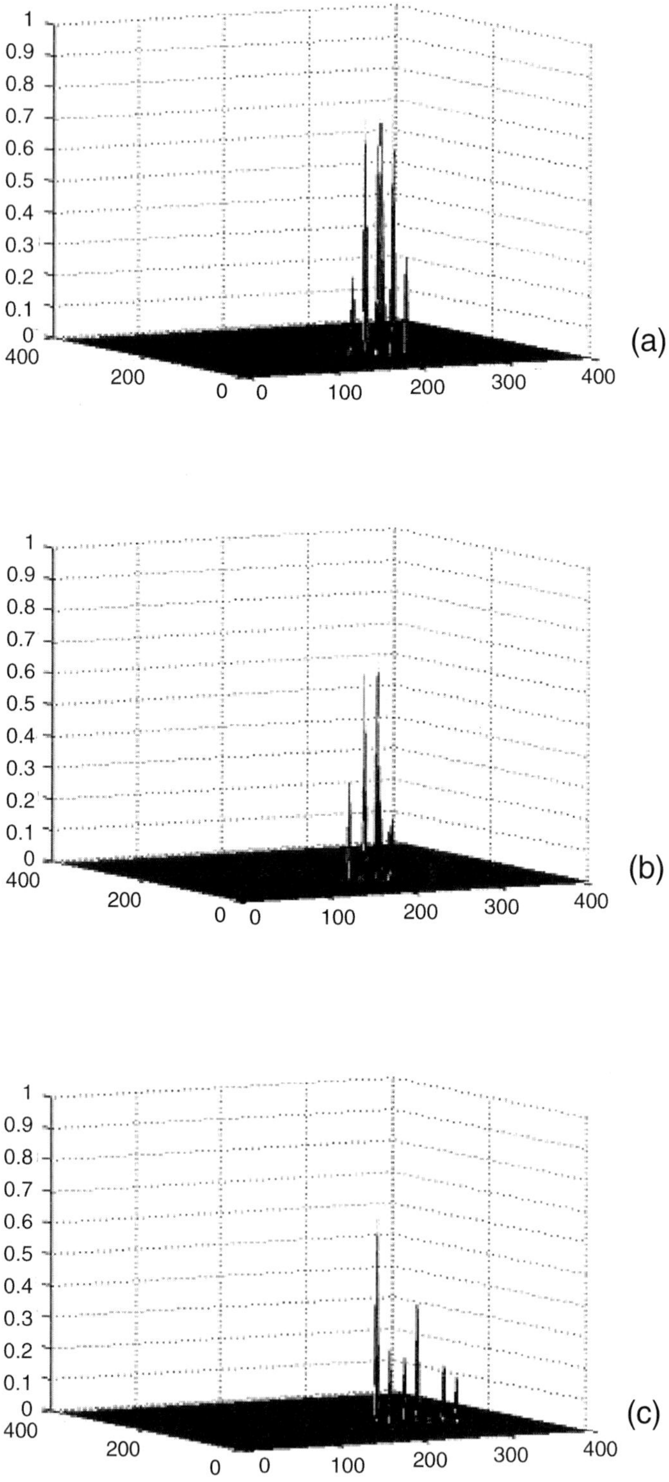

Fig. 6.17. Experimental results; (**a**) autocorrelation, cross-correlation between the reference and the rotated object with angles of (**b**) $-0.05°$; (**c**) $+0.05°$

graphic processor. Figure 6.17a shows the autocorrelation of the reference object. In this figure, there are several sharp peaks around the highest peak, which are caused by the correlation with the element image located in the center. To verify the rotational invariance property, correlations between the reference and slightly rotated objects are investigated. Figures 6.17b and c show results obtained with rotation angles of $\pm 0.05°$ degree, respectively. From the result, we can see that the position of the cross-correlation shifts shown in Fig. 6.17b and c are similar to results reported by Matoba et al. [42]. In this case, each sharp peak provides the spacing of the element image.

6.5 Summary

We have proposed and demonstrated a 3-D display system with integral imaging by use of a volume holographic memory system. The small multiple images formed by the microlens array are stored in the photorefractive crystal and reconstructed by using a phase-conjugate reading beam. We also analyze this optical system in terms of resolution, field of view, and speckle reduction. As a result, the system is able to display a 3-D object with an acceptable field of view and resolution. We also achieve a speckle-reduced image by using a rotating diffuser. We believe that this system can be extended to a 3-D real-object database. In addition to 3-D object display and 3-D video system, we have presented an all-optical 3-D object recognition system using a photorefractive volume holographic processor. We have presented experiments to illustrate the performance of the system using real 3-D objects. The experiments indicate that the 3-D object recognition system has high discrimination capability compared with its conventional 2-D correlator counterpart. This technique can be applied for recognition of 3-D objects, especially to detect small variations in a 3-D component. This 3-D recognition system is expanded to the integral imaging holographic processor, which is a useful technique for recognition of a rotated object.

References

1. T. Okoshi: *Three-dimensional Imaging Techniques* (Academic Press, New York 1971)
2. B. Javidi, E. Tajahuerce: Opt. Lett. **25**, 610 (2001)
3. N. Davis, M. McCormick, L. Yang: Appl. Opt. **27**, 4520 (1988)
4. B.P. Ketchel, G.L. Wood, R.J. Anderson, G.J. Salamo: Appl. Phys. Lett. **71**, 7 (1997)
5. H. Arimoto, B. Javidi: Opt. Lett. **26**, 157 (2000)
6. B. Javidi, F. Okano: *Three-dimensional Video and Display: Devices and Systems* (SPIE Press, Bellingham, WA 2001)

7. G. Lippmann: Comptes-Rendus **146**, 446 Seances Acad. Sci. (1908)
8. H.J. Caulfield (Ed.): *Handbook of Optical Holography* (Academic, New York 1979)
9. J.P. Huignard, J.P. Herriau, L. Pichon, A. Marrakchi: Opt. Lett. **5**, 436 (1980)
10. F. Wyrowski, O. Bryngdahl: J. Opt. Soc. Am. **A 5**(8), 1171 (1989)
11. H. Hoshino, F. Okano, H. Isono, I. Yuyama: J. Opt. Soc. Am. A **15**, 2059 (1998)
12. J. Arai, F. Okano, H. Hoshino, I. Yuyama: Appl. Opt. **37**, 2034 (1998)
13. H. Higuchi, J. Hamasaki: Appl. Opt. **17**, 3895 (1978)
14. F. Okano, H. Hoshino, J. Arai, I. Yuyama: Appl. Opt. **36**, 1598 (1997)
15. D. Psaltis, F. Mok: Sci. Am. **273**(11), 70 (1995)
16. A.D. McAulay: *Optical Computer Architectures* (Wiley, New York 1991)
17. E.G. Paek: *Optical Pattern Recognition with Microlasers* (NISTIR 6017, Gaithersburg 1998)
18. C. Warde, C.M. Schiller, J. Bounds, T.N. Horsky, G.A. Melnik, R.J. Dillon: Appl. Opt. **31**, 3971 (1992)
19. T. Okoshi: Appl. Opt. **10**, 2284 (1971)
20. J.W. Goodman: *Introduction to Fourier Optics* (McGraw-Hill, New York 1996)
21. S.H. Shin, B. Javidi: Appl. Opt. **41**(14), 2644 (2002)
22. X. An, D. Psaltis, G.W. Burr: Appl. Opt. **38**, 386 (1999)
23. D.L. Staebler, J.J. Amodei, W. Philips: 'Multiple storage of thick holograms in $LiNbO_3$.' In: *VII Int. Quantum Electronics Conf.* (Montreal 1972)
24. G.A. Rakuljic, B. Levya, A. Yariv: Opt. Lett. **17**, 1471 (1992)
25. K. Curtis, A. Pu, D. Psaltis: Opt. Lett. **19**, 993 (1994)
26. C. Denz, G. Pauliat, G. Roosen: Opt. Commun. **85**, 71 (1991)
27. A. Pu, R. Denkewalter, D. Psaltis: Opt. Eng. **36**, 2737 (1997)
28. M.A. Anastasio, X. Pan: JOSA Opt. Soc. Am. A, **17**, 391 (2000)
29. E. Tajahuerce, B. Javidi: Appl. Opt. **39**, 6595 (2000)
30. R. Bamler, J. Hofer-Alfeis: Opt. Acta. **29**, 747 (1982)
31. J. Rosen: Appl. Opt. **37**, 7538 (1998)
32. B. Javidi: *Smart Imaging System* (SPIE Press, Bellingham, WA 2001)
33. Y.B. Karasik: Appl. Opt. **36**, 7397 (1997)
34. B. Javidi, E. Tajahuerce: Opt. Lett. **25**, 610 (2000)
35. Y. Frauel, E. Tajahuerce, M. Castro, B. Javidi: Appl. Opt. **40**, 3887 (2001)
36. A.D. McAulay, J. Wang: Appl. Opt. **30**, 2885 (1991)
37. H.J. Coufal, D. Psaltis, G.T. Sincerbox (Ed.): *Holographic Data Storage* (Springer-Verlag, New York 2000)
38. S.H. Shin, B. Javidi: Opt. Lett. **26**, 1161 (2001)
39. Ph. Réfrégier: Opt. Lett. **15**, 854 (1990)
40. H. Hamam, H.H. Arsenault: Appl. Opt. **36**, 7408 (1997)
41. G. Barbastathis, M. Balberg, D.J. Brady: Opt. Lett. **24**, 811 (1999)
42. O. Matoba, E. Tajahuerce, B. Javidi: Appl. Opt. **40**, 3318 (2001)

Part II

Signal Processing, Coding, and Applications

7 Stereoscopic Video Compression Standard "MPEG-2 Multiview Profile"

H. Imaizumi and A. Luthra

7.1 Introduction

In this chapter, we describe a compression technology for stereoscopic television. Our brain acquires depth information from the subtle difference between the picture on the left retina and the picture on the right retina. It is well known that such a pair of pictures can be acquired with two cameras located at horizontally shifted positions. A simple solution to stereoscopic video transmission is the "simulcast" technique that is based on independently coding between the left view and the right view. However, this requires twice the conventional monoscopic digital transmission bandwidth. On the other hand, stereoscopic video compression exploits the interrelationship between the left view and the right view [6].

Efficient digital compression is one of the main issues of stereoscopic video so that a large volume of data can be easily stored or transmitted. Originally, the MPEG-2 video standard was primarily intended for coding of standard TV with good quality in the bit-rate range of 4 to 9 Mbit/s. However, the scope of the MPEG-2 video standard was expanded to include stereoscopic video as well as video of higher resolutions such as HDTV. The MPEG-2 video standard is partitioned into "Profiles" and "Levels" to support applications for a variety of video signals. The "Profile" specifies tools used for coding and the "Level" specifies the numerical parameters such as image size.

The Multiview Profile (MVP), which is one of the profiles of MPEG-2 video, is the compression standard for stereoscopic video. One of the main features of the MVP is the use of scalable coding tools to guarantee compatibility with the MPEG-2 Main Profile (MP), which is one of the most popular video compression standards used for digital television broadcasting, DVD video, etc. Scalable coding means two or more layers of coding. It consists of a base layer and one or more enhancement layers that contain incrementally more information about the video. In MVP one base layer and one enhancement layer is used. The left eye view is coded as the base layer and the right eye view as the enhancement layer. The base layer video is coded in conformance with the Main Profile. The enhancement layer video is coded with the scalable coding tools and exploits the correlation with the base layer video for efficient coding. Therefore, the conventional Main Profile decoder can decode the base layer video coded by a Multiview Profile encoder (backward compatibility), and the Multiview Profile decoder can decode a bitstream coded by a Main Profile encoder (forward compatibility).

7.2 Overview of the MPEG-2 Video Coding-Algorithm [1]

This section outlines the coding algorithm of the MPEG-2 Main Profile that is widely used for digital television, DVD video, etc.

The block diagram of the MPEG-2 video codec is shown in Fig. 7.1. One of the main features of the MPEG-2 video coding algorithm is the use of field/frame adaptive tools since MPEG-2 handles interlaced video signals. This is a major difference from the preceding video compression standards, such as ITU-T H.261 for videophones or MPEG-1 for video CD, which deal with only progressive (non-interlaced) video signals. The discrete cosine transform (DCT) is used to exploit the spatial redundancies and the motion compensator (MC) is used to exploit temporal redundancies. The quantization forces many of the DCT coefficients to zero and is then followed by variable length coding.

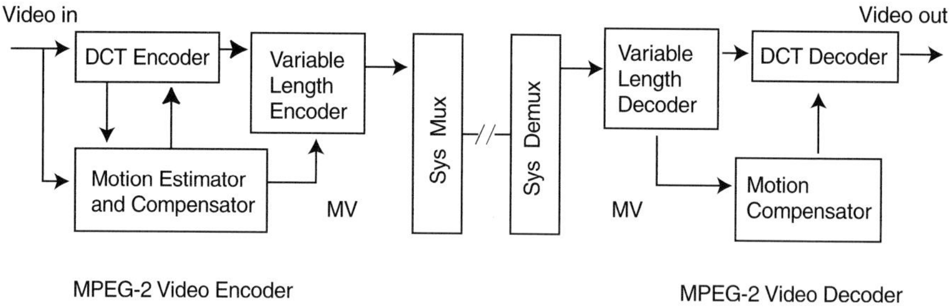

Fig. 7.1. A generalized block diagram for the MPEG-2 video coding algorithm

7.2.1 Interlaced Video

An interlaced raster scan is shown in Fig. 7.2. One of the principal benefits of interlaced scanning is that it reduces the scan rate with a reasonable tradeoff in the temporal and spatial (vertical) resolution. In an interlaced image, one frame consists of two fields sampled at different times. Two consecutive lines of a frame belong to alternate fields. The field scanned in the first half of a frame is called the first field and the field scanned in the next half is called the second field. The field containing the top scanning line and odd numbered lines of a frame is called the top-field or the odd-field, and the other field is called the bottom-field or the even-field. Since the vertical sampling position

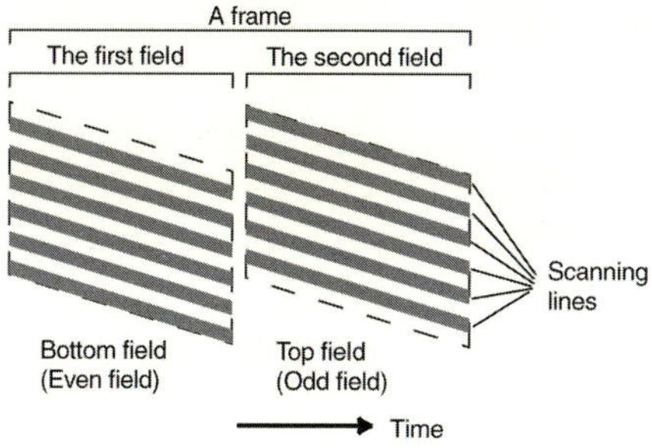

Fig. 7.2. Interlaced raster scan of an image

of the odd-field differs from that of the even-field, these fields have different parity.

7.2.2 Bidirectional Prediction

To exploit temporal redundancies, the macroblock-based motion compensation approach is adopted. A macroblock consists of a 16×16 luminance pixel block and its two associated 8×8 chrominance pixel blocks. A motion vector of up to half-pixel accuracy is used for predicting the displacement of each macroblock.

In bidirectional prediction, an average of two reference macroblocks can be used as the prediction. One is a displaced macroblock in the previous picture by a forward motion vector and the other is a displaced macroblock in the future picture by a backward motion vector. The reference macroblocks are shown in Fig. 7.3. The prediction error, which is the difference between the target pixel and its prediction, is coded and transmitted.

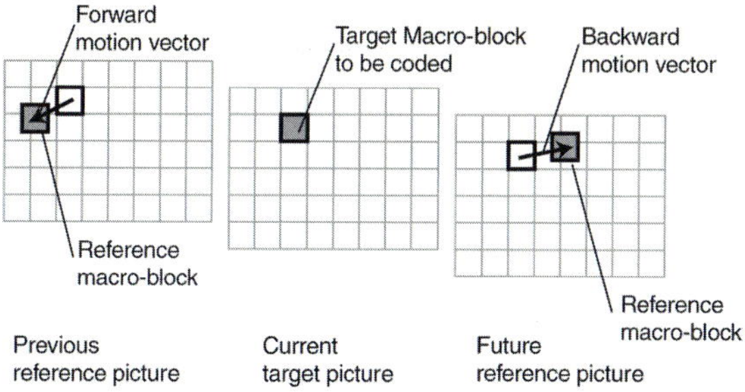

Fig. 7.3. Reference macroblocks for bidirectional prediction

7.2.3 Picture Coding Type

Pictures coded by using bidirectional prediction are called B-pictures. Pictures coded only by using forward prediction are called P-pictures. Pictures coded without using any prediction are called I-pictures. One of the three picture coding types is selected for each frame. Sequential pictures are coded by combining the three types of pictures. Figure 7.4 shows an example of the combination. I and P pictures are used as references for prediction. B-pictures are not allowed to be used as reference pictures.

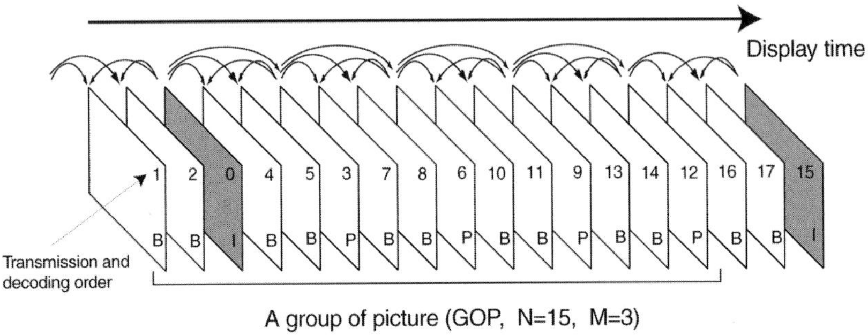

Fig. 7.4. Group of pictures with prediction distance, $M = 3$, and intradistance, $N = 15$. The prediction structure consists of three types of pictures, which are I-picture, P-picture and B-picture. Note that the transmission order of the pictures is not the same as the display order

A group of pictures (GOP) is a coding unit of pictures. The first coded picture of a GOP must be an I-picture in order to start decoding at the beginning of any GOP of a bitstream. The GOP structure contributes to random access as well as recovering from errors.

In B-pictures, one of four prediction modes is selected for each macroblock. The four prediction modes consist of a forward prediction, a backward prediction, an interpolation of both predictions where an average of the two reference macroblocks is taken, and intraframe coding without prediction. In P-pictures, either a forward prediction or intraframe coding is selected for each macroblock. In I-pictures, all macroblocks are coded by using only intraframe coding.

7.2.4 Picture Structure

MPEG-2 video is optimized for coding of interlaced video adaptive frame/field algorithm is used to dynamically select frame mode or field mode for each macroblock. Frame mode has the advantage of coding efficiency for still image areas and the field mode has the advantage of coding efficiency for moving image areas.

In MPEG-2, two picture structures are prepared for coding of interlaced video. With "frame pictures", each interlaced field pair is interleaved together into a frame. With "field pictures", individual constituent fields are divided into macroblocks separately. One of the two picture structures is chosen for coding of each frame.

In frame pictures, frame-motion compensation and field-motion compensation can be chosen for each macroblock. In field pictures, only field motion compensation can be used. In both types of picture structures, two fields of a frame must be coded with the same picture-coding type. In I-field pictures, however, the second field of the frame can be coded as either an I-field picture or a P-field picture.

7.2.5 Predictions for Frame Pictures

Three modes of predictions for frame pictures can be chosen for each macroblock.

The first, frame prediction for frame pictures, is shown in Fig. 7.5. The number of motion vectors is up to one for a macroblock of a P-frame and up to two for a macroblock of a B-frame.

Field prediction for frame pictures is shown in Fig. 7.6. A 16×16 pixel macroblock is divided into two 16×8 pixel blocks of the first field and the second field. Individual motion vectors are assigned for each block. Therefore, the number of motion vectors is two for a macroblock of a P-frame and four for a macroblock of a B-frame. The reference fields can be chosen from both the first field and the second field.

The third mode of prediction, called "dual prime" for P-pictures, is shown in Fig. 7.7. A target macroblock is predicted by an average of two reference fields. One motion vector and one small differential motion vector (DMV) are transmitted for prediction of a target macroblock. First, a target macroblock is split into top-field 16×8 pixels and bottom-field 16×8 pixels. For reference fields of the same parity with the split target blocks, the transmitted motion vector is used for the prediction. For reference fields of different parity with the target blocks, motion vectors scaled by the temporal distance and with DMV added are used for predictions. The dual prime prediction cannot be used if the previously coded picture was a B-picture or the first I-field picture of a frame.

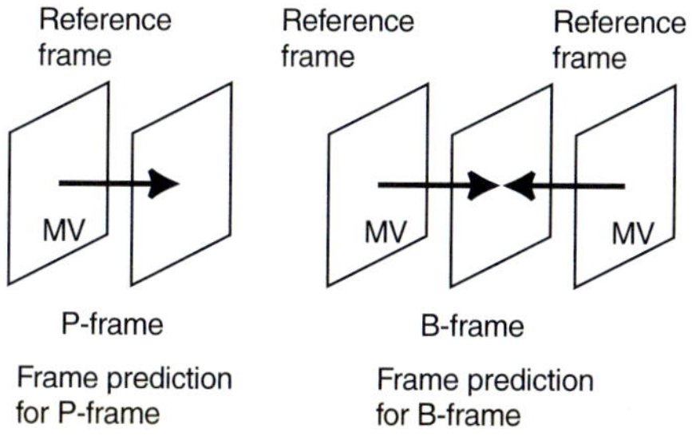

Fig. 7.5. Frame prediction for frame pictures

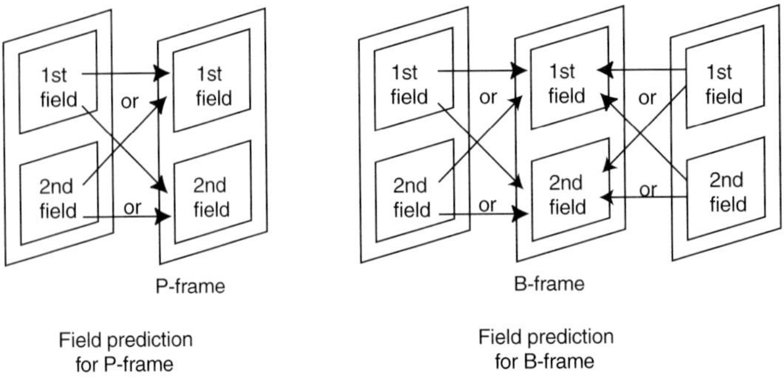

Fig. 7.6. Field prediction for frame pictures

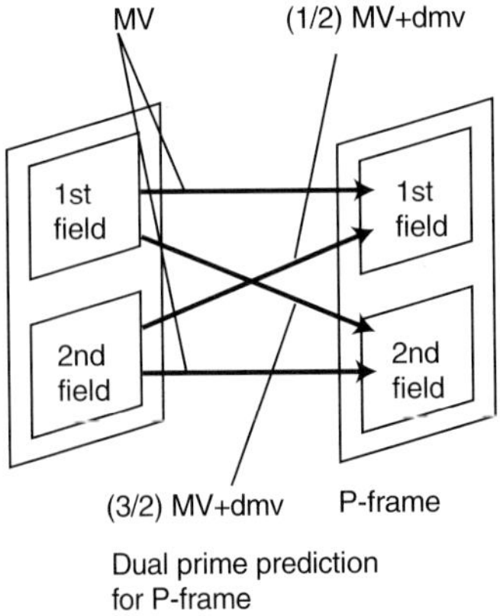

Fig. 7.7. Dual prime prediction for frame pictures

7.2.6 Predictions for Field Pictures

Three modes of predictions for field pictures can be chosen for each macroblock: 16×16 field prediction, 16×8 field prediction, and dual prime prediction. In the 16×16 field prediction, one motion vector is used for a macroblock of a P-field picture and two for a macroblock of a B-field picture. In the 16×8 field prediction, two motion vectors are used for a macroblock of a P-field picture and four for a macroblock of a B-field picture. The reference fields of P-field pictures are shown in Fig. 7.8. The reference fields of B-field pictures are the same as those of B-frame pictures shown in Fig. 7.6.

Dual prime prediction for field pictures is the same as the prediction for frame pictures, except that a predicted block consists of 16×16 pixels.

7.2.7 Field/Frame Adaptive DCT

A macroblock consists of four 8×8 DCT blocks. As shown in Fig. 7.9, the pixels that make up a DCT block are different for the frame DCT and the field DCT. In frame pictures, the frame DCT and the field DCT can be chosen

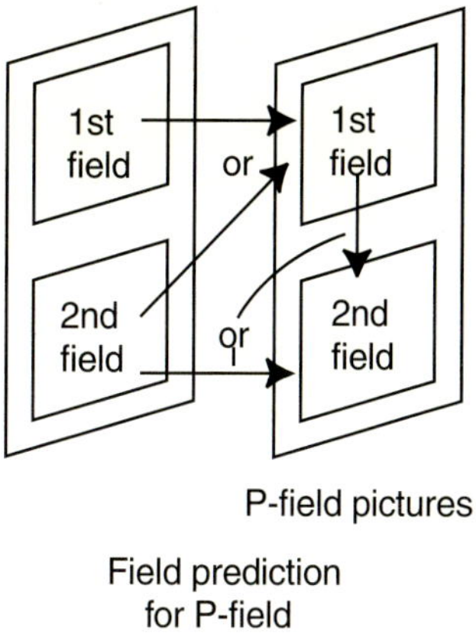

Fig. 7.8. Field prediction for P-field pictures

for each macroblock. On the other hand, only the field DCT can be used in field pictures.

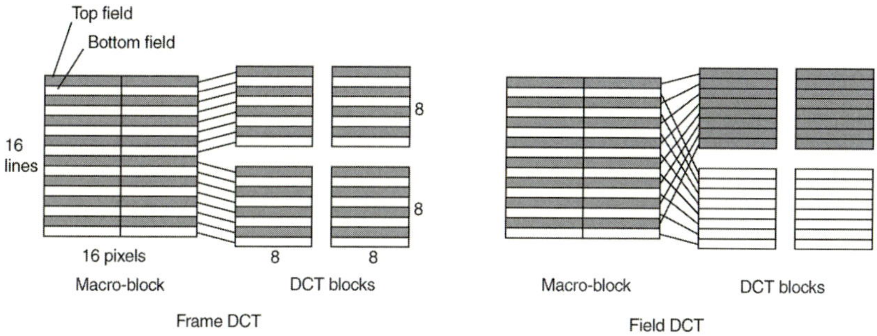

Fig. 7.9. Macroblock structures in frame DCT coding and field DCT coding

DCT coefficients followed by quantization are coded by using variable length coding and then transmitted.

7.3 Coding Algorithm for MPEG-2 Multiview Profile [2, 7]

The MPEG-2 video standard is partitioned into "Profiles" and "Levels" to support different applications. The "Profile" specifies tools used for coding and the "Level" specifies the numerical parameters. An actual specification is written as a combination of Profile and Level such as MP@ML (Main Profile at Main Level). The Multiview Profile (MVP), which is a video compression standard for stereoscopic television, is also one such profile. The MVP has the same four levels as MP: Low Level (maximum horizontal pixel size times maximum vertical pixel size: 352 × 288), Main Level (720 × 576), High 1440 Level (1440 × 1152) and High Level (1920 × 1152).

7.3.1 Compatibility of the Multiview Profile with the Main Profile

A block diagram for the MVP video encoder scheme is shown in Fig. 7.10. The base layer is assigned to a bitstream for the left view and the enhancement layer is assigned to a bitstream for the right view. The base layer video is coded in conformance with the MP. Therefore, the conventional MP decoder can decode the base layer video coded by a MVP encoder. The enhancement layer video is coded with a disparity compensation tool as well as a motion-compensation tool. A principal difference between the coding algorithm for the enhancement layer of MVP and that for other profiles is how the reference pictures are used for the predictions.

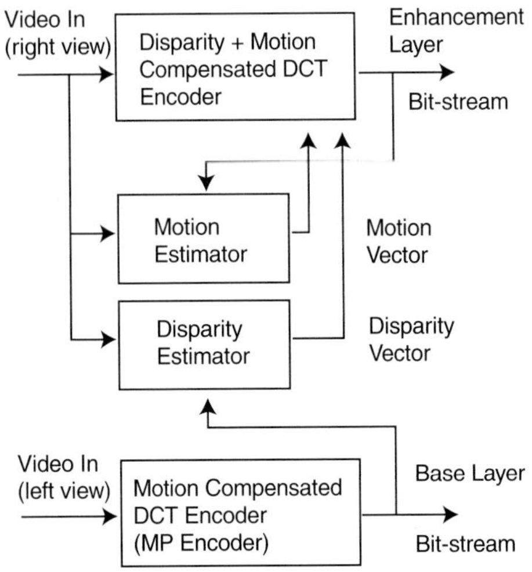

Fig. 7.10. A typical block diagram of a MVP encoder scheme. Two-layer coding provides a compatible stereoscopic video compression algorithm using disparity compensation and motion compensation

As described below, MVP uses a temporal-scalability tool. It allows two-way compatibility between MVP and MP. The temporal-scalability tools were originally prepared for enhancement of temporal resolution, for example, scalable coding between interlaced images and noninterlaced (progressive) images. At present, however, MVP is the only profile that uses the temporal scalability tool.

7.3.2 Predictions for the Multiview Profile

An example of the prediction structure for the MVP is shown in Fig. 7.11. As shown, the video sequences of the enhancement layer can be coded without I-pictures. This helps in reducing the bitrate, as I-pictures need more bits than B or P pictures. It is possible to reduce the bitrate for right-view sequences in the MVP and to allocate more bitrate for left-view sequences than the simulcast approach. This is why the picture quality of left-view sequences in the MVP is expected to be better than that in the simulcast approach.

The reference pictures for prediction of base-layer coding are the same as the MP. In enhancement-layer coding, I-picture, P-picture, and B-picture can be chosen for each picture. In I-pictures, only intraframe coding can be used. In P-pictures, forward prediction and intraframe coding can be chosen for each macroblock. The reference pictures for the forward prediction can be chosen for each picture from the base layer or the enhancement layer. The reference picture that is the most recent, or at the same time in display order, can be chosen from the base layer; or the reference picture that is the most recently decoded picture can be chosen from the enhancement layer. In B-pictures, three types of predictions and intraframe coding can be chosen for each macroblock. The reference picture for one of the predictions is the most recent, or at the same time, picture of the base layer in display order. The reference picture for another prediction is the most recently, decoded picture of the enhancement layer. The reference picture for the other prediction is the average of both of the reference pictures.

If the picture of the base layer is the first decoded I-picture of a GOP, the corresponding picture of the enhancement layer should be an I-picture or P-picture, the reference picture of which is from the base layer so as to start decoding from the GOP. In the example of Fig. 7.11, the picture type for the beginning of the GOP of the enhancement layer is a P-picture that refers to the base layer.

In addition, there are some restrictions for the prediction of field pictures so that pictures in the enhancement layer can be decoded immediately after

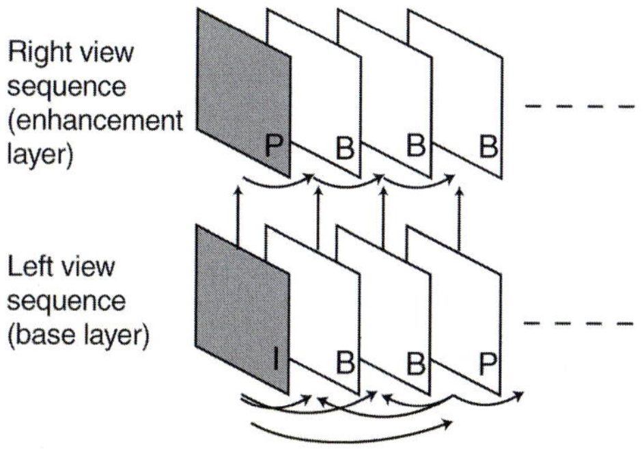

Fig. 7.11. Prediction structure for base and enhancement layers of the MVP

their corresponding required reference pictures have been decoded. In P-field pictures and B-field pictures, the second field of a frame in the base layer cannot be used for the reference picture for coding of the first field of a frame in the enhancement layer. This prevents the prediction from lower layer fields of the future. For the same reason, dual prime prediction from the base layer is partly restricted.

7.4 Camera Parameters

The Multiview Profile conveys the camera parameters, which are shooting geometry information and optical-lens parameters. Figure 7.12 shows the parameters to be transmitted, and Table 7.1 lists the ranges and resolutions of each parameter. It would be possible to interpolate a new scene from any viewpoint by calculating from the two transmitted views and the camera parameters. The camera parameters can be used to synthesize a 3-D-CG picture with the transmitted stereoscopic scenes.

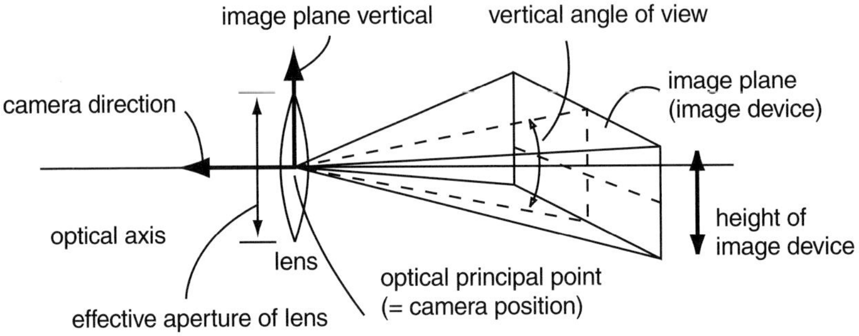

Fig. 7.12. Illustrated camera parameters conveyed by the MVP

The camera ID provides numbers for camera identification. The focal length divided by the F-number gives the effective aperture of the lens in Fig. 7.12.

7.5 Verification Test for Standardization for the Multiview Profile [3]

In this section, we describe the results of subjective assessment tests conducted as "verification tests" on the picture quality of the MPEG-2 Multi-

Table 7.1. Camera parameters conveyed by the MPEG-2 Multiview Profile

Parameter	Range	Resolution
Camera ID	0, 127	Integer
Height of image device	0.0, 4194.303 (mm)	0.001 (mm)
Focal length	0.0, 4194.303 (mm)	0.001 (mm)
F-number	0.0, 4194.303	0.001
Vertical angle of view	0, 180 (degree)	0.0001 (degree)
Camera position x		
Camera position y	−2,147,483.648, +2,147,483.647 (mm)	0.001 (mm)
Camera position z		
Camera direction x		
Camera direction y	−2,097,152, +2,097,151	Integer
Camera direction z		
Image plane vertical x		
Image plane vertical y	−2,097,152, +2,097,151	Integer
Image plane vertical z		

view Profile. Four organizations contributed to generate the test sequences by computer simulation in conformity with the MVP standard, and three organizations contributed to evaluate the picture quality of the sequences by stereoscopic viewing tests.

7.5.1 Bitstream Exchange [5]

Bitstream exchanges were conducted to verify that the MVP standard is fully consistent. Each company implemented its own encoder based on the MVP standard. Four bitstreams and the respective decoding sequences were exchanged with one another. The bitstreams encoded by one company's encoder were decoded by other companies' decoders. The decoded sequences were compared and were found to be perfectly consistent, thus verifying that there were no inconsistencies in the syntax or semantics of the MVP standard.

7.5.2 Test Conditions for the Subjective Assessment [4]

The picture quality of the MVP video was evaluated by subjective assessment tests. Three organizations contributed to this work. Each organization tested the picture quality of the compressed and then decompressed stereoscopic sequences by different stereoscopic display systems. The "simulcast approach" was also evaluated as a reference.

Table 7.2 provides an overview of the test conditions. The test setup was based on the documents of Rec. ITU-R BT.500-7. The double stimulus impairment scale method (variant II) was applied. A continuous scale was used in order to obtain more precise evaluations. Assessors were nonexpert viewers. The test sequences, which had the resolution of Rec. ITU-R BT.601, are summarized in Table 7.3.

Table 7.2. Overview of the subjective test conditions

Sequences	Street organ, flower pot, trapeze (525 lines, 30 Hz, 300 frames) fun fair (625 lines, 25 Hz, 250 frames)
Algorithms and bitrate (left view/ right view)	MVP@ML: 6/3 Mbit/s, 9/4 Mbit/s MP@ML ×2 (simulcast): 4.5/4.5 Mbit/s, 6.5/6.5 Mbit/s Source: original sequences without compression
Test method	The double stimulus impairment scale method (variant II) described in Rec. ITU-R BT.500-7, with continuous scale.
Stereoscopic display systems (picture size, viewing distance)	A: LCD projectors and polarizing eyeglasses (82 cm × 57 cm, 5H) B: Time sequential display display and LCD shutter eyeglasses (40.6 cm × 30.5 cm, 4H) C: Two-mirror (Wheatstone type) display system (19 cm × 14 cm, 5H)
Observers	Nonexpert viewers A: 19 observers, B: 18 observers, C: 24 observers

Table 7.3. Features of test sequences

Title of sequence	Features
Street organ	Objects: rapid walkers, a girl, a street organ Camera motion: rapid panning then slow panning The decoration of the street organ has a variety of depth variations
Flower pot	Objects: flowers and a flowerpot, a girl, background wall Camera motion: still, then slow panning The stalks of the flowers appear to have a fine texture
Trapeze	Objects: a trapeze artist Camera motion: slow panning The performer moves near and far very rapidly, but the moving area is not so large.
Fun fair	Objects: children and carousel Camera motion: still The carousel spins around very rapidly. There is a significant difference of average luminance between left and right sequences.

7.5.3 Results of the Subjective Assessment Tests

The results of the subjective assessment tests are shown in Fig. 7.13. The mean scores and the 95% confidence intervals of the means are indicated for each test condition. In the condition of the two-mirror system, evaluated test sequences were split into the first half of the sequence and the second half of the sequence, because of the limitation of display memory size at the test site.

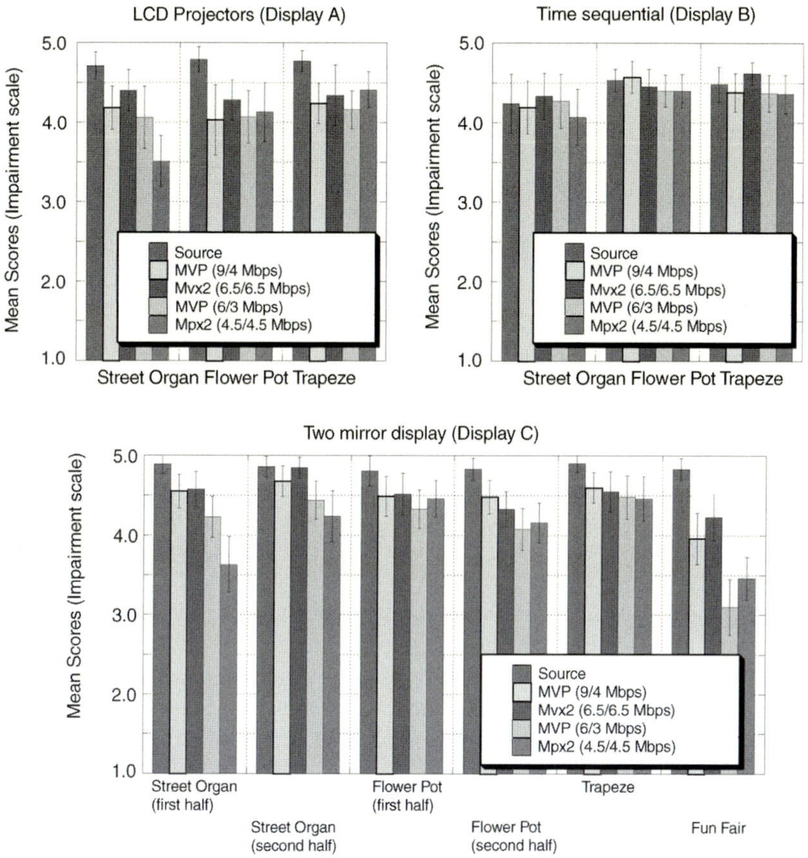

Fig. 7.13. Results of subjective assessment test of MVP. The mean scores and the 95% confidence intervals of mean scores are indicated for each test condition. In the condition of the two-mirror display system, test sequences "street organ" and "flower pot" were split into the first half of the sequences and the second half of the sequences

Key aspects of the test results are as follows:

- Within each of the four sequences, the mean score of the MVP sequence at a bitrate of 9/4 Mbps does not differ significantly from the mean score of the simulcast of MPs at a bitrate of 6.5/6.5 Mbps. Furthermore, the mean score of the MVP sequence at a bitrate of 6/3 Mbps does not differ significantly from the mean score of the simulcast of MPs at a bitrate of 4.5/4.5 Mbps, except the one pair of the sequences "street organ." For "street organ", the quality of MVP is superior to that of simulcast of MPs. These results show that differences in subjective evaluation between MVP and simulcast of MPs are very small at higher bitrate for pictures with slight motions ("flower pot" and "trapeze") and/or with significant luminance difference between left and right views ("fun fair").
- "Fun fair" is the sequence with the largest variation in mean scores. This sequence consists of large motion over most of the image area and most of the frames.
- No annoying coding artefacts were introduced by the MPEG-2 Multiview Profile coding scheme at the tested bitrates.

7.6 Conclusions

In this chapter, we introduced the algorithms of the MPEG-2 Multiview Profile, which is a video compression standard for stereoscopic television. Successful verification tests in the standardization process showed that the performance of the MVP was better than or equal to that of the simulcast approach.

The MPEG-2 Multiview Profile, which was established as an amendment of the MPEG-2 video standard in 1996 was integrated into the MPEG-2 video second edition "ISO/IEC 13818-2:2000".

Many encoder optimization procedures are not specified by the standard. Several performance-enhancement techniques have been developed and studied for the MVP. Examples include brightness balancing of two views for disparity estimation and compensation [8], picture quality of stereoscopic HDTV [9, 10], and rate control for stereoscopic video encoding [11]. These techniques further improve the coding efficiency of the MVP.

7.7 Acknowledgements

Test sequences were supplied by:

- CCETT
- NHK

Bitstream exchange was contributed by:

- Columbia University
- David Sarnoff Research Center
- General Instrument
- NHK

Subjective assessment tests were conducted by:

- CRC (Communication Research Centre, Ottawa)
- HHI (Heinrich-Herz-Institut, Berlin)
- NHK

We wish to thank the many experts in the field who contributed to the verification works for establishing the MPEG-2 Multiview Profile. They include Xuemin Chen, Tihao Chiang, Belle L. Tseng, Norihiro Suzuki, Ion-Paul Beldie, Kambiz Fazel, James Tam, Lew Stelmach, Eisuke Nakasu, Thomas Sikora, Laura Contin and Leonardo Chiariglione.

References

1. ISO/IEC JTC1/SC29/WG11: *Information Technology – Generic Coding of Moving Pictures and Associated Audio: Video* (2nd edn.) (ISO/IEC 13818-2:2000, 2000)
2. ISO/IEC JTC1/SC29/WG11: *Information Technology – Generic Coding of Moving Pictures and Associated Audio: Video*, ISO/IEC 13818-2 Amendment 3 (Multi-view Profile), MPEG96 N1366, Chicago, 1996
3. ISO/IEC JTC1/SC29/WG11: *Results of MPEG-2 Multi-View Profile Verification Test*", MPEG96 N1373, Chicago, 1996
4. ISO/IEC JTC1/SC29/WG11: *Test Procedures for Verification Testing of 13818-2/DAM3 (Multi-view Profile)*, MPEG96 N1227, Florence, 1996
5. Ajay Luthra, Hiroyuki Imaizumi: *Report of MPEG-2 Multi-View Profile Adhoc Group*, MPEG96 M0970, Tampere, 1996
6. Belle L. Tseng, Dimitris Anastassiou: 'Compatible video coding of stereoscopic sequences using MPEG-2's scalability and interlaced structure', *Int. Workshop on HDTV'94, Torino*, 1994
7. Ajay Luthra, Xuemin Chen: Proc. SPIE **3021**, Electric Imaging '97 (1997)
8. A. Puri, R.V. Kollarits, B.G. Haskell: Signal Proc.: Image Communi. **10**, 201 (1997)
9. H. Imaizumi, M. Sakai, K. Minomo, Y. Iwadate: *Bit-rate allocation for left and right view sequences on stereoscopic HDTV coding using MPEG-2 MVP@HL*, IPSJ Technical Report AVM 18-3, 1997 (in Japanese)
10. M. Katayama, H. Imaizumi, Y. Nojiri, S. Itou: *Subjective assessment of stereoscopic HDTV coding using MPEG-2 MVP@HL*, ITE Technical Report BSC 98-4, 1998 (in Japanese)
11. S. Naito, S. Matsumoto: Proc. SPIE **3653**, 1082 (1999)

8 The Ray-Based Approach to Integrated 3-D Visual Communication

T. Naemura and H. Harashima

8.1 Introduction

Further advancements in virtual reality and three-dimensional (3-D) imaging technologies are needed to achieve a high sense of presence and reality in the next-generation communications and information environment. At present, however, 3-D display technology is at the stage where a variety of systems are being studied [1–3], and the coming years should see amazing developments through advanced research. Under these circumstances, specialized 3-D information environments may be constructed for specific display systems leading to incompatibility in sending, receiving, and sharing 3-D data in the future. To prevent this problem from occurring, a flexible 3-D information environment that can support a variety of display systems must be constructed.

Along this line of thinking, the authors have already proposed the concept of "Integrated 3-D Visual Communication" [4]. This concept reflects the necessity of developing a neutral representation method (encoding method) independent of input/output systems (i.e. camera and display systems, referred to below as "3-D image formats") like stereo, multiview, and hologram, as shown in Fig. 8.1. Such a method deals with data directly associated with 3-D space in contrast to those making up individual 3-D image formats, and we refer to it as "spatial representation" in the following discussion.

Up to now, research into 3-D image-encoding technology has mainly targeted individual 3-D image formats [5, 6]. Here, methods that use structural models of objects that are conscious of spatial representation have also been studied [7–9]. Representation by structural models is highly compatible with computer graphics (CG) and is expected to be especially applicable to virtual reality (VR) applications. However, to appropriately estimate a structural model of an object from a 3-D image, a wide variety of problems that have turned up in the field of computer vision (CV) must be solved, and this cannot be easily achieved at present.

On the other hand, a technique using ray data has been proposed as a means of representing visual data in 3-D space [10–15]. This technique processes sets of ray data in contrast to conventional techniques that target image data, and is expected to achieve a high degree of freedom in processing by targeting ray data. It can be treated, moreover, as an especially useful concept in formulating a method of spatial representation that can comprehensively deal with a variety of 3-D image formats.

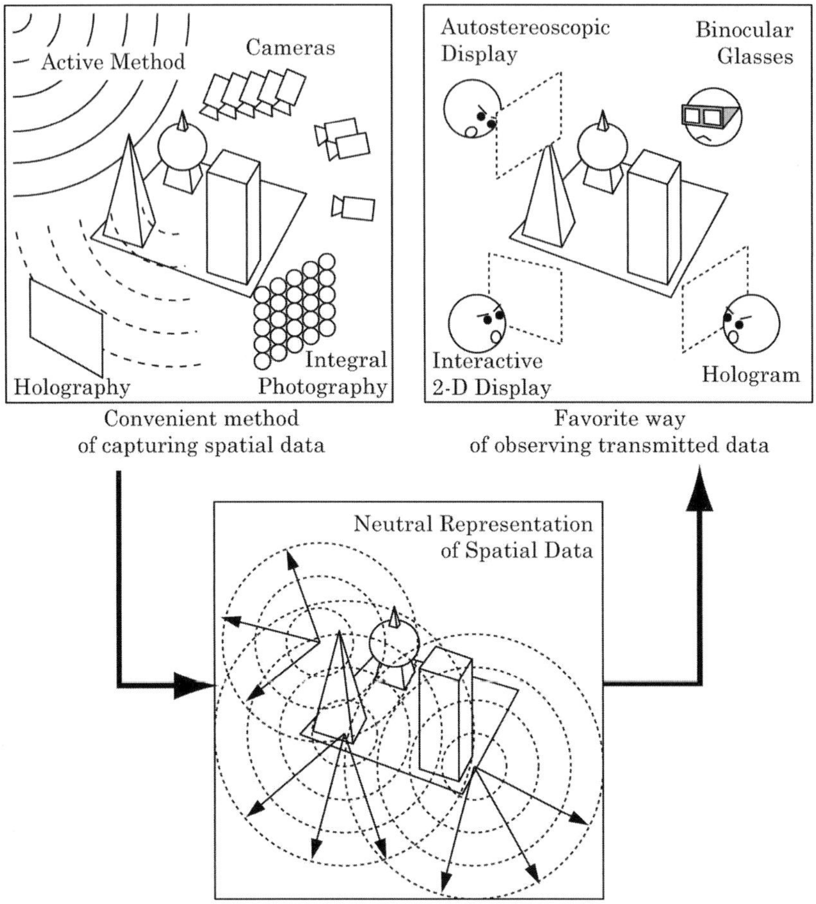

Fig. 8.1. Basic concept of "integrated 3-D visual communication" (revision of [16])

In this chapter, we first investigate an efficient method of representing ray data [16–19]. Here, by ignoring changes that accompany the propagation of light (such as interference and attenuation), we decrease the number of dimensions of data space for representing ray data and formalize a specific representation technique. We then show that spatial representation by this technique is equivalent to a method of representation based on sets of orthogonal (parallel) projection images independent of viewpoint. After this, we describe the construction of a system that can perform all processing from input to output in real time [20–22]. This system is a landmark development not only as a ray-based approach to representing visual data but also as an image-based rendering system. We then describe the application of this ray-based representation method to integral photography, a type of 3-D display technology, and show how its usefulness has been confirmed by experiment [23]. Next, with respect to the huge amount of data used in ray-based representation, we investigate how to compress this data efficiently and introduce the concept of a virtual object surface [24]. Finally, we report on a method for directly acquiring ray data using a telecentric lens [25].

8.2 Spatial Representation by Ray Data

"Visual data" means information on light rays that propagate through space. "Transmitting all information on light rays that fill 3-D space" therefore makes possible the "visual transmission of the 3-D space." With this in mind, we here outline a method of spatial representation focused on ray data [16, 17].

8.2.1 Representing 3-D Spatial Information by Ray Data

The set of all light rays that pass through point (X, Y, Z) in 3-D space is called a "pencil." 3-D space is filled with innumerable pencils, and the individual rays that make up a pencil are distinguished by direction (θ, ϕ). Accordingly, we let f denote ray data (brightness, color, etc.) in 3-D space and define f as the following function consisting of 5 parameters [26]:

$$\text{ray data} = f(X, Y, Z, \theta, \phi) \ . \tag{8.1}$$

In other words, 5 dimensions of data space are needed to record ray data. This data space is called the "plenoptic function" [26] or "ray space" [10].

Based on this function, pencil data at viewpoint $A(X_\mathrm{a}, Y_\mathrm{a}, Z_\mathrm{a})$ (all surrounding visual information in a QuickTimeVR panorama image [27]) can be written as follows:

$$f_\mathrm{a}(\theta, \phi) = f(X_\mathrm{a}, Y_\mathrm{a}, Z_\mathrm{a}, \theta, \phi) \ . \tag{8.2}$$

A 2-D image is a 2-D subspace of the 5-D ray space.

Capturing an image by a pinhole camera can be treated as an operation that obtains pencil data as image data. Conversely, given an observer at viewpoint A in a display system, stored pencil data may be read out and projected onto a picture screen. In this manner, capturing an image means the "sampling of pencil data," while image synthesis is achieved by the "reading out of pencil data." Also, by ignoring phase information of light, it becomes possible to not only synthesize pinhole images, but to also synthesize images virtually from $f(X, Y, Z, \theta, \phi)$ at various focal distances by supposing a lens system [28, 36]. In short, image data corresponding to any viewpoint, viewing direction, and focal distance are embedded in $f(X, Y, Z, \theta, \phi)$.

Spatial Representation for Restricted Viewpoints

When processing data obtained by moving a camera in the horizontal direction (along the line $(Y, Z) = (Y_a, Z_a)$, which is parallel to the X-axis), such as in stereo images or multiview images, the f function takes on the following form:

$$f_a(X, \theta, \phi) = f(X, Y_a, Z_a, \theta, \phi) . \tag{8.3}$$

In more general terms, an image sequence obtained by moving a camera along a straight line or curve $(X, Y, Z) = C(p)$ (including, for example, an image sequence of a walkthrough application) becomes a 3-D subspace of the 5-D ray space as follows:

$$f_C(p, \theta, \phi) = f(X, Y, Z, \theta, \phi)|_{(X,Y,Z)=C(p)} . \tag{8.4}$$

Likewise, an image sequence obtained on a curved surface $(X, Y, Z) = S(p, q)$ can be written as follows:

$$f_S(p, q, \theta, \phi) = f(X, Y, Z, \theta, \phi)|_{(X,Y,Z)=S(p,q)} . \tag{8.5}$$

This is a 4-D subspace of the 5-D ray space.

Accordingly, when camera position (observer viewpoint) is restricted to a curve $C(p)$ or a surface $S(p, q)$, the above technique decreases the number of parameters and represents spatial data in a more efficient manner.

Spatial Representation for Viewpoints with Greater Degrees of Freedom

We consider the case in which boundary surface S divides a 3-D space into two spaces. One is being described (described zone) and the other is the space in which the observer is present (visual zone), as shown in Fig. 8.2.

Here, assuming that "light travels in a straight line within the 3-D space without being affected by interference, attenuation, and the like," an image for a viewpoint situated at point O in the figure can be virtually synthesized by collecting ray data that passes through S and arrives at O. In other words, establishing the above assumption means that, for the ray data of (8.5), the observer can move to positions away from surface S and a 3-D degree of freedom can be obtained with respect to the viewpoint.

In this situation, where the viewpoint takes on a 3-D degree of freedom, the range of movement is, nevertheless, limited to either one of the two zones divided by boundary surface S. To put it another way, we can consider two ways of using (8.5) depending on which of the two zones is specified as the described zone, as depicted in Fig. 8.3. Figure 8.3a corresponds to object representation when objects enclosed in a zone are observed from the outside, and Fig. 8.3b corresponds to the environment representation when an observer, enclosed in a zone, observes the outside environment.

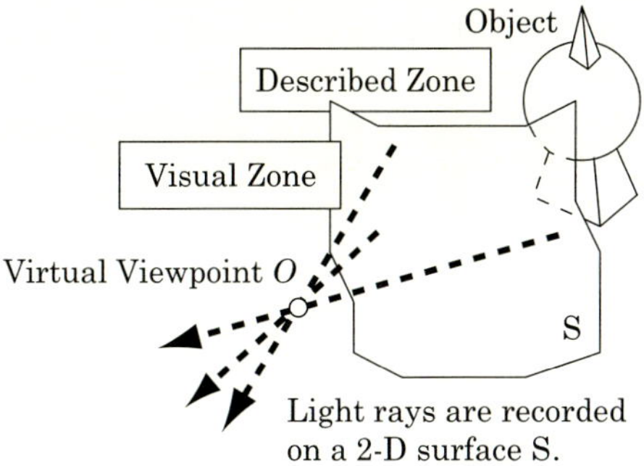

Fig. 8.2. Synthesis of virtual viewpoint images from light rays recorded on a surface (revision of [16])

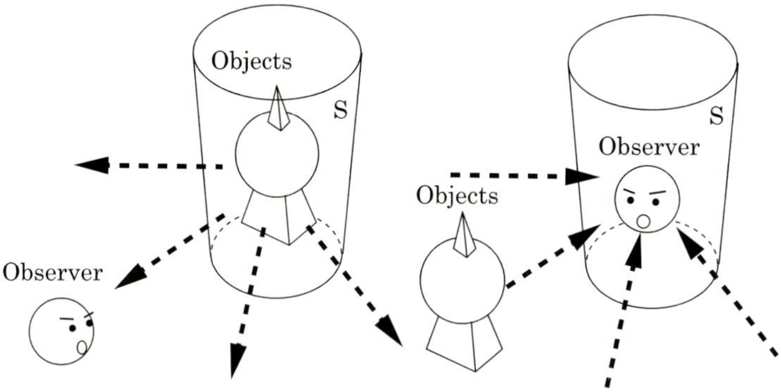

Fig. 8.3. Viewpoint-independent representation of 3-D spatial data (**a**) object representation applicable to multiplex hologram and so on; (**b**) environment representation applicable to the CAVE system and so on (revision of [16])

Accordingly, having 4 parameters as in (8.5) enables spatial representation with a sufficiently high degree of freedom. Compared to using 5 parameters, we can see that decreasing the number of dimensions of data space by one results in efficient spatial representation. The 4-D ray space was introduced to the field of visual communication in [10], and to the field of computer graphics in [14] as the "light field rendering."

We emphasize here that (8.5) corresponds to spatial representation dependent on the shape of surface S. For this reason, a conversion process will be required when camera arrangement S_i at the time of image capturing (input) differs from display shape S_o at the time of display (output). In short, conversion techniques dependent on both input and output must be defined exactly for the number of such combinations. This property detracts from the flexibility of a 3-D information environment.

This problem can be solved, however, by neutral representation of 3-D space independent of the boundary surface's shape. Here, by preparing a conversion from the input system to a neutral method of representation and a conversion from this neutral method of representation to the output system, any combination of input/output can be handled in a flexible manner. The following discusses a technique for achieving such neutrality independent of boundary-surface shape for the efficient 4-parameter spatial representation method described above.

8.2.2 Spatial Representation Independent of Boundary-Surface Shape

With reference to (8.5), we denote 4-parameter spatial representation independent of boundary-surface shape as follows:

$$f'(P,Q,\theta,\phi) = f(X,Y,Z,\theta,\phi)|_{(X,Y,Z)=S'(P,Q)} \,. \tag{8.6}$$

In this case, data of light propagating in a fixed direction $(\theta_\mathrm{a}, \phi_\mathrm{a})$ can be written as follows:

$$f'_\mathrm{a}(P,Q) = f'(P,Q,\theta_\mathrm{a},\phi_\mathrm{a}) \,. \tag{8.7}$$

Equation (8.7) corresponds to the recording of data for light rays oriented in the same direction, that is, data corresponding to an orthogonal projection of space, and the index of this data is given by the PQ coordinate system. In (8.5), as well, the pq coordinate system can be an index of orthogonal-projection data, but this system would be dependent on the shape of surface S.

With the above in mind, the purpose of this section is to define an easy-to-handle PQ coordinate system as an index to orthogonal-projection data. Differences in surface shape can then be absorbed by coordinate conversion between the pq coordinate system and PQ coordinate system.

Here, we define axis R coinciding with the direction of a light ray (optical axis) and denote its unit vector in the XYZ coordinate system as $\boldsymbol{r}$. Now, given two unit vectors $\boldsymbol{p}$ and $\boldsymbol{q}$ linearly independent of $\boldsymbol{r}$ and their corresponding axes P and Q, coordinate conversion from the XYZ coordinate system to the PQR coordinate system can be formalized. Specifically, vector $\boldsymbol{a} = (X_\mathrm{a}, Y_\mathrm{a}, Z_\mathrm{a})$ in the XYZ coordinate system converts as shown below to the PQR coordinate system:

$$\boldsymbol{a} = (X_\mathrm{a}, Y_\mathrm{a}, Z_\mathrm{a}) = P\boldsymbol{p} + Q\boldsymbol{q} + R\boldsymbol{r} \,. \tag{8.8}$$

Because changes to light during propagation are recorded along the R-axis, ignoring these changes would make values of the R coordinate meaningless. The PQ coordinate system, moreover, is an index of data for space projected in the $\boldsymbol{r}$ direction. The following investigates specific techniques for obtaining $\boldsymbol{p}$, $\boldsymbol{q}$, and $\boldsymbol{r}$.

Planar Recording of Ray Data

We first consider the case where the P and Q axes coincide with the X and Y axes, respectively, as shown in Fig. 8.4a.

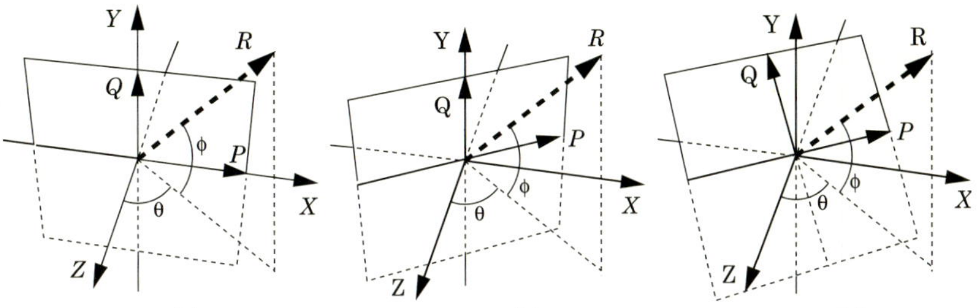

Fig. 8.4. Coordinate systems for recording ray data; (**a**) planar coordinate system; (**b**) cylindrical coordinate system; (**c**) spherical coordinate system (revision of [16])

Indicating ray direction as (θ, ϕ), we consider the coordinate systems shown. The $\boldsymbol{p}$, $\boldsymbol{q}$, and $\boldsymbol{r}$ unit vectors can be expressed as follows in terms of the XYZ coordinate system:

$$\begin{aligned}\boldsymbol{p} &= (1,\ 0,\ 0)\ ,\\ \boldsymbol{q} &= (0,\ 1,\ 0)\ ,\\ \boldsymbol{r} &= (\sin\theta\cos\phi,\ \sin\phi,\ \cos\theta\cos\phi)\ .\end{aligned} \tag{8.9}$$

Now, substituting these expressions for $\boldsymbol{p}$, $\boldsymbol{q}$, and $\boldsymbol{r}$ in (8.8), P and Q can be expressed as follows in terms of $(X_\mathrm{a}, Y_\mathrm{a}, Z_\mathrm{a})$:

$$\begin{aligned}P &= X_\mathrm{a} - Z_\mathrm{a}\tan\theta\ ,\\ Q &= Y_\mathrm{a} - Z_\mathrm{a}\tan\phi/\cos\theta\ .\end{aligned} \tag{8.10}$$

From these equations, the conversion from $f(X, Y, Z, \theta, \phi)$ to $f'(P, Q, \theta, \phi)$ can be defined. In other words, ray-data index (P, Q) can be determined from position coordinates $(X_\mathrm{a}, Y_\mathrm{a}, Z_\mathrm{a})$ in 3-D space. Accordingly, when data on light rays that pass through position $A(X_\mathrm{a}, Y_\mathrm{a}, Z_\mathrm{a})$ in 3-D space are needed, ray data recorded at position (P, Q) computed from (8.10) may be read out from $f'(P, Q, \theta, \phi)$.

This technique, however, cannot describe light rays that are parallel to the XY plane, that is, only rays that cross the XY plane can be recorded. For this reason, we call this technique "planar recording of ray data."

Cylindrical Recording of Ray Data

Next, we define the P, Q, and R axes as the result of rotating the X-, Y-, and Z-axes about the Y-axis by an angle of θ and then rotating only the

Z-axis about the X-axis by an angle of ϕ, as shown in Fig. 8.4b. In the same way as discussed above, we can derive the following equations. Here, $\boldsymbol{r}$ is the same as in planar recording:

$$\begin{aligned} \boldsymbol{p} &= (\cos\theta,\ 0,\ -\sin\theta)\ , \\ \boldsymbol{q} &= (0,\ 1,\ 0)\ , \qquad (8.11) \\ P &= X_{\mathrm{a}}\cos\theta - Z_{\mathrm{a}}\sin\theta\ , \\ Q &= -X_{\mathrm{a}}\sin\theta\tan\phi + Y_{\mathrm{a}} - Z_{\mathrm{a}}\cos\theta\tan\phi\ . \qquad (8.12) \end{aligned}$$

This technique cannot describe light rays that are parallel to the Y-axis, that is, only rays that cross the cylindrical surface parallel to the Y-axis can be recorded. This technique is therefore called "cylindrical recording of ray data."

Spherical Recording of Ray Data

Finally, we define the P, Q, and R axes as the result of rotating the X, Y, and Z axes about the Y-axis by an angle of θ and again about the X-axis by an angle of ϕ. In this case, the P, Q, and R axes are always perpendicular to each other. Again, similar to the above discussions, we derive the following equations:

$$\begin{aligned} \boldsymbol{p} &= (\cos\theta,\ 0,\ -\sin\theta)\ , \\ \boldsymbol{q} &= (-\sin\theta\sin\phi,\ \cos\phi,\ -\cos\theta\sin\phi)\ , \qquad (8.13) \\ P &= X_{\mathrm{a}}\cos\theta - Z_{\mathrm{a}}\sin\theta\ , \\ Q &= -X_{\mathrm{a}}\sin\theta\sin\phi + Y_{\mathrm{a}}\cos\phi - Z_{\mathrm{a}}\cos\theta\sin\phi\ . \qquad (8.14) \end{aligned}$$

In contrast to the above techniques, this one enables ray data in any direction to be recorded. In other words, all rays that cross the spherical surface defined about the origin can be recorded. We call this technique "spherical recording of ray data."

8.2.3 Overview of Spatial Representation System

A spatial representation system based on ray data can handle various input/output systems in a comprehensive manner through the following procedure.

1. Convert 3-D image input (e.g. stereo, multiview, hologram, or polygon formats) to ray data and store in $f'(P, Q, \theta, \phi)$.
2. Interpolate and fill gaps in $f'(P, Q, \theta, \phi)$.
3. Compress and transmit $f'(P, Q, \theta, \phi)$.
4. Read out and display appropriate ray data from $f'(P, Q, \theta, \phi)$ in accordance with the display system.

In general, it is not easy to obtain sufficient ray data by an actual camera system to fill $f'(P, Q, \theta, \phi)$ in a dense manner. In short, simply converting various types of 3-D image input to ray data can lead to gaps in $f'(P, Q, \theta, \phi)$. This is why interpolation and synthesis of ray data becomes necessary to fill these gaps. Specifically, virtually interpolated ray data will be added to actual ray data captured by the camera so as to generate a densely filled $f'(P, Q, \theta, \phi)$. The significance of this process is that a densely filled $f'(P, Q, \theta, \phi)$ so obtained represents a new world having aspects of both the real and virtual worlds.

The function $f(X, Y, Z, \theta, \phi)$ and (8.5) can be treated as a set of pencil data $f_{\mathrm{a}}(\theta, \phi)$ equivalent to a perspective projection image. On the other hand, the function $f'(P, Q, \theta, \phi)$ can be treated as the set of orthogonal projection $f'_{\mathrm{a}}(P, Q)$.

8.3 Construction of a Real-Time System

This section describes the constrution of a system that, based on the planar recording method described in Sect. 8.2.2, performs all processing from input (capture) to output (image synthesis) in real time [20–22].

This system is a landmark development not only as a ray-based approach to representing visual data but also as an image-based rendering system [11–15]. We call this real-time image-based rendering method "video-based rendering."

8.3.1 Transformation of the Planar-Recording Method

We consider the case in which the camera's optical axis is parallel to the Z-axis in the XYZ coordinate system. Letting F denote the distance from the camera's focus point to its pickup surface, the following relationship holds between the position (p_x, p_y) of a pixel on the captured image and the direction (θ, ϕ) of the light ray recorded on that position (See Fig. 8.5):

$$\begin{aligned} p_x &= -F \tan\theta \ , \\ p_y &= -F \tan\phi / \cos\theta \ . \end{aligned} \tag{8.15}$$

Here, we make the following assignments to express ray direction in terms of (x, y) instead of (θ, ϕ):

$$\begin{aligned} x &= p_x/F = -\tan\theta \ , \\ y &= p_y/F = -\tan\phi/\cos\theta \ . \end{aligned} \tag{8.16}$$

Furthermore, as the P and Q axes coincide with the X and Y axes, respectively, in the planar-recording method, $f'(P, Q, \theta, \phi)$ can be rewritten as follows:

$$\text{ray data} = f'(X, Y, x, y) \ . \tag{8.17}$$

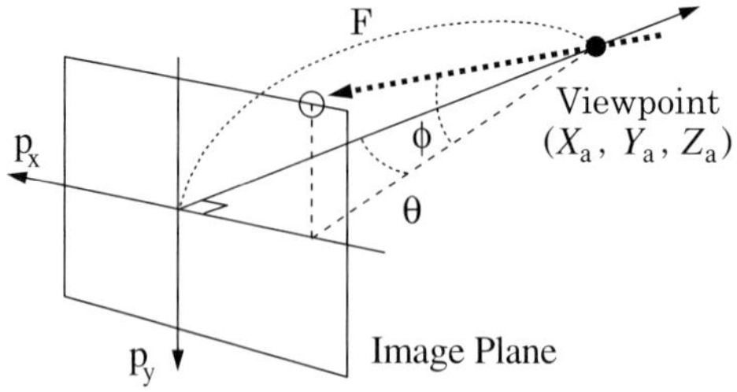

Fig. 8.5. Image coordinates (p_x, p_y) and directions of light rays (θ, ϕ) [18]

The following equations can now be derived from (8.10), (8.15), and (8.16), and we can see that corresponding points on the object form lines on the Xx and Yy planes (Consider X_a, Y_a, Z_a as constants):

$$X = X_a + Z_a\,x\,,$$
$$Y = Y_a + Z_a\,y\,. \tag{8.18}$$

This is a property conforming to the theory of epipolar plane images [29]. These expressions indicate that an object with a negative Z coordinate will be recorded on lines rising to the left on the Xx plane, as shown in Fig. 8.6.

Here, to obtain $f'(X, Y, x, y)$, cameras may be lined up on the XY plane and shots taken. On the other hand, to synthesize image $I_a(x, y)$ at viewpoint (X_a, Y_a, Z_a) from (8.17) and (8.18), ray data will have to be collected in the following manner:

$$I_a(x, y) = f'(X_a + Z_a\,x,\ Y_a + Z_a\,y,\ x,\ y)\,. \tag{8.19}$$

When $Z_a = 0$, $I_a(x, y) = f'(X_a,\ Y_a,\ x,\ y)$, which means that $f'(X, Y, x, y)$ is the set of images captured on the plane $Z = 0$. Equation (8.19) tells us that images at $Z \neq 0$ can be synthesized if a huge number of images can be captured on the $Z = 0$ plane.

8.3.2 Real-Time Interpolation of Light Rays

The purpose of this section is to find a means of performing processing from the acquisition of $f'(X, Y, x, y)$ to the synthesis of $I_a(x, y)$ in real time while varying viewpoint (X_a, Y_a, Z_a) interactively. In general, as a result of restrictions attached to camera spacing, $f'(X, Y, x, y)$ obtained in real time results in discrete sampling in the direction of the X- and Y-axes in comparison to sampling in the x- and y-directions. In order to interpolate light rays, the Lumigraph [15] utilizes a rough shape model, but it is not in real time.

Let the positions of K cameras be denoted by $(X_k, Y_k, 0)$ $(k = 0, \cdots, K-1)$ and the data space having gaps in real time by f_R. We first consider a simple interpolation technique that can be formalized as follows:

$$f(X, Y, x, y) = f_R([X], [Y], x, y) \tag{8.20}$$

where, for $k_{\rm opt}$ that satisfies the followings,

$$D_k(X,Y) \triangleq (X - X_k)^2 + (Y - Y_k)^2 \tag{8.21}$$

$$k_{\rm opt} \triangleq \arg\min_k D_k(X,Y) \tag{8.22}$$

we define the following expressions:

$$[X] \triangleq X_{\rm k_{opt}}, \quad [Y] \triangleq Y_{\rm k_{opt}} \,. \tag{8.23}$$

Interpolation results by the above technique for the Xx plane are shown in Fig. 8.6a. With this method, a discontinuous structure appears over the data space, which affects the quality of synthesized images at arbitrary viewpoints.

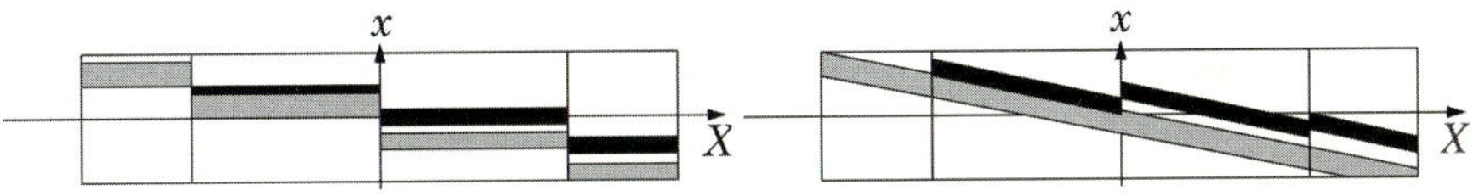

Fig. 8.6. Interpolation of a sparsely sampled data space: (**a**) $Z = \infty$; (**b**) $Z = Z_{\rm i}$(cont.) [20]

In response to the above problem, we now consider interpolation of the kind shown in Fig. 8.6b. This approximates the shape of the object as follows on plane $Z = Z_{\rm i}$.

$$\langle x \rangle \triangleq x - \frac{X - [X]}{Z_{\rm i}}, \quad \langle y \rangle \triangleq y - \frac{Y - [Y]}{Z_{\rm i}} \,. \tag{8.24}$$

Based on the above, the following approximation method can be given.

$$f(X,Y,x,y) = f_{\rm R}([X],[Y],\langle x \rangle,\langle y \rangle) \,. \tag{8.25}$$

From (8.25), we can see that the method of (8.20) corresponds to the case in which the objects lies at an infinite distance ($Z_{\rm i} = \infty$).

Here, by approximating the shape of the object at plane $Z = Z_{\rm i}$, interpolation as shown in Fig. 8.6b can be performed. Also, by creating an interface so that the value of $Z_{\rm i}$ can be varied interactively, it becomes possible to suppress distortion in the object of attention at the will of the observer.

8.3.3 Approximation Error

In the method described above, distortion occurs since the shape of the object is approximated at the $Z = Z_\mathrm{i}$ plane. When viewed from above the Xx plane, this distortion, at the sections where the switchover in $[X]$ occurs from X_k to X_{k+1}, takes on the form of large discontinuities in structure that worsen the farther an object lies away from $Z = Z_\mathrm{i}$. This situation can be expressed by a formula, as explained below.

Error Δx on the x-axis, when approximating an object that exists at $Z = Z_\mathrm{i} + \Delta Z$ at $Z = Z_\mathrm{i}$, can be expressed as follows from (8.24).

$$\begin{aligned}\Delta x &\stackrel{\triangle}{=} \left\{x - \frac{X - [X]}{Z_\mathrm{i} + \Delta Z}\right\} - \left\{x - \frac{X - [X]}{Z_\mathrm{i}}\right\} \\ &= (X - [X])\frac{\Delta Z}{Z_\mathrm{i}(Z_\mathrm{i} + \Delta Z)} \,. \end{aligned} \tag{8.26}$$

In other words, this error is propotional to $X - [X]$.

At this point, let us consider that the cameras are separated by a uniform interval D. The following therefore holds.

$$|X - [X]| \le \frac{D}{2} \,. \tag{8.27}$$

The equal sign corresponds to the section where $[X]$ switches over from X_k to X_{k+1}, in this expression. We can therefore write the following.

$$|\Delta x| \le \left|\frac{D\Delta Z}{2Z_\mathrm{i}(Z_\mathrm{i} + \Delta Z)}\right| \,. \tag{8.28}$$

Now, using (8.16), we consider converting error on the x-axis to error $\Delta\theta$ of angle θ that indicates the light-ray direction. In the discontinuous section on the Xx plane, we have both error from X_k and error from X_{k+1}, giving us the following.

$$\tan \Delta\theta = 2|\Delta x| = \left|\frac{D\Delta Z}{Z_\mathrm{i}(Z_\mathrm{i} + \Delta Z)}\right| \,. \tag{8.29}$$

As an example, let us consider an object with a thickness of 60 cm situated at a distance of 150 cm, that is, conditions of $Z_\mathrm{i} = -150$ and $-30 \le \Delta Z \le 30$. Given a camera interval of $D = 4\,\mathrm{cm}$, then, from (8.29), calculations give us $\Delta\theta = 0.0067\,\mathrm{rad}$ when $\Delta Z = 30$. This means that in the case of a wide-angle shot with a viewing angle, for example, of $\pi/2$ rad, and with resolution at a level of 235 pixels, interpolation can be performed without discontinuities being generated on the Xx plane. In particular, when $\Delta\theta$ becomes sufficiently small, no discontinuities will be generated even if the resolution of the Xx plane is increased. In other words, synthesis of a high-resolution image becomes possible.

Here, from (8.29), we can see that $\Delta\theta$ can be made small by

- Setting an object far away (making $|Z_{\mathrm{i}}|$ large) and/or
- Aligning cameras in a dense formation (making D small).

When setting an object at a distance, however, its image becomes small, which means that raising the resolution of the entire screen may not necessarily raise the resolution of the object. Accordingly, a method that sets the object far away is not a viable solution. On the other hand, a method that aligns the cameras in a dense formation, say $D = 1$ cm, would correspond to $\Delta\theta = 0.0017$ rad, making it possible to synthesize images without discontinuities up to a resolution of 942 pixels at a viewing angle of $\pi/2$ rad. In this case, an input bottleneck would occur due to camera resolution (e.g. 720 pixels). If, however, the viewing angle is narrow, say $\pi/8$ rad, then maximum resolution for $\Delta\theta = 0.0017$ rad becomes 235 pixels and input-camera resolution turns out to be higher.

We can summarize the above discussion as follows.

- The resolution of the input cameras is fixed, and varying the viewing angle changes the resolution of the light-ray direction.
- Based on (8.29), the configuration of the camera system determines $\Delta\theta$, which corresponds to the quantization step size of the light-ray direction. If $\Delta\theta$ were fixed, the number of light-ray directions increases as the viewing angle becomes wider.

A balance should therefore be achieved between the above variables when constructing a camera system.

8.3.4 System Configuration

Obtaining $f'(X, Y, x, y)$ in real time requires the simultaneous handling of multiview video input obtained from multiple cameras. In this chapter, we adopt the system configuration shown in Fig. 8.7.

The quad processor (QP) in the figure is a piece of equipment that integrates video images from four cameras and outputs a single video image on a screen partitioned into four sections. Consequently, video images from 16 cameras can be integrated into four images each having four sections by using four QP units. Then, by inputting these images into a fifth QP (QP E), the video images from 16 cameras can be integrated into a single image partitioned into 16 sections.

Here, using observer viewpoint $(X_{\mathrm{a}}, Y_{\mathrm{a}}, Z_{\mathrm{a}})$ and approximate distance Z_{i} to the subject, the computer processes the 16-section video input and synthesizes desired pictures. In this chapter, we handle video input as texture and achieve high-speed processing using a graphic engine.

8.3.5 Experimental Results

An external view of the camera array system that we constructed is shown in Fig. 8.8. It consists of 16 CCD cameras (MTV-2510EM) arranged on a board

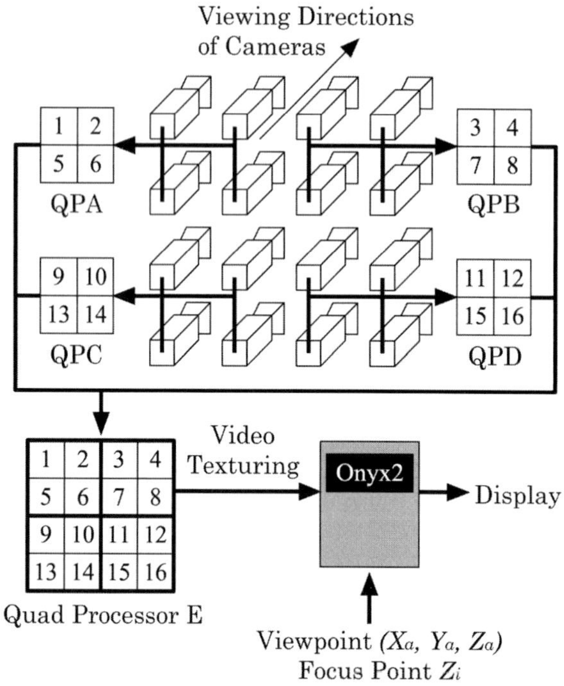

Fig. 8.7. System configuration for the real-time video-based rendering system [20]

Fig. 8.8. Camera array system [20]

in a 4-by-4 array with a camera spacing of 4 cm. This system can sample, at intervals of 4 cm, light rays passing through a 12 cm square plane.

Examples of moving pictures synthesized in real time are shown in Fig. 8.9. On the left are 16-section input images, and on the right are synthesized images achieved by a texture-mapping function targeting the areas enclosed by the white squares in the corresponding 16-section input image. These

Fig. 8.9. Results of real-time video-based rendering (*left*: 16-section input images, *right*: synthesized images) [20]

synthesized images are the result of virtually moving the viewpoint back and forth while looking between the walls from the fixed camera system. Specifically, in order from top to bottom in the figure, the camera moves backward (in a direction away from the character figures) in a virtual manner. As shown, the images of these figures become progressively smaller. Note that occlusion occurring between the wall in the foreground and an inner figure has been reproduced. This is a 3-D visual effect that cannot be obtained simply by zooming in or out. Object position Z_{i} can also be made to coincide with the position of the figures by mouse operation. Here, while errors occur in the foreground wall and background, favorable synthesis is achieved with regard to the area surrounding the figures.

For picture synthesis, we used Onyx2 Reality (R10000 195 MHz $\times$ 2) equipment manufactured by SGI and a video-texture function through DIVO (DIgital Video Option). Processing for extracting and aligning texture according to viewpoint was performed at 47 frames/s. This figure is greater than the 30 frames/s of video input, which indicates the possibility of even more complex processing.

8.3.6 Further Investigation

The second generation of video-based rendering systems has been also developed [21, 22, 34]. Figure 8.10 shows a new camera array system composed of 16 attachable camera head units [21]. We designed a connecting unit for a SONY XC-333 camera-head in order to realize more versatile camera-array configurations. As for the quad processor, we utilize SONY YS-Q430.

Fig. 8.10. Attachable camera head units and composed camera-array system (4×4, 3×5 and 1×16)

In order to suppress the discontinuities between stitched views on the right side of Fig. 8.9, the alpha-blending technique in the field of computer graphics can be utilized [21]. On the left side of Fig. 8.11, the areas enclosed by white squares correspond to those enclosed by white ones in Fig. 8.9. Broken lines indicate the areas used for blending with neighboring views. The right side of Fig. 8.11 shows the synthesized images in which the discontinuities are blurred.

The method described before approximates the shape of the scene by a plane $Z = Z_{\mathrm{i}}$. This will cause visible errors especially for the object apart

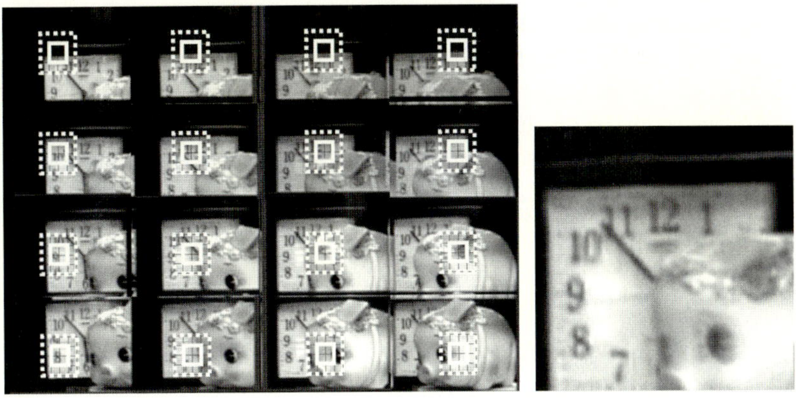

Fig. 8.11. Effect of alpha blending (*left*: 16-section input images, *right*: synthesized images)

from the plane $Z = Z_{\mathrm{i}}$ as shown in Fig. 8.12a, b and c. In order to extend the approximation method, we introduced a depth-measuring system (KOMATSU FZ930), which can estimate a rough 3-D shape of the scene from nine views captured by our camera array in real time [22]. We can utilize the 3-D shape dynamically updated in real time instead of a static single plane to interpolate light rays.

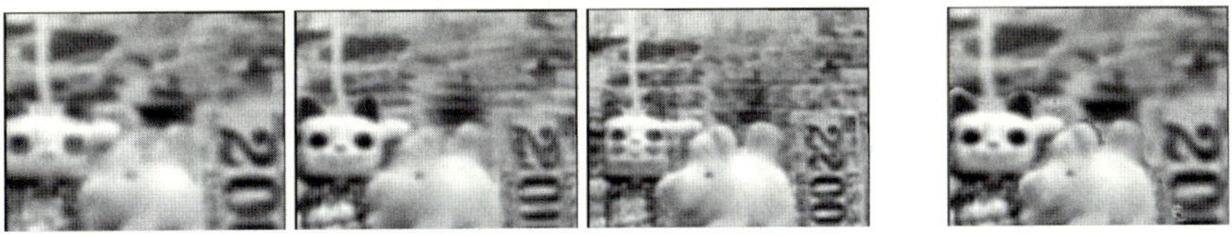

(a) (b) (c) (d)

Fig. 8.12. Effect of layered representation; (**a**) background layer where Z_{i} is adjusted to background; (**b**) middle object layer for a cat toy; (**c**) closer object layer for a pig toy; (**d**) composed image in which any object is in focus

In this experiment, the estimated shape is converted into a set of layers. We approximate the scene by three layers, that is, the scene is represented by three planes corresponding to background, a toy cat and a toy pig, respectively. Figure 8.12d shows the result of superimposing the layers. We can see that the visible errors in Fig. 8.12a, b and c are effectively suppressed in Fig. 8.12d thanks to the layered representation.

8.4 Acquisition of Dense Data Space Utilizing Integral Photography

In Sect. 8.3, we reported on the real-time application of the planar-recording method using a camera array. In this section, we describe a technique that samples ray data at a much greater density than a camera array by using integral photography (IP) as the input format [23].

8.4.1 Using a Real-Time IP System

The NHK Science and Technical Research Laboratories has achieved a real-time full-color IP system using a bundle of optical fibers and a high-definition camera [30, 31]. In this system, the high-definition camera picks up the inverted image of a target object as elemental images, the number of which corresponds exactly to the number of lenses (optical fibers). An example is shown in Fig. 8.13.

Fig. 8.13. Example of an IP image [23]

In this section, we examine a system that feeds an image like the one shown here to a computer and synthesizes a new image interactively according to the position of the observer without the use of special optics. The method of Sect. 8.3 and the method described here are compared in Table 8.1. In either case, we have a method that deals with function $f'(X, Y, x, y)$ of the

Table 8.1. Comparisons of methods in Sect. 8.3 and 8.4

Method	No. of views (X, Y resolution)	No. of pixels in elemental image (x, y resolution)
Sect. 8.3	4×4	About 180×120
Sect. 8.4	54×57	About 20×20

planar-recording method. There is a difference, however, in how they sample ray data.

In short, these two input systems differ only in the way they sample ray data, and the algorithm of the ray-representation method can be equally applied to both. The experimental results presented below were obtained by applying the algorithm described in Sect. 8.3 to the IP image (Fig. 8.13) captured by a high-definition camera.

8.4.2 Experimental Results

Figure 8.14a shows the result of simulating IP playback optics by computer with respect to the input image of Fig. 8.13. The number of elemental images in this setup (54×57) corresponds to the resolution of the synthesized image. Figure 8.14b, on the other hand, shows the result of applying the method of Sect. 8.3 to IP input. Depth parameter Z_{i} has been adjusted for the dog in the front. Picture quality partially improves according to object depth.

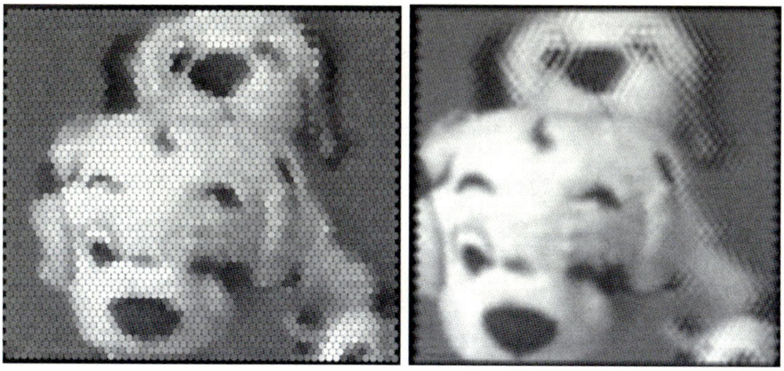

Fig. 8.14. Results of image synthesis from IP input; (**a**) simulating IP playback optics; (**b**) proposed method [23]

Figure 8.15 shows results for different types of source pictures. These results demonstrate that the method of Sect. 8.3 is effective for objects that reflect in a shiny and complex manner and for moving objects like people.

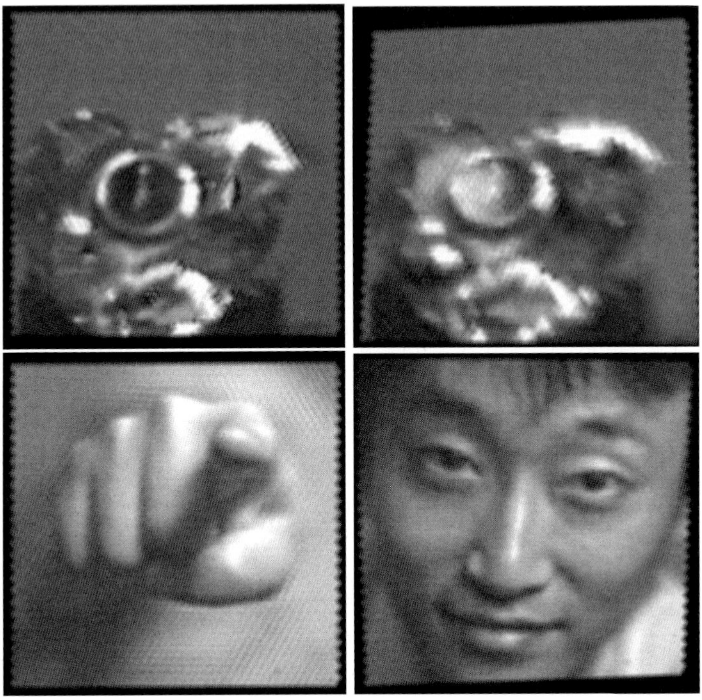

Fig. 8.15. Other results for objects that reflect in a shiny and complex manner and for moving objects like people [23]

The above images were synthesized in real time while interactively varying viewpoint and depth parameter Z_i of the object.

8.5 Compression of Ray Data

An important issue in the ray-representation method is how to efficiently compress the huge volume of spatial data present [24, 32, 33, 35].

8.5.1 4-D Discrete Cosine Transform

Compression using the discrete cosine transform (DCT) has found wide use in the field of picture encoding. Here, in contrast to two dimensions in the case of pictures, we can consider four-dimensional DCT targeting the function $f'(P, Q, \theta, \phi)$. In particular, by defining 4-D zigzag scanning and the like for DCT coefficients, this approach has been found to be an effective method of compressing ray data [32].

8.5.2 Improving Compression Efficiency by Redefining the Boundary Surface

Let us take another look at the 4-D data space $f'(X,Y,x,y)$ described in Sect. 8.3. We know that data on light rays passing through plane $Z = 0$ will be recorded in $f'(X,Y,x,y)$. If we now consider light rays that pass though plane $Z = Z_{\mathrm{a}}$, we can derive an expression for 4-D data space $g(X,Y,x,y)$ in the same manner as (8.19), as shown below:

$$g(X,Y,x,y) = f'(X + Z_{\mathrm{a}}x,\ Y + Z_{\mathrm{a}}y,\ x,\ y)\ . \tag{8.30}$$

In other words, the boundary surface may be redefined while preserving ray data.

Furthermore, by making use of the fact that redundancy in ray data becomes easier to use by selecting an appropriate value for Z_{a}, it becomes possible to further improve compression by 4-D DCT. In general, a greater improvement in compression efficiency by 4-D DCT is observed when setting Z_{a} near the object and recording ray data in $g(X,Y,x,y)$ as compared to defining a boundary surface far from the object and configuring 4-D data space $f'(X,Y,x,y)$.

8.5.3 Virtual Object Surface [24,35]

Expanding on the above ideas, we can do more than just bring the boundary surface closer to the object. We can also define a surface or layer approximating the shape of the object and thereby make it possible to perform even more efficient data compression. This is equivalent to enclosing the object in a 3-D display (a surface for which the manner of viewing changes according to viewpoint). In the following, this kind of surface that approximates the shape of the object to handle ray data is called a "virtual object surface."

When recording and transmitting visual information of an object using a virtual object surface, there are two types of data that must be conveyed:

- data on the shape of the virtual object surface, and
- data on the light rays that pass through the virtual object surface.

From the standpoint of data compression, we can consider that

- a virtual-object-surface shape that closely approximates the shape of the object results in high correlation among the rays at a single point on the virtual object surface and simplifies the compression of ray data.

On the other hand, we can expect that

- the amount of data needed to describe the shape of a virtual object surface to increase as that shape becomes more complex.

In other words, we have a tradeoff relationship, and there exists an optimal degree of shape approximation that minimizes the total amount of data.

8.5.4 Experimental Results

We compare compression efficiency for ray data under a variety of conditions as described below (Fig. 8.17).

(a) Each image of a multiview picture is compressed by JPEG.
(b) A multiview picture is treated as a moving picture and compressed by MPEG2.
(c) Compression with respect to $f'(X, Y, x, y)$ is performed by 4-D DCT.
(d) Z_a is optimally set and compression with respect to $g(X, Y, x, y)$ is performed by 4-D DCT.
(e) Each layer in the layered model shown in Fig. 8.16 is compressed by 4-D DCT as a virtual object surface.
(f) Texture data in the layered model is treated separately in the technique described in (e).

Fig. 8.16. Example of layered model [24]

Conditions (a) and (b) constitute an experiment with the aim of comparing conventional techniques with the proposed method. Conditions (c), (d), and (e), on the other hand, constitute an experiment to compare various techniques that use the 4-D DCT method described earlier. Condition (f), meanwhile, has been added to these experiments as a technique for separately transmitting texture in the layered model beforehand considering the difficulty of interactive image display if 4-D DCT is time consuming.

Experimental results are shown in Fig. 8.18. The horizontal axis shows total code length required to transmit the multiview picture, and the vertical axis shows average SNR at various capture points. The amount of data in the multiview picture before encoding is 437 Mb.

For case (d) where a single-plane model is used as a virtual object surface, the point corresponding to the highest compression rate was searched for while varying the distance (disparity) from the camera plane to the virtual

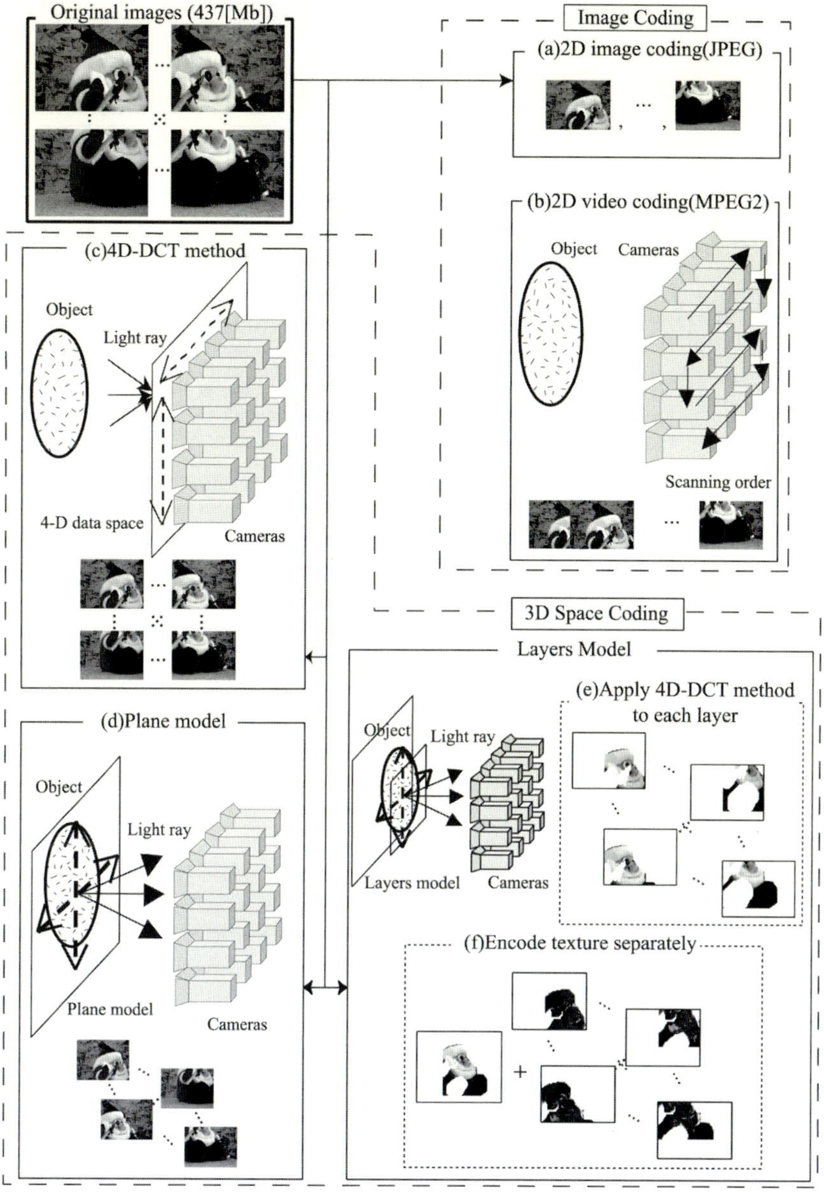

Fig. 8.17. Relationship between various experiments (revision of [24])

object plane. This point was found at a disparity of 40 pixels, which represents a positional relationship in which the virtual object surface pierces the object.

The original images used in these experiments exhibit a disparity of from 30 to 40 pixels since the camera interval is coarse compared to the image

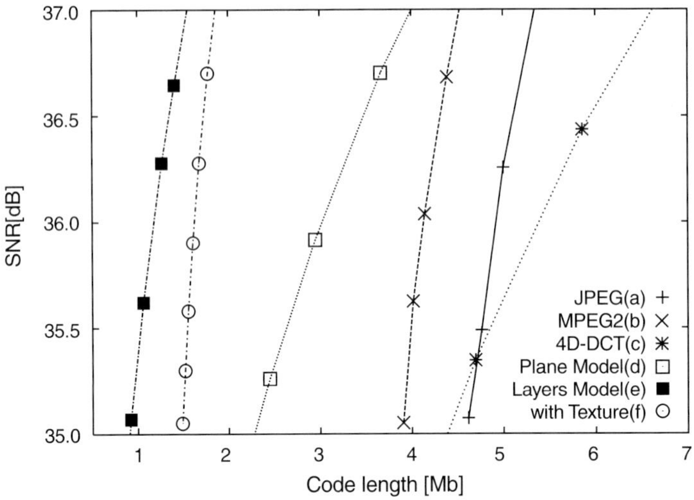

Fig. 8.18. Results of light-ray data compression (revision of [24])

resolution. In other words, if coordinate X in 4-D data space $f'(X, Z, x, y)$ changes by 1 pixel, the corresponding image coordinate x will shift from 30 to 40 pixels. This greatly exceeds the 8-pixel block size at the time of coding, which means that redundancy in space itself cannot be sufficiently used in technique (c) that directly codes the camera plane. In actuality, techniques (a) and (b) that apply JPEG and MPEG2, respectively, exhibit higher compression rates than technique (c), which is thought to be due to the above reason.

However, even in the case of technique (d) that uses a simple plane, the use of a virtual object surface greatly improves the compression rate obtaining better results than techniques (a) and (b) that apply image coding. Also, when using layered model (e), compression rate improves even further resulting in a compression rate of 345:1 at 36.3 dB. Technique (f) that codes texture separately, on the other hand, results in a compression rate worse than that of technique (e). This is due to the overhead generated by separate coding of texture.

Although technique (f), in which overhead is generated, is inferior to technique (e) in terms of data compression, it can perform texture mapping at high speed with respect to the shape of the virtual object surface, and makes interactive-image synthesis possible. For example, images with quality like that shown in Fig. 8.16 can be synthesized before 4-D DCT operations complete. At present, high-speed image synthesis can be expected through the use of hardware-based texture mapping. As a result, an observer can select a viewpoint interactively with the same ease as 3-D computer graphics, and in addition, can appreciate an image of photo quality that reflects the result of 4-D DCT operations by halting the video for a few seconds. Here, layer and texture code length was 1.03 Mb for an SNR of 29.6 dB.

When applying 2-D image coding like JPEG to each image in a set of multi-images, the correlation between images cannot be used at all and effective compression cannot be expected for the most part. On the other hand, when using 2-D video coding like MPEG2, coding that employs the correlation between consecutive frames can be performed. However, only the correlation between adjacent images in relation to camera positions can be used. Moreover, for multiview images featuring large disparity between images, it is difficult to determine the motion (disparity) vector with good accuracy, and as a result, disparity compensation prediction does not function effectively.

Space coding, in contrast, applies 4-D DCT to 4-D data space, which makes it possible to use the correlation between adjacent images not only in the horizontal direction but also in the vertical and diagonal directions, and even between images separated by greater distances. Use of a virtual object surface, moreover, enables images with even large disparity to be handled. For these reasons, the proposed technique can produce superior results, as shown in Fig. 8.18.

We have therefore confirmed that using redundancy in 4-D data space makes for efficient compression surpassing that performed on individual 2-D pictures.

8.6 Sampling Ray Data by a Telecentric Lens

As described in Sect. 8.2, 4-D data space $f'(P, Q, \theta, \phi)$ can be treated as a set of orthogonal projection images $f'_{\mathrm{a}}(P, Q)$. Accordingly, the use of a telecentric lens that can directly capture orthogonal projection images makes it possible to obtain a 4-D data space directly [25]. While Sects. 8.3, 8.4, and 8.5 have focused on techniques based on the planar-recording method, this section describes an experiment using the spherical recording method that can handle orthogonal projection images without distortion.

8.6.1 Capturing Orthogonal Projection Images

In the experiment, orthogonal projection images of a set of objects are first captured from multiple directions. Some of these images are shown in Fig. 8.19. Specifically, these are images of two cubes of the same size arranged one behind the other. It can be seen that both objects in a pair are displayed with the same size regardless of the distance from the lens. This confirms that the telecentric lens is taking orthogonal projection images. These images can now be used to configure 4-D data space by spherical recording.

Fig. 8.19. Examples of orthogonal projection images [25]

8.6.2 Synthesis of Perspective Views

Figure 8.20 shows the results of synthesizing perspective views of the objects described above at various viewpoints by reading out appropriate ray data from 4-D data space. We can see here how two objects in a pair are now displayed with different sizes in accordance with their relative positions.

Fig. 8.20. Results of synthesizing perspective projection images from a set of orthogonal projection images [25]

Capturing a large volume of orthogonal projection images enables perspective views to be synthesized at a variety of viewpoints.

8.7 Conclusions

We have presented a technique of representing 3-D spatial information based on ray data. Three methods of recording ray data were formalized here: planar recording, cylindrical recording, and spherical recording. In comparison with past techniques that use perspective projection images, this technique uses orthogonal projection images that enable spatial representation independent of viewpoint. The effectiveness of the proposed technique has been demonstrated experimentally by synthesizing 3-D computer graphics using, for example, a real-time video-based rendering system and integral photography. We have also proposed the concept of a virtual object surface and demonstrated its effectiveness in compressing ray data. Finally, the use of a telecentric lens was examined as a way of directly obtaining ray data, and its effectiveness was also demonstrated by experiment.

This ray representation method is a simple and straightforward approach to the problem of how to represent 3-D space, an issue shared by various fields including 3-D image communications, computer graphics, and virtual reality. In future research, we plan to continue our efforts in systematizing basic theory and to further develop various elemental technologies associated with input systems (camera array, integral photography, and telecentric lenses), compression and transmission methods (virtual object surface, 4-D DCT), and interactive display systems (image-based rendering).

Acknowledgements

We would like to extend our appreciation to NHK Science and Technical Research Laboratories for Fig. 8.13, an IP image taken by them, and to Tsukuba University for use of their multiview image database in the experiment of Sect. 8.5. Finally, we wish to thank our students; Junji Tago for his contribution in Sect. 8.3.6, Tatsuya Yoshida in Sect. 8.4, Takahide Takano in Sect. 8.5, and Keisuke Takeuchi in Sect. 8.6.

References

1. T. Okoshi: *Three-Dimensional Imaging Techniques* (Academic Press, New York 1976)
2. S. Benton: 'The Second Generation of the MIT Holographic Video System'. In: *TAO 1st Int. Symp. on Three Dimensional Image Communication Technologies, Tokyo, December 6–7, 1993*, ed. by J. Tsujiuchi, J. Hamasaki, M. Wada (Telecommunication Advancement Organization of Japan, Tokyo 1993) pp. S-3-1-1–S-3-1-6
3. K. Perlin, S. Paxia, J.S. Kollin: 'An Autostereoscopic Display'. In: *SIGGRAPH'00, 26th Annual Conf. on Computer Graphics, Los Angeles, July 23–28, 2000* (ACM Press, New York 2000) pp. 319–326
4. H. Harashima: 'Three-Dimensional Image Coding, Future Prospect'. In: *Picture Coding Symposium of Japan (PCSJ'92), Japan, October 7–9, 1992*, 9–12 (in Japanese)
5. T. Motoki, H. Isono, I. Yuyama: Proc. IEEE **83**, 7, pp. 1009–1021 (1995)
6. T. Naemura, M. Kaneko, H. Harashima: Inst. Electron. Inform. Commun. Eng. (IEICE) Trans. Inf. Syst. **E82-D**, 3, 558–567 (1999)
7. T. Fujii, H. Harashima: Inst. Electron. Inform. Commun. Eng. (IEICE) Trans. Inf. & Syst. **E77-D**, 9, 987–995 (1994)
8. J.R. Ohm, M.E. Izquierdo: IEEE Trans. Circuits Syst. Video Technol. **7**, 5, 801–811 (1997)

9. D. Tzovaras, N. Grammalidis, M.G. Strintzis: IEEE Trans. Circuits Syst. Video Technol. **7**, 2, 312–327 (1997)
10. T. Fujii: A Basic Study on the Integrated 3-D Visual Communication. Ph.D Thesis, Course of Elec. Eng., The Univ. of Tokyo (1994) (in Japanese)
11. A. Katayama, K. Tanaka, T. Oshino, H. Tamura: 'A Viewpoint Independent Stereoscopic Display Using Interpolating of Multi-Viewpoint Images'. In: *Stereoscopic Displays and Virtualreality Systems II, San Jose, February 7–9, 1995*, ed. by S.S. Fisher, J.O. Merritt, M.T. Bolas (SPIE, Washington 1995) **2409**, 11–20
12. L. McMillan, G. Bishop: 'Plenoptic Modeling: An Image-Based Rendering System'. In: *SIGGRAPH'95, 22nd Annual Conf. on Computer Graphics, Los Angeles, August 6–11, 1995* (ACM Press, New York 1995) pp. 39–46
13. T. Yanagisawa, T. Naemura, M. Kaneko, H. Harashima: 'Handling of 3-Dimensional Objects in Ray Space'. In: *Inform. Syst. Soc. Conf. Inst. Electron. Inform. Commun. Eng. (IEICE), Tokyo, September 5–8, 1995* (IEICE, Tokyo 1995) D-169 (in Japanese)
14. M. Levoy, P. Hanrahan: 'Light Field Rendering'. In: *SIGGRAPH'96, 23rd Annual Conf. on Computer Graphics, New Orleans, August 4–9, 1996* (ACM Press, New York 1996) pp. 31–42
15. S. Gortler, R. Grzeszczuk, R. Szeliski, M. Cohen: 'The Lumigraph'. In: *SIGGRAPH'96, 23rd Annual Conf. on Computer Graphics, New Orleans, August 4–9, 1996* (ACM Press, New York 1996) pp. 43–54
16. T. Naemura: Ray-Based Coding of Real Space and Its Application to Shared Space Communication. Ph.D Thesis, Course of Elec. Eng., The Univ. of Tokyo (1996) (in Japanese)
17. T. Naemura, M. Kaneko, H. Harashima: Signal Process.: Image Commun. **14**, 21–37 (1998)
18. T. Naemura, H. Harashima: 'The Ray-Based Approach to Augmented Spatial Communication and Mixed Reality'. In: *Mixed Reality – Merging Real and Virtual Worlds*, ed. by Y. Ohta, H. Tamura (Ohmsha, Tokyo, Springer-Verlag, Berlin 1999), pp. 165–182
19. T. Naemura, H. Harashima: 'Ray-Based Approach to Integrated 3D Visual Communication'. In: *Three-Dimensional Video and Display: Devices and Systems (Critical Reviews of Optical Science and Technology), Boston, November 5–6, 2000*, ed. by B. Javidi, F. Okano (SPIE, Washington 2000) **CR76**, pp. 282–305
20. T. Naemura, H. Harashima: 'Real-Time Video-Based Rendering for Augmented Spatial Communication'. In: *Visual Commun. Image Process. (VCIP'99), San Jose, January 25–27, 1999*, ed. by K. Aizawa, R.L. Stevenson, Y. Zhang (SPIE, Washington 1998) **3653**, pp. 620–631
21. J. Tago, T. Nitta, T. Naemura, H. Harashima: 'Advances in VBR(1) – Multi-View Input System and Real-Time Ray Interpolation–'. In: *Annual Conf. of Inst. Electron. Inform. Commun. Eng. (IEICE), Shiga, March 26–29, 2001* (IEICE, Tokyo 2001) A-16-18 (in Japanese)
22. J. Tago, T. Nitta, T. Naemura, H. Harashima: 'Advances in VBR(2) – Dynamic Layered Representation by Real-Time Measurement'. In: *Annual Conf. of Inst. Electron. Inform. Commun. Eng. (IEICE), Shiga, March 26–29, 2001* (IEICE, Tokyo 2001) A-16-19 (in Japanese)
23. T. Naemura, T. Yoshida, H. Harashima: Opt. Express **8**, 255–262 (2001), http://www.opticsexpress.org/oearchive/source/30085.htm

24. T. Takano, T. Naemura, H. Harashima: Inst. Electron. Inform. Commun. Eng. (IEICE) Trans. **J82-D-II**, 1804–1815 (1999) (in Japanese)
25. K. Takeuchi, T. Naemura, H. Harashima: J. Inst. Image Inform. Telev. Eng. **54**, 10, 1474–1477 (2000) (in Japanese)
26. E. Adelson, J. Bergen: 'The Plenoptic Function and the Elements of Early Vision'. In: *Computer Models of Visual Processing*, ed. by M. Landy, J. Movshon (MIT Press, Cambridge 1991), chap. 1
27. S.E. Chen: 'QuickTimeVR – an Image-Based Approach to Virtual Environment Navigation –'. In: *SIGGRAPH'95, 22nd Annual Conf. on Computer Graphics, Los Angeles, August 6–11, 1995*, (ACM Press, New York 1995) pp. 29–38
28. T. Fujii, T. Kimoto, M. Tanimoto: J. Inst. Image Inform. Telev. Eng. **52**, 3, 356–363 (1998) (in Japanese)
29. R.C. Bolles, H.H. Baker, D.H. Marimont: Int. J. Computer Vision, **1**, 7–55 (1987)
30. F. Okano, H. Hoshino, J. Arai, I. Yuyama: App. Opt. **36**, 7, 1598–1603 (1997)
31. H. Hoshino, F. Okano, H. Isono, I. Yuyama: Opt. Soc. Am. A **15**, 8, 2058–2065 (1998)
32. T. Takano, T. Naemura, M. Kaneko, H. Harashima: J. Inst. Image Inform. Telev. Eng. **52**, 9, 1321–1327 (1998)
33. M. Magnor, B. Girod: IEEE Trans. Circuits Systems Video Technol., **10**, 3, 338–343 (2000)
34. T. Naemura, J. Tago, H. Harashima: IEEE Computer Graphics and Applications, **22**, 2, 66–73 (2002)
35. T. Takano, T. Naemura, H. Harashima: Systems and Computers in Japan, **32**, 12, 47–59 (2001)
36. A. Isaksen, L. MacMillan, S.J. Gortler: 'Dynamically Reparameterized Light Fields'. In: *SIGGRAPH'00, 27th Annual Conf. on Computer Graphics, Los Angeles, July 23–28, 2000* (ACM Press, New York 2000) pp. 297–306

9 Information Flow in Streaming 3-D Video

D.J. Brady, S. Feller, D. Kammeyer, E. Cull, L. Fernandes, R. Stack, and R. Brady

9.1 Introduction

Radical improvements in electronic sensor and processor capabilities in the past decade have destabilized basic definitions of imaging in general and three-dimensional (3-D) imaging in particular. Conventionally, imaging refers to analog focal or holographic systems that integrate information acquisition and processing. Increasingly aggressive digital processing, however, diminishes the processing role in the sensor head. Particularly in sensor arrays, there is often no need for a well-formed "image" in analog space.

The divide between digital and analog systems is particularly pronounced for multidimensional imaging. Holographic and stereoscopic sensors record the illusion of 3-D scenes, but do not in fact construct 3-D models. Tomographic and other 3-D scene analysis schemes create true 3-D digital models from sensor array data. In most cases, however, users do not demand and cannot process full 3-D models.

3-D video systems may be categorized in the following hierarchy:

1. *Integrated capture and display systems.* These systems capture a stereo view and project that view for the user. Such systems include holographic and autostereoscopic recording and display solutions. These systems may be represented by the block diagram shown in Fig. 9.1.
2. *Discrete capture and display systems.* These systems include stereo camera recording/holographic or stereo display pairs. A block diagram for this approach is shown in Fig. 9.2.
3. *Capture, inversion and display systems.* As exemplified by tomographic imagers, these systems form complete 3-D models of the scene and reproject these models for viewers. Figure 9.3 is a block diagram for this approach.

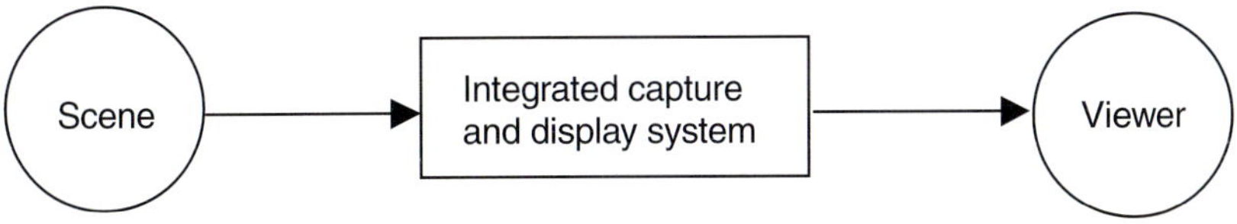

Fig. 9.1. An integrated stereoscopic capture and display system

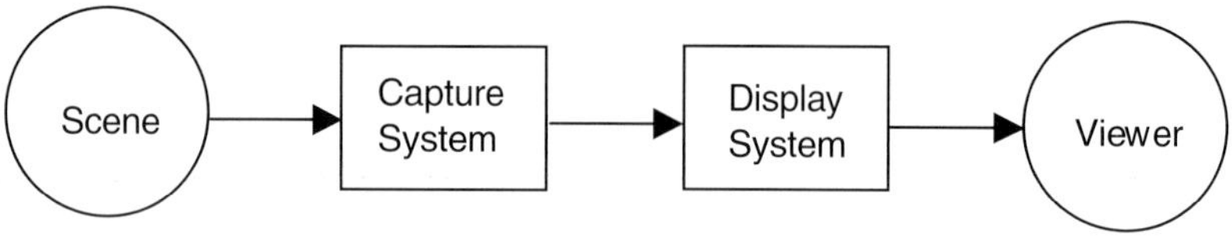

Fig. 9.2. A discrete capture and display system

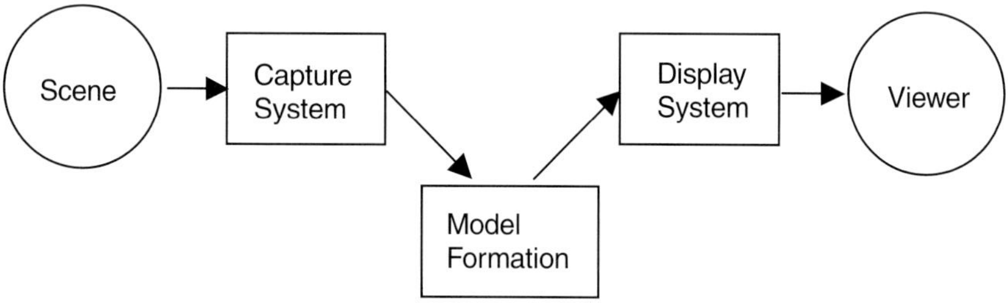

Fig. 9.3. A capture, inversion, and display system

Approach 1 was developed in the age of analog processing. Ubiquitous digital processing power makes 2 and 3 increasingly attractive. The fundamental attraction of these two approaches is that they support multiple users. 2 and 3 are also more adaptive and allow separate optimization of capture and display parameters.

Our focus in this chapter is multiuser functionality. We are interested in "streaming" 3-D video, meaning video that is mapped in real time across networks. In this context "video" refers to real-world images rather than computer generated graphics. (Although we allow for the possibility that computed views may be used to augment, approximate or otherwise improve the visual illusion of displayed real-world scenes. Just as digital imaging is blurring the distinction at the analog/digital interface between image formation and image processing, digital display is blurring the distinction at the digital/analog interface between image projection and image creation.)

Approaches 2 and 3 become considerably more complex in multiuser environments. The primary issue is *at what level should branching into multiuser pathways begin?* Approach 2, for example, might be constructed based on the single capture system block diagram shown in Fig. 9.4 or the multiple user and multiple capture system shown in Fig. 9.5. A leading example of the Fig. 9.4 approach would be 3-D television, in which a single stereo view is shown to many users. The Fig. 9.5 approach has been less commonly implemented, but has been developed on the Argus Imaging Array, as described below.

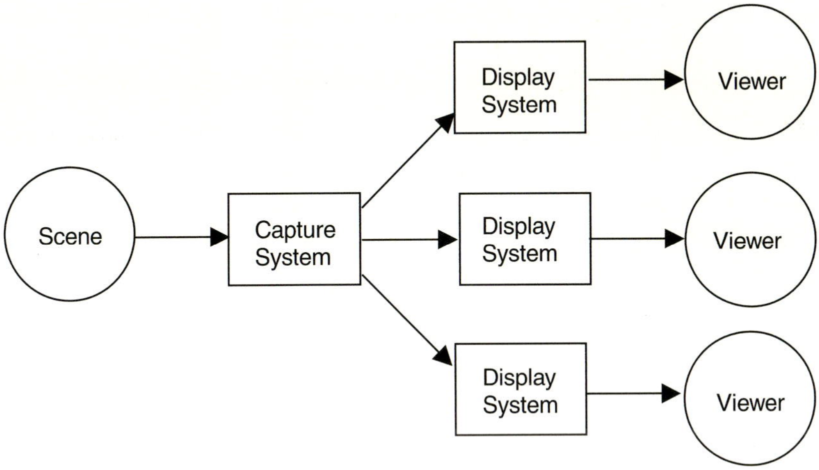

Fig. 9.4. A multiuser capture and display system

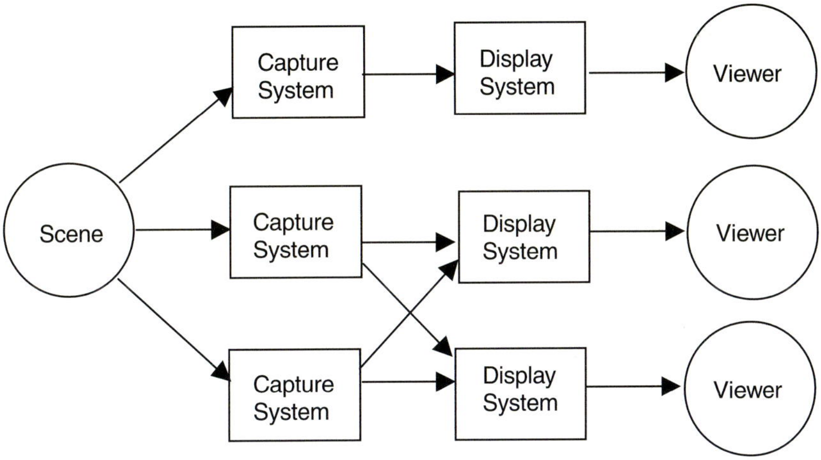

Fig. 9.5. A multiuser, multicapture system

Image-inversion systems also correspond to a variety of potential information flow maps. One can imagine complex interconnectivities between distributed capture systems, full and partial inversion systems and distributed displays.

While considerable analysis has been applied to localized capture of stereo and tomographic data, to the inversion of this data for scene models and to the reprojection and display of partially and fully inverted data, relatively little attention has been devoted to how networks of sensors and display systems might be integrated in interactive 3-D video. To address the need for analysis of data flow on distributed sensor networks, we constructed a distributed sensing and display testbed at the University of Illinois.

The Argus system has since been installed and upgraded at Duke University.

This chapter describes our 3-D video testbed, which we have named Argus. According to the Greek Mythology link www.has.brown.edu/~maicar

Argus has been called "the all-seeing because he had eyes in the whole of his body. Others say he had one hundred eyes in his head and that they slept two at a time in turn while the rest remained on guard.

Our Argus is a distributed supercomputer with 64 distributed digital cameras. This chapter describes the hardware architecture of this system and presents results from two preliminary tests of 3-D image streaming under diverse conditions.

9.2 Argus Hardware

Argus is composed of two major parts. The first is the digital hardware. The second is the sensor space, which is essentially a studio surrounded by cameras. The Argus computer system is a computational cluster of 32 dual-processor slave computers controlled by a master node. Each slave node contains two 400 MHz Pentium processors while the master computer contains two 450 MHz processors.

Communication on the computational cluster is achieved via a Gb switch. Each slave node has a 100 Mb connection to the local network while the master node has a Gb optical link. This networking paradigm was chosen because it allows algorithms developed for Argus to take advantage of recent advances in distributed processing and real-time computing. The computational resources of the entire system can be shared across the system as in a Beowulf class computer cluster described by [1, 2]. A diagram of the system interconnectivity is illustrated in Fig. 9.6.

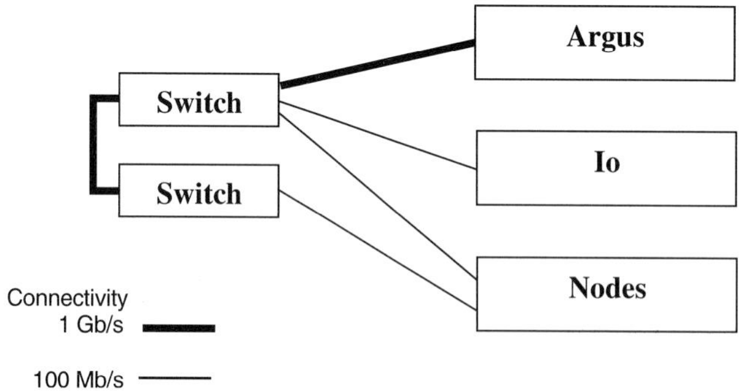

Fig. 9.6. Argus connectivity. Io is the file server. Argus is the master node

The sensor space consists of sixty-four cameras equally spaced around a fourteen-foot diameter ring. While the cameras are typically mounted in a planar configuration, the height of the cameras can be changed to allow for dual-plane, helical, and other variations. The frame of the imaging space is constructed out of steel pipe while the cameras are mounted on eight forty-five degree arcs that can be set at variable heights. The cameras circuitry is enclosed in a 2.5 by 3 inch metal box. The structure and layout of the imaging system can be seen in Fig. 9.7.

The lighting system for Argus was developed to provide consistent luminescence throughout the imaging space. To eliminate all light coming from outside the system, the interior of the sensor space is covered with black cloth with holes cut for the lenses of the cameras. Black carpet is used to limit reflection from the floor. In addition to the existing fluorescent room lights, we added ten low-wattage halogen bulbs as high as possible above the sensor array to provide even illumination across the sensor space.

The sensors used in Argus are grayscale CMOS focal planes with 2-mm lenses. The output from the cameras is an NTSC signal with 320×240 pixel resolution. The signal from each camera is transmitted to a node in the computational cluster over a standard coaxial cable. Each node in the com-

Fig. 9.7. The Argus sensor space. The small silver boxes are the cameras. In use, the image cage is draped with black curtains and the floor is covered with black carpet

putational cluster contains two WinTV capture cards for converting analog sensor data into a digital image stream. The cameras share a common power supply and control system to provide frame synchronization capabilities and centralized control.

A significant challenge in the construction of Argus was the accurate alignment of the sensors. Initially, the cameras were all placed at the same height with a laser leveling system. The direction of the cameras was determined by adjusting the alignment until an image of the laser at the source directly overlapped an image of the laser spot at the sensor. Due to the imprecise method of mounting the cameras, physical sensor alignment was a significant challenge, as we were only able to align the cameras to within two pixels of the ideal location.

To compensate for the limitations in the physical alignment procedure, we place a point source in the center of the imaging space to provide a reference for each of the cameras. We are able to digitally align the cameras to within a pixel of their ideal orientation by translating the images until the point source is centered. While this procedure is able to calculate the orientation of the sensor, we are developing alignment procedures to completely determine the position as well as the orientation of the sensors.

9.3 Algorithms

Argus is capable of generating both stereo pair views of the image space (treating the sensor array as a discrete mesh as in Fig. 9.5) and complete 3-D models of the space (as depicted in Fig. 9.3). Each type of data has different characteristics and performance capabilities in distributed-display applications. This section describes both data types and discusses the advantages and disadvantages of each.

Showing each eye separate images with slightly shifted perspectives can produce the illusion of three-dimensionality. This slight shift of viewpoint mimics the way we view the real world where the shift is determined by the separation distance between the eyes. The end result is a very convincing 3-D image. The images shown to the left and right eye are considered a stereo pair. Stereoscopes have used this technique for years to give the viewer a sense of depth from pairs of traditional images.

In the Argus project, stereo pairs are produced using images captured from two adjacent cameras. These images are then digitally aligned to make up for any errors in the physical camera placement. A shift is also necessary to prevent the images from converging to produce a cross-eyed effect. Since the system consists of a large number of cameras in a circle, a stereo pair can be produced at all points of interest around the exterior of the space. An example of a stereo pair streamed from Argus is shown in Fig. 9.8.

As the stereo-pair data is being streamed, the system is able to store images from the cameras locally on the cluster. This allows a user to control

Fig. 9.8. Stereo pair images of a dancer in Argus. Notice how the perspective is slightly different between the left and right views

the timeliness of the system. At the display end, time can be stopped, slowed down, sped up, or even reversed, much like the controls on a VCR. The ability to control time allows a flexible way to view the space and makes this type of data an attractive method for analyzing time-dependent object motion.

The basic stereo-pair-generation algorithm requires minimal computation. This quickly changes, however, as we look at improving the capabilities of the system. One advantage of having sensors embedded in the computational cluster is the ability to implement complex algorithms quickly. We can take advantage of advances in parallel computing to provide increased data processing while maintaining high-speed data streams. Multiuser support and intermediate-view generation are two potential improvements that will take advantage of the computational resources of the system.

In addition to generating stereo-pair data to create stereoscopic views, the Argus system uses cone-beam tomography to generate a complete 3-D model of the imaging space. In tomographic applications, the space is represented digitally by a 3-D array of voxels, or volume elements. In cone-beam tomography, the value of each voxel is an indication of the probability that the corresponding physical space to the voxel is filled. These voxel arrays are viewed using CRUMBS [3] or similar 3-D imaging tools.

The process of generating tomographic data in cone-beam tomography is best described in terms of ideal pinhole cameras. In this ideal system, each point in the imaging plane corresponds to the intensity along a ray formed between the point and the pinhole, projecting into space. Since all of the rays pass through the pinhole, the locus of the rays forms a code, thus the concept of cone-beam imaging. In this form of tomography, a 3-D model is generated by backprojecting the rays from a series of pinhole cameras into a voxel space in an additive manner. If the background of the images is removed prior to the backprojection, the intensity of each voxel indicated the probability of an object being at that position in physical space. An intensity threshold can then be applied to the voxel array, leaving only the voxels with high intensity that define objects in the real space. Although the pinhole camera is an idealized concept, a camera with a sufficiently small aperture is a close approximation, and the general approach is still valid.

Tomographic setups such as our cone-beam approach normally have simple matrix solutions. Unfortunately though, these matrices have millions of rows and columns, and are thus infeasible to invert quickly. Several methods have been devised to simplify the computation, including Feldkamp's [4] algorithm, which, although it is an inexact solution, provides good quality and can be efficiently computed. Feldkamp's algorithm consists of two major steps. The first step is a filtering step that reduces the background components of the images while enhancing the objects of interest. This can be efficiently achieved by means of a fast Fourier transform. The backprojection step is comparatively computationally expensive for voxel arrays nontrivial sizes due to the size of the voxel array and the number of rays that must be projected into that space.

In order to approach real-time performance, we developed a parallel implementation of Feldkamp's cone-beam algorithm to harness the computational capabilities of the Beowulf cluster using the MPICH [5, 6] implementation of the message passing interface (MPI) standard. For each voxel array we generate, each node captures an image from a camera, performs the filtering step, and then begins backprojecting. Since in Feldkamp's algorithm the final volume is the sum of the contributions from each camera, each node backprojects its discretely sampled image into a subset of the volume. These volumetric subsets are passed from node to node until all of the voxel array has been filled. At this point, the volume is a sum of the contributions from each of the nodes and the process is complete. Figure 9.9 shows a ray projection of a reconstructed 3-D volume in Argus. For reference, Fig. 9.10 shows a single-camera view of the same scene.

One of the challenges of streaming volumes, as opposed to stereo, is that the display hardware must be much more sophisticated. Figure 9.9 was constructed using SGI hardware at the National Center for Supercomputing Applications. The level and cost of this hardware (a multiple processor Onyx system with specialty 3-D projection hardware) is well beyond the typical capabilities of commercial viewers. Of course, the 3-D projection capacity of desktop displays is developing rapidly. In any case, it is difficult to visualize the utility of this data from projections like Fig. 9.9. The real value is understood when interactively rotating and exploring the reconstructed volume in a CAVE [7] virtual reality space or a similar facility.

The speed at which data can be transmitted out of the Argus is a major factor in the utility of the system. Although we are constrained by the network bandwidth between Argus and a remote site, some work can be done locally to help improve performance, both in terms of overall throughput and timeliness at the receiving end.

Fig. 9.9. Reconstructed view from a 3-D tomographic model of the dancer

Fig. 9.10. Raw camera view of a dancer in the Argus imaging system

The throughput of the system can be drastically improved if a compression algorithm is applied to the data. A simple compression algorithm such as run-length encoding (RLE) can typically reduce the size of a volumetric dataset by a factor of four. The stereo pair data was only reduced moderately with compression. Since speed is a major consideration for this system, the amount of compression performed needs to be balanced with the impact the compression algorithm will have on the computational resources of the system.

The timeliness of the system can be enhanced by generating only as many frames as can be sent over the network. A performance hit occurs when too many frames are generated by the system. For example, if the system is able to generate frames 1.5 times faster than they can be transmitted, it is possible that the network will have to wait on a frame to complete if it is overwriting the latest frame. Although this overlap is typically not too great, this can affect how the system performs.

9.4 Performance

As described in Sect. 9.3, we have implemented both stereo pair display and cone-beam tomography algorithms on Argus. The relative advantages of the two systems depend on the viewer display and network performance. For a network of users that have access to the Argus system over a high-bandwidth, low-latency network it often makes more sense to directly calculate views and to map individual views across the network to users. For users connected over unreliable networks with considerable latency, it may be more efficient to calculate, compress, and transmit a global world model. This qualitative view is based on experience while streaming data across a LAN, between North Carolina and Illinois, and between the United States and Japan.

The rate that stereo pair data can be streamed is determined by the network bandwidth and the size of the frames. Each stereo-pair frame is a combination of two 320×240 images, or 153,600 bytes. A network bandwidth of 4.6 Mb/s is necessary to transfer the entire data stream without dropping frames. A 10 Mb/s network is able to sustain the maximum data rates produced by this system. In typical situations, the stereo-pair reconstruction is slowed to transmit 6–8 stereo view pairs/s over the network. Each pair is a 153.6 kb data set. Sustained network bandwidth of just over 1 Mb/s is required for this application.

For relatively local demonstrations, real-time operation is attractive. Stereo views can be calculated at video rate, allowing real-time interactive stereo visualization of the Argus studio. At remote sites, in contrast, the latency of the network makes real-time interactivity difficult. While streaming to Japan for example, the latency was 6 s. This latency is due to packet switching rather than network transmission time and is determined by logical network distance rather than physical distance. Between North Carolina and Illinois, the experimental latency was about 2 s. Latency is typically well under a second across the local network.

Our cone beam implementation is able to reconstruct only 1.2 volumes per second. At this rate, the data rate required for transmission of the volumes is 40 Mb/s. Simple RLE compression reduces the size of the volumes by a factor of four, so ten megabits is enough to provide this level of performance. Internally, however, RLE is not used during the computation of the volumes, so the aggregate bandwidth required inside the cluster to perform the computation is 1.3 Gb/s. Our network can handle greater data rates than this. However, as the network load increases, the processors are forced to spend more time handling network traffic, and therefore, are able to do less computation.

In our reconstructions, features become discernible in the volumes when they are 3–4 pixels in size on the cameras. This does not compare favorably to other optical tomography experiments [8]. There are several reasons for this. First, the quality of cameras being used is quite low. The consistency in intensity response is poor between cameras, and therefore contributions from each camera are not weighted equally. In other experiments, a single camera was used and the object was rotated, so this was not an issue. This problem could be corrected by measuring the response of each camera to a known test object. Second, the registration is not perfect. This could be corrected by using more sophisticated test objects and antialiasing techniques to obtain subpixel registration. Thirdly, the lenses suffer from manufacturing inconsistencies, and have curvature of field. Furthermore, the lenses form a nonorthoscopic image that is not corrected for in our software. This causes some images to be quite blurry, and distortions to be introduced onto all of the images.

The advantage of transmitting the volume reconstruction is that remote users can interact with the 3-D space in real time on their own display facilities, rather than waiting for Argus to respond to requests for new views. In contrast with the network-limited performance of the stereo-view system, real-time volume display tends to be limited by the model generation rate. Each volume frame is a data cube of two million bytes. Transmission at the generation rate requires ten megabits per second. Currently, our software only handles 128×128 images, however this is not a fundamental limitation of the algorithm.

9.5 Conclusions

We have described demonstrations of two of the many possible topologies for streaming 3-D video from a distributed sensing and processing array to distributed users. We found advantages of source-side stereo projection included:

- Lower computational loads at both the data capture and display interfaces allow higher-quality images to be streamed for stereo-pair systems.
- Stereo projection requires fewer assumptions on scene interpretation and introduces fewer system artifacts.
- Stereo processing is much faster at the source end, allowing faster interactivity and real-time display for local users.

Advantages of tomographic reconstruction and volume transmission included:

- High latency inhibits interactivity with source-side stereo projection. This problem is resolved with display-side display projection.
- Aggregate bandwidth to multiple remote users can be reduced while allowing independent view selection for each user with volume transmission.
- While tomographic backprojection does not automatically reduce the data volume of the joint set of images taken by the camera array, compression of the reconstructed volume is more computationally straightforward than joint compression of a complete set of images. Thus, transmission of the reconstructed volume tends to be more efficient than transmission of the complete raw data set.

While the two approaches we have demonstrated here do not by any means represent a complete spectrum of possible schemes for streaming 3-D video from a sensor space, they do represent benchmarks for what can be done. We are most pleased by how well processing is integrated into the physical structure of the sensor space on this system.

To our knowledge, Argus is the first demonstration of a distributed supercomputer with embedded sensor resources. Many additional possible streaming topologies can be implemented on this system. In practice, we expect neither of the benchmarks we have described here will represent an ideal solution. Various approaches to joint real-time compression across the array and embedded processing on the network to maximize the data flow and the visual experience of users might be attempted. Since our experience has shown that the ideal approach for one network topology differs from the ideal for other topologies, we expect that adaptive algorithms will be particularly attractive for this application. Algorithms might particularly focus on the user topology and on user display resources. Since a single user is able to process the full sensor data set, the underlying problem should be viewed as a mapping between distributed capture and display systems, as in Fig. 9.5. In contrast with Fig. 9.5, however, the ideal system will include joint processing nodes that integrate data from multiple capture heads.

The problem of streaming 3-D video comes down to tradeoffs between the cost of sensing and the cost of computation. Our stereo-pair transmission approach emphasizes sensing over computation. The tomographic approach emphasizes computation. The ideal system will adaptively exploit available sensor and processing resources based on the immediate requirements of the user.

9.6 Acknowledgements

Argus is an Air Force Office of Scientific Research project supported by the program *Distributed Optoelectronic Processing for Multidimensional Digital Imaging*, grant number F49620-99-1-0229.

References

1. C. Reschke, T. Sterling, D. Ridge, D. Savarese, D. Becker, P. Merkey: 'A Design Study of Alternative Network Topologies for the Beowulf Parallel Workstation.' In: *Proc. High Performance and Distributed Computing 1996*
2. D. Ridge, D. Becker, P. Merkey, T. Sterling Becker, P. Merkey, 'Beowulf: Harnessing the Power of Parallelism in a Pile-of-PCs.' In: *Proc. IEEE Aerospace 1997*
3. R. Brady, J. Pixton, G. Baxter, P. Moran, C. Potter, B. Carragher, A. Belmont: 'CRUMBS: A Virtual Environment Tracking Tool for Biological Imaging.' In: *Proc. IEEE Symp. on Frontiers in Biomedical Visualization, Atlanta, GA, October 30 1995*, pp. 23–32
4. L.A. Feldkamp, L.C. Davis, J.W. Kress: J. Opt. Soc. Am. **1**(6), 612 (1984)
5. W. Gropp, E. Lusk, N. Doss, A. Skjellum: Parallel Comput. **22**(6), 789 (1996)
6. W. Gropp, E. Lusk: *User's Guide for MPICH, a Portable Implementation of MPI* (Mathematics and Computer Science Division, Argonne National Laboratory 1996)
7. T.A. DeFanti, C. Cruz-Neira, D.J. Sandin: IEEE Spectrum **30**, 28 (1993)
8. D.L. Marks, R.A. Stack, D.J. Brady, D. Munson, R.B. Brady: Science **284**, 2164 (1999)

10 Three-Dimensional Image Capturing and Representation for Multimedia Ambiance Communication

T. Ichikawa, S. Iwasawa, K. Yamada, T. Kanamaru, T. Naemura, K. Aizawa, S. Morishima, and T. Saito

10.1 Multimedia Ambiance Communication

Demand is growing for more realistic three-dimensional (3-D) image spaces that can enhance interpersonal communication. Existing 3-D image space systems are mainly based on computer graphics (CG); examples include Cybertown [1] and InterSpace [2]. Unfortunately, the image spaces created by existing CG techniques are unnatural and certainly not attractive. Now, studies are increasingly focused on ways of creating a sense of presence by making use of actual images through image-based rendering [3–11].

Nevertheless, a photo-realistic 3-D image space based on actual images has yet to be achieved, and attempts to create such spaces are still unsatisfactory for communication with a shared environment. Prior research has introduced several interesting approaches, but a truly effective photo-realistic 3-D image space has not yet been created for interpersonal communication within a shared environment. Our goal is to create a photo-realistic image space that looks natural and inviting while minimizing the computer power needed. This is accomplished by layering different types of images to create the 3-D image space. The laws of perspective and the characteristics of human visual perception are used to alter the actual images to make them more realistic [12]. Our approach is to realize what we call "multimedia ambiance communication" [13].

Multimedia ambiance communication can be made possible by creating an immersive environment consisting of a 3-D image space stereoscopically projected on a curved screen. The concept of a layered structure defined for painting, such as long-range, mid-range, and short-range views, can be applied to a 3-D image space. New techniques, such as two-plane expression, high-quality panorama image generation and setting representation for image processing, 3-D image representation and generation for photo-realistic 3-D image space, have been developed. Also, we propose a life-like avatar within the 3-D image space. To obtain the characteristics of the user's body, a human subject is scanned using a Cyberware whole-body scanner. The output from the scanner, a range image, is a good start for modeling the avatar's geometric shape. A generic human surface model is fitted to the range image. The obtained model is topologically equivalent even if our method is applied to another subject. If a generic model with motion definitions is employed, common motion rules can be applied to all models made from the generic model.

10.2 3-D Image Space with Three Layers

A 3-D image space of a location is established as a surrounding environment on a computer, and users from more than one location share this image space for communication and shared work. The concept of a layered structure, as defined for painting, using long-range, mid-range, and short-range views, can be applied to a 3-D image space that uses images.

For long-range objects like the sky, mountains, or forests, which appear to be at a distance from the user, a 3-D structure is not needed. A 2-D scene expression should be sufficient. However, mid-range objects like trees and monuments look slightly different, depending on which eye is used to view them. A pseudo-3-D scene expression or a stereoscopic scene expression that enables perception of whether objects are in front of or behind other objects, as well as occlusion, and that provides rough depth information should be sufficient. Finally, for short-range objects, a 3-D structure is considered necessary to provide depth information so that such objects can be viewed from any direction. In communication, however, the priority is to convey visual information about the people of the other party, such as facial expressions and gestures. And unlike for objects, the inability to view people from any direction is not a serious handicap.

Furthermore, images such as the ground or the surface of a body of water, which extend from long- and mid-range views up to short-range views, must appear to be continuous. The characteristics of human visual perception for different viewing ranges are summarized in Table 10.1.

10.3 Immersive Environment for Ambiance Communication

10.3.1 Configuration of a 3-D Image Space Display Unit

We developed a 3-D image-space display unit that uses three projectors and a curved screen. This display unit can ensure sufficient brightness at the space corresponding to the user location. Although the curved screen degrades the view angle compared with CAVE [14], it enables the camera capturing the user's image to be placed to the left, to the right, above, or below the screen. A rear-type projector is even more suitable as a display unit for a face-to-face communication environment. The main specifications of the unit are given in Table 10.2 and the exterior is shown in Fig. 10.1.

Table 10.1. 3-D layer-structured image space

	Condition	Types of images required
Long-range views (Landscapes)	It is possible to perceive that it is a distant landscape. It can be perceived as a continuous space.	2-D scene expression. Dynamic images of actual panoramic images with a wide field of vision angle.
Mid-range views (Objects and human avatars)	Objects appear of disappear when the location of a visual point/direction changes. Whether objects are in front of or behind other objects can be perceived.	Pseudo-3-D objects. Stereoscopic images. Sparse multi-view images.
Short-range view (Objects and human avatars)	Objects are visible from the perimeter. People do not have to be visible from all directions because the objective is communication.	3-D objects (based on CG). Dense multi-view images.

Table 10.2. Specification for 3-D image space display unit

Radius	320 cm
Horizontal view angle	About 90°
Horizontal pixels	1024 pixels/screen
Horizontal length	About 500 cm
Vertical view angle	About 24°
Vertical pixels	768 pixels
Vertical length	About 140 cm
Height of viewpoint	About 130 cm

10.3.2 Use of a Computer for Creating and Displaying a 3-D Image Space

A graphics computer (SGI Onyx2 Infinite Reality, 8 R10K CPUs) is used to display images in real time with tracing a viewers line of vision. It also performs high-level graphic and image processing, such as integrating images

Fig. 10.1. 3-D image space display unit

and CG. This has been combined with the curved display unit described in Sect. 10.3.1 to form a test bed for ambiance communication.

10.3.3 Video Camera System Design for Capturing the Images of the Ambient Environment

We have also developed a three-camera video system for capturing high-quality images to be used for the surrounding environment. The three-camera system features accurate frame synchronization between the three channels, and a battery for use outdoors. It can be used to capture actual moving panoramic images with the three cameras or stereoscopic images with two or three cameras. The center channel is used for easing occlusion finding, segmentation, and stereo matching. The adjustment ranges and direct reading accuracy of the pan heads and stage are described in Table 10.3. Figure 10.2 shows the cameras, pan heads, and stage of the system.

Table 10.3. Adjustment range of pan-head parameters and direct reading accuracy

	Range	Accuracy
Pan	90° to 45° (for left and right)	0.1°
	±30° (center)	0.1°
Tilt	±15°	0.1°
Roll	±5°	0.1°
Height	±5 mm	0.1 mm
Baseline	56 to 500 mm (from center)	1 mm

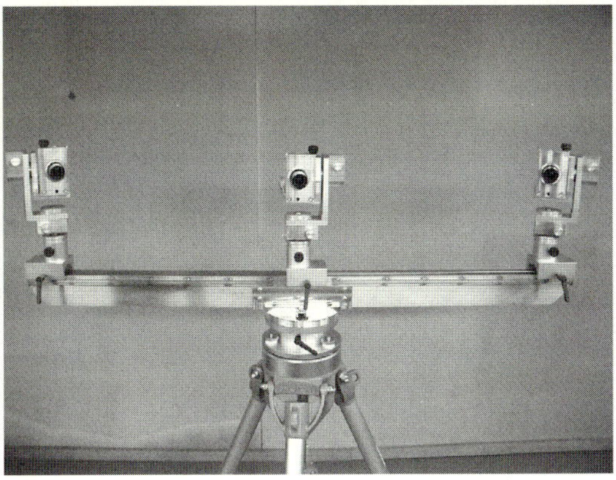

Fig. 10.2. Three-camera system

10.4 Expression for 3-D Image Space

10.4.1 Two-Plane Expression for Image Space Consisting of Long-Range View and the Ground or the Surface of a Body of Water

As an example of a two-plane expression, let us look at a case in which there is a sky and forest in a long-range view, and the ground extends to a point as shown in Fig. 10.3. In this case, a pseudo-3-D image expression can be composed of two virtual planes. The image of the long-range view and ground is divided into two parts. The ground part is projected onto the horizontal virtual plane perpendicular to the vertical virtual plane for the long-range view. This expression not only gives a more realistic setting of the ground, but also commands a more natural view when synthesized with other objects, and gives potentialities of deformations for a number of purposes. Figure 10.4 shows an example of the result of this expression method. The method is suitable for panoramic and static 3-D image spaces. As the boundary between the long-range view and the ground is not quite a straight line, and the boundary is the same as that for a long-range view that reaches a distance of nearly one hundred meters or so, the image appears natural.

The projection is done according to the following process. From the viewpoint O in Fig. 10.5, the long-range view image at the right end is perceived, so as to have the same vertical view angle θ_{v} as the camera's. The variable L is the distance between O and F, the center of the image. Let v be the height of the image, p the height of the ground/water part, and θ_{f} the angle between the image and the projection plane, all of which are known parameters. The ground/water part CE is projected onto CD, and B on CE is projected onto A. The length from the top of the image to B is j, and the length between C and A is y. Finally, θ is the angle between OF and OB, θ_d is the angle between EC and EO, and θ_e is the angle between BC and BO. The relation between y and j is described by the following equation:

$$
\begin{aligned}
y &= \frac{j - v + p}{\frac{1}{\tan\theta_f} + \frac{1}{\tan\theta_e}} \cdot \frac{1}{\sin\theta_f} \\
&= \frac{j - v + p}{\frac{1}{\tan\theta_f} + 1/\tan\left(\frac{\pi}{2} - \arctan\frac{j - v/2}{L}\right)} \cdot \frac{1}{\sin\theta_f} \\
&= \frac{j - v + p}{\frac{1}{\tan\theta_f} + 1/\tan\left(\frac{\pi}{2} - \arctan\frac{j - v/2}{\frac{v/2}{\tan(\theta_v/2)}}\right)} \cdot \frac{1}{\sin\theta_f}\,. \qquad (10.1)
\end{aligned}
$$

10.4.2 Setting Representation for Mid-range Views

The mid-range views cover a distance of over several meters and do not require perfect 3-D information. Instead, an image can be represented more simply using depth information by approximate representation on a plane. Planar images, for example, can be used like stage settings to approximate

Fig. 10.3. An example of an image consisting of a sky, forest and ground

Fig. 10.4. An example of the result of the expression method

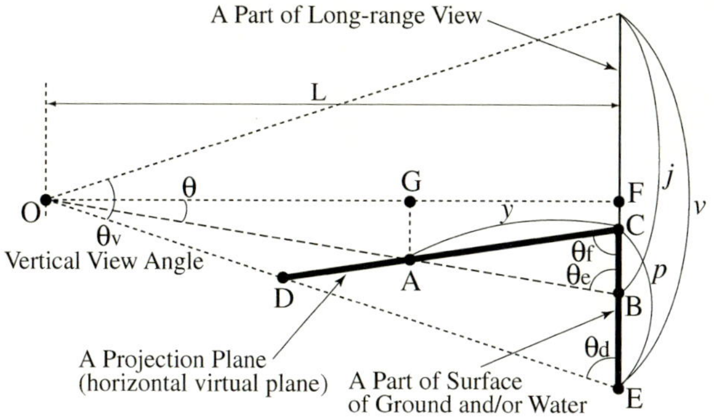

Fig. 10.5. The expression method on a horizontal virtual plane

objects and describe the 3-D structure. Although this method does not describe a perfect 3-D structure, it is effective as a simplified means of describing space in mid-range views. Figure 10.6 shows this concept of setting representation. We considered two methods of obtaining the parameters of plane for representing a setting. In one method we obtained the parameters using stereoscopic image sequences captured by the three-camera system. In the other method we obtained the parameters using a rough depth map measured by the laser scanner.

Setting Representation Using Stereoscopic Image Sequences. Mid-range objects are considered to be represented approximately by a plane. An equation of a plane is $k_1X + k_2Y + k_3Z = 1$. Using the brightness values of the left- and right-side images, k_1, k_2, and k_3 are determined by using a method of reiterative calculations that minimizes the estimated error in disparity. In this method we can obtain parameters regardless of whether mid-range objects move or not.

Setting Representation Using Rough Depth Map. The plane parameters cannot be obtained when the value of the tolerance diverges, so we used a rough depth map to obtain the plane with certainty. To construct a 3-D image space consisting of a surrounding environment, we need to measure a depth map over a wide angle, even if the depth map is not accurate. To do this, we used an LMS-Z210 3-D laser scanner as specified in Table 10.4. Also, the laser scanner can measure rough image data. The measurement took about 50 seconds for 555 points/80 degrees horizontally and 278 points/40 degrees vertically, and so the image is limited as a still image in this case.

Measurement of range data and capturing image data. We measured range data and texture data using the following system, as shown Fig. 10.7. It includes a 3-D laser scanner, laptop computer, and digital camera. The laptop computer controls the 3-D laser scanner. The center of the mirror in the 3-D

Fig. 10.6. Setting Representation

laser scanner and the center of the lens in the digital camera are at the same height, which means that the optical axis of the 3-D laser scanner is identical to that of the digital camera. The digital camera is mounted on a camera base and can turn in the polar direction at the center of the lens. The range and intensity data measured by the LMS-Z210 are shown as follows. The intensity is the degree of intensity of the reflection beam off objects, and has a value from 0 to 255. Figure 10.8 shows an example of range data and Fig. 10.9 shows an example of intensity data. Figure 10.10 shows an example of image data captured by the digital camera (SONY Cybershot F55K).

Estimation of plane parameters. A plane equation is represented as follows:

$$k_1X + k_2Y + k_3Z = 1 . \tag{10.2}$$

We obtain the plane parameters of the house in Fig. 10.11. These parameters, k1, k2, and k3, are obtained by minimizing the mean square error of the range data. In the case of the wall indicated in black in Fig. 10.11, the parameters are

$$k_1 = -0.0343,\ k_2 = 0.0244,\ k_3 = 0.0011 .$$

The intercepts of the X, Y, and Z axes are calculated as -29.15 ($Y = 0$ and $Z = 0$), 40.98 ($X = 0$ and $Z = 0$), and 909.09 ($X = 0$ and $Y = 0$) respectively, that is, the intercept of the Z axis is a large value, so the plane of the black wall is vertical to the ground. So we have the correct plane parameters.

10.5 Expression for a Lifelike Avatar

We come up with a lifelike avatar that can express facial expressions and reflects an external appearance of a person. We expect that it would be

Fig. 10.7. Range and image data measurement system

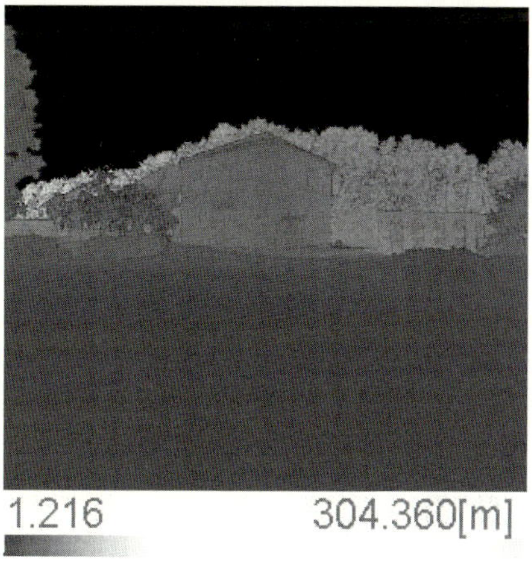

Fig. 10.8. Example of range data

Fig. 10.9. Example of intensity data

Fig. 10.10. Example of image data

Fig. 10.11. Example of a plane on a house

possible to use the template model of a person and fit it with the range data of the user to represent the person. This would be applied to any persons independently of their appearance. Figure 10.12 gives an overview of how this lifelike avatar would be created. As a 3-D-body template model, we employed wire-frames of a human body. Range and texture image data of the user's body are obtained by using the 3-D range finder, Cyberware Whole Body Scanner, and we used it as a geometric guide to fit the template model. Finally, we have the desired human model for a lifelike avatar.

10.6 Construction of 3-D Image Space

We constructed a 3-D image space using the techniques described in Sect. 10.4. Figure 10.13 shows an example of this space. The long-range view – the sky, forest, and ground – are represented by a two-plane model and are expressed with a photo-realistic panoramic image. The setting representation represents the mid-range object – the house. The plane parameters of the house are

Table 10.4. Technical specification of LMS-Z210 laser range finder

Mesurement range	2 to 350 m
Line-scan mechanism	Rotating polygonal mirror
Scan angle range	Maximum 80°
Scan speed	Up to 37.5 scan/second
Fram-scan mechanism	Rotating optical head
Scan angle range	Maximum 333°
Scan speed	Up to 15 deg/sec
Measurement accuracy	±2.5 cm
Measurement resolution	2.5 cm

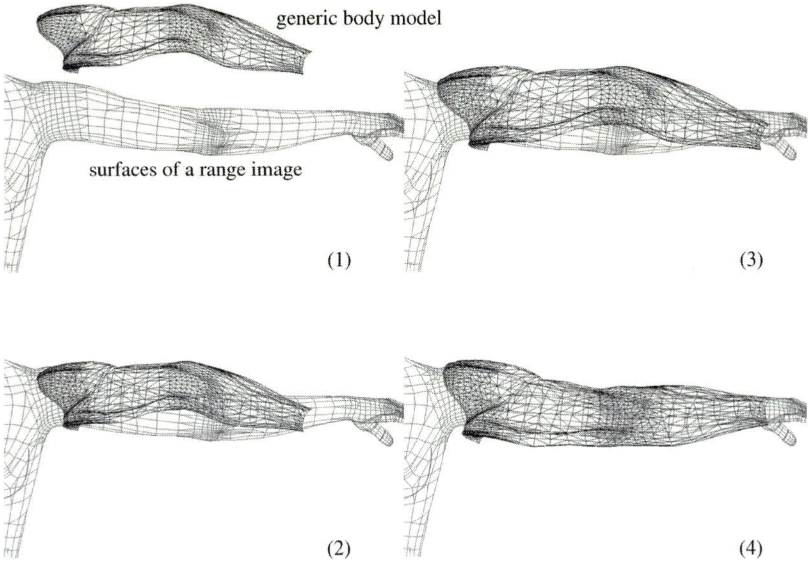

Fig. 10.12. Body-fitting process

obtained by the method mentioned above, and the image data of the house are matched to the planes of house. The house is expressed with a photo-realistic image. The result is that we feel as if we are a part of this 3-D image space and we can move around in it as if we were actually there. A mid-range object (the tree) and short-range objects (the flowers) are constructed by conventional CG techniques.

Fig. 10.13. Photo-realistic 3-D image space using the proposed method

10.7 Conclusions

Multimedia ambiance communication is a meaning of shared-space communication. Multimedia ambiance communication makes use of actual images, combined with the laws of perspective, as the visual characteristics of human beings, to establish a photo-realistic 3-D image space that provides a natural surrounding environment that users feel a part of. We have developed the concept of a layered structure, as used in painting, that consists of long-range, mid-range, and short-range views. For each view, including, for example, the ground, the water surface, and objects, scene expressions have been considered. We have constructed the test bed for multimedia ambiance communication, consisting of a high-speed graphic computer and a curved screen stereoscopically projected from the rear of the screen, and developed the three-camera system for capturing environment images. Also, we have proposed and developed a 3-D image space model, image processing for long-range view including of ground and for high quality panorama image, and a setting representation for mid-range view. Furthermore, a modeling method for a lifelike avatar within the 3-D image space is proposed.

Acknowledgements

We thank Museum Meiji-mura in Japan for their cooperation with measurements and capturing images.

References

1. Cybertown: http://www.cybertown.com/
2. InterSpace: http://www.is.ntts.com/
3. P.E. Debevec, C.J. Taylor, J. Malik: 'Modeling and Rendering Architecture from Photograph: A Hybrid Geometry- and Image-based Approach'. In: *SIGGRAPH 1996 Conf. Proc.*, August 4–9, 1996, New Orleans, Louisiana, ed. by Holly Rushmeier (ACM, New York 1996) pp. 11–20

4. S.M. Seitz, C.R. Dyer: 'View Morphing: Synthesizing 3-D Metamorphoses Using Image Transforms'. In: *SIGGRAPH 1996 Conf. Proc.*, August 4–9, 1996, New Orleans, Louisiana, ed. by Holly Rushmeier (ACM, New York 1996) pp. 21–30
5. M. Levoy, P. Hanrahan: 'Light Field Rendering'. In: *SIGGRAPH 1996 Conf. Proc*, August 4–9, 1996, New Orleans, Louisiana, ed. by Holly Rushmeier (ACM, New York 1996) pp. 31–42
6. S.J. Gortler, et al.: 'Lumigraph'. In: *SIGGRAPH 1996 Conf. Proc.*, August 4–9, 1996, New Orleans, Louisiana, ed. by Holly Rushmeier (ACM, New York 1996) pp. 43–54
7. T. Naemura, M. Kaneko, H. Harashima: 'Orthographic Approach to Representing 3-D Images and Interpolating Light Rays for 3-D Image Communication and Virtual Environment'. Signal Processing: Image Communication, **14** (1998) pp. 21–37
8. R. Szeliski, H. Shum: 'Creating Full View Panoramic Image Mosaic and Environment Maps'. In: *SIGGRAPH 1997 Conf. Proc.*, August 3–8, 1997, Los Angeles, California, ed. by Turner Whitted (ACM, New York 1997) pp. 251–258
9. P. Rademacher, G. Bishop: 'Multiple-Center-of-Projection Images'. In: *SIGGRAPH 1998 Conf. Proc.*, July 19–24, 1998, Orlando, Florida, ed. by Michael F. Cohen (ACM, New York 1998) pp. 199–206
10. Y. Yu, J. Mark: 'Recovering Photometric Properties of Architectural Scenes from Photographs'. In: *SIGGRAPH 1998 Conf. Proc.*, July 19–24, 1998, Orlando, Florida, ed. by Michael F. Cohen (ACM, New York 1998) pp. 207–217
11. J. Shade, S. Gortler, L. He, R. Szeliski: 'Layered Depth Images'. In: *SIGGRAPH 1998 Conf. Proc.*, July 19–24, 1998, Orlando, Florida, ed. by Michael F. Cohen (ACM, New York 1998) pp. 231–242
12. J. Gibson: *The Ecological Approach to Visual Perception* (Houghton Mifflin, Boston 1979)
13. T. Ichikawa, T. Yoshimura, K. Yamada, T. Kanamaru, H. Suga, S. Iwasawa, T. Naemura, K. Aizawa, S. Morsihima, T. Saito: 'Multimedia Ambiance Communication based on Actual Moving Pictures'. In: *Proc. 1999 IEEE Int. Conf. on Image Processing, Vol. III*, October 25–28, 1999, Kobe, Japan (IEEE, New York 1999) pp. 36–40
14. C. Cruz-Neira, T.A. DeFanti, D.J. Sandin: 'Surround-Screen Projection-based Virtual Reality: The Design and Implementation of CAVE'. In: *SIGGRAPH 1993 Conf. Proc.*, August 1–6, 1993, Anaheim, California, ed. by Stephen Spencer (ACM, New York 1993) pp. 132–142

11 Three-Dimensional Display and Recognition System Based on Multiple Perspectives Integral Imaging

O. Matoba, E. Tajahurece, and B. Javidi

11.1 Introduction

For implementation of ultimate television and robot vision, there is growing interest in developing three-dimensional (3-D) displays and 3-D information processing systems [1–13]. Integral photography is one of the popular methods of displaying 3-D objects [2–5]. In integral photography or integral imaging, a 3-D object is projected onto a two-dimensional (2-D) elemental image array that contains different perspectives by a lens array, and then 2-D elemetal images are recorded in a film or a charge coupled device (CCD) sensor. In the reconstruction of the 3-D object, the illumination of the developed film or a spatial light modulator that displays the signal detected by the CCD can reconstruct the 3-D object because the light is traced back in the same way as in the recording. The detection by CCD sensors can make it possible to send the data through communication networks to remote users. The remote users can reconstruct the 3-D object optically [2, 3] or numerically [4–6]. In this chapter, we present a method to display a 3-D object using integral imaging in a computer [4–6]. The numerical reconstruction can improve the image quality by enhancing contrast and reducing the noise. It is possible to reconstruct the images viewed from different angles by proper sampling of the electronically recorded multiple perspectives.

We also present an optical system to recognize 3-D objects by using perspectives of the 3-D objects based on integral imaging [8]. These perspectives have 3-D information of the object as described above. Thus it is possible to develop a 3-D object recognition system by correlation between the perspectives of a reference 3-D object and an input 3-D object. The correlation of two-dimensional perspectives is performed optically in a conventional joint transform correlator. We can develop a real-time optical recognition system because all the processes can be performed optically without converting into electrical signals. Although the microlens array limits the space–bandwidth product of the system, many objects of interest are far from the lens array and are small in terms of the space–bandwidth product. The proposed technique might be more appropriate for these objects. We present both numerical and experimental correlation results corresponding to the recognition of a 3-D object.

11.2 Three-Dimensional Display Based on Integral Imaging

Figure 11.1 shows a setup of the 3-D display based on integral imaging. A 3-D object is illuminated by incoherent light and then the diffracted light and the reflected light are projected by a microlens array into 2-D elemental images that show different perspectives. The microlens array used in the experiments has 60×60 lenses in an area of 25 mm square. The distance between the 3-D object and the microlens array is 50 mm. To form the image array of the object onto the CCD camera, a camera lens with a focal length of 50 mm is inserted between the CCD camera and the microlens array. The magnification factor of the elemental image array formed by the camera lens is adjusted such that the size of the elemental image array becomes almost the same as the size of a tip of the CCD camera.

A die with dimensions 15 mm $\times$ 15 mm $\times$ 15 mm is used as a 3-D object. Elemental image arrays of the die are shown in Fig. 11.2. Figure 11.2a shows 19×19 elemental images out of 60×60 elements. Figure 11.2b is an enlarged image array of Fig. 11.2a. Since the CCD camera has 2029×2044 pixels, each image element is recorded with about 34 square pixels. In the CCD, each pixel of the elemental image array is stored in a computer as 10 bits of data.

We describe a method to reconstruct the 3-D image in the computer using the recorded elemental image array. The numerical reconstruction of the 3-D images is realized by simply extracting points periodically from the measured elemental image array as shown in Fig. 11.3. Each elemental image contains 34×34 pixels. Therefore, we extract the pixel at every 34th pixel in the vertical and horizontal lines of the elemental image array. These extracted pixels reconstruct an image viewed from a particular angle. To reconstruct images viewed from other angles, the position of the pixels to be extracted are shifted stepwise.

A picture of the original object and three examples of reconstructed images are shown in Fig. 11.4. Small dots seen in the reconstructed images are

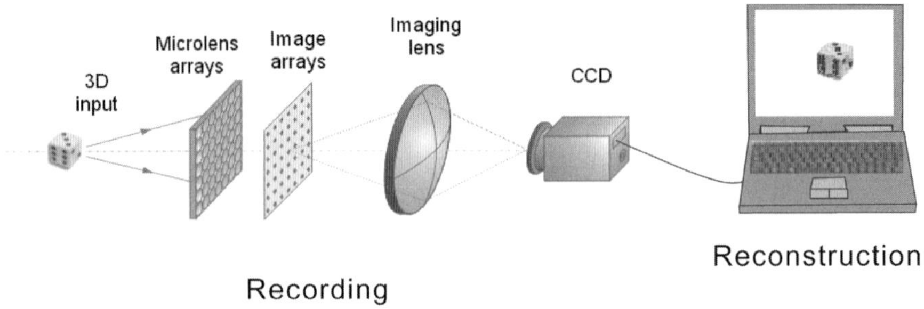

Fig. 11.1. Setup of a computer-based 3-D display based on integral imaging

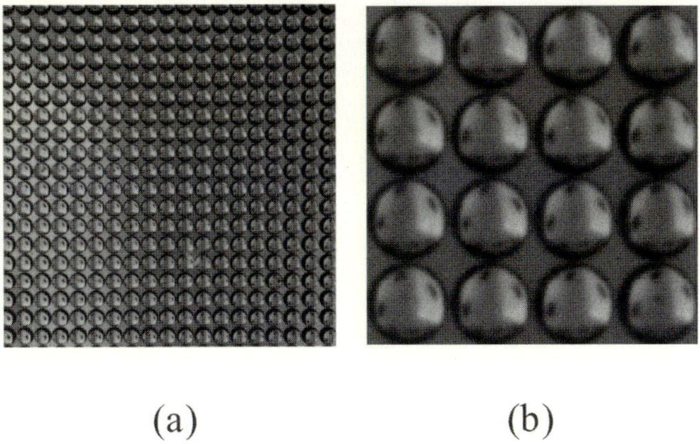

Fig. 11.2. (a) 2-D perspectives of the die; (b) enlarged images of (a)

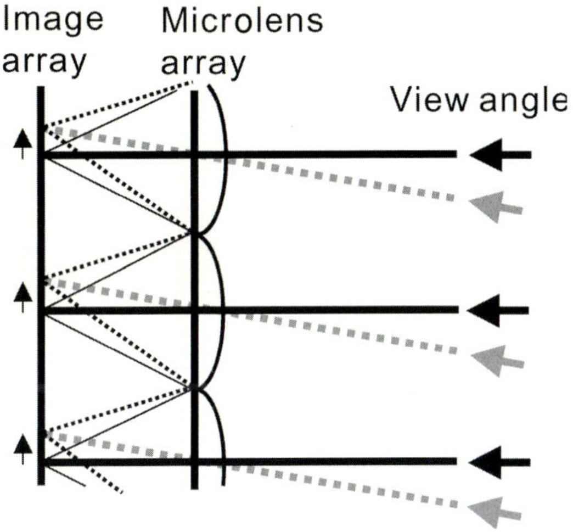

Fig. 11.3. Position of points to be focused by different view angles

the result of defects and dead lenses in the microlens array. Appropriate image processing will improve their quality. The resolution of the reconstructed image is determined by the resolution of the CCD camera and the number of lenses in the microlens array. The number of pixels of the reconstructed image is the same as the number of lenses in the microlens array. Therefore, the reconstructed images shown in Figs. 11.4 contain 60×60 pixels. Although only three reconstructed images are shown in this paper, a sequence of images can be reconstructed by changing the view angle stepwise.

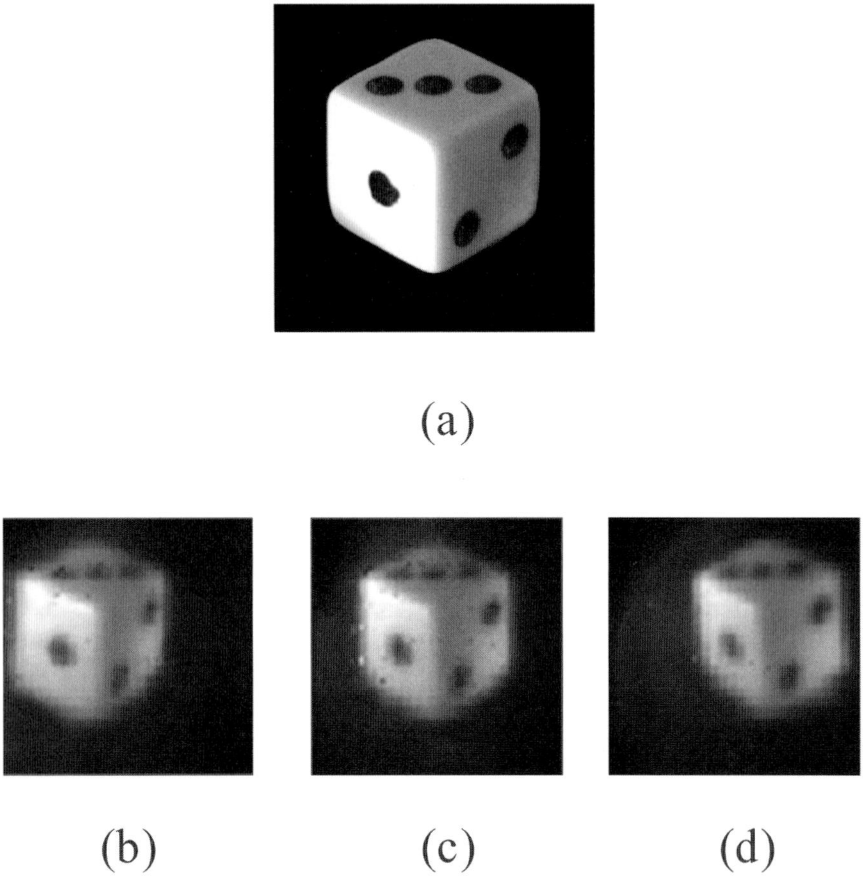

Fig. 11.4. (**a**) Die used as a 3-D object; (**b**)–(**d**) reconstructed images viewed from three different angles

11.3 Three-Dimensional Object Recognition System Based on Integral Imaging

11.3.1 System

Using perspectives of the 3-D object, we can develop the 3-D object recognition system by correlation between perspectives of a reference 3-D object and an input 3-D object. Figure 11.5 shows a schematic diagram of the 3-D object recognition system. The system consists of two subsystems; one is a projection system using a lens array, and the other is a joint transform correlator. First, the input and the reference 3-D objects are projected onto arrays of 2-D elemental images by respective lens arrays under incoherent illumination. In analogy with integral photography recording methods, each elemental image contains a different 2-D perspective of the 3-D object. This means that the depth information of the 3-D object is distributed among the 2-D elemental images. In the second stage, both arrays of elemental images are imaged onto an optically addressed spatial light modulator (SLM) to convert them into coherent images. This conversion allows us to use incoherent light in the recording process to avoid coherent noise. Finally, a joint transform correlator is employed to perform the correlation between all the input and the reference elemental images simultaneously. Using a Fourier-transforming lens, the joint power spectrum is recorded at another optically addressed SLM. After taking another Fourier transformation, the correlation between the input and the reference images is obtained at a CCD.

We explain the operation of the 3-D object recognition system. Let us consider $S_{ij}(x, y)$ and $R_{ij}(x, y)$ to be the (i, j) elemental images of an input and a reference 3-D object, respectively. The spatial coordinates are defined in Fig. 11.5. At the input plane, the input data to be used in the joint transform correlator are written as

$$I(x,y) = \sum_{k=-N_x}^{N_x} \sum_{l=-N_y}^{N_y} S_{kl}(x,y) \oplus \delta(x - k\Delta_x - kM\Delta_x, y - l\Delta_y - lM\Delta_y) + \sum_{k=-N_x}^{N_x} \sum_{l=-N_y}^{N_y} R_{mn}(x,y) \oplus \delta(x - k\Delta_x - kM\Delta_x + \alpha, y - l\Delta_y - lM\Delta_y + \beta) \;, \quad (11.1)$$

where $\oplus$ denotes convolution, $\delta(\bullet)$ is a delta function, M is the magnification of the imaging system in the lens arrays, and Δ_x and Δ_y are the periods of the lens arrays along the x and y axes, respectively. In (11.1) α and β are the distances between the object and the reference images along the x

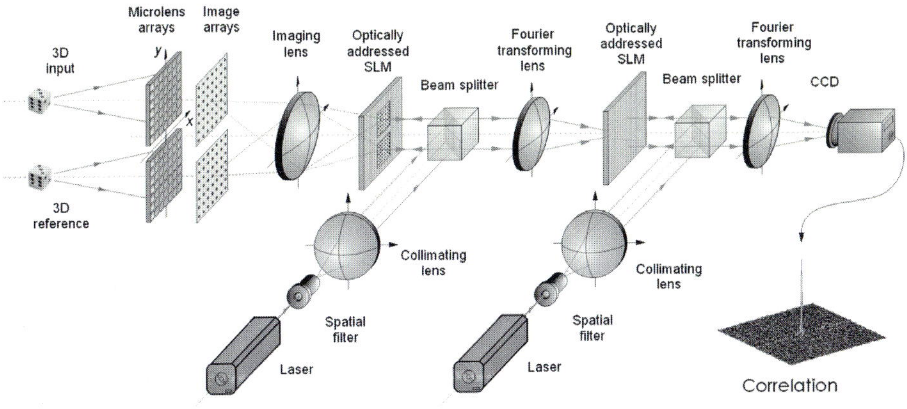

Fig. 11.5. Schematic diagram of a real-time 3-D object recognition system

and the y axes, respectively. The number of microlenses along the x and y directions is given by $2\ N_x+1$ and $2\ N_y+1$, respectively. The spatial shifts, $M\ \Delta_x$ and $M\ \Delta_y$, are introduced in (11.1) to take into account the displacement of each elemental image with respect to the optical axis of the corresponding microlens. After the incoherent-to-coherent conversion, the input data described by (11.1) are used as the input in the joint transform correlator architecture. The joint power spectrum, $\tilde{J}_{\mathrm{c}}(\nu,\eta)$, is given by

$$\tilde{J}_{\mathrm{c}}(\nu,\eta)=\left|\tilde{I}(\nu,\eta)\right|^2\ , \tag{11.2}$$

where

$$\begin{aligned}\tilde{I}(\nu,\eta)&=\iint I(x,y)\exp\left\{-\mathrm{j}\frac{2\pi}{\lambda f}(x\nu+y\eta)\right\}\mathrm{d}x\,\mathrm{d}y\\&=\sum_{k=-N_x}^{N_x}\sum_{l=-N_y}^{N_y}\tilde{S}_{kl}(\nu,\eta)\exp\left[-\mathrm{j}\frac{2\pi}{\lambda f}\{k(1+M)\Delta_x\nu\right.\\&\quad\left.+l(1+M)\Delta_y\eta\}\right]+\sum_{m=-N_x}^{N_x}\sum_{n=-N_y}^{N_y}\tilde{R}_{mn}(\nu,\eta)\\&\quad\exp\left\{-\mathrm{j}\frac{2\pi}{\lambda f}\left[\{m(1+M)\Delta_x-\alpha\}\nu+\{n(1+M)\Delta_y-\beta\}\eta\right]\right\}\ .\end{aligned} \tag{11.3}$$

In (11.3), λ is the wavelength of the laser source, f is the focal length of the Fourier-transforming lens, and $\tilde{S}_{kl}(\nu,\eta)$ and $\tilde{R}_{mn}(\nu,\eta)$ are the Fourier transforms of $S_{kl}(x,y)$ and $R_{mn}(x,y)$, respectively. Substituting (11.3) in (11.2), we take into account only the following term, which gives the correlation:

$$\begin{aligned}\tilde{J}_{\mathrm{c}}(v,\eta)&=\sum_{k=-N_x}^{N_x}\sum_{l=-N_y}^{N_y}\sum_{m=-N_x}^{N_x}\sum_{n=-N_y}^{N_y}\tilde{S}_{kl}(v,\eta)\tilde{R}^*_{mn}(v,\eta)\\&\quad\times\exp\left\{-\mathrm{j}\frac{2\pi}{\lambda f}[\{(k-m)(1+M)\Delta_x+\alpha\}v\right.\\&\quad\left.+\{(1-n)(1+M)\Delta_y+\beta\}\eta]\right\}\ ,\end{aligned} \tag{11.4}$$

where the asterisk denotes the complex conjugate. After taking the Fourier-transform of $\tilde{J}_{\mathrm{c}}(v,\eta)$, we obtain the following correlation results;

$$\begin{aligned}\tilde{J}_{\mathrm{c}}(x,y)&=\sum_{k=-N_x}^{N_x}\sum_{l=-N_y}^{N_y}\sum_{m=-N_x}^{N_x}\sum_{n=-N_y}^{N_y}S_{kl}(x,y)\\&\quad\otimes R_{mn}(x-(k-m)(1+M)\Delta_x-\alpha,\\&\quad y-(l-n)(1-M)\Delta_y-\beta)\ ,\end{aligned} \tag{11.5}$$

where $\otimes$ denotes correlation. Equation (11.5) shows that all the crosscorrelations between the elemental images of the input and the reference 3-D objects are superimposed according to the spatial displacement of the corresponding elemental images. When the input is the same as the reference, the autocorrelation between the corresponding perspectives, i.e. when $k = m$ and $l = n$, is superimposed at the same position (α, β) on the output plane. If the input object shifts in the x–y plane, the elemental images also shift according to the shift of the input. From (11.5), the shift of the elemental images causes the displacement of the autocorrelation peak. Thus, the proposed system can detect the spatial shift of the object along the x and the y axes. Note that moving the 3-D object along the optical axis causes the spatial shift of the elemental images along the radial direction. It would be necessary to include other techniques to calculate the correlation in this case.

The proposed system can detect a rotated 3-D object. Let us consider that an input object is a reference object rotated slightly on the y axis. Figure 11.6 shows the projection of the input into elemental images. Each lens sees the input 3-D object from a different angle. Figure 11.6a describes the reference recording process while Fig. 11.6b corresponds to that of the rotated reference. For simplicity, only a 2-D input image and 1D elemental images are represented in Fig. 11.6. The bottom lens faces a surface of the input object in Fig. 11.6a. If the 3-D object is rotated through an angle $\theta = (\Delta/a)$, as shown in Fig. 11.6b, the adjacent lens faces the same surface

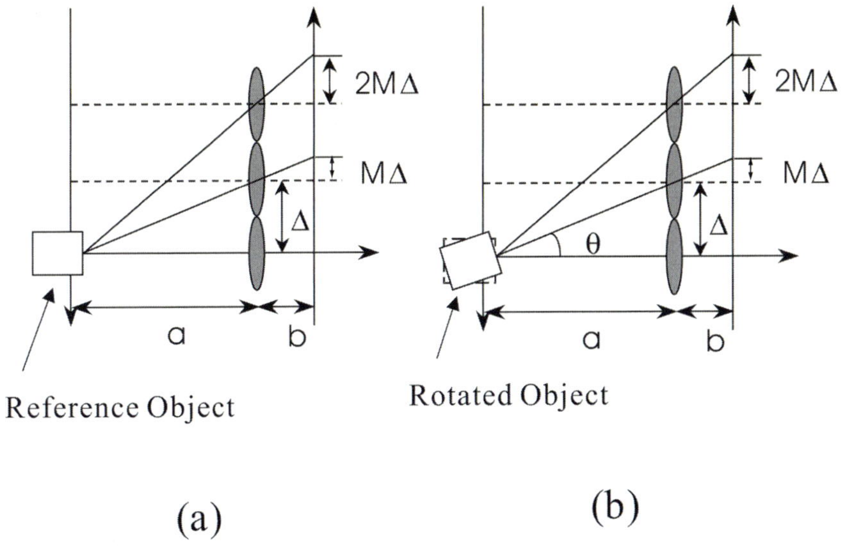

Fig. 11.6. Projection of the 3-D input into 2-D perspectives: (**a**) without and (**b**) with rotation

of the input object. Thus, the rotation of the 3-D object results only in the spatial shift of the elemental images.

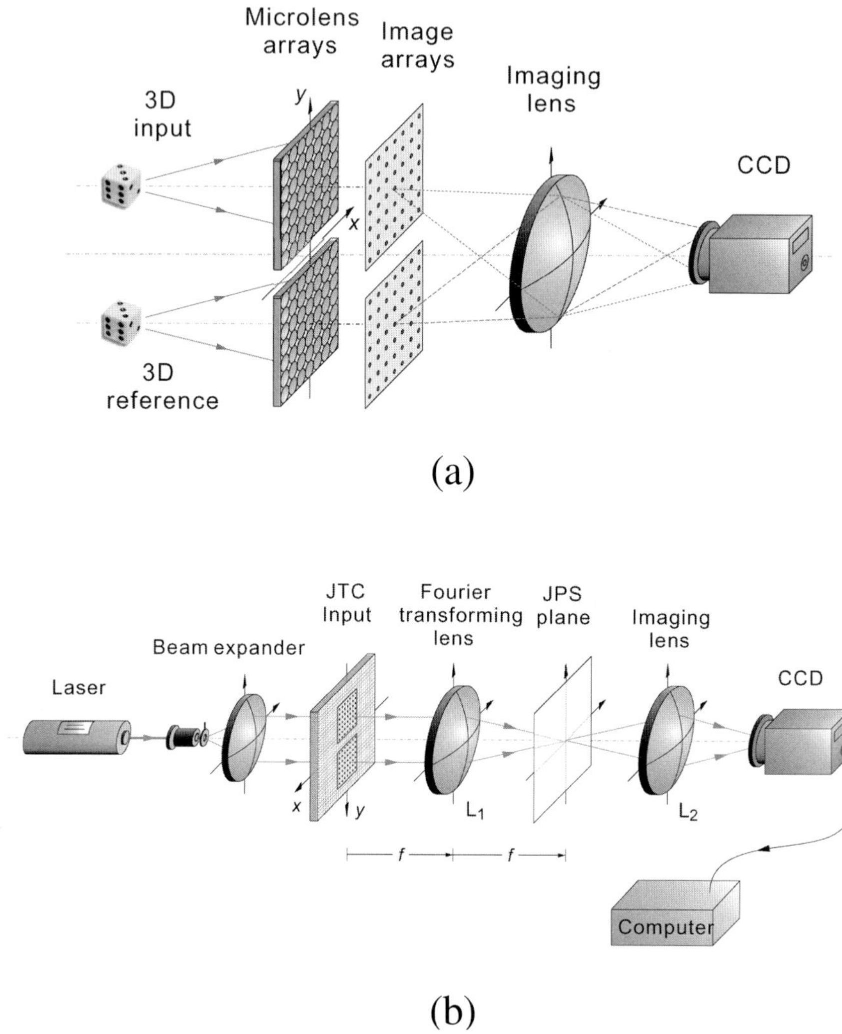

Fig. 11.7. Experimental setup: (**a**) recording of 2-D perspectives and (**b**) a joint transform correlator; JTC and JTS denote the joint transform correlator and the joint power spectrum, respectively

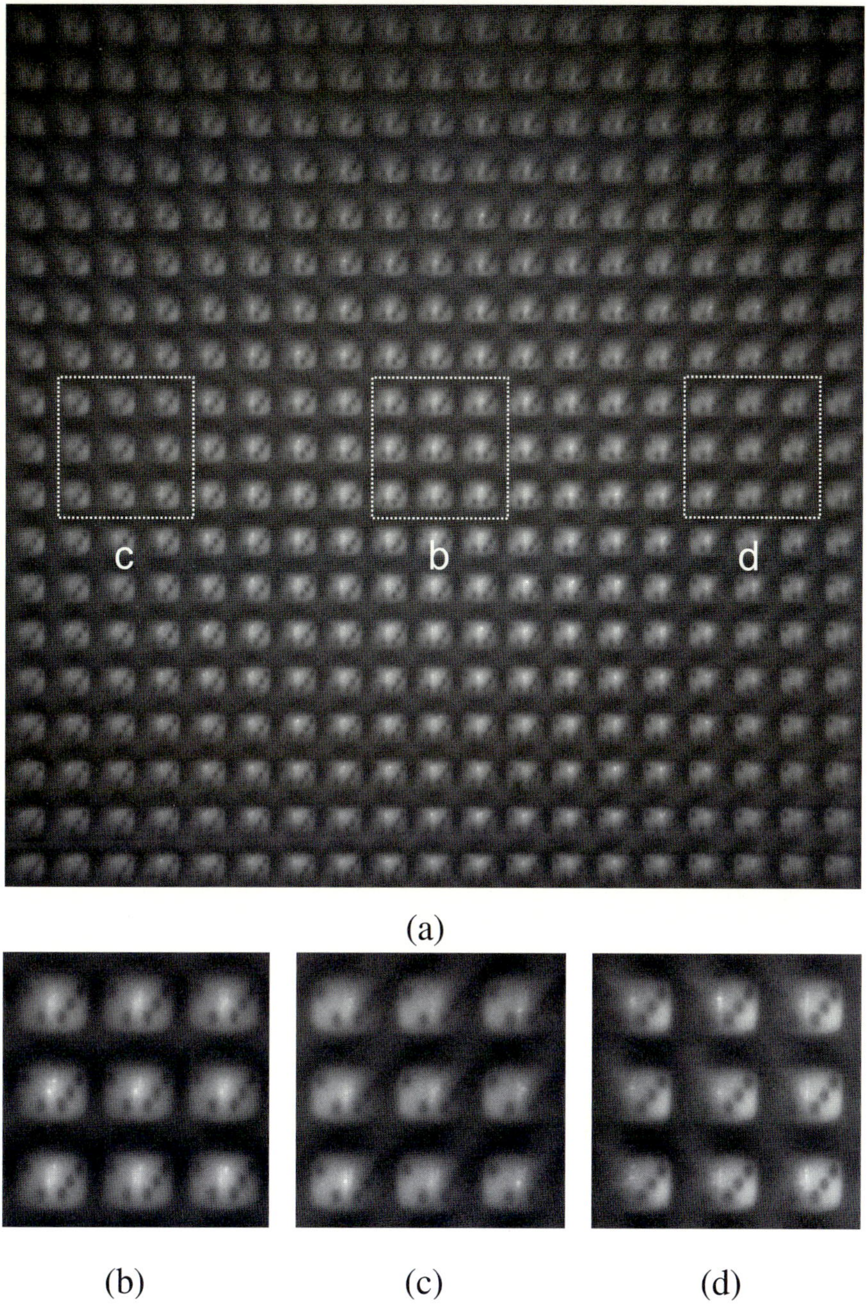

Fig. 11.8. 2-D perspectives of the reference die: (**a**) central 19×19 images of the die; (**b**), (**c**), and (**d**) enlarged images of central, left, and right regions of (**a**), respectively

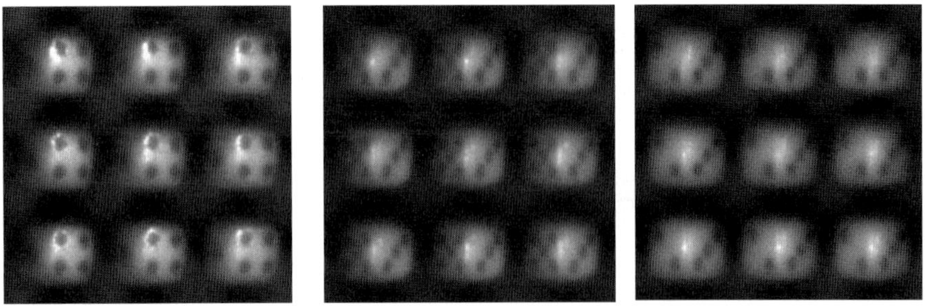

Fig. 11.9. 2-D perspectives of (**a**) input object with completely different faces of the die, (**b**) input object with a similar face of the die, and (**c**) rotated reference with an angle of 0.26°

11.3.2 Numerical and Experimental Results

In Fig. 11.7 we show the experimental setup to demonstrate the proposed system. Up to now there has been no commercially available optically addressed SLM that has a number of pixels large enough to record the necessary number of elemental images of the input and the reference 3-D object. Thus, in this chapter, to implement the system in Fig. 11.5, we use a photographic film instead of an optically addressed SLM to record the elemental images.

Figure 11.7a shows the recording system of the elemental images. A xenon lamp illuminates the input and the reference 3-D objects. Each 3-D object is imaged on different perspectives according to the different positions of the microlenses in the array and then is recorded by a megapixel CCD camera. The lens array consists of 38×38 lenses with a period of 250 μm and a focal length of 1.3 mm. After digital negative conversion of the recorded images, we take pictures of the images in negative films. This process creates a positive image in the film. After development, the film is placed in the input plane of the joint transform correlator architecture as shown in Fig. 11.7b. A He–Ne laser beam is used as a coherent source to illuminate the input, and the reconstructed elemental images are Fourier-transformed by lens L1. The focal length of L1 is 380 mm. Lens L2 reduces the size of the joint power spectrum that is recorded by another megapixel CCD. We take a second Fourier transform of the joint power spectrum in a computer and obtain the correlation of the object and the reference. Note that we can also take the second Fourier transform using a simple optical system to develop a real-time recognition system.

Figure 11.8a shows a partial set of elemental images, 19×19 images, of the reference 3-D object. In these experiments, two cubic dice with dimensions 4.6 mm $\times$ 4.6 mm $\times$ 4.6 mm are used as input and reference 3-D objects. Figures 11.8b–d show enlarged images of central, left, and right regions of the reference elemental images array. Each elemental image records a different

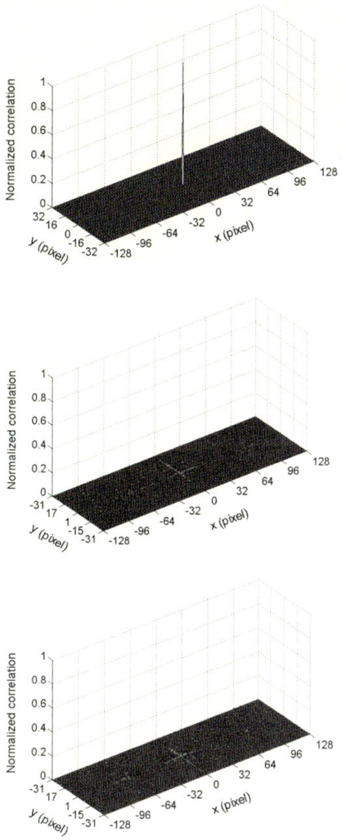

(a)

(b)

(c)

Fig. 11.10. Numerical results of (**a**) autocorrelation of the reference die, (**b**) crosscorrelation between the reference die and a die with completely different faces of the die, and (**c**) crosscorrelation between the reference and a die with a similar face of the die

perspective according to its position in the array of lenses. Figure 11.9a, b are the elemental images obtained by using a die with completely different faces and a die with similar faces to the reference, respectively. Figure 11.9c is a rotated reference object with an angle of 0.26°. The die is rotated on the y axis.

We analyze the system performance of the proposed system with a digital implementation. After obtaining the elemental images with the system shown in Fig. 11.7a, the correlation based on the joint transform correlator is calculated in a computer. Figure 11.10a shows the normalized autocorrelation of the reference object. In this figure, only a window centered with the autocorrelation peak is displayed to extract the correlation result. There is a very sharp autocorrelation peak as a result of the addition of all the autocorrelations of

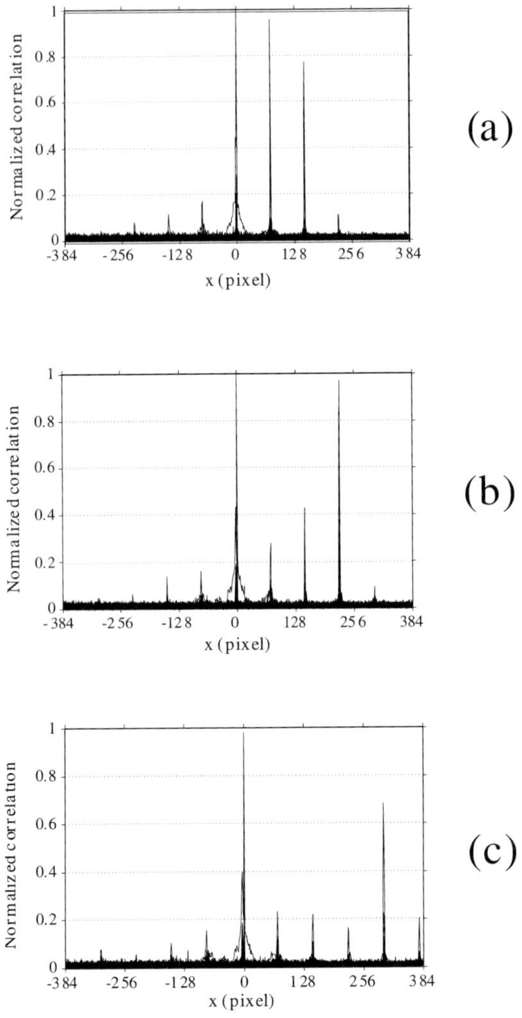

Fig. 11.11. Crosscorrelations between the reference and the rotated references with angles of (**a**) 0.26°, (**b**) 0.52°, and (**c**) 0.78°

the elemental images in the same position. In the correlation calculation, the joint power spectrum is binarized to enhance the correlation results [14, 15]. Figures 11.10b, c are crosscorrelations between the reference die and a die with completely different faces and between the reference die and a die with a similar face, repectively. But these peaks are 19 and 14.6 times smaller than the autocorrelation peak and thus we can recognize the original reference 3-D object by thresholding. The results in Figs. 11.10b, c are normalized with

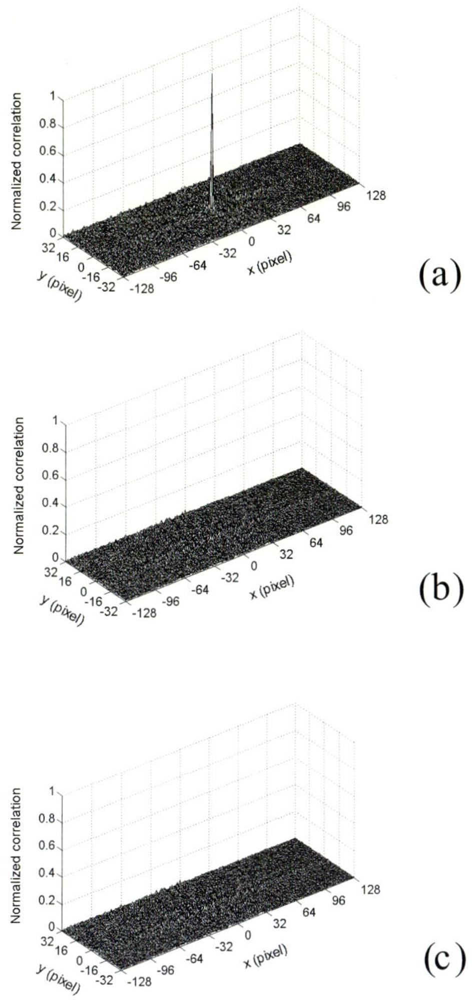

Fig. 11.12. Experimental results: (**a**) autocorrelation of the reference; (**b**) crosscorrelations between completely difference faces of the reference die; and (**c**) crosscorrelations between similar faces of the reference die

the value of the autocorrelation peak. Note that even when we use a very similar object, with one of the faces identical to one of the reference, the crosscorrelation peak is smaller than the autocorrelation peak. This result shows that the proposed system can discriminate the difference between the reference and similar inputs.

It is possible to show that the proposed optical system is also able to recognize slightly rotated objects. Figures 11.11a,b are crosscorrelations be-

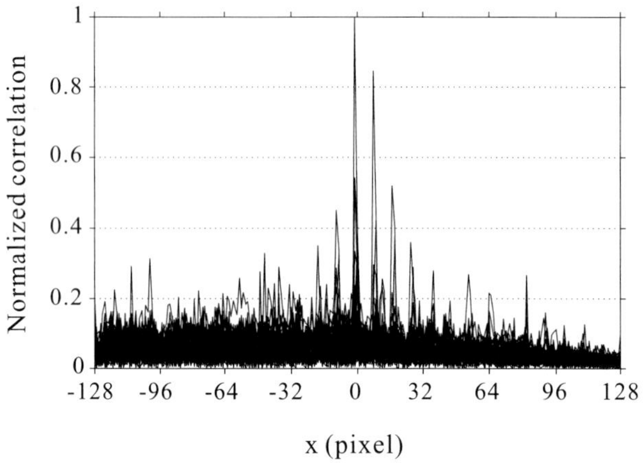

Fig. 11.13. Experimental result of crosscorrelation between the reference and the rotated reference with an angle of 0.26°

tween the reference and the rotated references with angles of 0.26°, 0.52°, and 0.78°, respectively. The peak at the center ($x = 0$) is the autocorrelation of the die shape. We can see that the second highest peak is shifted as the rotation angle is increased. The position of the second highest peak gives us information about the rotation angle of the reference.

We demonstrate the previous results by performing the correlation experiments optically with the system in Fig. 11.7b. Figure 11.12a shows the autocorrelation of the reference. We can see a sharp peak. In the crosscorrelations between Fig. 11.8a and Fig. 11.9a, and between Fig. 11.8a and Fig. 11.9b, there are peaks at the same position as the autocorrelation peak, as shown in Fig. 11.12b, c, respectively. But these peaks are 11.5 and 10 times smaller than the autocorrelation peak. These peaks are due to the correlation of the die shape as described in the numerical experiments. From these results we conclude that the system can recognize the 3-D object by thresholding.

Figure 11.13 shows the crosscorrelation between Fig. 11.8a and Fig. 11.9c that is optically obtained. There are many peaks, as shown in Fig. 11.11. The second highest peak is shifted 9 pixels from the autocorrelation peak. The position of the peak gives information about the rotation angle of the reference.

11.4 Conclusions

We have presented a computer-based 3-D display and a 3-D object recognition system based on intergral imaging. In the 3-D display system, a 3-D object is projected into an array of 2-D perspectives and then is recorded by a CCD sensor. By properly sampling the perspectives in the computer, we can reconstruct the images viewed from different angles. The computer reconstruction can improve the image quality by image processing. The experimental results showed that the 3-D die was successfully reconstructed. We have also presented a 3-D object recognition system that uses perspectives of the 3-D objects. In the recognition system, a 3-D object recognition is performed by correlation between all the perspectives of a reference 3-D object and an input 3-D object in a conventional optical 2-D correlator. The experimental results showed that the system was able to recognize 3-D objects even when we use an input object very similar to the reference. We also showed that the the system can recognize a slightly rotated 3-D object. When an optically addressed spatial light modulator is available, the whole process can be implemented in real time.

References

1. B. Javidi: *Smart Imaging Systems* (SPIE Press, Washington 2001)
2. F. Okano, H. Hoshino, J. Arai, I. Yuyama: Appl. Opt. **36**, 1598 (1997)
3. F. Okano, J. Arai, H. Hoshino, I. Yuyama: Opt. Eng. **38**, 1072 (1999)
4. H. Arimoto, B. Javidi: Opt. Lett. **22**, 157 (2001)
5. H. Arimoto, B. Javidi: 'Integral Three-dimensional Imaging with Digital Image'. In: Proc. SPIE **CR 76**, eds. by B. Javidi and F. Okano, pp. 117–130 (2001)
6. H. Arimoto, B. Javidi: 'Integral Three-Dimensional Imaging with Computer-Based Algorithms.' In: Proc. Annual Meeting IEEE Lasers Electro-Optical Society (IEEE LEOS) 2000, Vol. 2, The Westin Rio Mar Beach, Nov. 13–16, 2000; Conference chair A. Owyoung; Rio Grande, Puerto Rico, Nov. 13, 2000; pp. 752–752
7. T. Naemura, T. Yoshida, H. Harashima: Opt. Express **8**, 255 (2001)
8. O. Matoba, E. Tajahuerce, B. Javidi: Appl. Opt. **40**, 3318 (2001)
9. A. VanderLught: IEEE Trans. Inf. Theory **IT-10**, 139 (1964)
10. C.S. Weaver, J.W. Goodman: Appl. Opt. **5**, 1248 (1966)
11. J. Rosen: Opt. Lett. **22**, 964 (1997)
12. B. Javidi, E. Tajahuerce: Opt. Lett. **25**, 610 (2000)
13. J.J. Esteve-Taboada, D. Mas, J. García: Appl. Opt. **38**, 4760 (1999)
14. B. Javidi: Appl. Opt. **28**, 2358 (1989)
15. B. Javidi: 'Nonlinear Joint Transform Correlators'. In: *Real-time Optical Information Processing*, ed. by B. Javidi, J.L. Horner (Academic Press, San Diego 1994) pp. 115–183

12 Imaging Architectures with Volume Holographic Elements

G. Barbastathis

12.1 Introduction

Imaging remains one of the primary applications of optical science. An imaging system receives emitted or scattered radiation from the environment, and transforms it in order to recover the structure of an object of interest within the scene. Therefore, imaging may be defined as the solution to an inverse problem of light propagation. In this paper we review optical imaging with emphasis on the acquisition of scene information in three or more dimensions (including, for example, color), and present several imaging architectures where volume holograms are used as imaging elements.

Until the development of CCD cameras in the early to mid-1970s, the almost universal purpose of imaging systems had been the production of geometrically exact projections of the three-dimensional (3-D) world onto 2-D sensor arrays, which were permanent (such as photographic film) or instantaneous (such as an observer's retina). Remarkable exceptions such as computed tomography and radio astronomy, known earlier, had limited use or no use in the domain of optical frequencies. However, the gradual increase in the availability of digital cameras, ample digital computing power to process the images, and digital networks to distribute them produced a revolutionary shift in optical imaging design. The necessity to produce "images" on the detector plane that are physically analogous to the imaged objects became secondary, as long as the detector captured sufficient information to allow the recovery of the objects after appropriate processing operations. This paradigm shift had two apparent additional benefits. Firstly, it simplified the design of optical elements, since several geometric imaging imperfections (e.g. defocus) could be corrected digitally, to some extent. Secondly, and most importantly, digital processing allowed the user to recover *more* object information from the sensor data. For example, by processing several 2-D intensity patterns one may recover surface topography data about opaque objects ($2 + 1/2$-D images) or volumetric structure data (3-D images) and even spectral data in addition (4-D images) about semi-transparent objects. Examples of this principle in action are triangulation-based imaging systems, the confocal microscope, and coherence imaging systems.

The general availability of digital processing is undoubtedly the principal reason for the increased power of imaging systems. It would be desirable to have optical elements available with equally general behavior to maximize design flexibility. Lenses, for example, are rather limited to low-power polynomial surface shapes due to manufacturing constraints present even in modern techniques such as injection molding and servo-controlled grinders. Diffractive optical elements achieve a much broader range of responses, but the transformations they can effect are still limited to two dimensions. The most general optical elements available are volume holograms, which can be thought of as self-aligned 3-D stacks of diffractive elements operating on the incident field coherently as it propagates through. The purpose of this paper and some other recent work [1, 2] is to begin the exploration of volume holographic elements as enabling new functionalities and optimization schemes in imaging systems.

Volume holography was introduced in a seminal paper by van Heerden [3] as an extension of the well-known results about Bragg diffraction to the domain of optical wavelengths. Since then, volume holograms have been used in data storage [3–6], optical interconnects and artificial neural networks [7, 8], and communications [9–13]. To date commercial applications of volume holograms are for spectral filtering [14] and 3-D storage devices. While volume holograms can replicate the function of imaging system components, such as lenses, beam splitters, spatial or spectral filters, holograms do not outperform conventional components for these functions. Volume holography as a tool is extremely attractive in emerging digital imaging systems, however, because volume holograms have more design degrees of freedom per unit system aperture than any other optical component. Design complexity allows volume holograms to extract more sophisticated features from fields, enabling sensor design to target features for object reconstruction.

In this paper we overview imaging systems from the perspective of information transformations (Sect. 12.2), discuss the advantages of using holographic elements in high-performance imaging systems (Sect. 12.3), and present the basic volume diffraction theory (Sect. 12.4) upon which all volume holographic applications are based. The 90° holographic imaging geometry and its properties are presented in Sect. 12.5.

12.2 Field- and Intensity-Transforming Imaging Systems

As mentioned already, imaging is an inversion operation which recovers the structure of a light source (i.e. the object) from the radiation received in an observation space. The source may be primary or secondary, depending on the application. The transformations needed to recover the object can be effected by a variety of instruments. We classify these as *field-transforming elements*, which include traditional optical elements, such as prisms, lenses, and holograms, as well as *intensity-transforming elements*, which are primarily implemented by digital processing. The latter possibility became widely available through the proliferation of reasonably priced high-performance computers and digital cameras (CCD since the 1970s and CMOS more recently).

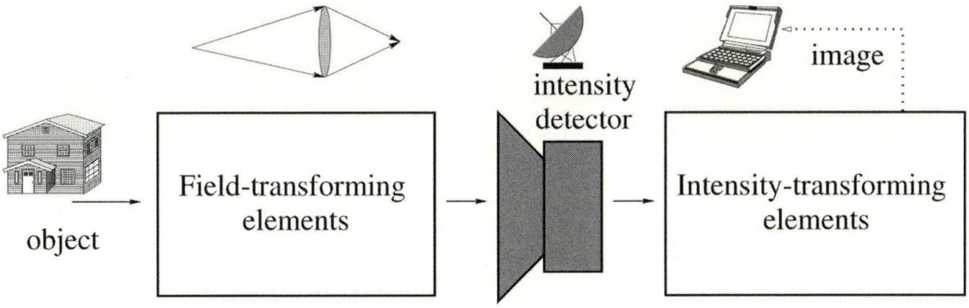

Fig. 12.1. General classification of elements composing an imaging system

A generic imaging system is shown in Figure 12.1. Field-transforming elements are the first to receive the radiation from the object. The field produced by this stage is converted to an electrical signal by the intensity detector. The electrical signal is then fed to the intensity-transforming stages, which produce the final image.

As the respective names suggest, there is a fundamental difference in the nature of the transformations that can be effected by field- and intensity-transforming imaging elements. Field-transforming elements operate directly on the electromagnetic field that composes the optical wave. For coherent fields, this property allows field-transforming elements to modify the amplitude as well as the phase of the (deterministic) input wave. For example, a spherical lens imposes a quadratic phase modulation in the paraxial approximation. In the case of partially coherent fields, field-transforming elements operate rather on the random process which represents one realization of the optical field (different in every experiment); as a result, these elements are capable of modifying all the moments of the random field process.

At optical frequencies, the detection step (i.e. the conversion of the electromagnetic field to an electrical signal, such as a voltage) is fundamentally limited to return the intensity of the field. For coherent fields, the intensity is proportional to the modulus of the complex amplitude; i.e. phase information is lost. For partially coherent fields, the only information available past the detector is the first moment (intensity) of the coherence function. In both cases, the intensity is a measure of the *average energy* (normalized over the unit area) carried by the optical field, after cancelling out field fluctuations. It is well known that the intensity detection step is not by any means destructive for other types of information carried by optical fields. For example, interferometry exploits the detectors' quadratic nonlinearity law to map the phase of the field onto the intensity for coherent fields, and the phase of the second-order correlation function onto the intensity for partially coherent fields.

Therefore, the designer has considerable freedom in selecting the location of the detection step in an optical imaging system and, subsequently, the division of labor between the field- and intensity-transforming parts of the system. In practice, some configurations work better than others, depending on the application. The three examples below illustrate this point.

The confocal microscope, invented by M. Minsky [15] and popularized by Wilson and Sheppard [16, 17], is an example of an imaging system which is primarily based on field transformations. It operates by the lowest-dimensional measurements possible, i.e. point measurements, and constructs a 3-D image by scanning the volume of the specimen and obtaining the emitted intensity values one point at a time. The geometry of the optical system is such that light emitted locally from a very small portion of the object only is allowed to reach the detector. The rest of the light is rejected by the detector pinhole. The proportional light contribution to a single measurement as a function of the object coordinates is equivalent to the 3-D point-spread function (PSF) of the system; it can be calculated accurately under various aberration conditions using Fourier optics [18, 19]. Clearly, the function of partitioning the source is performed in the field domain by elements such as the objective lenses and the detector pinhole. The role of intensity transformations is confined to storing the scanned data in a 3-D matrix for subsequent display, a role which is relatively minor as far as the imaging operation is concerned. Confocal microscopy has been implemented in many different variants for improved light efficiency or resolution, e.g. differential interference [20], fluorescence [21], two-photon [22], etc.; it has been spectacularly successful, primarily in various applications of biological and biomedical imaging.

Coherence imaging is an example of computational imaging that relies on global, rather than local, measurements. It is based on a fundamental result, derived independently by van Cittert and Zernicke [23, 24], which states that the degree of statistical correlation of the optical field in the far zone, expressed as a complex function over the exit pupil of the imaging system, is the Fourier transform of the object intensity distribution. Therefore, the object can be recovered by measuring the coherence function through interferometry, and then inverse-Fourier transforming the result. The application of the van Cittert–Zernicke theorem in the radio frequency spectral region is the basis of radio astronomy [25], which yields by far the most accurate images of the most remote cosmic objects. The most common formulation of the theorem relates the mutual coherence in a plane at infinity to a two-dimensional source intensity distribution, but extentions to 3-D sources have been derived by various authors [26–29]. The far-field version of the extended van Cittert–Zernicke theorem was recently implemented experimentally [30–32]. A full generalization of the theorem has also been developed and experimentally implemented to allow Fresnel zone reconstruction in projective coordinates [33]. In most coherence imaging systems, the "work balance" of the imaging operation is shifted almost exclusively to intensity transforms (from the intensity one derives the correlation function and then Fourier-transforms digitally to retrieve the source itself). The role of field transforms is limited to "folding" the field in order to generate a self-interference pattern which maps the field autocorrelation onto the intensity detector [33–35].

Extended depth of field imaging with the use of cubic phase masks [36, 37] is an example of a system where the roles of field and intensity transforms are more balanced than in the two systems described previously. The cubic phase mask "scrambles" the optical field, thereby forming a strongly blurred intensity image in the focal plane. However, application of the inverse deblurring digital transform leads to recovery of the object(s) in the input field with relative insensitivity to their depth. Thus, a clever combination of field and intensity transforms leads to a significant advantage in improving the perceived quality of the recovered digital images.

Optical coherence tomography [38] is an interesting hybrid between confocal microscopy and coherence imaging. It should be classified as a field-transforming technique, since it uses field decorrelation due to decoherence to achieve depth slicing (in the confocal microscope, the slicing function is performed by the pinhole). An extension, spectroscopic optical coherence tomography [39] exploits the phase information in the recorded fringes to extract in addition the spectral composition of the object through a Fourier transform. This version assigns increased importance to the intensity-transforming operation, thereby increasing the amount of information that it returns about the object.

Summarizing the above descriptions, their marked design differences also indicate the diversity in their domain of application. Confocal microscopy and optical coherence tomography are excellent for imaging thick specimens such as tissue and other semi-transparent materials. In these cases, the pure version of non-confocal (global) coherence imaging would have limited fringe visibility due to increased quasi-constant background radiation. On the other hand, for objects that can be modelled as a superposition of discrete, mutually incoherent radiators distributed in 3-D space, or for opaque objects, coherence imaging overperforms other techniques due to the reduced cost of the required optical elements and the reduced scanning time. Cubic phase maskbased imaging is appropriate for digital photography kinds of applications, where defocus-free 2-D projections of 3-D objects are sufficient, but it can be combined with scanning to return $2 + 1/2$-D data as well [37].

12.3 Holography in Imaging

A class of optical elements that allow considerable flexibility in their optical response is, of course, holograms. A hologram is determined either by the profiles of the two optical beams that interfere to record it, or can be fabricated by etching a waveform on a suitable material (typically glass). In either case, several state-of-the-art devices are available for determining the hologram response. For example, in the former case, spatial light modulator (SLM) technology allows spatial amplitude and phase modulation of the optical beams to a resolution down to 10 μm; in the latter case, photolithography, and electron beam patterning have been used to generate very sophisticated diffractive optical elements for communications, display, and other applications. To date, holograms have not been very popular as optical elements in practical imaging systems. A notable exception is holographic interferometry [40] and two-wavelength interferometry [41, 42], where the hologram does not function as a fixed imaging element, but rather as a sophisticated detector that captures phase properties of the object. Bertero and collaborators [43–49] have proposed a method of superresolving confocal microscopy using diffractive elements calculated based on singular system theory.

Volume holograms for complex field transformation and feature extraction have been highly developed in the context of holographic storage and interconnection. Digital data storage, where each stored hologram corresponds to a page of information, is the most straighforward application. Despite this apparent simplicity, the system geometry is extremely important for the capacity and function of data storage. On the next level of complexity, artificial neural networks have used holographic mappings for dendritic interconnections. Before the spectacular improvement of VLSI technology in the 1990s [50, 51], volume holograms were considered as a primary contender for the efficient storage and implementation of the massive interconnections needed for complex pattern-recognition tasks. Many of the design considerations from data storage and neural net systems can be applied to the design of holograms for imaging applications. As will become evident in the remainder of this chapter, some of the fundamental properties of holographic storage techniques, in particular the spatial selectivity and degeneracies of spherical reference volume holograms [52], can be applied verbatim to imaging. Even though the architectures we study here are different from disk geometries [52], the similarity simplifies our intuitive understanding of the problem.

We propose to use volume holograms as optical imaging elements for one main reason: a volume element provides a larger number of degrees of freedom in defining the optical response, compared to a surface element (e.g. a thin hologram) of the same aperture. This is intuitively obvious from dimensional arguments, and was proven formally in [53, 54] using the modal properties of electromagnetic fields. We will not repeat the formal arguments here, but point out the desirable and undesirable features of volume holography that should be taken into account in the design process. The main price to pay for the advanced design flexibility is that the control problem of defining the hologram response (i.e. "programming" the volume hologram) becomes considerably more difficult and is accomplished at the expense of diffraction efficiency [54]. One interesting consequence of the degrees of freedom afforded by volume holograms is that the shift invariance of these elements can be limited in at least one dimension. This is a useful property for 3-D imaging systems, e.g. the optical-sectioning ability of confocal microscopes can be thought of as resulting from the shift variance of the pinhole response.

Summarizing the design parameters of volume holograms:

(a) the required imaging degrees of freedom should be achieved by using as small a number of holograms as possible, in order to maintain high individual diffraction efficiency for each hologram;
(b) the recording of volume holograms is an expensive, material-limited process that should not be performed in real time; it is better, therefore, to use volume holograms as fixed elements that have been predesigned, fabricated in the factory, and delivered to the user, rather than as dynamic elements modifiable in real time.

What function should the volume hologram perform inside an optical imaging system? Unlike other optical elements, volume holograms can perform *several* functions. For example, we have implemented experimentally a confocal microscope [2] where the hologram performs the function of a matched filter for depth selectivity, replacing the pinhole. In the remainder of this paper, we present the basic theory and point-spread function calculations for an extension of the confocal microscope idea to spatio-spectral imaging of incoherent objects.

12.4 Volume Diffraction Theory

A volume hologram is a spatial modulation of the refractive index of a dielectric material, expressed as a function $\Delta\varepsilon(\boldsymbol{r})$. The modulation is effective for values of the space coordinate $\boldsymbol{r}$ satisfying $\boldsymbol{r} \in V_{\mathcal{H}}$, where $V_{\mathcal{H}}$ is the volume occupied by the holographic material. When the hologram is illuminated by a probe field $E_{\mathrm{p}}(\boldsymbol{r})$, the diffracted field $E_{\mathrm{d}}(\boldsymbol{r}')$ is found as the solution to Maxwell's equations in an inhomogeneous medium, with modulated refractive index. The solution is simplified if we assume that the magnitude of the modulation is much smaller than the unmodulated refractive index ε_0:

$$|\Delta\varepsilon(\boldsymbol{r})| \ll \varepsilon_0, \qquad \boldsymbol{r} \in V_{\mathcal{H}}, \tag{12.1}$$

because the weak diffraction approximation (also known as "Born's approximation") can then be applied. The diffracted field is given by [55]

$$E_{\mathrm{d}}(\boldsymbol{r}') = \iiint_{V_{\mathcal{H}}} E_{\mathrm{p}}(\boldsymbol{r})\Delta\varepsilon(\boldsymbol{r}) \times \frac{\exp\{\mathrm{i}k|\boldsymbol{r}-\boldsymbol{r}'|\}}{|\boldsymbol{r}-\boldsymbol{r}'|}\mathrm{d}^3\boldsymbol{r}, \tag{12.2}$$

where $k = 2\pi/\lambda$ is the wavenumber, and the last term in the integrand is recognized as the scalar Green's function for free space. The derivation of (12.2) from Maxwell's equations is beyond the scope of this paper.

Equation (12.2) has the following simple interpretation. Assume that the volume grating is composed of infinitessimal scatterers, the strength of the scatterer located at $\boldsymbol{r} \in V_{\mathcal{H}}$ being $\Delta\varepsilon(\boldsymbol{r})$. Then the diffracted field is the coherent summation of the fields emitted by all the scatterers when

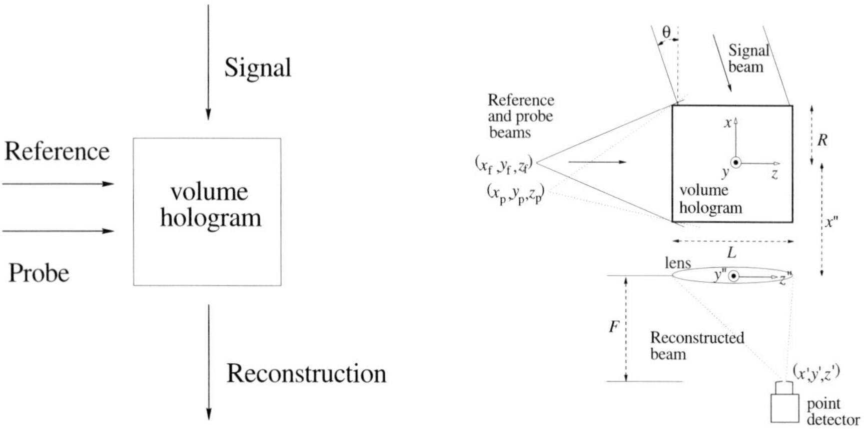

Fig. 12.2. The 90° holographic recording geometry (*left*, schematic; *right*, parameter definitions)

they are excited by the incident field $E_\mathrm{p}(\boldsymbol{r})$. Naturally, this picture ommits higher-order scattering, i.e. fields generated when the field scattered from one infinitesimal scatterer reaches other infinitesimal scatterers. This omision, though, is consistent with the weak scattering approximation, which says that these higher-order effects are even weaker, and, therefore, negligible.

Expression (12.2) is computationally efficient when spherical waves are involved in the recording of the hologram. For other types of fields, a representation of the diffracted field and the grating in wavevector space works better [55], but is beyond the scope of this paper. We will here limit our discussion to the "90°-geometry" (Fig. 12.2), where the reference and signal beam are incident on two normal faces of a cube-like recording medium. The probe beam is incident from the same direction as the reference, and the reconstruction appears as a continuation of the signal.

12.5 Volume Holographic Imaging in the 90°-Geometry

12.5.1 Optical Slicing Through Bragg Mismatch

The recording and probing geometry is shown in Figure 12.2. The holographic medium is disk-shaped with radius R in the x, y plane and thickness L along the z direction. The reference beam is a spherical wave at wavelength λ_f produced by a point source at $\boldsymbol{r}_\mathrm{f} = x_\mathrm{f}\hat{\boldsymbol{x}} + y_\mathrm{f}\hat{\boldsymbol{y}} + z_\mathrm{f}\hat{\boldsymbol{z}}$. We express this wave in the paraxial approximation as

$$E_\mathrm{f}(\boldsymbol{r}) = \exp\left\{ \mathrm{i}2\pi\frac{z - z_\mathrm{f}}{\lambda_\mathrm{f}} + \mathrm{i}\pi\frac{(x - x_\mathrm{f})^2 + (y - y_\mathrm{f})^2}{\lambda_\mathrm{f}(z - z_\mathrm{f})} \right\} . \tag{12.3}$$

Consistent with the paraxial approximation, the hologram radius R is assumed to be smaller than any longitudinal distance that the fields propagate over; i.e. $R \ll |z_{\mathrm{f}}| + L/2$.

The signal beam is a plane wave incident at angle $u \ll 1$ with respect to the $\hat{\boldsymbol{x}}$ axis, expressed as

$$E_{\mathrm{s}}(\boldsymbol{r}) = \exp\left\{-\mathrm{i}2\pi\left(1-\frac{u^2}{2}\right)\frac{x}{\lambda} + \mathrm{i}2\pi u\frac{z}{\lambda}\right\} . \tag{12.4}$$

The modulation of the material refractive index resulting from exposure to beams E_{f}, E_{s} is given by

$$\Delta\varepsilon(\boldsymbol{r}) = |E_{\mathrm{f}}(\boldsymbol{r}) + E_{\mathrm{s}}(\boldsymbol{r})|^2 . \tag{12.5}$$

Out of the four terms in the interference pattern, we will insert only $E_{\mathrm{f}}^*(\boldsymbol{r})E_{\mathrm{s}}(\boldsymbol{r})$ in the volume diffraction equation (12.2) for the remainder of this paper. The remaining three terms are Bragg-mismatched and do not diffract significantly.

The probe field is a spherical wave at wavelength λ_{p} emanating at $\boldsymbol{r}_{\mathrm{p}} = x_{\mathrm{p}}\hat{\boldsymbol{x}} + y_{\mathrm{p}}\hat{\boldsymbol{y}} + z_{\mathrm{p}}\hat{\boldsymbol{z}}$. The expression for the probe field is

$$E_{\mathrm{p}}(\boldsymbol{r}) = \exp\left\{\mathrm{i}2\pi\frac{z-z_{\mathrm{p}}}{\lambda_{\mathrm{p}}} + \mathrm{i}\pi\frac{(x-x_{\mathrm{p}})^2+(y-y_{\mathrm{p}})^2}{\lambda_{\mathrm{p}}(z-z_{\mathrm{p}})}\right\} . \tag{12.6}$$

For an observation point $\boldsymbol{r}''$ near the $\hat{\boldsymbol{x}}$ axis (see Fig. 12.2), the diffraction integral in the paraxial approximation is given as

$$E_{\mathrm{d}}(\boldsymbol{r}'') = \iiint E_{\mathrm{p}}(\boldsymbol{r})\Delta\varepsilon(\boldsymbol{r})\exp\left\{\mathrm{i}2\pi\frac{x''-x}{\lambda_{\mathrm{p}}} + \mathrm{i}\pi\frac{(y''-y)^2+(z''-z)^2}{\lambda_{\mathrm{p}}(x''-x)}\right\} \times \mathrm{circ}\left(\frac{\sqrt{x^2+y^2}}{R}\right)\mathrm{rect}\left(\frac{z}{L}\right)\mathrm{d}^3\boldsymbol{r} . \tag{12.7}$$

Performing the Fourier transform in y'', z'', and subsequently the integrals in cylindrical coordinates [1], we obtain a result of the form

$$\tilde{E}_{\mathrm{d}}(\boldsymbol{r}') = 2\pi R^2 \int_{-L/2}^{L/2} \exp\left\{\mathrm{i}\pi C(z)\right\}\mathcal{L}\left(2\pi A(z)R^2, 2\pi B(z)R\right)\mathrm{d}z , \tag{12.8}$$

where $\mathcal{L}(u,v)$ is the field description near the Gaussian focal point of an aberration-free lens, and the coefficients are given by:

$$A(z) = \frac{1}{\lambda_{\rm p}(z - z_{\rm p})} - \frac{1}{\lambda_{\rm f}(z - z_{\rm f})}\,, \tag{12.9}$$

$$B_x(z) = -\frac{x_{\rm p}}{\lambda_{\rm p}(z - z_{\rm p})} + \frac{x_{\rm f}}{\lambda_{\rm f}(z - z_{\rm f})} + \frac{y'^2 + z'^2}{2\lambda_{\rm p} F^2} - \frac{u^2}{2\lambda_{\rm f}} + \frac{1}{\lambda_{\rm f}} - \frac{1}{\lambda_{\rm p}}\,, \tag{12.10}$$

$$B_y(z) = -\frac{y_{\rm p}}{\lambda_{\rm p}(z - z_{\rm p})} + \frac{y_{\rm f}}{\lambda_{\rm f}(z - z_{\rm f})} - \frac{y'}{\lambda_{\rm p} F}\,, \tag{12.11}$$

$$C(z) = \frac{x_{\rm p}^2 + y_{\rm p}^2}{\lambda_{\rm f}(z - z_{\rm p})} - \frac{x_{\rm f}^2 + y_{\rm f}^2}{\lambda_{\rm f}(z - z_{\rm f})} + 2\left(\frac{u}{\lambda_{\rm f}} - \frac{z'}{\lambda_{\rm p} F} + \frac{1}{\lambda_{\rm p}} - \frac{1}{\lambda_{\rm f}}\right) z\,. \tag{12.12}$$

The diffracted power received by a fixed, infinitesimally small detector pixel located at $(x'/F, y'/F)$ when $\lambda_{\rm p} = \lambda_{\rm f}, y_{\rm p} = 0$ are kept fixed and $x_{\rm p}$, $z_{\rm p}$ are allowed to vary is plotted in Fig. 12.3. The reason for the drop in diffracted power shown in Fig. 12.3 is, naturally, Bragg mismatch.

The easiest physical explanation of Bragg mismatch in this case is, going back to (12.8), to think of the diffracted field as the coherent superposition of fields produced by a continuum of lenses distributed along the hologram thickness and focused at the same Gaussian point image. The continuum of focused fields add constructively when the hologram is Bragg matched, i.e. when the incident field is matched to the holographic fringes. This happens when the probe field is an exact replica of the reference field, and in some other cases described in more detail in the next section. When the hologram is Bragg mismatched, the superimposed lens fields go out of phase, and the total diffracted intensity drops drastically. This mechanism of Bragg mismatch provides "optical slicing" in volume holographic imaging systems. It should

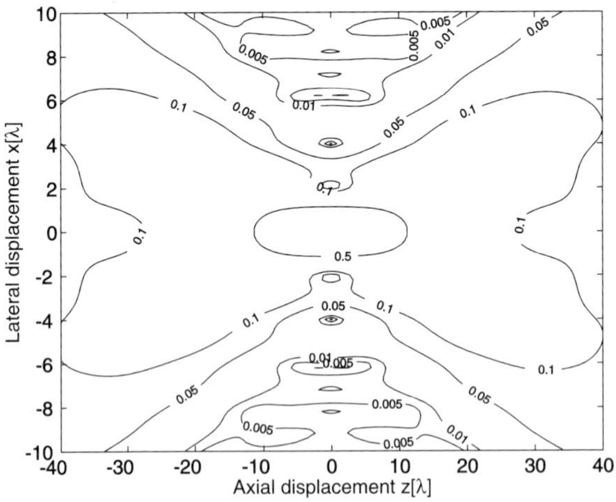

Fig. 12.3. Intensity contour plot of the field diffracted by a point source at position $\boldsymbol{r}_{\rm p} = (x_{\rm p}, 0, z_{\rm p})$ in the geometry of Figure 12.2 with $R = 750\lambda$, $L = 1500\lambda$. The point detector is located at the focal point $x'/F = y'/F = 0$

be compared with the function of a pinhole in confocal microscopy systems, and decoherence in optical coherence tomography systems.

Using information theory arguments, we have proven that optical slicing by a volume hologram produces images of superior quality compared to a confocal microscope, provided that the hologram exceeds a critical diffraction efficiency [56, 57]. Even though an exhaustive discussion of holographic image formation is beyond the scope of this paper, it is worth noting that the improvement over confocal microscopy is due to the matched-filtering properties of volume holograms. These properties are known well and have been exploited repeatedly in optical pattern recognition [58–60]; we instead use matched filtering for imaging. A comparison of volume holographic imaging with optical coherence tomography has not yet been carried out in the literature.

12.5.2 Holographic Degeneracies and Spatio-Spectral Tomography

As mentioned in the previous section, Bragg mismatch is avoided if the probe location and wavelength change simultaneously to match the fringe orientation and spacing of the volume grating. From the diffraction integral (12.8), we can see that the condition for obtaining a strong diffracted field is equivalent to setting the arguments of $\mathcal{L}$ as well as the varying portion of C equal to zero. Applying these conditions to (12.9–12.12) we obtain the *degeneracy surface* of the hologram as

$$z_\mathrm{p} = \frac{z_\mathrm{f}}{\mu}, \tag{12.13}$$

$$\frac{x_\mathrm{p}}{z_\mathrm{p}} = -\mu\left(1 - \frac{x_\mathrm{f}}{z_\mathrm{f}} - \frac{u^2}{2}\right) + \left(1 - \frac{y'^2 + z'^2}{2F^2}\right), \tag{12.14}$$

$$\frac{y_\mathrm{p}}{z_\mathrm{p}} = \mu\frac{y_\mathrm{f}}{z_\mathrm{f}} + \frac{y'}{F}, \tag{12.15}$$

where the parameter $\mu = \lambda_\mathrm{p}/\lambda_\mathrm{f}$ is obtained as the root of the quadratic equation

$$D\mu^2 + G\mu - H = 0 \tag{12.16}$$

with magnitude closest to 1. The coefficients of the quadratic are given by

$$D = 1 + \frac{x_\mathrm{f}^2 + y_\mathrm{f}^2}{z_\mathrm{f}^2} - 2\frac{x_\mathrm{f}}{z_\mathrm{f}} - u^2, \tag{12.17}$$

$$G = \left(1 + \frac{x_\mathrm{f}}{z_\mathrm{f}}\right)^2 + \left(\frac{y_\mathrm{f}}{z_\mathrm{f}} + \frac{y'}{F}\right)^2 + \frac{z'^2}{F^2} + (u-1)^2 - 2\left(1 + \frac{x_\mathrm{f}^2 + y_\mathrm{f}^2}{z_\mathrm{f}^2}\right), \tag{12.18}$$

$$H = \left(1 - \frac{z'}{F}\right)^2. \tag{12.19}$$

Now suppose that the volume hologram is probed by an extended, incoherent, polychromatic object. The intensity of the diffracted field is observed by a planar two-dimensional detector array located at the focal plane of the diffracted field, as shown in Figure 12.4a. One may equivalently think of the probe as the incoherent superposition of elementary broadband point sources $\{(x_{\mathrm{p}}, y_{\mathrm{p}}, z_{\mathrm{p}})\,; \lambda_{\min} \le \lambda_{\mathrm{p}} \le \lambda_{\max}\}$, distributed within the volume $V_{\mathcal{P}}$ of the probe. The resulting diffracted field is the incoherent superposition of the fields that each elementary source would diffract if it probed the hologram individually. According to the above remarks, the elementary sources lying on the degeneracy surface (i.e. the elementary probe sources whose coordinates satisfy (12.13–12.16) result in a strong diffracted field at separable locations on the detector plane, whereas Bragg mismatch suppresses the field due to all other elementary probe sources. In other words, the field reaching the detector plane in Figure 12.4a is primarily due to a narrow spatio-spectral slice of the probe distributed around the degeneracy surface. The degeneracy surface of the hologram (in 4-D space: 3 spatial dimensions plus 1 spectral dimension) corresponds to the probe slice that is mapped onto the planar detector. The shape of a typical degeneracy surface for this geometry is given in Figure 12.4b.

12.5.3 Point-Spread Function and Resolution of Volume Holographic Imaging

Like other optical imaging elements, the volume hologram performs a low-pass filtering operation on the probe object due to the finite angular aperture

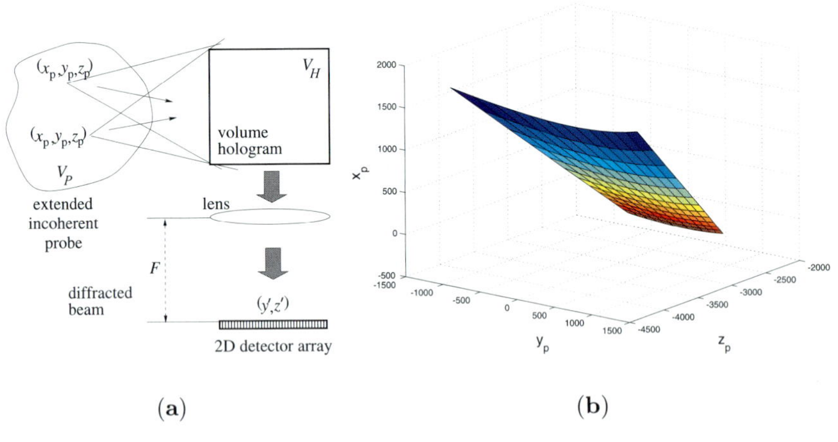

Fig. 12.4. (**a**) Volume holographic imaging setup in the 90° geometry. (**b**) Spatio-spectral slice of the object shown in (**a**) extracted by the volume hologram, according to (12.13–12.15) for $\boldsymbol{r}_{\mathrm{f}} = (100, -100, -2500)\,\lambda$, $u = 0$, $-0.3 \le x'/F, y'/F \le 0.3$. Blue corresponds to $\mu = 0.619$ and purple to $\mu = 1.247$

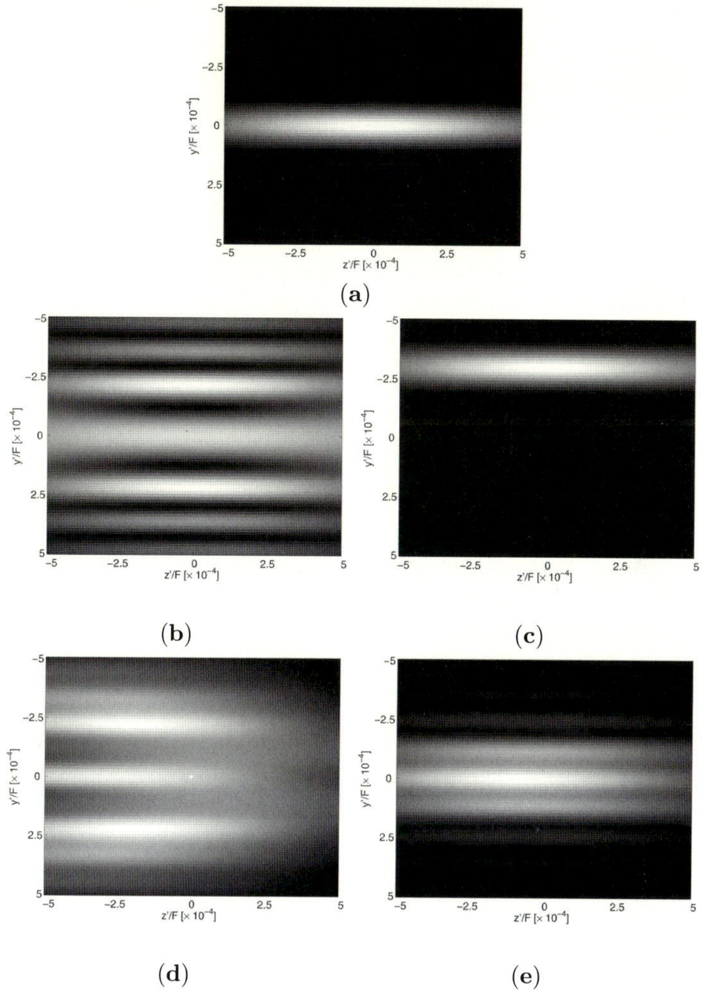

Fig. 12.5. (**a**) Bragg-matched ($\boldsymbol{r}_\mathrm{p} = \boldsymbol{r}_\mathrm{f}$, $\lambda_\mathrm{p} = \lambda_\mathrm{f}$) impulse response $h(y', z')$ of the volume hologram. The peak intensity at $(y' = 0, z' = 0)$ is normalized to 1. (**b**) Impulse response $h(y', z')$ for $\boldsymbol{r}_\mathrm{p} = \boldsymbol{r}_\mathrm{f} + 7.5\lambda_\mathrm{f}\hat{\boldsymbol{x}}$, $\lambda_\mathrm{p} = \lambda_\mathrm{f}$. The peak intensity is 0.0015. (**c**) Impulse response $h(y', z')$ for $\boldsymbol{r}_\mathrm{p} = \boldsymbol{r}_\mathrm{f} + 7.5\lambda_\mathrm{f}\hat{\boldsymbol{y}}$, $\lambda_\mathrm{p} = \lambda_\mathrm{f}$. The peak intensity is 0.96. (**d**) Impulse response $h(y', z')$ for $\boldsymbol{r}_\mathrm{p} = \boldsymbol{r}_\mathrm{f} + 7.5\lambda_\mathrm{f}\hat{\boldsymbol{z}}$, $\lambda_\mathrm{p} = \lambda_\mathrm{f}$. The peak intensity is 0.03. (**e**) Impulse response $h(y', z')$ for $\boldsymbol{r}_\mathrm{p} = \boldsymbol{r}_\mathrm{f}$, $\lambda_\mathrm{p} = \left(1 + 10^{-4}\right)\lambda_\mathrm{f}$. The peak intensity is 0.05

that it subtends relative to the probe. In the 90° geometry, the imaging system aperture is determined by the lateral radius R of the hologram. The resolution is related to the intensity distribution $h(y', z')$ that a point source

$\{(x_p, y_p, z_p); \lambda_p\}$ creates in the detector plane when it is imaged through the system.

The response of the hologram is found by direct integration of the diffraction equation (12.8). Numerical results are given in Figure 12.5. Note the sharp drop in the peak diffracted intensity at the detector plane, which is observed when the probe is displaced in the $\hat{\boldsymbol{x}}$ or $\hat{\boldsymbol{z}}$ directions or if its wavelength λ_p is altered relative to the reference source. On the other hand, displacement in the $\hat{\boldsymbol{y}}$ direction does not significantly reduce the diffraction efficiency, since the degeneracy surface is almost aligned with the y axis (see Fig. 12.4b.)

12.6 Conclusions

The use of volume holograms for optical imaging presents tremendous potential for the capture of rich visual information. The early results presented here and in earlier work [1, 2] demonstrate the feasibility of volumetric imaging spectrometry, i.e. four-dimensional (4-D) imaging using this approach. The applications are broad, encompassing target acquisition and recognition, bio-imaging, and chemical sensing.

This work was supported by the NSF CAREER Award and the Air Force Research Laboratories (AFRL/MNG, Eglin AFB.) The author is grateful to A. Sinha, W. Liu, D. Psaltis, M. Balberg, and D.J. Brady for helpful discussions and suggestions.

References

1. G. Barbastathis, M. Balberg, D.J. Brady: Opt. Lett. **24**, 811 (1999)
2. G. Barbastathis, D.J. Brady: Proc. IEEE **87**, 2098 (1999)
3. P.J. van Heerden: Appl. Opt. **2**, 393 (1963)
4. D. Psaltis: Byte **17**, 179 (1992)
5. J.F. Heanue, M.C. Bashaw, L. Hesselink: Science **265**, 749 (1994)
6. D. Psaltis, F. Mok: Sci. Am. **273**, 70 (1995)
7. Y.S. Abu-Mostafa, D. Psaltis: Sci. Am. **256**, 66 (1987)
8. J. Hong: Opt. Quant. Electr. **25**, S551 (1993)
9. D.J. Brady, A. G.-S. Chen, G. Rodriguez: Opt. Lett. **17**, 610 (1992)
10. P.-C. Sun, Y. Fainman, Y.T. Mazurenko, D.J. Brady: Proc. SPIE **2529**, 157 (1995)
11. P.-C. Sun, Y.T. Mazurenko, W.S.C. Chang, P.K.L. Yu, Y. Fainman: Opt. Lett. **20**, 1728 (1995)
12. K. Purchase, D. Brady, G. Smith, S. Roh, M. Osowski, J.J. Coleman: Proc. SPIE **2613**, 43 (1996)
13. D.M. Marom, P.-C. Sun, Y. Fainman: Appl. Opt. **37**, 2858 (1998)
14. G.A. Rakuljic, V. Levya: Opt. Lett. **18**, 459 (1993)
15. M. Minsky: Microscopy apparatus. US Patent No. 3 013 467 (1961)

16. T. Wilson, C.J.R. Sheppard: *Theory and Practice of Scanning Optical Microscopy* (Academic Press, 1984)
17. T. Wilson (Ed.): *Confocal Microscopy* (Academic Press, 1990)
18. C.J.R. Sheppard, A. Choudhury: Opt. Acta **24**, 1051 (1977)
19. C.J.R. Sheppard, C.J. Cogswell: J. Microsc. **159**, 179 (1990)
20. C.J. Cogswell, C.J.R. Sheppard: J. Microsc. **165**, 81 (1992)
21. I.J. Cox, C.J.R. Sheppard, T. Wilson: Optik **60**, 391 (1982)
22. P.T.C. So, T. French, W.M. Yu, K.M. Berland, C.Y. Dong, E. Gratton: Bioimaging **3**, 49 (1995)
23. P.H. Van Cittert: Physica **1**, 201 (1934)
24. F. Zernike, Proc. Phys. Soc. **61**, 158 (1948)
25. V. Schooneveld: 'Image Formation from Coherence Functions in Astronomy'. In: *IAU Colloquium*, Groningen, 1978, Vol. 49, ed. by D. Reidel
26. A.J. Devaney: J. Math. Phys. **20**, 1687 (1979)
27. W.H. Carter, E. Wolf: Opt. Acta **28**, 227 (1981)
28. I.J. LaHaie: J. Opt. Soc. Am. A **2**, 35 (1985)
29. A.M. Zarubin: Opt. Commun. **100**, 491 (1992)
30. J. Rosen, A. Yariv: Opt. Lett. **21**, 1011 (1996)
31. J. Rosen, A. Yariv: J. Opt. Soc. Am. A **13**, 2091 (1996)
32. J. Rosen, A. Yariv: Opt. Lett. **21**, 1803 (1996)
33. D. Marks, R. Stack, D.J. Brady: Appl. Opt. **38**, 1332 (1999)
34. K. Itoh, Y. Ohtsuka: J. Opt. Soc. Am. A **3**, 94 (1986)
35. D.L. Marks, R.A. Stack, D.J. Brady, D.C. Munson, Jr., R.B. Brady: Science **284**, 2164 (1999)
36. E.R. Dowski, W.T. Cathey: Appl. Opt. **34**, 1859 (1994)
37. D.L. Marks, R. Stack, D. Brady, J. van der Gracht: Opt. Lett. **24**, 253 (1999)
38. D. Huang, E.A. Swanson, C.P. Lin, J.S. Schuman, W.G. Stinson, W. Chang, M.R. Hee, T. Flotte, K. Gregory, C.A. Puliafito, J.G. Fujimoto: Science **254**, 1178 (1991)
39. U. Morgner, W. Drexler, F.X. Kartner, X.D. Li, C. Pitris, E.P. Ippen, J.G. Fujimoto: Opt. Lett. **25**, 111 (2000)
40. P.K. Rastogi (Ed.): *Holographic Interferometry* (Springer-Verlag, Berlin, Heidelberg 1994)
41. J.C. Wyant: Appl. Opt. **10**, 2113 (1971)
42. C. Polhemus: Appl. Opt. **12**, 2071 (1973)
43. M. Bertero, P. Brianzi, E.R. Pike: Inverse Probl. **3**, 195 (1987)
44. M. Bertero, P. Boccacci, M. Defrise, C. De Mol, E.R. Pike: Inverse Probl. **5**, 441 (1989)
45. M. Bertero, P. Boccacci, R.E. Davies, E.R. Pike: Inverse Probl. **7**, 655 (1991)
46. M. Bertero, P. Boccacci, R.E. Davies, F. Malfanti, E.R. Pike, J.G. Walker: Inverse Probl. **8**, 1 (1992)
47. J.G. Walker, E.R. Pike, R.E. Davies, M.R. Young, G.J. Brakenhoff, M. Bertero: J. Opt. Soc. Am. A **10**, 59 (1993)
48. J. Grochmalicki, E.R. Pike, J.G. Walker, M. Bertero, P. Boccacci, R.E. Davies: J. Opt. Soc. Am. A **10**, 1074 (1993)
49. M. Bertero, P. Boccacci, F. Malfanti, E.R. Pike: Inverse Probl. **10**, 1059 (1994)
50. C. Mead: Proc. IEEE **78**, 1629 (1990)
51. C.A. Mead: Analog Int. Circ. & Signal Proc. **6**, 9 (1994)
52. G. Barbastathis, M. Levene, D. Psaltis: Appl. Opt. **35**, 2403 (1996)

53. D. Psaltis, D. Brady, X.G. Gu, S. Lin: Nature **343**, 325 (1990)
54. D. Brady, D. Psaltis: J. Opt. Soc. Am. A **9**, 1167 (1992)
55. H. Coufal, D. Psaltis, G. Sincerbox (Eds.): *Holographic Data Storage* (Springer Verlag, Heidelberg 2000)
56. G. Barbastathis, A. Sinha: Trends in Biotechnology **19**, 383 (2001)
57. G. Barbastathis, A. Sinha, M.A. Neifeld: 'Information-Theoretic Treatment of Axial Imaging'. In: *OSA Topical Meeting on Integrated Computational Imaging Systems, Albuquerque, NM, 2001*, pp. 18–20
58. H.-Y.S. Li, Y. Qiao, D. Psaltis: Appl. Opt. **32**, 5026 (1993)
59. B.L. Volodin, B. Kippelen, K. Meerholz, B. Javidi, N. Peyghambarian: Nature **383**, 58 (1996)
60. H.S. Seung, B. Javidi: Opt. Lett. **26**, 1161 (2001)

13 Compression of Digital Holograms for Three-Dimensional Video

T.J. Naughton, Y. Frauel, O. Matoba, B. Javidi, and E. Tajahuerce

13.1 Introduction

Digital holography has been used for three-dimensional (3-D) measurement and inspection. The field of image recognition [1–3] has also taken advantage of this technique, and 3-D object recognition through digital holography [4–7] is now an established approach among the general class of 3-D object recognition techniques [4–15]. It has recently been suggested that digital holograms could be streamed over a network to effect a form of 3-D video [16, 17]. A real-time optical reconstruction method for digital holograms has been proposed and demonstrated [18]. Our digital holograms are created through phase-shift interferometry [19, 20]. Each digital hologram encodes multiple views of the object from a small range of angles. A particular view of the object can be constructed by extracting the appropriate window of pixels from the hologram and applying a numerical propagation technique [4, 20]. In advance of knowing which of the 4 M pixels are required for particular views, each hologram requires 65 Mbytes of storage in its native double-precision format (5 s of transmission time over a 100 Mbit/s network connection). This is too slow for real-time object reconstruction, and impractical for any type of holographic video streaming. We would like to adapt and apply data compression techniques [21] to these holograms to permit more efficient storage and transmission.

Algorithms for 3-D data compression do exist, such as the wavelet technique of Bilgin et al. [22] or the vector quantization technique of Qian et al. [23] for hyperspectral imagery. However, these techniques are not suitable for our purposes. Our holograms are actually 2-D complex-valued images and differ from a multispectral image, for example, in that there is little correlation between the (real and imaginary) components of each pixel.

One problem with digital holograms is that they contain speckle. Compression of noise, including speckle noise, has been investigated previously. Vago et al. [24] have studied the compression of data from a speckle interferometry application and found that a low-pass filter served to selectively reduce unwanted speckle noise. Shahnaz et al. [25] have applied JPEG (Joint Photographic Experts Group) [26, 27] compression to images with speckle noise and found that its compression performance suffers greatly in the presence of speckle. A homomorphic transform-based technique in the image plane seems to be the best method for removing this noise [25]. These techniques have been applied to real-valued images with simulated speckle noise. Murtagh et al. [28] have investigated compression of real data from astronomical images. By modeling and removing noise they have been able to increase compression rates by a factor of 6 over JPEG. We have found that applying median filtering and mean filtering (separately or combined) to the reconstructed image provides a good tradeoff between image detail and robustness to speckle noise [6, 16–18]. However, it is not possible to remove the speckle in the hologram itself because it actually carries some information.

Holographic data compression is a new field of research. Nomura et al. [29] have investigated quantization of real-valued holograms for encryption applications. Compression rates of 4.0 [see (13.3) for definition of compression rate] for real-valued holograms of binary 2-D inputs with minimal reconstruction error were achieved. They have found that baseline JPEG (the standard JPEG implementation) performs poorly for such holograms. However, the digital holograms of interest in this paper are of 3-D objects and are composed of complex-valued pixels, which means they cannot be processed directly with standard image-compression tools. Very recently, modifications to standard transform techniques, including the discrete cosine transform (DCT) [30] at the heart of baseline JPEG, for the compression of complex-valued digital holograms have been proposed [16]. The positive results from this work concur with previous studies with 2-D images that show that careful manipulation of the quantized cosine coefficients can improve image quality over baseline JPEG [31].

The first stage in the evolution of a video compression system is the compression of individual holographic frames [16, 17]. In this chapter, we demonstrate several lossless and lossy approaches for compressing digital holograms, calculate each of their compression rates, and predict the average performance of combining particular approaches into a complete compression code. Industry-standard lossless data compression algorithms (Lempel-Ziv, Lempel-Ziv-Welch, and Huffman coding) were employed. The lossy techniques were based on subsampling, quantization, discrete Fourier transformation, and phase-only reconstruction. When evaluating lossy techniques, we do not examine the error in the decompressed hologram directly. We are not interested in compression noise or artifacts (such as blocking effects) that appear in the decompressed hologram, only in how lossy compression affects subsequent object reconstruction. For this reason we use an object plane metric, with normalized RMS error, to measure the integrity of the decompressed hologram. For various degrees of speckle reduction, we quantify the number of Fourier coefficients that can be removed from the hologram domain, and the lowest level of hologram domain quantization achievable, without incurring significant error in the reconstructed object domain.

In Sect. 13.2, we describe phase-shift digital holography and describe our experimental setup. In Sect. 13.3, we apply standard lossless data compression techniques to the digital holograms. The simplest form of lossy compression, that of resampling or hologram resizing, is examined in Sect. 13.4, and quantization is examined in Sect. 13.5. In Sect. 13.6, we use a discrete Fourier transform (DFT) technique to selectively remove Fourier coefficients from 8×8 pixel nonoverlapping blocks of the hologram, and in Sect. 13.7 we investigate the use of a phase-only hologram in the reconstruction process.

13.2 Phase-Shift Digital Holography

We record digital holograms with an optical system based on a Mach–Zehnder interferometer (see Fig. 13.1). A linearly polarized argon ion (514.5 nm) laser beam is divided into object and reference beams, both of which are expanded and spatially filtered. The first beam illuminates a reference object placed at a distance $d = 350\,\text{mm}$ from a 10-bit 2028×2044 pixel Kodak Megaplus CCD camera. We refer to the complex amplitude distribution in the plane of the object as $U_0(x, y)$. The reference beam passes through half-wave plate RP_1 and quarter-wave plate RP_2. The linearly polarized beam can be phase-modulated by rotating the two retardation plates. Through permutation of the fast and slow axes of the plates we can achieve phase shifts of 0, $\pi/2$, π, and $3\pi/2$. The reference beam combines with the light diffracted from the object and forms an interference pattern in the plane of the camera. At each of the four phase shifts we record an interferogram. We use these four real-valued images to compute the camera-plane complex field $H_0(x, y)$ by phase-shift interferometry [19]. We call this computed field a digital hologram. Our holograms have dimensions 2028×2044 pixels and are originally in floating point representation with 8 bytes of amplitude information and 8 bytes of phase information for each pixel.

A digital hologram $H_0(x, y)$ contains sufficient amplitude and phase information to reconstruct the complex field $U(x, y, z)$ in a plane in the object

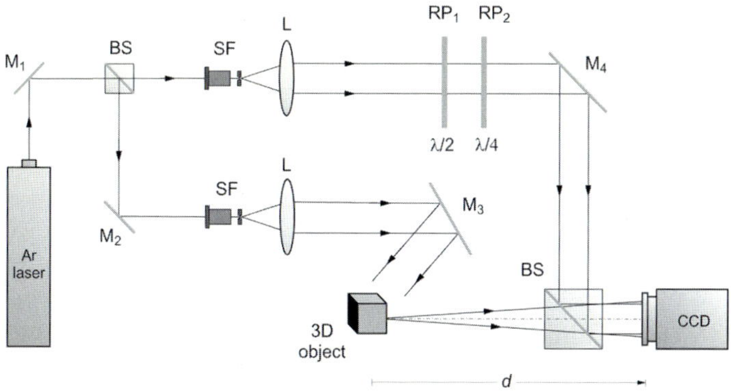

Fig. 13.1. Experimental setup for phase-shift digital holography: M, mirror; BS, beam splitter; SF, spatial filter; L, lens; RP, retardation plate

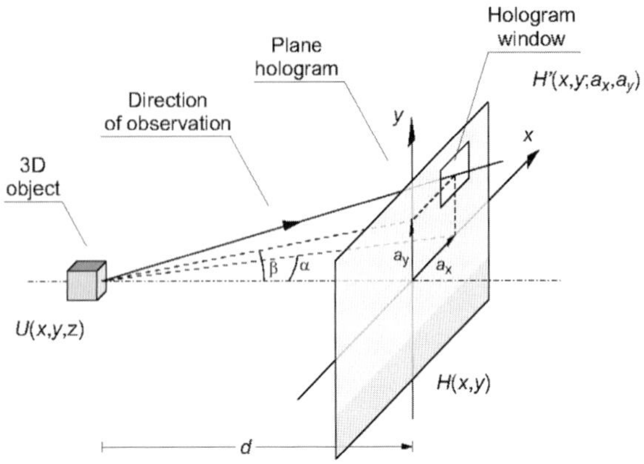

Fig. 13.2. Relationship between the hologram window and angle of view

beam at any distance z from the camera [4, 20]. This can be calculated using the Fresnel–Kirchhoff formula:

$$U(x, y, z) = H_0(x, y) \star h(x, y, z) , \tag{13.1}$$

where

$$h(x, y, z) = -\frac{\mathrm{i}}{\lambda z} \exp\left(\mathrm{i}\frac{2\pi}{\lambda}z\right) \exp\left(\mathrm{i}\pi\frac{(x^2 + y^2)}{\lambda z}\right) \tag{13.2}$$

is the point-spread function for free space, λ is the wavelength of the illumination, and $\star$ denotes a convolution operation. At $z = d$, and ignoring errors in digital propagation due to discrete space (pixelation) and rounding, the digital reconstruction $U(x, y, z)$ approximates $U_0(x, y)$. Furthermore, as with conventional holography [32] a windowed subset of the hologram can be used to reconstruct a particular view of the object, as shown in Fig. 13.2.

In principle, it would be possible to reconstruct the object with only one of the four interferograms. Indeed, it is possible to simulate the readout of the hologram by multiplying it by the reference wave and then applying (13.1) to simulate the propagation. This would give an image of the object as the first diffraction order. However, there are two reasons why we cannot use this method in our experiments. Firstly, we do not know precisely the reference beam. We designed it to be close to a uniform plane wave, but undoubtedly it will contain defects that we have no knowledge of. The second reason is that the size of the camera pixels (9 µm × 9 µm) limits the frequency of the spatial carrier. Practically, this means that the reference beam has to be almost parallel to the object beam in order for the fringes to be larger than the pixels. When reading the hologram, all of the diffraction orders are thus superimposed and it is not possible to separate the image of the object from the unwanted orders. For these reasons we must use the four-interferogram phase-shift technique to compute the complex amplitude of the object beam.

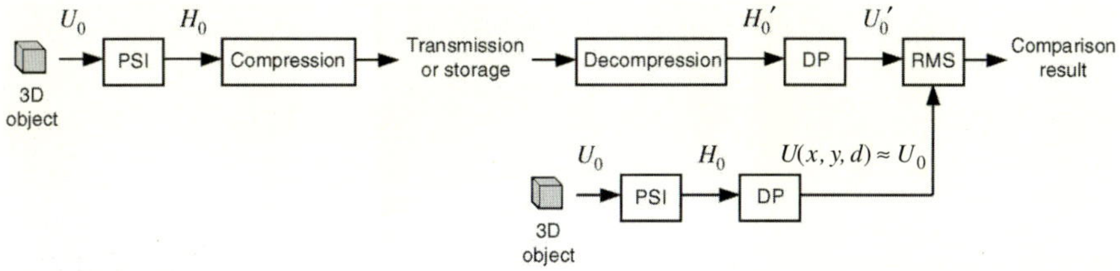

Fig. 13.3. Illustration of the problem statement: digital hologram H_0 must be compressed and transmitted such that decompressed and reconstructed U_0' compares closely with the approximation of the original complex amplitude distribution $U(x, y, d)$. PSI, image capture and phase-shift interferometry stage; DP, digital propagation (reconstruction) stage; RMS, calculation of RMS difference

The problem we address in this chapter can be stated as follows. A digital hologram H_0 of some object U_0 is to be compressed and transmitted from sender to receiver (see the illustration in Fig. 13.3). At the receiver, the hologram is decompressed as H_0' and an object U_0' reconstructed by numerical propagation. We wish to determine the most effective way of compressing H_0 such that U_0' is reconstructed with minimal deviation from U_0.

13.3 Lossless Holographic Data Compression

Lossless data compression techniques are used in situations where the data must be faithfully decompressed, such as in text compression. If we use lossless techniques, we are assured that U_0' will be identical to U_0, apart from rounding and pixelation errors. The set of 3-D objects used in the compression experiments is shown in Fig. 13.4. The digital holograms were treated as binary data streams. Three of the most common industry-standard compression techniques were chosen: Huffman coding [33], Lempel–Ziv coding (LZ77) [34], and Lempel–Ziv–Welch coding (LZW) [35, 36].

13.3.1 Huffman Coding

Huffman coding [33], an entropy-based technique, is one of the oldest and most widely used compression methods. It replaces each symbol in the input by a codeword, assigning shorter codewords to more frequent symbols. Huffman coding is often used for text compression and is used as the final stage in baseline JPEG compression [26, 27]. Huffman coding is an example of a variable-length code, replacing input blocks of fixed size by variable-length strings of bits.

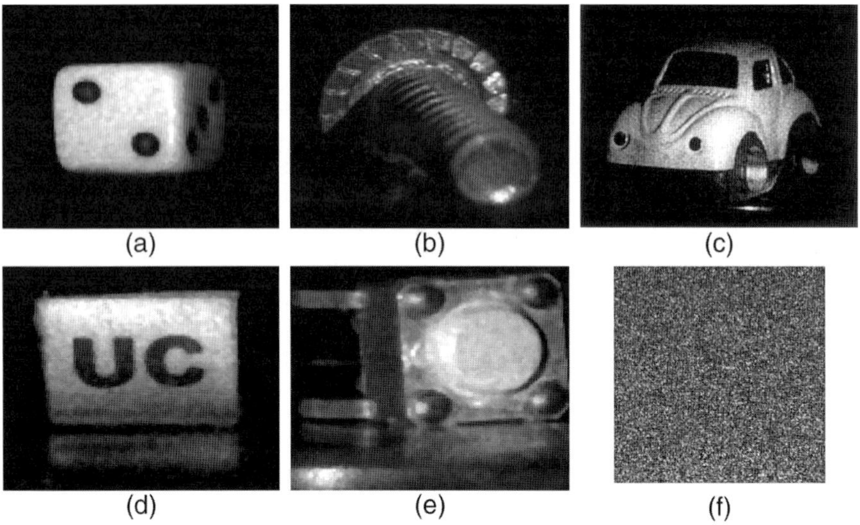

Fig. 13.4. The set of holograms used in these experiments: (**a**) through (**e**) are the amplitudes of the reconstructed wavefronts for holograms No. 1 through No. 5, respectively; (**f**) is the amplitude of an example 512×512 subset of digital hologram No. 1

13.3.2 LZ77

The LZ77 algorithm [34] takes advantage of repeated substrings in the input data. In contrast to Huffman coding, a variable-length string of input symbols is replaced by a fixed-size codeword (a reference to the previous occurrence of that string). The size of this codeword limits its sliding history window (the number of previously seen symbols that the algorithm can refer to). Variants of LZ77 apply additional compression to the output of the substitution stage. A statistical compressor such as Huffman is often chosen to take advantage of the unequal probabilities of output codewords.

13.3.3 LZ78/LZW

The LZ77 algorithm has a shortcoming in that the way it references repeated substrings makes it biased towards local redundancy. In LZ78 coding [35], a dictionary (or lookup table) of previously encountered substrings is maintained to make more efficient use of the reference codewords. This substantially reduces the number of comparisons required during each encoding step. The LZW algorithm [36] is a popular refinement of LZ78, permitting variable-sized codewords and the deletion of old strings from the dictionary, as implemented in the GIF (Graphics Interchange Format) algorithm.

13.3.4 Lossless Compression Results

The result of using these industry-standard lossless algorithms is shown in Table 13.1. Treating the holograms (each a sequence of pairs of amplitude and phase values) as binary data streams achieves compression rates in the range 1.0 to 1.98, where the compression rate r is calculated from

$$r = \frac{\text{uncompressed size}}{\text{compressed size}} , \qquad (13.3)$$

and where a rate of 1.0 was used when no compression was achieved or when the processed hologram was actually larger in size. By the term compression rate we indicate the number of bits of uncompressed data that are effectively communicated with a single bit of compressed data. As shown in Table 13.1, on average, each bit of compressed data encodes 1.33 bits of uncompressed holographic data with LZ77 coding, and 1.04 bits with LZW and Huffman coding. A compression rate of 2.75 would be expected when compressing English text. Given their size (65 Mbytes) it is worrying how poorly the holograms compress using these techniques. Their performance is possibly due to the noisy influence of speckle in the hologram data (see Fig. 13.4f). In general, the fact that the Huffman algorithm is outperformed by the Lempel–Ziv algorithms is no surprise. However, that the simpler and locally-biased LZ77 consistently outperforms LZW could tell us something about the nature of the holographic data and about the local characteristics of its speckle content: that in fact there is some local redundancy that has been exploited by the sliding window of LZ77.

Table 13.1. Compression with LZ77, LZW, and Huffman coding, when the hologram is treated as a single binary data stream with alternating amplitude and phase angle components; c. r., compression rate

Hologram No.	Size (kB)	LZ77 (kB)	LZW (kB)	Huffman (kB)	LZ77 c. r.	LZW c. r.	Huffman c. r.
1	64 769	52 038	64 769	62 236	1.24	1.00	1.04
2	64 769	62 353	64 769	62 298	1.04	1.00	1.04
3	64 642	32 718	54 766	61 784	1.98	1.18	1.05
4	64 769	54 923	64 769	62 262	1.18	1.00	1.04
5	64 769	53 608	64 769	62 267	1.21	1.00	1.04
Averages:					1.33	1.04	1.04

Table 13.2. Compression with LZ77, LZW, and Huffman coding, when the hologram is treated as two separate real and imaginary data streams; c. r., compression rate

Hologram No.	Size (kB)	LZ77 (kB)	LZW (kB)	Huffman (kB)	LZ77 c. r.	LZW c. r.	Huffman c. r.
1	64 769	9356 + 9127	8931 + 8791	13 899 + 13 079	3.50	3.65	2.40
2	64 769	29 890 + 29 752	32 385 + 32 385	30 987 + 31 009	1.09	1.00	1.04
3	64 642	7254 + 8180	7747 + 8318	27 154 + 27 282	4.19	4.02	1.19
4	64 769	9465 + 9125	9047 + 8805	13 815 + 13 027	3.48	3.63	2.41
5	64 769	9227 + 9061	8839 + 8718	13 439 + 13 082	3.54	3.69	2.44
Averages:					3.16	3.20	1.90

13.3.5 Intermediate Coding

For some compression algorithms, the way data is represented can have a large impact on how well it will be compressed. We investigate three additional representations (or intermediate codings) for the digital holographic data. First, the pixel values are divided into separate amplitude and phase data streams to exploit the possible redundancy between neighboring amplitude values. This produced slight but consistent improvements in compression rate for each hologram and for each compression algorithm, increasing the range to 1.0 – 2.26 and the LZ77 average to 1.46 [16]. (Compression rates were not affected if the data was concatenated into one stream prior to compression, rather than compressing the two streams separately.) For the other representations, the holographic data was transformed from the (amplitude, phase) domain to the equivalent, neglecting rounding errors, (real, imaginary) domain. Treating the hologram as a binary data stream of pairs of real and imaginary values results in an improvement of almost a factor of two [16], achieving lossless compression rates in the 1.0 – 4.3 range. Separating the pixel values into separate real and imaginary data streams gives comparable results (an average compression rate of 3.2 for LZW) as shown in Table 13.2. The real and imaginary data streams can also be concatenated and compressed as one larger data stream for the same overall compression gain.

From Table 13.2, it is also evident that both the real and imaginary streams are equally difficult to compress. Since Huffman's algorithm exploits the frequency of occurrence of binary words rather than their relative positioning, it is no surprise that the changes in representation proposed here did not have an impact on the Huffman compression results. For algorithms that exploit the repeated occurrence of substrings (LZ77 and LZW), compression rates can be doubled by transforming the digital hologram into a binary stream of real and imaginary values, rather than leaving it as the amplitude and phase values produced by default by phase-shift interferometry.

Although compression rates of over 4.0 could be achieved for some holograms, others created with the same experimental setup and phase-shift interferometry algorithms could not be compressed at all. If lossless compression of digital holograms is required, then these results indicate that the hologram should be stored as a binary data stream of separate real and imaginary matrices, and compressed with LZW for best average performance (with an expected average compression rate of 3.2). Although LZ77 still achieves the maximum compression of any hologram, on average LZW is marginally better, making it the method of choice.

13.3.6 Lossless and Lossy Image Compression

Standard lossless image compression algorithms can be applied to the digital holograms. This would require some intermediate processing to convert each complex-valued hologram into two positive real-valued images. The performance of GIF and TIFF-LZW (Tagged Image File Format) compression can be gauged from the results for the LZW algorithm in Table 13.2. The performance of portable network graphics (PNG) can be gauged by looking at the results for the LZ77 algorithm. If a higher rate of compression is required then a lossy form of compression will have to be applied.

We expect a lossy strategy to give good results because although 16 bytes are used to store a digital holographic pixel, the holograms have been constructed from four intensity images, each with only 10 bits of resolution. A digital hologram cannot be treated as a real-valued image as in conventional lossy image compression. A change locally in a digital hologram will, in theory, affect the whole reconstructed object. This is demonstrated in Sect. 13.4 by showing that the slightest resizing of the hologram can dramatically influence reconstruction quality. The criteria for grading the performance of lossy compressors will also not be the same as in conventional image compression. The JPEG algorithm [26, 27] was defined largely by the qualitative opinions of human observers on the errors introduced to the decompressed image. Our grading system will have to be different. The errors introduced into the digital hologram as a result of lossy compression are not of direct concern; it is errors in the reconstructed object, loss of viewing angle, and so on, that are of most interest. Lossy systems are investigated next.

13.4 Compression by Resampling

The simplest and most common form of lossy compression is that of resampling. In general, in optical information processing, a single image might be resampled many times: sampling by the spatial frequency of the CCD sensor, sampling of the camera signal by the interface card, sampling by the pixel resolution of a monitor screen or spatial light modulator (SLM), and so on. A convenient (albeit application-dependent) method of compressing a standard real-valued image is to resample or resize it as small as is possible without corrupting features of interest. Subsampling is often used in conventional video streaming applications in order to maintain a constant frame rate over a fluctuating bandwidth connection.

It has been shown that digital holograms are unduly sensitive to resampling [16, 17]. In a 3-D object-recognition application it was shown that resizing the hologram to 0.97 of its previous side-length caused the normalized crosscorrelation peak height to fall dramatically to the order of 10^{-2} of its former height [16, 17]. This sensitivity is due in some part to speckle. With the slightest modification to the digital hologram a completely different speckle pattern is reconstructed, and the similarity between reconstructions drops considerably. In order to reduce the effect of speckle we discard the phase information in the reconstructed object wavefront and apply a median filtering operation. Our justification for keeping only the object plane amplitude information is based on this information's dependency on both the amplitude and phase of the hologram plane: if the amplitude information in the object plane has been reconstructed correctly, this indicates that sufficient amounts of both amplitude and phase information were preserved in the hologram

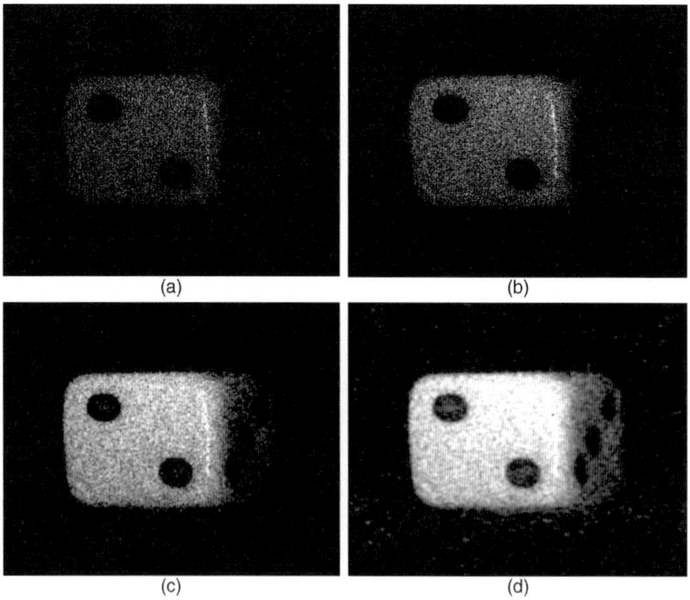

Fig. 13.5. Increasing the level of median filtering in hologram No. 1: (**a**) with no median filtering; (**b**) with a 3×3 pixel median filtering neighborhood; (**c**) with a 7×7 pixel neighborhood; (**d**) with an 11×11 pixel neighborhood

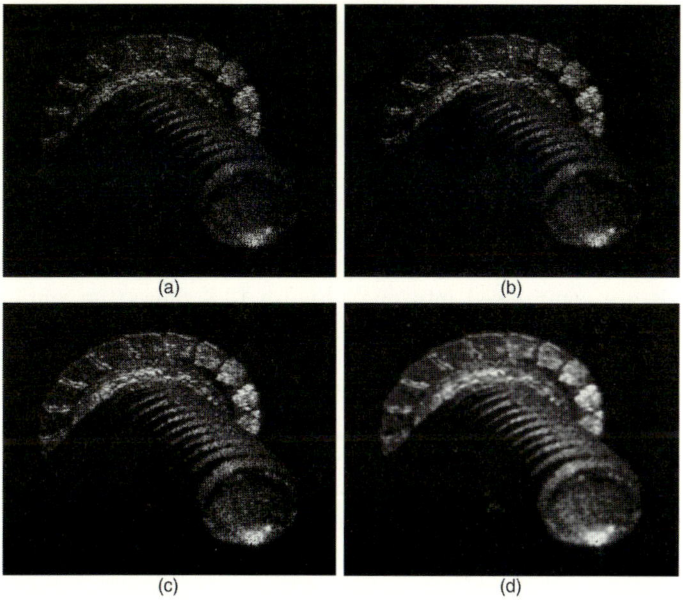

Fig. 13.6. Increasing the level of median filtering in hologram No. 2: (**a**) with no median filtering; (**b**) with a 3×3 pixel median filtering neighborhood; (**c**) with a 5×5 pixel neighborhood; (**d**) with an 11×11 pixel neighborhood

plane during compression. The appropriate level of median filtering will be dependent on the application. Figures 13.5 and 13.6 show the effect of increasing the median filtering neighborhood in the reconstructed object plane for holograms No. 1 and No. 2, respectively. As the level of median filtering increases, the gross structure of the object is enhanced and detail (including speckle) is removed.

For the following figures, we examine the effect of resizing a digital hologram when only the amplitude information of the reconstructed wavefront is examined. To ensure that the object is reconstructed at the same distance from the hologram plane the complex-valued holograms are returned to full size after the loss due to compression is introduced. (We assume that this resizing operation does not introduce additional error.) The objects were reconstructed, their phase component removed, and the object amplitudes evaluated in terms of the normalized RMS difference D, calculated from

$$D = \frac{1}{P_{U_0}} \sqrt{\frac{1}{N_x N_y} \sum_{m=0}^{N_x-1} \sum_{n=0}^{N_y-1} \left| U_0(m,n) - U_0'(m,n) \right|^2} \,, \tag{13.4}$$

where (m, n) are discrete spatial coordinates in the object plane, and N_x and N_y are the number of samples in the x and y directions, respectively. P_{U_0} is the power per pixel in the uncompressed object amplitudes and is defined as

$$P_{U_0} = \sqrt{\frac{1}{N_x N_y} \sum_{m=0}^{N_x-1} \sum_{n=0}^{N_y-1} |U_0(m,n)|^2} \,. \tag{13.5}$$

Figure 13.7a contains plots of the normalized RMS difference for hologram No. 1. For each curve, the hologram is resampled with nearest neighbor interpolation before reconstruction, the phase information at the object plane is removed, and the amplitude median filtered using one of three neighborhoods. The large artifacts in each plot occur at exactly a side-length resizing of 0.5 (when four pixels are compressed into one). Figure 13.8 shows the amplitude of the reconstructed object for selected amounts of resampling with nearest neighbor interpolation. Figure 13.7b contains plots of the normalized RMS difference for hologram No. 1 for three different interpolation strategies (nearest-neighbor, bilinear, and bicubic) and with or without 11×11 pixel median filtering. Figure 13.9 shows the amplitude of the reconstructed object for selected amounts of resampling with bicubic interpolation. As the level of median filtering is increased (as the neighborhood is increased) the hologram tolerance to resampling increases. Specific applications will dictate when too much filtering obscures salient details in the reconstructed object. Resampling, if it is acceptable for a particular application, can therefore improve the compression rate. Furthermore, since resampling simply reduces the number of pixels and (possibly) changes their values rather than compresses the underlying data format, the lossless techniques of Sect. 13.3 can be applied as a final step.

Figure 13.7 shows that even in the presence of measures introduced to reduce the effects of speckle, such as removing the phase and applying me-

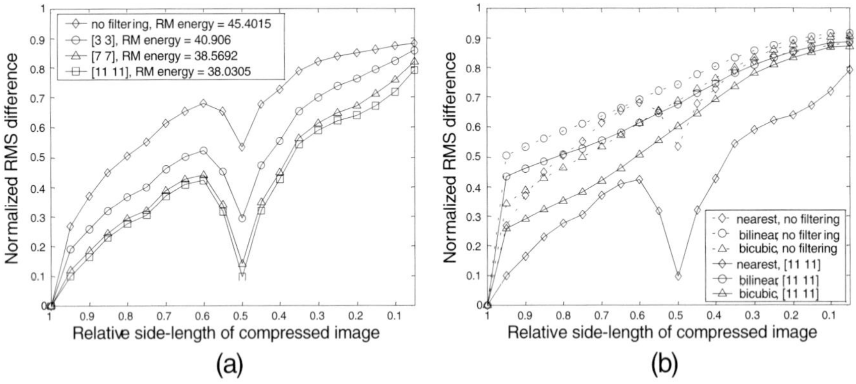

Fig. 13.7. Resizing hologram No. 1 and using only the amplitude information in the reconstructed object plane to measure error: (**a**) normalized RMS difference for nearest-neighbor interpolation with median-filtering neighborhoods of 3×3, 7×7, and 11×11 pixels, and without median filtering; (**b**) normalized RMS difference for three interpolation strategies (nearest-neighbor, bilinear, and bicubic)

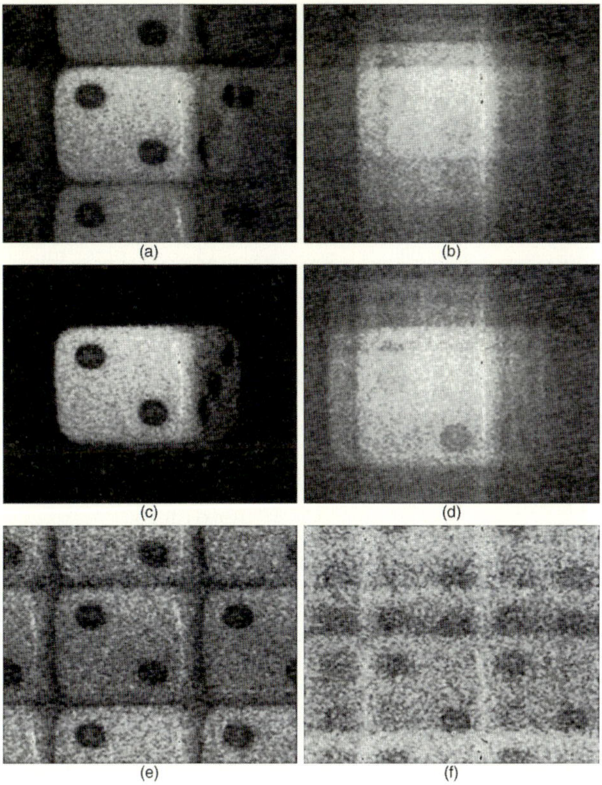

Fig. 13.8. The effect on reconstructed object amplitude of resampling in the hologram domain with nearest-neighbor interpolation, and 11×11 pixel median filtering, for hologram No. 1. Reconstructions can be compared with the original in Fig. 13.5d. Resizing to relative side-lengths of (**a**) 0.75; (**b**) 0.55; (**c**) 0.5; (**d**) 0.45; (**e**) 0.25; (**f**) 0.2

dian filtering, the RMS errors grow quickly with resizing. Nearest-neighbor interpolation introduces fewest errors in reconstruction; however, bicubic interpolation might be preferable because it degrades more gracefully as the undersampling increases. Taking a quantitative example from Fig. 13.7b, if 11×11 pixel median filtering is acceptable, then resizing to a side-length of 0.75 with bicubic interpolation will reduce the number of pixels by a factor of 1.78. This, combined with the lossless techniques of Sect. 13.3, would give an average compression rate of 5.7. This compression, however, comes at the cost of an unsatisfactory 0.38 normalized RMS error. The technique of resampling achieves its best performance for a side-length resizing of 0.5, 11×11 pixel median filtering, and nearest-neighbor interpolation, where a compression rate of 12.8 can be averaged for a normalized RMS error of less than 0.1.

13.5 Quantization

The technique of complex amplitude quantization is analogous to spatial-domain nearest-neighbor resampling. By reducing the number of possible

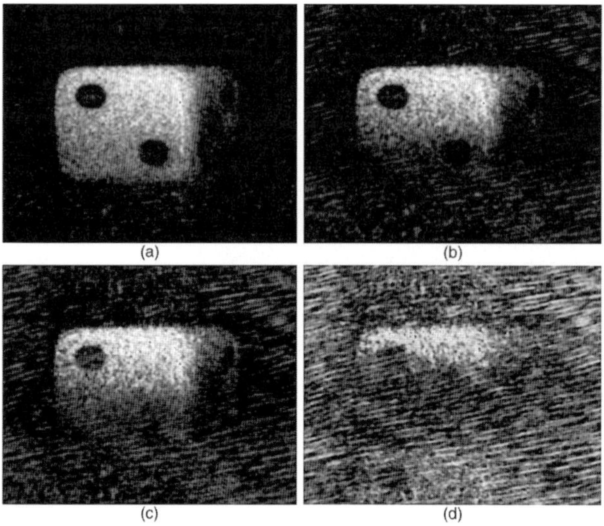

Fig. 13.9. The effect on reconstructed object amplitude of resampling in the hologram domain with bicubic interpolation, and 11×11 pixel median filtering, for hologram No. 1. Reconstructions can be compared with the original in Fig. 13.5d. Resizing to relative side-lengths of (**a**) 0.8; (**b**) 0.5; (**c**) 0.45; (**d**) 0.25

values (or levels) available to each pixel we reduce the number of bits required to describe it. Such a technique would be anticipated to provide a compression factor of at least 6.4 (8 bytes/10 bits) with minimal losses since images with only 10 bits of intensity resolution were used in the phase-shift interferometry stage. Experiments to measure the dependency of the object reconstruction integrity on the number of quantization levels in digital holograms have been performed [16, 17]. Figure 13.10a shows a plot of normalized RMS difference against number of bits per data value for hologram No. 1, for each of five

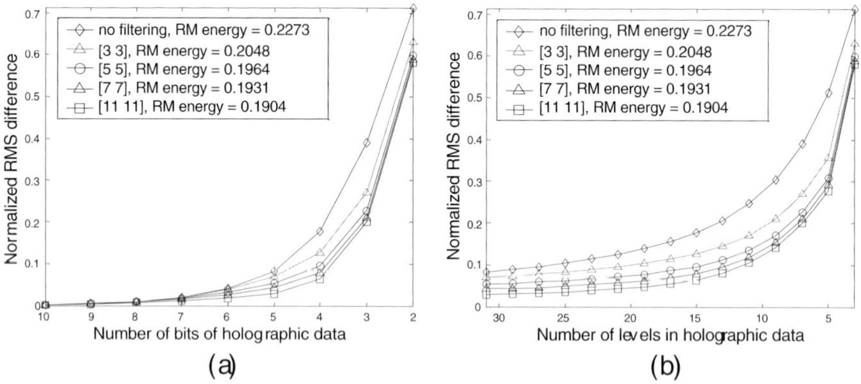

Fig. 13.10. Normalized RMS difference plotted against quantization for hologram No. 1: (**a**) number of bits in each of the real and imaginary values; (**b**) linear scale plot of the number of levels in each of the real and imaginary values, from 5 bits (31 levels) to 2 bits (3 levels)

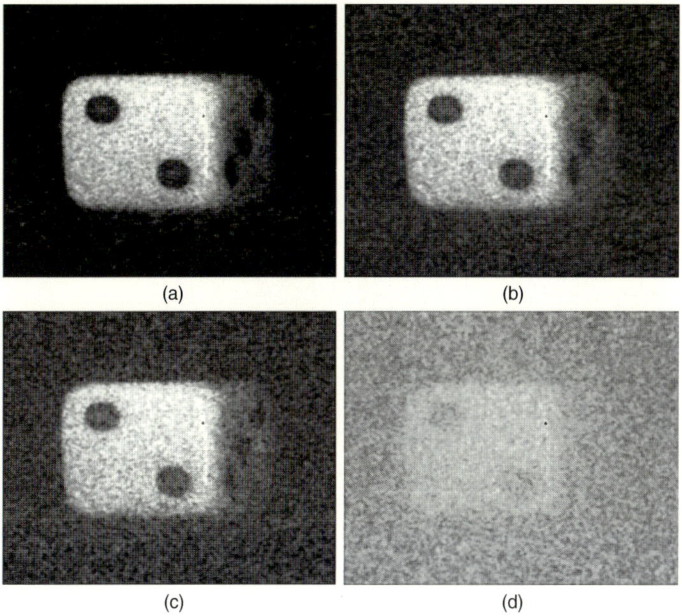

Fig. 13.11. Reconstructed amplitudes for hologram No. 1 for various numbers of quantization levels and with 11 × 11 pixel median filtering (reconstructions can be compared with Fig. 13.5d): (**a**) 4 bits (15 quantization levels); (**b**) 3 bits (7 quantization levels); (**c**) 5 quantization levels; (**d**) 2 bits (3 quantization levels)

median-filtering neighborhoods from 1×1 pixels (no filtering) to 11×11 pixels. The digital holograms were stored in real-imaginary format; each holographic pixel requires two such data values. For a finer view of this dependency, Fig. 13.10b shows a plot of the normalized RMS difference against the number of individual quantization levels. The quantization levels were chosen to be symmetrical about zero; as a result n bits encode $2^n - 1$ levels. For example, two bits encode levels $\{-1, 0, 1\}$, 3 bits encode levels $\{-3, -2, -1, 0, 1, 2, 3\}$, and so on. The number of levels 31, 15, 7, and 3, correspond to 5, 4, 3, and 2 bits of resolution, respectively. Figure 13.11 shows the reconstructed object amplitudes for hologram No. 1 for selected numbers of quantization levels.

Figures 13.12 and 13.13 contain the results of applying quantization to hologram No. 2. Similar compression performances were recorded. Recall that this is the hologram that could not be compressed with the lossless techniques. Note also that for hologram No. 2 a median-filtering neighborhood of only 3×3 pixels performs as well as a larger filter at high compression rates.

Between 6 bits and 7 bits (corresponding to compression rates of 16.0 and 12.8, respectively) need to be retained in each of the real and imaginary components for low (< 0.02) reconstruction losses. With 4 bits and with moderate amounts of median filtering, a normalized RMS error of less than 0.1 can be

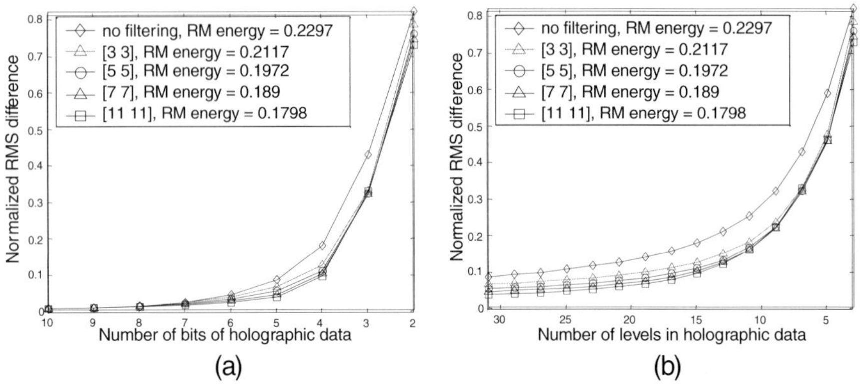

Fig. 13.12. Normalized RMS difference plotted against quantization for hologram No. 2: (**a**) number of bits in each of the real and imaginary values; (**b**) linear scale plot of the number of levels in each of the real and imaginary values, from 5 bits (31 levels) to 2 bits (3 levels)

achieved, corresponding to a compression rate of 16. Applying an entropy coder (such as Huffman or a run-length technique) to the quantized pixel values could further increase these compression rates. Quantization appears to be a promising technique. A simplified form of Fourier-domain quantization is studied in the next section.

13.6 DFT-Based Compression

The JPEG algorithm [26, 27] performs most of its compression during a quantization stage. It uses the DCT [30] to allow it to perform quantization in the spatial-frequency domain. The JPEG standard is defined for real-valued images only, but we adapt it to digital holograms, taking instead the DFT of each nonoverlapping block of 8×8 pixels. Rounding errors aside, this will not reduce the amount of information in the image but will tend to concentrate the majority of the hologram information into a few DFT coefficients in each block. By quantizing particular coefficients in each block we reduce the length of its bit description and thus allow an entropy coder to further compress the coefficients. In our simplified version of the technique, we set a fixed number of the smallest DFT coefficients in each block to zero [16, 17]. This involves sorting the values in each 8×8 DFT and setting the $(64 - n)$ lowest-valued coefficients to zero, where n is a positive integer in the range $1-64$ denoting the number of DFT coefficients to be retained in each block. Sorting of the DFT coefficients was performed by amplitude first, and then (if necessary) by phase angle. Upon decompression (a blockwise inverse DFT) the hologram was used to construct object U_0' for comparison with U_0. Once

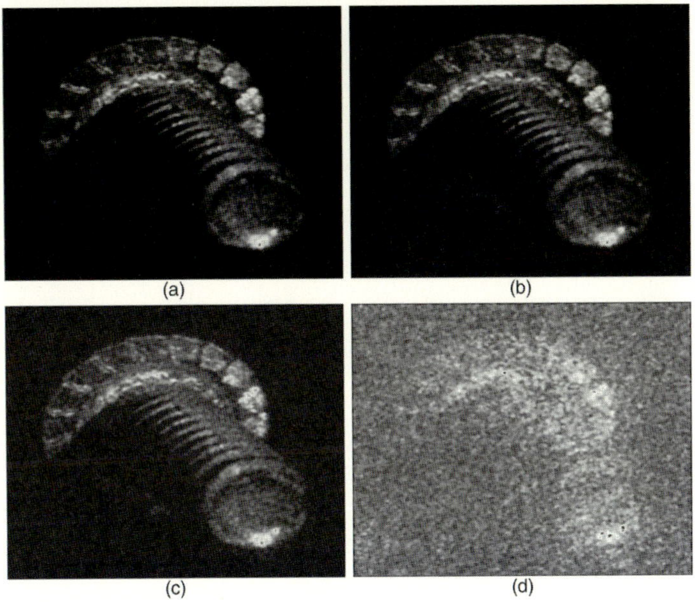

Fig. 13.13. Reconstructed amplitudes for hologram No. 2 for various numbers of quantization levels and with 11 × 11 pixel median filtering (reconstructions can be compared with Fig. 13.6d): (**a**) 4 bits (15 quantization levels); (**b**) 3 bits (7 quantization levels); (**c**) 5 quantization levels; (**d**) 2 bits (3 quantization levels)

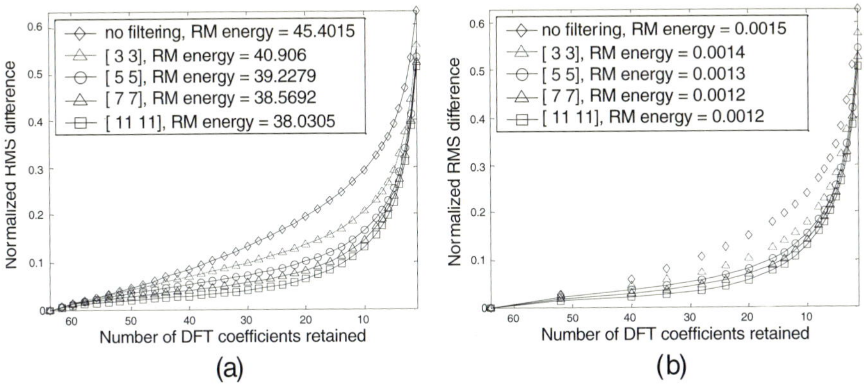

Fig. 13.14. Normalized RMS difference as a function of DFT coefficients retained, for various median-filter neighborhoods: (**a**) for hologram No. 1, and (**b**) for hologram No. 2

again, the phase of the object wavefront was discarded and the amplitude median-filtered prior to comparison to lessen the effects of speckle.

Figure 13.14 shows the normalized RMS difference for hologram Nos. 1 and 2, for various values of n and for different median-filtering neighborhoods.

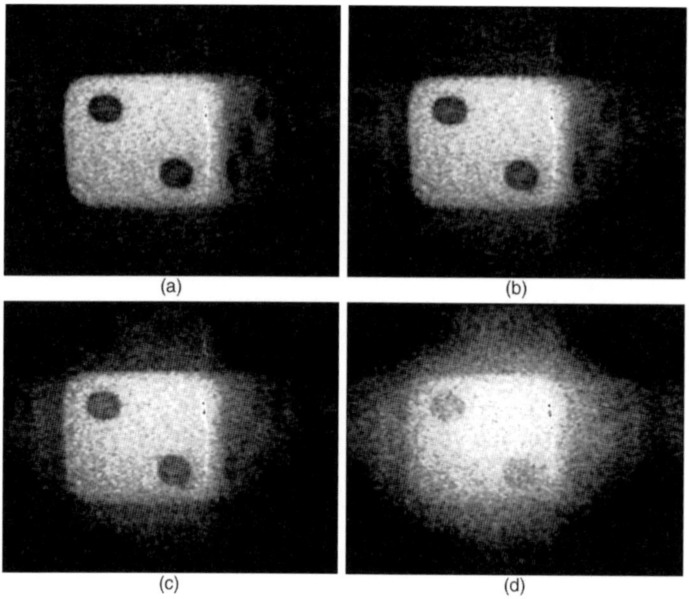

Fig. 13.15. Reconstructed object amplitudes (11×11 median-filtered) for selected numbers of DFT coefficients retained in each 8×8 pixel block of hologram No. 1 (reconstructions can be compared with Fig. 13.5d): (**a**) 10 coefficients (15.6% of all coefficients retained); (**b**) 5 coefficients (7.8%); (**c**) 3 coefficients (4.69%); (**d**) dc coefficient only (1.56%)

Even though hologram No. 2 could not be compressed with the lossless techniques, it does not appreciably degrade faster than hologram No. 1 under this lossy algorithm. Figures 13.15 and 13.16 show the reconstructed object amplitudes for selected values of n. The results are consistent over both holograms: with 11×11 median filtering, as many as 92% of the DFT coefficients can be removed for moderate reconstruction quality (a normalized RMS error of 0.22; see Figs. 13.15b and 13.16b). This corresponds to a compression rate of 12.8. For RMS errors of less than 0.1, no more than 78% of the coefficients can be removed, giving a compression rate of 4.6. The remaining nonzero-valued DFT coefficients are stored with 8 bytes (for each real and imaginary value) and so the quantization technique of Sect. 13.5 or the lossless techniques of Sect. 13.3 could be used to further increase these compression rates.

13.7 Phase-Only Digital Hologram

We have found that a digital hologram's amplitude values contribute less to the final reconstructed object than its phase distribution [18]. In general, compression rates can be doubled by equalizing the amplitude values in

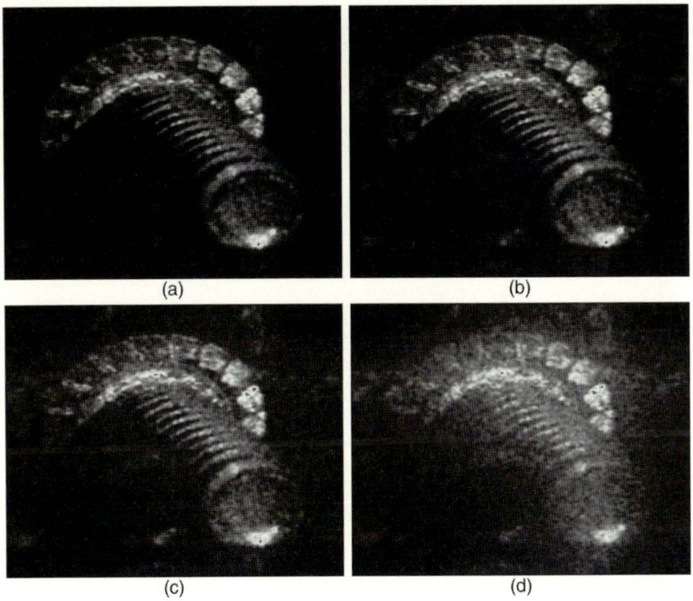

Fig. 13.16. Reconstructed object amplitudes (11×11 median-filtered) for selected numbers of DFT coefficients retained in each 8×8 pixel block of hologram No. 2 (reconstructions can be compared with Fig. 13.6d): (**a**) 10 coefficients (15.6% of all coefficients retained); (**b**) 5 coefficients (7.8%); (**c**) 3 coefficients (4.69%); (**d**) dc coefficient only (1.56%)

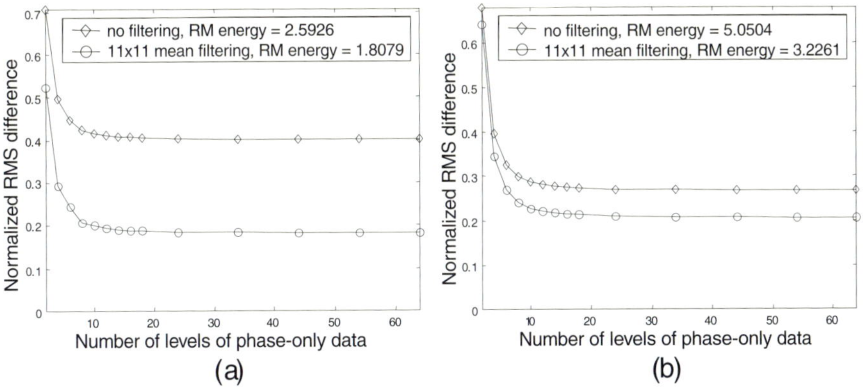

Fig. 13.17. Normalized RMS difference as a function of quantization in a phase-only digital hologram: (**a**) for hologram No. 1, and (**b**) for hologram No. 2

a digital hologram, and can be further increased by quantizing the remaining phase values. In Fig. 13.17 we examine the reconstructed error as a function of phase-only quantization. The intensity of the reconstructed object from

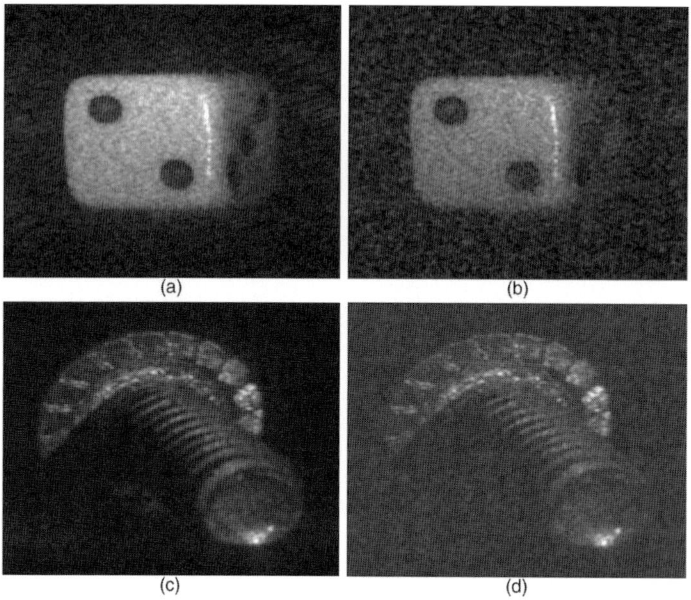

Fig. 13.18. Reconstructed object amplitudes (11×11 mean-filtered) for selected phase-only quantizations: (**a**) using 4-bit and (**b**) 1-bit (binary) phase-only data from hologram No. 1; (**c**) using 4-bit and (**d**) 1-bit phase-only data from hologram No. 2

the phase-only hologram was compared with the intensity of an object reconstructed from a fully complex version of the hologram. In order to take account of the change in hologram energy caused by amplitude equalization the fully complex and phase-only reconstructions U_0 and U_0', respectively, were first rescaled so that each had, on average, an amplitude value of 1.0 per pixel. The rescaled objects, U_r and U_r', were calculated from

$$U_\mathrm{r} = U_0 \times \left(\frac{1}{N_x N_y} \sum_{m=0}^{N_x-1} \sum_{n=0}^{N_y-1} U_0(m,n) \right)^{-1} ,$$
$$U_\mathrm{r}' = U_0' \times \left(\frac{1}{N_x N_y} \sum_{m=0}^{N_x-1} \sum_{n=0}^{N_y-1} U_0'(m,n) \right)^{-1} , \quad (13.6)$$

where once again (m,n) are discrete spatial coordinates in the object plane, and N_x and N_y are the number of samples in the x and y directions, respectively. Mean filtering was then applied to the rescaled reconstructions prior to calculating the normalized RMS difference.

The reconstructed objects with this method do suffer from high reconstruction errors due to speckle noise, but the method has a major advantage in that it is somewhat invariant to reductions in quantization (note the almost constant error above 16 levels in Fig. 13.17 and the comparable quality between reconstructions in Fig. 13.18). This makes the method suitable for implementation with inexpensive liquid crystal SLMs [18]. If the application will permit normalized RMS errors of over 0.2, high compression rates can be achieved with quantized phase-only digital holograms: the objects in Figs. 13.18a and 13.18c were each reconstructed from a hologram with a compression rate of 32.

13.8 Conclusions

We have investigated various techniques for the compression of digital holograms created by phase-shifting interferometry. With industry-standard lossless data-compression techniques an average lossless compression rate of 3.2 can be expected. This rate was achieved by an intermediate coding of separated real and imaginary components before application of the LZW algorithm. Lossy resampling techniques, combined with phase-removal and median filtering to lessen the effects of speckle, were examined in terms of the normalized RMS reconstruction error. Hologram resampling resulted in a high degradation in the reconstructed image quality, but for resizing to a side-length of 0.5 (and in the presence of a high degree of median filtering) a compression rate of 12.8 could be achieved. Nearest-neighbor interpolation provided the lowest (albeit still high) reconstruction errors of the three interpolation strategies investigated.

Quantization proved to be a very effective technique. Each real and imaginary component could be reduced from its original 8 bytes to 4 bits while maintaining an acceptable reconstruction quality, resulting in a compression rate of 16. The technique based on the removal of DFT coefficients achieves approximate compression rates of up to 4.6 for reasonable reconstruction integrity, depending on the degree of median filtering allowed. It is anticipated that this can be improved further by applying lossless compression or quantization to the remaining DFT coefficients.

Based on a compression rate of 10.7 (6-bit quantization and a normalized RMS error of between 0.018 and 0.02, depending on the hologram chosen) complex-valued holographic video frames of dimensions 640×640 pixels could be streamed over a 100 Mbit/s connection at a rate of 20 Hz or frames with 1024×1024 pixels at 8 Hz. If the reconstruction quality can be sacrificed (permitting normalized RMS errors of 0.2, for example) a combination of phase-only holograms and 4-bit quantization could be used to stream 1024×1024 pixel frames at 24 Hz over the same connection. Further compression gains are anticipated in future work to apply interframe compression strategies to the holographic data (in the same way that the MPEG algorithm removes redundancy from its stream of JPEG frames).

Acknowledgements

T.J. Naughton wishes to acknowledge support from Enterprise Ireland, the Department of Electrical and Computer Engineering, University of Connecticut, and the Department of Computer Science, National University of Ireland, Maynooth.

References

1. A. VanderLugt: IEEE Trans. **IT-10**, 139 (1964)
2. J.W. Goodman: *Introduction to Fourier Optics*, 2nd edn. (McGraw-Hill, New York 1996)
3. A.D. McAulay: *Optical Computer Architectures* (Wiley, New York 1991)
4. B. Javidi, E. Tajahuerce: Opt. Lett. **25**, 610 (2000)
5. E. Tajahuerce, O. Matoba, B. Javidi: Appl. Opt. **40**, 3877 (2001)
6. Y. Frauel, E. Tajahuerce, M.-A. Castro, B. Javidi: Appl. Opt. **40**, 3887 (2001)
7. Y. Frauel, B. Javidi: Opt. Lett. **26**, 1478 (2001)
8. A. Pu, R.F. Denkewalter, D. Psaltis: Opt. Eng. **36**, 2737 (1997)
9. R. Bamler, J. Hofer-Alfeis: Opt. Acta **29**, 747 (1982)
10. J. Rosen: Appl. Opt. **37**, 7538 (1998)
11. J.J. Esteve-Taboada, D. Mas, J. García: Appl. Opt. **38**, 4760 (1999)
12. J. Guerrero-Bermúdez, J. Meneses, O. Gualdrón: Opt. Eng. **39**, 2828 (2000)
13. Ph. Réfrégier, F. Goudail: 'Statistical Processing of Polarization Diversity Images'. In: *Optoelectronic Information Processing: Optics for Information Systems, Valencia, Spain, June 2001*, Proc. SPIE **CR81** (SPIE Press, Bellingham, Washington 2001) p. 262
14. B. Javidi (Ed.): *Smart Imaging Systems* (SPIE Press, Bellingham, Washington 2001)
15. B. Javidi (Ed.): *Image Recognition, Algorithms, Systems, and Applications* (Marcel-Dekker, New York 2002)
16. T.J. Naughton, Y. Frauel, B. Javidi, E. Tajahuerce: 'Compression of Digital Holograms for Three-Dimensional Object Recognition.' In: *Algorithms and Systems for Optical Information Processing V, San Diego July 2001*, Proc. SPIE **4471** (SPIE Press, Bellingham, Washington 2001) p. 281
17. T.J. Naughton, Y. Frauel, B. Javidi, E. Tajahuerce: Appl. Opt., September 2001 to appear vol. 41, no. 17, 10 June 2002
18. O. Matoba, T.J. Naughton, Y. Frauel, N. Bertaux, B. Javidi: Appl. Opt., not published yet
19. J.H. Bruning, D.R. Herriott, J.E. Gallagher, D.P. Rosenfeld, A.D. White, D.J. Brangaccio: Appl. Opt. **13**, 2693 (1974)
20. I. Yamaguchi, T. Zhang: Opt. Lett. **22**, 1268 (1997)
21. M. Rabbani: *Selected Papers on Image Coding and Compression*, SPIE Milestone Series **MS48** (SPIE Press, Bellingham, Washington 1992)
22. A. Bilgin, G. Zweig, M.W. Marcellin: Appl. Opt. **39**, 1799 (2000)
23. S. Qian, A.B. Hollinger, D.J. Williams, D. Manak: Opt. Eng. **35**, 3242 (1996)
24. J. Vago, H. Vermeulen, A. Verga: Opt. Eng. **36**, 3052 (1997)
25. R. Shahnaz, J.F. Walkup, T.F. Krile: Appl. Opt. **38**, 5560 (1999)

26. G.K. Wallace: Commun. ACM **34**, 30 (1991)
27. A. Léger, T. Omachi, G.K. Wallace: Opt. Eng. **30**, 947 (1991)
28. F. Murtagh, J.-L. Starck, M. Louys: Int. J. Imaging Syst. Technol. **9**, 38 (1998)
29. T. Nomura, A. Okazaki, M. Kameda, Y. Morimoto, B. Javidi: 'Digital Holographic Data Reconstruction with Data Compression'. In: *Algorithms and Systems for Optical Information Processing V, San Diego, July 2001*, Proc. SPIE **4471** (SPIE Press, Bellingham, Washington 2001) p. 235
30. N. Ahmed, T. Natarajan, K.R. Rao: IEEE Trans. **C-32**, 90 (1974)
31. E.Y. Lam, J.W. Goodman: Appl. Opt. **37**, 6213 (1998)
32. H.J. Caulfield: *Handbook of Optical Holography* (Academic Press, New York 1979)
33. D.A. Huffman: Proc. IRE **40**, 1098 (1952)
34. J. Ziv, A. Lempel: IEEE Trans. **IT-23**, 337 (1977)
35. J. Ziv, A. Lempel: IEEE Trans. **IT-24**, 530 (1978)
36. T.A. Welch: IEEE Computer **17**, 8 (1984)

14 Three-Dimensional Image Security

B. Javidi, E. Tajahuerce, and O. Matoba

14.1 Introduction

Data security has become an important field of research in the last few years, and new techniques of pattern recognition, encryption, and verification, or improvements of existing methods, are introduced every year. In this direction, optical processing systems have shown some advantages over purely electronic processing [1–10]. Optical processors provide a large degree of freedom to secure data, high space–bandwidth product, and high speed of operation. On top of that, optical systems constitute an unavoidable step in data security applications when we are dealing with information already contained in the optical domain like two-dimensional (2-D) images or three-dimensional (3-D) objects.

In general, the information to be secured or verified by optical means is encoded as a 2-D image using amplitude, phase, polarization or wavelength modulation of light and optically or optoelectronically processed. One important approach to encrypt this information is to use random phase codes to modify the Fraunhofer or Fresnel diffraction patterns of the 2-D input image [1, 2, 11–26]. This method is similar to that used for securing or multiplexing holographic memories [27–35]. In general, the encrypted image contains both amplitude and phase and, therefore, holographic recording may also be required [36]. This fact makes it difficult to transmit the encrypted information over conventional communication channels. On the other hand, several digital holography methods have been applied to record fully complex information with an intensity recording device, such as a CCD camera [37–44]. Among them, digital phase-shifting interferometry stands out as a versatile and efficient technique [45–49]. After transmission, the holograms are reconstructed by using a digital computer or optically.

Several techniques combining optical encryption and digital holography have already been reported. They record electronically the holographic information associated with the diffraction patterns of a 2-D image by using off-axis [50] and in-line [51, 52] digital holography. Encryption is performed by modifying the diffraction pattern with random phase-codes. The encryption is performed immediately and directly on the complex information, and the information can be transmitted through conventional communication channels. Decryption at a remote site requires no more computation than the usual image reconstruction procedure. Therefore, there is a potential speed advantage over full-digital encryption techniques. The original information is then reconstructed digitally by simulating free-space propagation.

In this chapter we review a method which extends the previous technique to optically encrypt 3-D information by means of digital holography [53]. A phase-shifting interferometer records the phase and amplitude information generated by a 3-D object at a plane located in the Fresnel diffraction region by using an intensity-recording device. This information is encrypted optically with the Fresnel diffraction pattern generated by a random phase mask and stored electronically. To perform decryption, the key is also electronically recorded by phase-shifting interferometry. The encrypted hologram can be transmitted electronically but can only be decoded with the proper key.

After decryption, images of the 3-D object focused at different planes can be generated digitally [53]. Furthermore, the method allows us to reconstruct the 3-D object with different perspectives from a single encrypted image. It is important to note that this technique does not require us to send the key exclusively through a digital communication channel. Instead, a copy of the random phase key itself can be sent to the authorized user. This results in improved security over pure electronic encryption techniques because random phase masks are difficult to counterfeit. However, the transmission of successive encrypted images can be performed rapidly and efficiently by electronic means.

Finally, we show that the field diffracted by the 3-D object can be reconstructed optically after decryption with the key [54]. We have proposed a method that uses phase-only information of the decrypted complex field from the digital hologram to reconstruct the 3-D object. The proposed method is suitable for the real-time optical reconstruction of the 3-D object because phase-only reconstruction allows us to use commercially available display devices such as liquid crystal spatial light modulators (SLMs). In this way, we take advantage of minimal optical power loss in the reconstruction process.

An experimental demonstration of the encryption process is presented. We show that the decryption of the hologram is only possible with the proper key. The digital reconstruction of different views of the 3-D object is also demonstrated experimentally. Finally, the reconstruction of 3-D objects using phase-only information is evaluated numerically and experimentally by using a liquid crystal spatial light modulator.

In Sect. 14.2 we describe the basic principle of the method to encrypt 3-D information using phase-shifting interferometry and random phase codes. The method to obtain the key and to decrypt the hologram is explained in Sect. 14.3. In Sect. 14.4 we discuss the technique for the digital reconstruction of the decrypted information, showing that it is possible to generate different perspectives of the input 3-D object. In Sect. 14.5 we describe the method to reconstruct optically the decrypted 3-D object. Finally, Sect. 14.6 summarizes the conclusions.

14.2 Three-Dimensional Object Encryption With Digital Holography

The basic principle of an optical security system based on random phase encoding was proposed by Javidi and Horner and later improved by Réfrégier and Javidi [1, 2]. Two random phase masks, one in the input plane and other in the Fourier plane of an optical Fourier processor are used to convert the input image into wide-sense stationary white noise. In this section we show how this technique can be adapted to optically encrypt the complex amplitude distribution generated by a 3-D object illuminated with coherent light. The phase and amplitude of the diffraction pattern is measured using a phase-shifting interferometer. Encryption is accomplished by locating a random phase mask at the reference beam of the interferometer.

The outline of the optical system to encrypt 3-D information is shown in Fig. 14.1. It is based on a Mach–Zehnder interferometer architecture. The first beam splitter (BS) divides a collimated light beam from an argon laser into the reference and the object beams. The object beam illuminates the 3-D input after reflection in a couple of mirrors (M). Assuming that the incident light is diffracted only once, we can describe the opaque object as a 3-D continuum distribution of point sources with relative amplitude $U_{\mathrm{O}}(x, y, z)$, were x and y are transversal coordinates and z is the paraxial distance to the output plane. In this way, the complex amplitude distribution at the output plane, $U_{\mathrm{H}}(x, y)$, can be evaluated with the following 3-D integral of superposition:

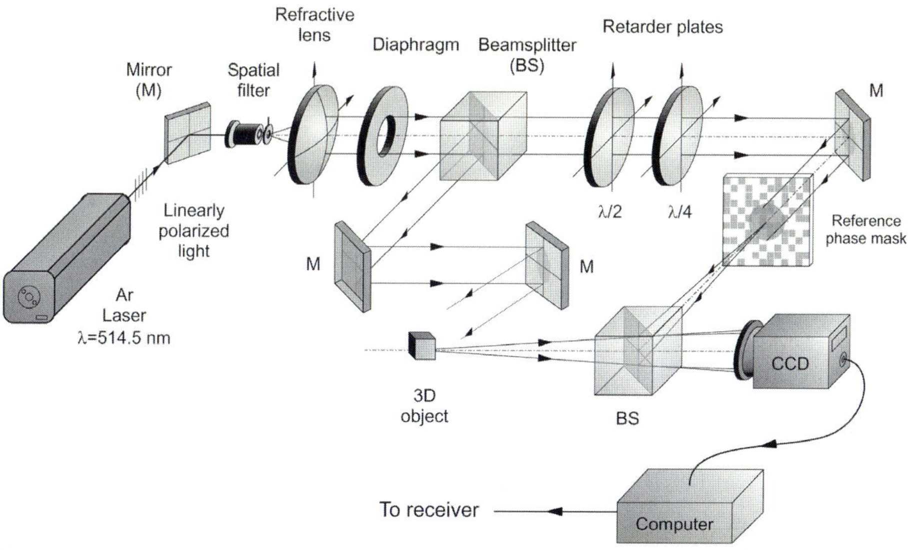

Fig. 14.1. Phase-shifting holographic system to encrypt 3-D information

$$U_{\mathrm{H}}(x,y) = A_{\mathrm{H}}(x,y)\exp\left[\mathrm{i}\,\phi_{\mathrm{H}}\,(x,y)\right] = \frac{1}{\mathrm{i}\,\lambda}\int\int_{-\infty}^{\infty}\int U_{\mathrm{O}}(x',y';z)\ \times$$
$$\times\ \frac{1}{z}\ \exp\left(\mathrm{i}\frac{2\pi}{\lambda}\,z\right)\exp\left\{\mathrm{i}\frac{\pi}{\lambda\,z}\left[(x-x')^2+(y-y')^2\right]\right\}\ \mathrm{d}x'\,\mathrm{d}y'\,\mathrm{d}z\ , \tag{14.1}$$

where λ is the wavelength of the incident light. In (14.1), $A_{\mathrm{H}}(x,y)$ and $\phi_{\mathrm{H}}(x,y)$ are the amplitude and phase, respectively, of the complex amplitude distribution at the output plane generated by the object beam. Note that in this approximation, neglecting secondary diffraction, the object can be considered also as a continuum of 2-D images at different distances z to the output plane. We will show in Sect. 14.3 that by measuring the amplitude $A_{\mathrm{H}}(x,y)$ and phase $\phi_{\mathrm{H}}(x,y)$ of the diffraction pattern in (14.1) it is possible to reconstruct the amplitude distribution generated by the 3-D object in different planes, in particular those in which the object appears in focus, computing an inverse Fresnel-transform operation either optically or numerically.

The parallel reference beam in Fig. 14.1 travels through two retarder plates, one quarter-wave plate and one half-wave plate, is reflected by a mirror, and is diffracted by a random phase mask. The diffracted light is reflected by a second beam splitter and interferes with the light diffracted by the 3-D object in the CCD detector. The system is aligned in such a way that, without the phase mask, the reference beam generates a plane wave traveling perpendicular to the CCD sensor after reflection in the second beam splitter. The light provided by the argon laser is linearly polarized. In this way, by suitably orienting the phase retarders, the phase of the reference beam can be shifted to different values. This is achieved by aligning successively the different slow and fast axes of the phase retarders with the polarization of the incident light. In this way, the phase of the reference beam can be modulated with phase shifts $\Delta\varphi_1 = 0$, $\Delta\varphi_2 = -\pi/2$, $\Delta\varphi_3 = -\pi$, and $\Delta\varphi_4 = -3\pi/2$.

The phase mask, with a random phase distribution $\Phi(x,y)$, and located at a distance $L = L_1 \mathrm{r} L_2$ from the detector (see Fig. 14.1), generates a complex field for each phase shift $\Delta\varphi_{\mathrm{p}}$ given by the following Fresnel integral:

$$U_{\mathrm{R}}(x,y;\Delta\varphi_{\mathrm{p}}) = \exp\left[\mathrm{i}\,\Delta\varphi_{\mathrm{p}}\right]\ \exp\left[i\frac{\pi}{\lambda\,L}\ (x^2+y^2)\right]$$
$$\times\int_{-\infty}^{\infty}\int \exp\left[\mathrm{i}\phi(x',y')\right]\ \exp\left[\mathrm{i}\frac{\pi}{\lambda\,L}\ (x'^2+y'^2)\right]$$
$$\times\exp\left[-\mathrm{i}\frac{2\pi}{\lambda L}(xx'+yy')\right]\ \mathrm{d}x'\ \mathrm{d}y'\ , \tag{14.2}$$

In the previous equation some constant factors have been neglected. Note that (14.2) can be written in convolution form as [8]

$$U_{\mathrm{R}}(x,y;\Delta\varphi_{\mathrm{p}}) = A_{\mathrm{R}}(x,y)\ \exp\left[\mathrm{i}\left(\varphi_{\mathrm{R}}(x,y)+\Delta\varphi_{\mathrm{p}}\right)\right] =$$
$$= \exp\left(\mathrm{i}\,\Delta\varphi_{\mathrm{p}}\right)\ \exp\left[\mathrm{i}\Phi\ (x,y)\right]\otimes\exp\left[\mathrm{i}\frac{\pi}{\lambda L}\ (x^2+y^2)\right]\ . \tag{14.3}$$

In this equation, $A_\mathrm{R}(x,y)$ and $\varphi_\mathrm{R}(x,y)$ are the amplitude and phase, respectively, of the complex amplitude distribution at the output plane generated by the reference beam when both fast axes of the retarder plates are aligned with the direction of polarization, i.e. when $\Delta\varphi_\mathrm{p}=0$. Since $\Phi(x,y)$ is a random phase distribution, from (14.3) we conclude that $A_\mathrm{R}(x,y)$ and $\varphi_\mathrm{R}(x,y)$ will be random-noise-like functions.

The complex light field at the output plane can be evaluated with digital phase-shifting interferometry by recording intensity patterns with different phase shifts of the reference beam [45]. The intensity at the output plane is given by the coherent superposition of (14.1) and (14.3), i.e.

$$I_\mathrm{p}(x,y) = [A_\mathrm{H}(x,y)]^2 + [A_\mathrm{R}(x,y)]^2 + \\ + 2A_\mathrm{H}(x,y)\, A_\mathrm{R}(x,y)\, \cos[\phi_\mathrm{H}(x,y) - \varphi_\mathrm{R}(x,y) - \Delta\varphi_\mathrm{p}] \,. \tag{14.4}$$

By recording four interference patterns $I_\mathrm{p}(x,y)$ with the reference beam phase-shifted by the previous four values of $\Delta\varphi_\mathrm{p}$, it is possible to obtain the encrypted phase and amplitude. This is known as the four-step method; see, for example, [48]. From (14.4) it is straightforward to show that the measured phase is given by

$$\phi_\mathrm{E}(x,y) = \phi_\mathrm{H}(x,y) - \varphi_\mathrm{R}(x,y) = \arctan\left(\frac{I_4 - I_2}{I_1 - I_3}\right) \,. \tag{14.5}$$

The amplitude can be calculated from the following equation:

$$A_\mathrm{E}(x,y) = A_\mathrm{H}(x,y)\, A_\mathrm{R}(x,y) = \frac{1}{4}\left[(I_1 - I_3)^2 + (I_4 - I_2)^2\right]^{1/2} \,. \tag{14.6}$$

For simplicity, in (14.5) and (14.6) we have omitted the spatial dependence of $I_\mathrm{p}(x,y)$. The two functions given by (14.5) and (14.6), i.e. the amplitude $A_\mathrm{E}(x,y)$ and phase $\phi_\mathrm{E}(x,y)$, constitute the encrypted image.

Without knowledge of the random amplitude $A_\mathrm{R}(x,y)$ and phase $\varphi_\mathrm{R}(x,y)$ generated by the reference beam, which act as two keys for decryption, it is very difficult to recover the amplitude $A_\mathrm{H}(x,y)$ and phase $\phi_\mathrm{H}(x,y)$ necessary to reconstruct the images of the 3-D object by inverse Fresnel propagation. Note that the random phase mask and its 3-D position act also as a key. Therefore, the input information is encrypted such that it can only be decrypted with the reference complex amplitude distribution or using the same reference phase mask $\phi_\mathrm{R}(x,y)$ located in the proper 3-D position. In contrast with other phase encryption methods, not only the phase but also the amplitude of the diffraction pattern of the input is modified by the introduction of the random phase mask in the reference beam. Furthermore, instead of the Fraunhofer diffraction pattern, an arbitrary Fresnel diffraction pattern of the input object is used for encryption. Therefore, phase-retrieval algorithms are difficult to apply. Besides, by dealing with Fresnel diffraction patterns, the intensity of the output images is properly adapted to the dynamic range of the detector. In this way, the introduction of a second random phase mask in the object beam is not necessary.

In summary, the 3-D input information can be transmitted securely by using conventional communication channels. We only need to send the amplitude $A_{\mathrm{E}}(x,y)$ and phase $\phi_{\mathrm{E}}(x,y)$ of the encrypted digital hologram coded as two conventional image distributions. We show next that the digital hologram can be decrypted in a simple way with the proper key and that it is possible to recover the input both digitally or optically.

14.3 Key Recording and Decryption of the Digital Hologram

To obtain the key amplitude and phase functions, $A_{\mathrm{R}}(x,y)$ and $\varphi_{\mathrm{R}}(x,y)$, the 3-D object is removed and, with the help of one mirror, the object beam propagates through the second beam splitter parallel to the axis of the detector, as is shown in Fig. 14.2. In this way, the reference and object parallel beams interfere in-line into the output plane. With this optical system, we are able to record the other four intensity patterns $I'_{\mathrm{p}}(x,y)$ by phase-shifting the reference beam. Now, the key phase $\phi_{\mathrm{K}}(x,y)$ is given by

$$\phi_{\mathrm{K}}(x,y) = \phi_{\mathrm{C}} - \varphi_{\mathrm{R}}(x,y) = \arctan\left(\frac{I'_4 - I'_2}{I'_1 - I'_3}\right), \tag{14.7}$$

while the amplitude $A_{\mathrm{K}}(x,y)$ is calculated from

$$A_{\mathrm{K}}(x,y) = A_{\mathrm{C}} A_{\mathrm{R}}(x,y) = \frac{1}{4}[(I'_1 - I'_3)^2 + (I'_4 - I'_2)^2]^{1/2} . \tag{14.8}$$

The parameters ϕ_{C} and A_{C} are the constant phase and amplitude of the object beam and can be substituted by 0 and 1, respectively. Note that we only need to do this operation once, to obtain the key, but many 3-D input objects can be encrypted with the same encrypting key. If necessary, the key can be modified very simply by changing the random phase mask or its 3-D position. The key can be sent to the authorized user in the form of two conventional images containing the 2-D distributions $A_{\mathrm{K}}(x,y)$ and $\phi_{\mathrm{K}}(x,y)$. Alternatively, we can send a copy of the phase mask itself and its 3-D position. The operation of decryption is performed digitally by combining the encrypted hologram, constituted by $A_{\mathrm{E}}(x,y)$ and $\phi_{\mathrm{E}}(x,y)$, with the key functions $A_{\mathrm{K}}(x,y)$ and $\phi_{\mathrm{K}}(x,y)$ in the following way:

$$\phi_{\mathrm{D}}(x,y) = \phi_{\mathrm{E}}(x,y) - \phi_{\mathrm{K}}(x,y) , \tag{14.9}$$

and

$$A_{\mathrm{D}}(x,y) = \begin{cases} \dfrac{A_{\mathrm{E}}(x,y)}{A_{\mathrm{K}}(x,y)}, & \text{if } A_{\mathrm{K}}(x,y) \neq 0 \\ 0 & \text{otherwise} \end{cases} . \tag{14.10}$$

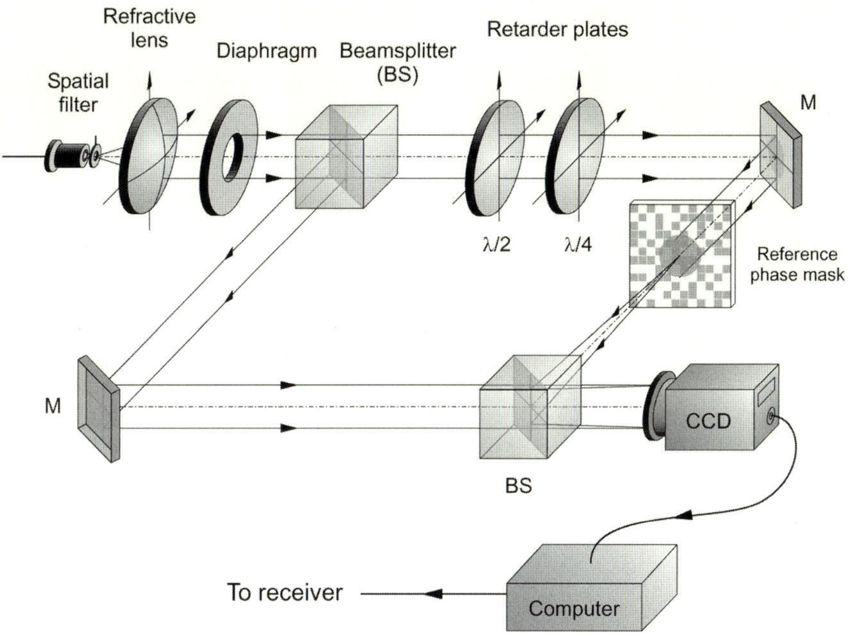

Fig. 14.2. Optical system to measure the key

The decrypted functions $A_D(x,y)$ and $\phi_D(x,y)$ constitute the amplitude and phase of the decrypted Fresnel hologram $U_D(x,y) = A_D(x,y)\exp[i\,\phi_D(x,y)]$. Except for constant factors, the complex function $U_D(x,y)$ reproduces the Fresnel hologram of the 3-D object, $U_H(x,y)$, and thus by free-space propagation of $U_D(x,y)$ the amplitude distribution of the input object can be reconstructed.

The operation of decryption can also be done directly from the intensity measurements in the following way:

$$\phi_D(x,y) = \arctan\left[\frac{(I_4 - I_3)(I_1' - I_3') - (I_1 - I_3)(I_4' - I_2')}{(I_4 - I_2)(I_4' - I_2') - (I_1 - I_3)(I_1' - I_3')}\right], \tag{14.11}$$

and

$$A_D(x,y) = \left[\frac{(I_1 - I_3)^2 + (I_4 - I_2)^2}{(I_1' - I_3')^2 + (I_4' - I_2')^2}\right]^{1/2}. \tag{14.12}$$

In this case, the two functions $I_{13}(x,y) = [I_1(x,y) - I_3(x,y)]$ and $I_{42}(x,y) = [I_4(x,y) - I_2(x,y)]$ can constitute the encrypted information, in the form of two irradiance distributions, while the functions $I_{13}'(x,y) = [I_1'(x,y) - I_3'(x,y)]$ and $I_{42}'(x,y) = [I_4'(x,y) - I_2'(x,y)]$ can act as the key.

14.4 Digital Reconstruction of the 3-D Object

After decryption, a direct digital reconstruction of the input object can be performed by computing a Fresnel integral numerically. Let $U_{\mathrm{D}}(m,n)$ be the discrete amplitude distribution of the decrypted digital hologram, where m and n are discrete coordinates along the orthogonal directions x and y, respectively. In this way, $x = m\ \Delta x$ and $y = n\ \Delta y$, where Δx and Δy are the resolutions of the CCD detector. The discrete complex amplitude distribution $U_{\mathrm{O}}(m',n')$ at a plane orthogonal to the digital hologram, located at a distance d, is given, aside from constant factors, by the following discrete Fresnel transformation:

$$U_{\mathrm{O}}(m',n') = \exp\left[\frac{-\mathrm{i}\pi}{\lambda d}(\Delta x'^2 m'^2 + \Delta y'^2 n'^2)\right] \sum_{m'=0}^{N_x-1} \sum_{n'=0}^{N_y-1} U_{\mathrm{D}}(m,n)$$
$$\times \exp\left[\frac{-\mathrm{i}\pi}{\lambda d}(\Delta x^2 m^2 + \Delta y^2 n^2)\right] \exp\left[-\mathrm{i}2\pi\left(\frac{m'm}{N_x} + \frac{n'n}{N_y}\right)\right] . \qquad (14.13)$$

In (14.13), m' and n' are discrete coordinates at the reconstruction plane, $\Delta x'$ and $\Delta y'$ denote the resolution at this plane, and N_x and N_y are the number of pixels of the detector along the x and y axis, respectively. It can be shown that the resolutions $\Delta x'$ and $\Delta y'$ along the horizontal and vertical directions in the object plane are given by

$$\Delta x' = \frac{\lambda d}{N_x \Delta x} \quad \text{and} \quad \Delta y' = \frac{\lambda d}{N_y \Delta y} . \qquad (14.14)$$

In this way, the resolution improves as the number of samples in the hologram plane increases.

It is important to note that different regions of the hologram record light arising from different perspectives of the 3-D object. Therefore, as in conventional analogue holography, we can reconstruct different views of the 3-D object by using different windows in the decrypted hologram, as shown in Fig. 14.3, and illuminating them with tilted plane waves. To this end, let us consider that the input object to be encrypted was located at a distance d from the CCD.

To reconstruct a particular view of the decrypted 3-D object, first, we must consider only a rectangular window, $U'_{\mathrm{D}}(m,n;a_x,a_y)$, of the digital hologram $U_{\mathrm{D}}(m,n)$ centered at the proper location. The parameters a_x and a_y denote the coordinates of the center of the window. In this way, the information contained in this region of the hologram corresponds to a direction of observation that subtends angles α and β with the optical axis given by

$$\alpha = \frac{\alpha_x \Delta_x}{d} \quad \text{and} \quad \beta = \frac{\alpha_y \Delta_y}{d} . \qquad (14.15)$$

Second, we must consider that the hologram is illuminated by a parallel light beam directed towards the 3-D object, i.e. a light beam tilted by angles α

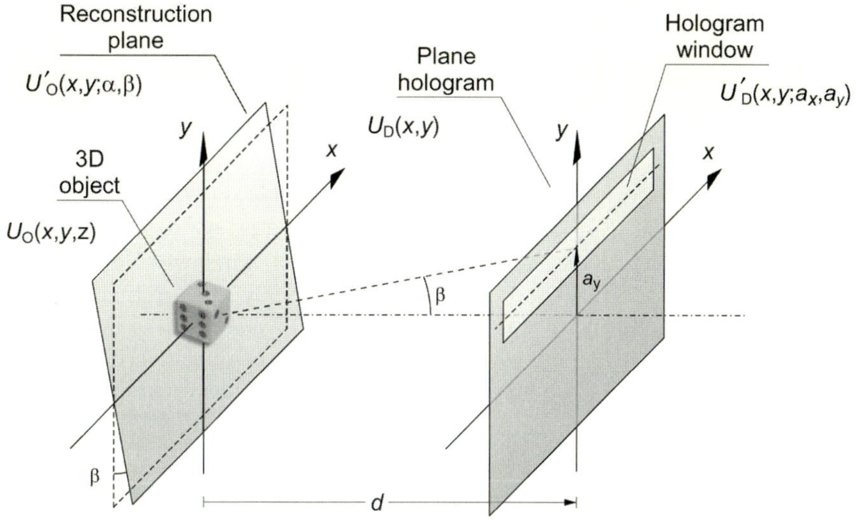

Fig. 14.3. Location of the window in the digital hologram to generate different vertical perspectives of the 3-D object

and β with respect to the optical axis given by (14.15). In this way, the perspective of the object will remain centered in the reconstruction of the hologram. The angular range achieved will be limited only by the size of the detector.

From the above considerations, the rectangular region of the hologram to be used for reconstruction must be defined as

$$U'_{\mathrm{D}}(m,n;a_x,a_y) = U_{\mathrm{D}}(m,n)\ \mathrm{rect}\left(\frac{m-a_x}{b_x},\frac{n-a_y}{b_y}\right) \times \exp[\mathrm{i}2\pi(a_x m + a_y n)]\ , \tag{14.16}$$

where $\mathrm{rect}(x,y)$ is the rectangle function and b_x and b_y denote its transverse size. The linear phase factor in (14.16) simulates the effect of a tilted plane wave incident on the hologram. Now, the discrete complex amplitude distribution $U'_{\mathrm{O}}(m,n;\alpha,\beta)$ at a plane located at a distance d and tilted by angles α and β with respect the hologram plane can be computed using the following equation:

$$U'_{\mathrm{O}}(m',n';\alpha,\beta) = \exp\left[\frac{-\mathrm{i}\pi}{\lambda d}(\Delta x'^2 m'^2 + \Delta y'^2 n'^2)\right] \sum_{m'=0}^{N_x-1}\sum_{n'=0}^{N_y-1} U'_{\mathrm{D}}\left(m,n;\frac{\alpha d}{\Delta x},\frac{\beta d}{\Delta y}\right)\exp\left[\frac{-\mathrm{i}\pi}{\lambda d}(\Delta x^2 m^2 + \Delta y^2 n^2)\right] \times \exp\left[-\mathrm{i}2\pi\left(\frac{m'm}{N_x}+\frac{n'n}{N_y}\right)\right]\ . \tag{14.17}$$

Note that the introduction of the linear phase factor in (14.16) is equivalent to a tilted plane wave illuminating the hologram in (14.17).

The geometry of the configuration for the particular case of a vertical variation of the perspective (i.e. restricted to $a_x = 0$) is that depicted in Fig. 14.3. The maximum angle of view, $\beta_{max} = (N_y - b_y)\Delta y/d$, is a function of the CCD size and the window size. This angle range can be increased by increasing the size of the CCD or by decreasing the size of the window used for reconstruction. However, as was pointed out in (14.14), the resolution of the reconstructed image decreases on selecting smaller windows. Nevertheless, It is important to realize that the hologram window does not need to be square. Therefore, when we want to obtain a set of perspectives in only one direction, it is possible to improve the resolution of the object reconstruction by increasing the size of the hologram window in the transverse direction, as is shown in Fig. 14.3.

The operation in (14.17) can be efficiently computed using a fast Fourier transform algorithm [55]. Thus, different perspectives can be generated at high speed. Points on the surface of the object at distances z from the hologram different from d will appear defocused in the reconstructed image. Nevertheless, the planes of reconstruction can also be changed easily in the computer, starting from the same digital hologram. Note also that the field of focus can be increased by diminishing the size of the hologram window at the expense of a reduction of resolution.

In our experimental verification, we performed the encryption of a 3-D object with the phase-shifting interferometer depicted in Fig. 14.1. A mega-pixel camera with pixel size equal to 9×9 μm was used to record the interference patterns between the object and reference beams with a dynamic range equivalent to 256 grey levels. The object has the shape of a cube with 10 mm lateral size. In Fig. 14.4 we show the axial view of the object generated from a non-encrypted digital hologram using (14.13). The object was located at a distance $d = 570$ mm from the CCD and no phase mask was used at the reference beam. The resolution of the reconstructed image is limited by the size of the CCD, which acts as the aperture in the reconstruction process, as is shown in (14.14). As in conventional coherent holographic techniques, the quality of the image is also affected by speckle. The speckle size is, roughly speaking, equal to $1.2\,\lambda/NA$, where NA is the numerical aperture of the system. Thus the size of the speckle is approximately equal to the resolution of the reconstructed image given by (14.14).

In Fig. 14.5 we show the amplitude and phase of the encrypted hologram, computed from (14.5) and (14.6), when the random phase mask is located as shown in Fig. 14.1. The phase mask is a plastic diffuser of randomly varying thickness with a correlation length of 6 μm in both the x and y directions. Figure 14.6 is a picture of the amplitude and phase key generated by the random phase mask at the output plane obtained by using the optical system in Fig. 14.2. From the encrypted and the key intensity patterns we obtained

Fig. 14.4. Image of the 3-D object to be encrypted obtained by phase-shifting interferometry with the system in Fig. 14.1 without a reference phase mask

the decrypted hologram by using (14.9) and (14.10). The decrypted image shown in Fig. 14.7a was generated using the entire decrypted hologram. For comparison, in Fig. 14.7b we show a decryption performed without the correct key. Noise in the decrypted image arises from errors in the evaluation of the Fresnel diffraction pattern generated by the phase mask in the CCD. These errors are due to the limited size of the detector pixels and the grey-level quantization.

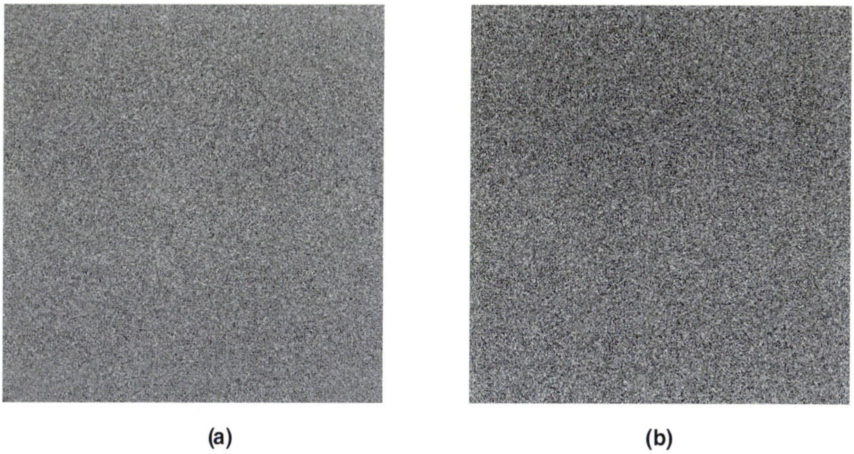

Fig. 14.5. Gray-level representation of (**a**) the encrypted amplitude and (**b**) the encrypted phase of the 3-D object in Fig. 14.4 by using a random reference phase mask

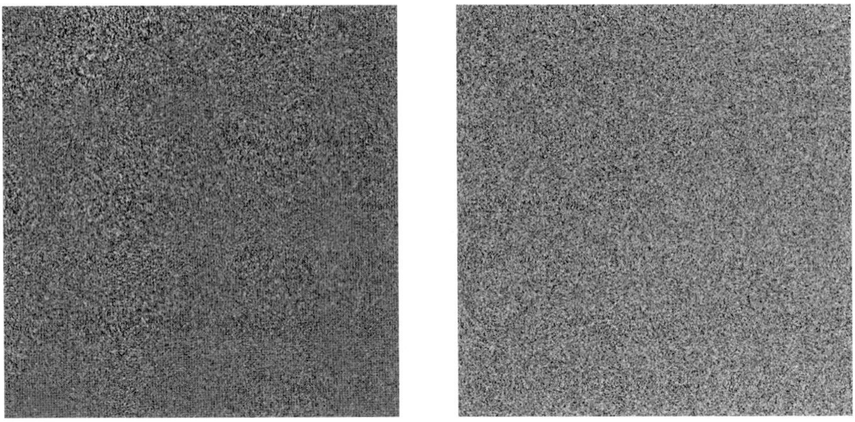

(a) (b)

Fig. 14.6. Representation of (**a**) the key amplitude and (**b**) the key phase of the random phase mask used to encrypt the information in Fig. 14.5

(a) (b)

Fig. 14.7. (**a**) Result of decryption of the encrypted information contained in Fig. 14.5 by using the key in Fig. 14.6; (**b**) Incorrect decryption using a wrong phase key

Finally, in Fig. 14.8 we show different perspectives of the decrypted 3-D object reconstructed from the decrypted digital Fresnel hologram. We used a rectangular window to limit the hologram with a vertical size of 256 pixels and horizontal size equal to the size of the CCD. The size reduction is necessary to reconstruct different perspectives from different regions of the hologram. But this also reduces the resolution and, consequently, the quality of the images with respect to that obtained using the whole digital hologram. Thus, we only reduce the size of the window in the direction for which we

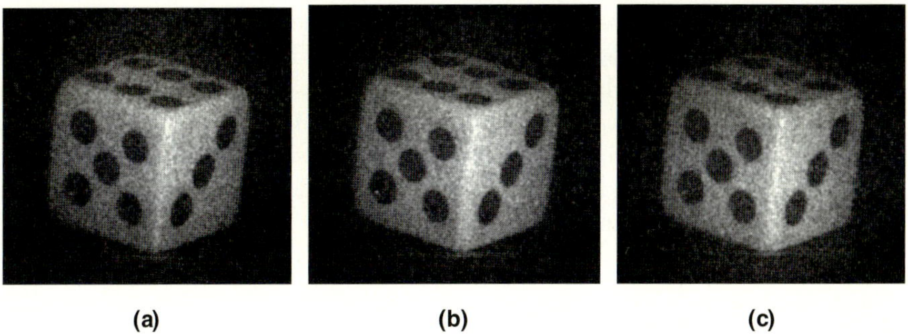

Fig. 14.8. Different perspectives of the 3-D object obtained after decryption. Angles of view are: (**a**) $\beta = 0.7°$; (**b**) $\beta = 0°$; and (**c**) $\beta = -0.7°$

want to achieve different views of the 3-D object. In this case, the window was displaced only in the vertical direction, generating an angular difference of 0.7° between the different perspectives labeled a, b, and c in Fig. 14.8. Any other view angle, limited only by the size of the CCD, can be reconstructed in the same way.

14.5 Optical Reconstruction

In this section we show that a real-time 3-D display using phase-only information of the complex field from a Fresnel digital hologram is able to reconstruct the input 3-D object. This is possible because the phase distribution of the hologram is a wavefront containing the 3-D position information of the point sources constituting the input object. Phase-only reconstruction allows us to use commercially available display devices such as liquid crystal spatial light modulators (SLMs) because these devices can only be operated, in principle, in either the phase or amplitude modulation mode. Furthermore, using only this phase information, we can reduce by half the storage requirements of the digital hologram.

With this technique it is possible to reconstruct optically a 3-D object which has been previously encrypted at a remote site. First, the Fresnel digital hologram of the 3-D object is encrypted and recorded optically as shown in Fig. 14.1 by using the technique described in Sect. 14.2. Both the digital hologram and the key are transmitted digitally to the authorized user. Second, a digital decryption process, as explained in Sect. 14.3, provides a Fresnel digital hologram of the input 3-D object if the proper key is used. Finally, the phase information of this decrypted digital hologram can be encoded in the SLM in order to reconstruct optically the original input object.

As a first approach, in order to verify experimentally the ability of this method to reconstruct optically a decrypted digital hologram, we recorded a Fresnel digital hologram of a 3-D input object and reconstructed it with a phase-only SLM. Figure 14.9a and b show a numerically reconstructed 3-D object using complex information and phase-only information, respectively, from its Fresnel digital hologram. We use a screw as a 3-D object in this experiment, which is located at a distance of 390 mm from the CCD. We can see that the phase-only information can reconstruct the 3-D object successfully. However, speckle noise influences both 3-D reconstructed images. To reduce the speckle noise, we apply a low-pass filter (mean filtering) to the reconstructed 3-D objects. The low-pass filtered images are presented in Fig. 14.9c and d. The sizes of the windows for the mean filter in Fig. 14.9c and d are 11×11 pixels and 21×21 pixels, respectively. In this way speckle noise is removed to a large extent by low-pass filtering. The intensity levels are scaled to enhance the contrast.

In practical systems, there is a limit on the number of phase levels that can be displayed in phase-only spatial light modulators (P-SLMs). Object reconstructions from phase-only information are compared with intensity images reconstructed from fully complex information, for different numbers of quantization levels. For these calculations, we use a mean filter with 11×11 pixels. Figure 14.10 shows the reconstructed 3-D object: without quantization reduction in Fig. 14.10a, and with quantization reductions to 4 bits and 1 bit, in Fig. 14.10b and c, respectively. Note that we can discern the 3-D objects

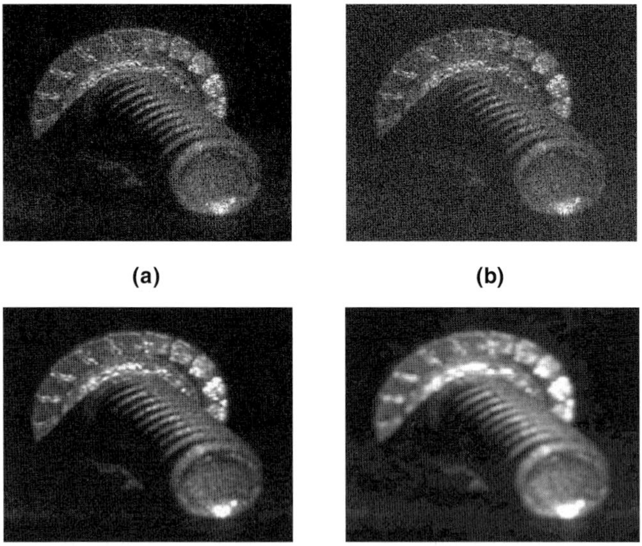

(a) (b) (c) (d)

Fig. 14.9. Reconstruction of the input 3-D object from (**a**) fully complex field and (**b**) phase-only information. Low-pass filtered images after reconstruction using phase-only information with a mean filter of (**c**) 11×11 pixels and (**d**) 21×21 pixels

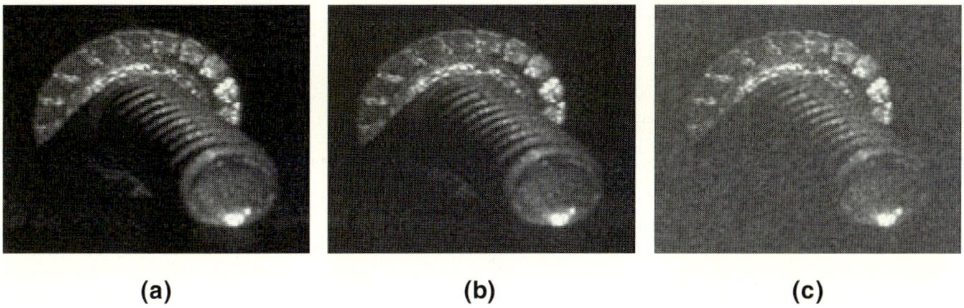

Fig. 14.10. Reconstructed 3-D object without quantization reduction in (a), and with quantization reductions to 4 bits in (b), and 1 bit in (c)

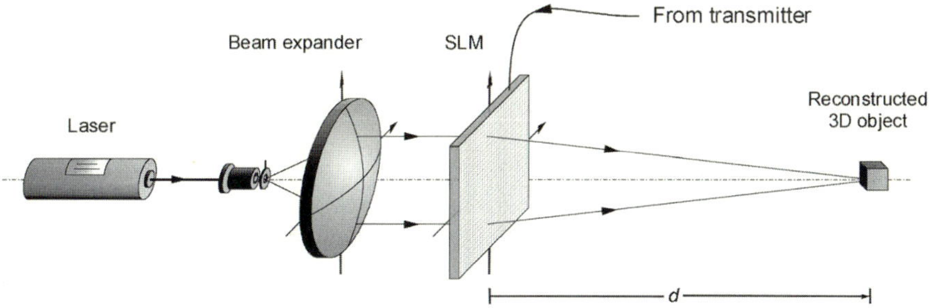

Fig. 14.11. System for 3-D object optical reconstruction using phase-only information from the digital hologram

even when the phase level of the P-SLM is binary. As few as 16 levels (4 bits) are sufficient to reconstruct the 3-D objects without significant loss of detail.

The optical setup to reconstruct 3-D objects is shown in Fig. 14.11. A He–Ne laser operated at a wavelength of 632.8 nm is used. After collimation and beam-width expansion the He–Ne laser beam illuminates a liquid crystal SLM with 1024×768 pixels. The size of each pixel is approximately $18\,\mu m \times 18\,\mu m$. This SLM does not have enough pixels to display the complete 2044×2028 pixel phase-only hologram, and so a 1024×768 pixel window is presented. By measuring the phase retardation of the SLM in a Mach–Zehnder interferometer, we determined that the maximum amount of phase retardation possible with our SLM is 0.6π. The phase retardation is almost linearly proportional to the signal level from the computer. Figure 14.12 shows the reconstructed 3-D object. Figure 14.12a–c show reconstructions at distances of 135 mm, 155 mm, and 165 mm from the SLM, respectively. Note that Fig. 14.12b shows the object with better focusing. From these figures, we can see that the images of the screw are reconstructed at different planes. The calculated position of the in-focus reconstruction is 126 mm. This systematic error is caused by the use of a different readout wavelength, different pixel size, and a readout wavefront that is not exactly the same as a plane wave.

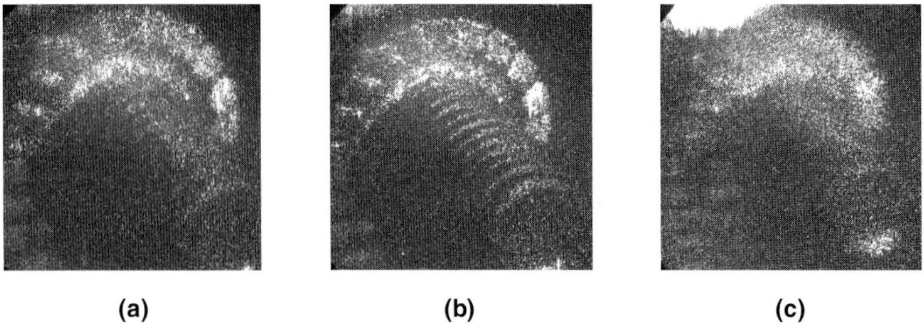

Fig. 14.12. Experimental results; (**a**), (**b**) and (**c**) are reconstructed images of the screw where the CCD is located at 135 mm, 155 mm, and 165 mm from the SLM, respectively

14.6 Conclusions

We have described an optoelectronic technique based on digital phase-shifting holography for 3-D security applications. Information codified as a 3-D image is optically processed and captured electronically. The encrypted data can be transmitted securely through conventional digital communication channels and reconstructed digitally. Views of the decrypted 3-D object can be reconstructed at different axial distances and with different perspectives.

Since an optical system is required to record the digital hologram, optical encryption with a random phase mask represents a convenient way to secure 3-D information. Furthermore, our technique combines the high space–bandwidth product and speed of optical processing with the advantages of electronic transmission, storage, and postprocessing. On the other hand, the method allows us to avoid electronic transmission of the key if this is desired to increase security. Besides, other electronic encryption techniques can also be applied to the digital hologram.

After decryption, the reconstruction of the 3-D object can be performed optically by encoding the decrypted digital hologram in a spatial light modulator. In our experimental verification we have shown that it is possible to optically reconstruct a Fresnel digital hologram of a 3-D object by using only the phase of the complex distribution constituting the digital hologram.

14.7 Acknowledgments

E. Tajahuerce acknowledges partial support from the Dirección General de Investigación Científica y Técnica, Spain, under project PB98-1049-C02-01.

References

1. B. Javidi, J.L. Horner: Opt. Eng. **33**, 1752 (1994)
2. Ph. Réfrégier, B. Javidi: Opt. Lett. **20**, 767 (1995)
3. Ph. Lalanne, H. Richard, J.C. Rodier, P. Chavel, J. Taboury, K. Madani, P. Garda, F. Devos: Opt. Comm. **76**, 387 (1990)
4. N. Yoshikawa, M. Itoh, T. Yatagai: Opt. Lett. **23**, 1483 (1998)
5. R.L. van Renesse: *Optical Document Security* (Artech House, Boston 1998)
6. A.D. McAulay: *Optical Computer Architectures* (John Wiley, New York 1991)
7. H.-Y. Li, Y. Qiao, D. Psaltis: Appl. Opt. **32**, 5026 (1993)
8. J.W. Goodman: *Introduction to Fourier Optics* (McGraw-Hill, New York 1996)
9. C.L. Wilson, C.I. Watson, E.G. Paek: Proc. SPIE **3073**, 373 (1997)
10. A. Pu, R. Denkewalter, D. Psaltis: Opt. Eng. **36**, 2737 (1997)
11. R.K. Wang, I.A. Watson, C.R. Chatwin: Opt. Eng. **35**, 2464 (1996)
12. B. Javidi, A. Sergent, G. Zhang, L. Guibert: Opt. Eng. **992**, 992 (1997)
13. F. Goudail, F. Bollaro, B. Javidi, Ph. Réfrégier: J. Opt. Soc. Am. A **15**, 2629 (1998)
14. G. Unnikrishnan, J. Joseph, K. Singh: Appl. Opt. **37**, 8181 (1998)
15. B. Javidi, E. Ahouzi: Appl. Opt. **37**, 6247 (1998)
16. J.L. Horner, B. Javidi, (Eds.): *Optical Engineering Special Issue on Optical Security,* Vol. 38 (SPIE, Bellingham 1999)
17. O. Matoba, B. Javidi: Opt. Lett. **24**, 762 (1999)
18. O. Matoba, B. Javidi: Appl. Opt. **38**, 7288 (1999)
19. P.C. Mogensen, J. Glückstad: Opt. Lett. **25**, 566 (2000)
20. B. Javidi, T. Nomura: Opt. Eng. **39**, 2439 (2000)
21. G. Unnikrishnan, J. Joseph, K. Singh: Opt. Lett. **25**, 887 (2000)
22. Z. Zalevsky, D. Mendlovic, U. Levy, G. Shabtay: Opt. Commun. **180**, 15 (2000)
23. T. Nomura, B. Javidi: Opt. Eng. **39**, 2031 (2000)
24. B. Zhu, S. Liu, Q. Ran: Opt. Lett. **25**, 1159 (2000)
25. E. Tajahuerce, J. Lancis, B. Javidi, P. Andres: Opt. Lett. **26**, 678 (2001)
26. S. Liu, Q. Mi, B. Zhu: Opt. Lett. **26**, 1242 (2001)
27. T.F. Krile, M.O. Hagler, W.D. Redus, J.F. Walkup: Appl. Opt. **18**, 52 (1979)
28. J.E. Ford, Y. Fainman, S.H. Lee: Opt. Lett. **15**, 1088 (1990)
29. C. Denz, G. Pauliat, G. Roosen, T. Tschudi: Opt. Commun. **85**, 171 (1991)
30. H. Lee, S.K. Jin: Appl. Phys. Lett. **62**, 2191 (1993)
31. J.F. Heanue, M.C. Bashaw, L. Hesselink: Appl. Opt. **34**, 6012 (1995)
32. Y.H. Kang, K.H. Kim, B. Lee: Opt. Lett. **22**, 739 (1997)
33. C. Denz, K.O. Mueller, F. Visinka, T.T. Tschudi: Proc. SPIE **3802**, 142 (1999)
34. C.C. Sun, W.C. Su, B. Wang, Y. Ouyang: Opt. Commun. **175**, 67 (2000)
35. X. Tan, O. Matoba, T. Shimura, K. Kuroda: Appl. Opt. **40**, 4721 (2001)
36. H.J. Caulfield (Ed.): *Handbook of Optical Holography* (Academic Press, London 1979)
37. L. Onural, P.D. Scott: Opt. Eng. **26**, 1124 (1987)
38. U. Schnars: J. Opt. Soc. Am. A **11**, 2011 (1994)
39. G. Pedrini, Y.L. Zou, H.J. Tiziani: J. Mod. Opt. **40**, 367 (1995)
40. U. Schnars, W.P.O. Jüptner: Appl. Opt. **33**, 179 (1994)
41. Y. Takaki, H. Kawai, H. Ohzu: Appl. Opt. **38**, 4990 (1999)
42. J.C. Marron, K.S. Schroeder: Appl. Opt. **31**, 255 (1992)
43. U. Schnars, T.M. Kreis, W.P.O. Jüptner: Opt. Eng. **35**, 977 (1996)

44. E. Cuche, F. Bevilacqua, C. Depeursinge: Opt. Lett. **24**, 291 (1999)
45. J.H. Bruning, D.R. Herriott, J.E. Gallagher, D.P. Rosenfeld, A.D. White, D.J. Brangaccio: Appl. Opt. **13**, 2693 (1974)
46. K. Creath: In *Progress in Optics,* Vol. XXVI, ed. by E. Wolf (North-Holland, Amsterdam 1988) pp. 349–393
47. J. Schwider: In *Progress in Optics,* Vol. XXVIII, ed. by E. Wolf (North-Holland, Amsterdam 1990) pp. 271–359
48. I. Yamaguchi, T. Zhang: Opt. Lett. **22**, 1268 (1997)
49. T. Zhang, I. Yamaguchi: Opt. Lett. **23**, 1221 (1998)
50. B. Javidi, T. Nomura: Opt. Lett. **25**, 29 (2000)
51. E. Tajahuerce, O. Matoba, S.C. Verrall, B. Javidi: Appl. Opt. **39**, 2313 (2000)
52. S. Lai, M.A. Neifeld: Opt. Commun. **178**, 283 (2000)
53. E. Tajahuerce, B. Javidi: Appl. Opt. **39**, 6595 (2000)
54. O. Matoba, T.J. Naughton, Y. Frauel, N. Bertaux, B. Javidi: Appl. Opt. Lett. **27**, 321 (2002)
55. J.W. Cooley, J.W. Tukey: Math. Comput. **19**, 297 (1965)

15 3-D Display and Interaction Technologies for Desktop Computing

S. Pastoor and J. Liu

15.1 Need for a Next-Generation Human–Computer Interface

Rapid progress in the enabling technologies has provided the basis for universal data and communication networks such as the Internet. Standardization allows the exchange of various types of information (e.g. JPEG for still images, MPEG for motion sequences and HTML/VRML for complex hypermedia objects), so that increasingly powerful networked multimedia systems become available. The end users are confronted with a range of new possibilities such as access to huge external data bases delivering information for daily use or for entertainment (encyclopedic information, video games, video-on-demand), access to numerous transactional services (home banking, e-commerce), as well as video communication and tele-cooperation in distributed work groups. In professional applications, networked computers are increasingly being used in order to achieve cost-effective production and accelerated product development thanks to new, flexible working structures, and in order to monitor and optimally control business processes.

It is now widely accepted that exploiting the tremendous potential of the emerging information technologies means that new user interface technologies and new interaction paradigms need to be developed, enabling the intentions of a user to be communicated to the computer in a more natural and intuitive way. A natural and intuitive user interface takes advantage of human propensities, physical analogies with familiar devices, and conceptual analogies with cultural standards [1] and, hence, allows users to infer the appropriate interactions with the computer without prior learning.

When the affordances of an interactive object are perceptually obvious, it is easy for a user to know how to interact with it. The use of easily understandable visual icons in graphical user interfaces has, for example substantially reduced the amount of knowledge people have to remember about the interface: in order to interact with the computer, users need primarily to learn some basic rules about how to interact with the simulated world of objects displayed on the computer screen [2]. Hence, graphical user interfaces based on windows, icons, menus, and a pointing device have considerably simplified the use of computers compared with purely text-oriented input and output techniques so that they have become commonplace now.

On the other hand, given that humans have a highly evolved sensory system (hearing, vision, taste, smell, touch, etc.), capable of gathering very detailed information about the environment, and comparably advanced means to communicate their intentions, there should be a great potential to further improve user interfaces by exploiting these capabilities. Moreover, human-to-human communication is continuous, i.e. we constantly receive information and show reactions on several "output" channels simultaneously, whereas human–computer interaction is basically performed in a sequential manner. Compared to the current state-of-the-art, much more information-rich ways to show what is happening in the system and much richer styles of interaction between a user and a computer are imaginable. In this context, multimodal interface design aims to make the most appropriate and effective use of the variety of human sensory and effector channels in combination. "The future of computing will be more like flight simulators or video games. Users will fly through large amounts of information, be able to select and filter information, combine components, and create meaningful results" [3].

The need for novel human–computer interface concepts became obvious for the first time in tasks where it was essential to see the structural relationships in large, hierarchically organized data bases. In this context, the advantage of three-dimensional representations for the visualization of relationships and patterns in complex information was clearly shown.

Compared to hierarchically organized data bases (such as the data stored on the hard disk of a PC), in networked systems like the Internet users are confronted with a considerably more complex situation. Regarding the structuring of information, the Internet is an example of a (very large) hypermedia system: the data and services available are distributed to nodes which are connected to open, non-hierarchical network structures via direct links, thus forming a virtual information space. This enables free, nonlinear access. The user can select any entry point in the network, whilst each node offers various alternatives for continuing navigation. The nodes can contain conventional, independent media objects or elements with modular design, leading to subsequent nodes of information. Besides static types of information (text, graphics and still images), dynamic time-based multimedia information (animation, moving pictures and sound) is presented at the user interface. In principle, this applies both to the activated media objects themselves and to the control elements for selecting, positioning, zooming, editing, etc. (e.g. animated icons showing the dynamic actions conveyed).

Empirical user studies with these types of systems have been carried out since the end of the 1980s [4]. Due to a lack of standardized testing methods, as well as the huge differences in the design of the systems tested, the results cannot be easily compared. All that is currently certain is that from a user point of view, some severe problems or risks exist which are difficult to accept and for which a convincing solution has yet to be found.

The following risks are immanent in the hypermedial structuring of non-hierarchical data bases such as the Internet [5]:

1. User disorientation ("getting lost in hyperspace"): the complexity and the unlimited scope of a non-hierarchically structured, open network means that it is impossible to offer a comprehensive overview. The incomplete visibility of relevant contexts and relationships hinders the generation of an adequate conceptual user-model of the system (i.e. the way in which the system structure and functionality is understood by the user) and subsequent goal-oriented intentional actions based on this model [1].[1] In order to improve orientation, it should be clear which sources of information and which processing functions are available when entering the system, which is the current position in the network, which path the user has taken to get there, which branching alternatives exist – both forwards and backwards – and how the user can switch between the different forms of use (e.g. reading instead of editing, exploratory navigation as opposed to specific search).
2. Increased "cognitive overhead": nonlinear access results in unusual selection decisions in terms of the desired sequence, execution, and type of presentation of the media objects available. Each leap between the nodes interrupts reading and may demand adjustment to new functionalities.

The integration of telecommunication and telecooperation into networked hypermedia systems brings two further risks [4, 6]:

3. Intrusiveness: this refers to the interference in the apparently delicate balance between the user's desire for a highly sensory communication situation (social presence), on the one hand, and the need for distance and control, on the other, as well as the loss of spontaneity and discretion with the transmission of informal communication.
4. Loss of personal control ("diffuse domain control"): telecooperation requires a fusing of the borders between one's own (private) working areas and the publicly accessible working areas. This could result in withdrawal or explicit territorial behavior instead of control by way of subtle social behavior.

In general, the control of privacy is an important issue whenever data is represented in a shared environment. For example, the privacy issue is most salient in the concept of ubiquitous computing [7]. This concept foresees a world where a variety of interconnected stationary and mobile

[1] According to Norman, the success of any form of man–machine interaction depends on adherence to elementary principles of design: the "visibility" of the system concept and the possible user actions, the immediate "feedback" on the effects, the possibility of developing a consistent "conceptual model" as well as the implantation of familiar tool functions and properties – even in a wide sense ("affordances", "constraints", "mappings").

(portable/wearable) computers is imbedded in a highly perceptive environment. Video cameras and other types of sensors will track any movement and even sense the emotional reactions of the users [8]. Based on this information, intelligent software agents are designated to individually address users and to adapt the user interface to individual needs at any time and any place. Moreover, these agents will be permitted to act autonomously on behalf of the user – within certain limits, which have to be set and controlled by the user.

In practice, all of the above-mentioned problems are prone to be linked with extremely negative emotional components such as frustration and helplessness. Counteractive measures aimed at providing users with quick cognitive orientation, personal control and sovereign handling of the computer environment without restricting functionality from the very start are considered to be the prime requirement for the design of future user interfaces [9].

In the following chapter we are going to overview current approaches to novel, information-rich multimodal interfaces.

We will show that there is a clear tendency in graphic interface design to make use of the highly evolved human capabilities to remember and evaluate spatial and temporal relationships in visualized information. Key issues in multimodal interaction are the integration of various interaction styles as well as the development of the component technologies required for nonintrusive perceptual user interfaces (PUIs, Sect. 15.2). In Sect. 15.3 we will review in more detail our own approach to a multimodal desktop computer which has a three-dimensional user interface displayed on an autostereoscopic three-dimensional (3-D) display and a range of video-based tracking components forming the PUI (the mUltimo3-D system). Following the principle of a user-centered design, we have repeatedly evaluated and redesigned the mUltimo3-D system (Sect. 15.4).

15.2 Multimodal Human–Computer Interaction

Humans are able to perceive information via multiple input channels using the senses of sight, hearing, touch, smell, taste, and balance. Perception via the visual and auditive modalities generally outperform the haptic, olfactory and gustatory modalities in terms of the bandwidth, and spatial and temporal resolution of the information transmitted. That is one of the reasons why visual and auditory displays have been dominant in human–computer interaction. Various output channels, such as speech, gestures, and written text, are used by humans to communicate with other people. The concept of multimodal interaction means not simply to provide the display and sensory techniques that enable such interaction with a computer (which is of course not simple from an engineering viewpoint). The real challenge is to fuse the various modalities in an intelligent way [10]. This is the task of interface software agents.

According to George G. Robertson [11], the interest in multimodal interfaces is based on the assumptions that (1) a much higher bandwidth is possible with these systems, (2) different modalities excel at different tasks (what may change over time and user), (3) errors and disfluencies can be reduced dramatically (due to the statistical advantage of combining multiple observations), and (4) multimodal interfaces are more natural so that they are more engaging. Multimodal output leverages the human senses and the perceptual system, because we can perceive multiple things at once; multiple input leverages the human motor capabilities and communication skills, because we can do multiple things at once.

Figure 15.1 shows a simple model of multimodal human–computer interaction. The role of the computer is considered as that of a (increasingly intelligent) tool helping the user to solve tasks. The user is regarded as the dominant and decisive part in human–computer interaction. In a broad sense, the user's role is to develop concepts and intentions, to communicate his/her goals to the computer, to develop problem solving strategies with the aid of the computer, and to supervise the actions taken and results delivered by the computer. This strategy is based on the assumption that humans are highly efficient in the way they define tasks, look at problems to be solved from different perspectives and at different levels of abstraction, and combine different sources of information in order to draw conclusions. Computers are strong in using lexical knowledge, can quickly and reliably solve tasks, if the strategy has already been delivered (programmed) by the user, and will presumably develop increased cognitive and problem-solving capabilities in the future. Hence, in this model computers should have some kind of intelligence in order to be able to interpret the user's intentions and to adequately support the user with the resources available (this is commonly referred to as the task of intelligent software agents [13]). Various input and output channels are imaginable to communicate intentions and results.

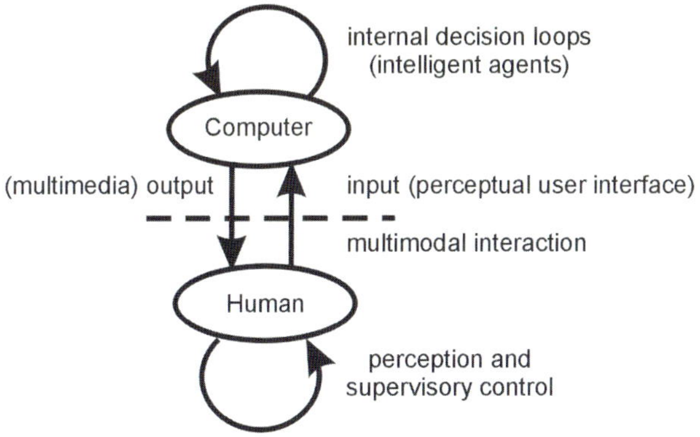

Fig. 15.1. Basic processes and channels in human–computer interaction (adapted from [12])

Humans have developed various modalities to communicate with other humans and to interact with their physical environment. Current research in human–computer interaction aims at improving the computer's perceptive faculties so that users can communicate their intentions in the most natural way (perceptual user interface, evaluating visual, auditive, and tactile information). In this chapter we will focus on the computer side of this interaction model, i.e. on the computer output (graphical user interface, sound events, and haptic feedback) and on the multimodal input.

15.2.1 Toward Perceptual User Interfaces

The core aspect in the development of the first graphic user interfaces was to create electronic counterparts of the physical objects in an office and to allow equivalent actions with the electronic representations on the screen [2]. Hence, the icons on the screen of a PC monitor represent familiar objects such as printed papers, folders, filing cabinets, and a paper bin. The overall organizing metaphor presented on the screen is the picture of a typical desktop. The mouse was developed to enable actions equivalent to physically handling documents, i.e. to select a document, to open, close, copy, move it, and to delete it by dropping it into a trash bin. The interface concept takes Norman's design principles into account ("What you see is what you get") and is called the WIMP concept, referring to the basic modules of Windows framing related information objects, Icons, Menus, and a Pointing device. The WIMP interface was basically developed in the 1960s and 70s and implemented for the first time in a product in 1981 by Xerox PARC.

Hence, the current user-interface metaphor was born of a concept design reflecting the generation and management of written paper documents, i.e. in a "paper world". On the other hand, studies on how electronic information systems are used in today's office have shown that the desktop metaphor is no longer adequate when the organization of information becomes the central aspect of computer use [14]. During the last decade, research work has started on alternatives to the desktop metaphor right up to a generation change (post-WIMP interface). Habit, repeated face-lifting, and the fact that the explosive expansion of Internet use began just recently may be some reasons for the continued existence of WIMP systems. The following is intended to show some of the main streams which may determine the components of a post-WIMP interface.

An alternative to displaying file structures in the form of hierarchical file folders is to chronologically arrange them in the sequence in which they were generated or as they arrived via the network as e-mails, downloads, or Web pages (Lifestreams [15]). A series of overlapping documents become larger in perspective, the nearer they are to the user. The newest document is the largest and appears in the front whilst the oldest documents disappear into the depths of the screen. The user can navigate back and forth in the stream and pick out and look at any document. An essential aspect when redesigning graphic interfaces is that context information must be made visible in order to simplify the finding and combining of information for the user. Lifestreams are one step in this direction since they refer to the user's time context when creating or processing a document, i.e. by integrating the human's excellent ability to remember time-based relationships into the visualization concept.

Direct manipulation of screen objects [9] is a fundamental feature of graphic user interfaces. With a view to information retrieval (browsing, navigating), Shneiderman has developed this principle to a mantra: "Overview first, then zoom and filter, and finally details on demand." The "overview" and "zoom" aspects have triggered the development of Zooming User Interfaces (ZUIs).[2] Figure 15.2 shows, for example, the Pad++ Web Browsers [16] with a sequence of linked Web pages. When browsing through the links, the path is indicated by lines. The currently focused node is enlarged. Surrounding nodes are reduced to thumbnails – they offer context information and show the current position in the Web. In the graphic interfaces in "Lifestreams" and "Pad++ Web Browsers", the size cue and the effect of occlusion are used to indicate distance: "smaller" and "occluded" means "farther away" (in terms of time in "Lifestreams" and in terms of location in "Pad++ Web Browsers").

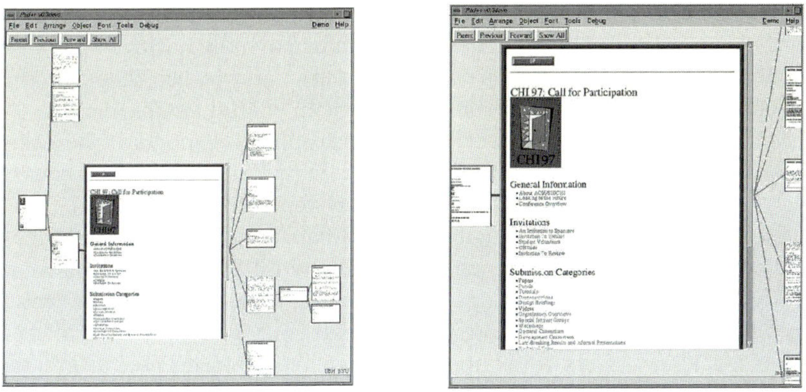

Fig. 15.2. Zooming User Interface (Pad++ Web Browser [16], Courtesy of Ben Bederson, University of Maryland). The current document is enlarged. The path through the network is clearly shown by links to thumbnails of the documents already visited

[2] According to Shneiderman, "direct manipulation" means direct interaction with visually perceived information objects. This requires complete visibility of the search area, the possibility to quickly explore, the availability of tools for pointing, selecting and dragging, and the immediate feedback on the consequences of an action (with the option to undo the respective action).

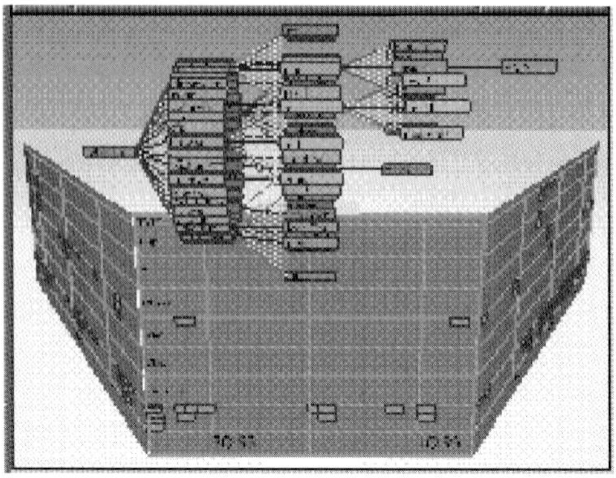

Fig. 15.3. User interface with animated 3-D visualization of hierarchically structured data (Cone Tree Browser, Courtesy George Robertson and Stuart Cart, Xerox Corp., and ACM Publications) and perspective display of context information in the peripheral wings of a Perspective Wall. When a data object is selected, the corresponding "pages" of the cone tree or the respective side of the wall are turned to the front [20]. From S.K. Card et al. *The Information Visualizer, An Information Workspace*, CHI'91, pp. 181–186 © (1991) ACM, Inc. Reprinted by permission

The continuous scaling of size and occlusion in order to create the impression of spatial perspective appears to be the logical next step, which means the step into the third dimension. Humans have the natural ability to unconsciously perceive various spatial depth cues[3] and to use these for the purpose of conscious orientation and movement in spatial environments.

Recent approaches more intensively use 3-D graphical elements at the interface; these elements are actually flat but they are supplemented by perspective, shading, texture gradient, and other monocular depth cues ($2^1/_2$-D) [17–19].[4] For example, the Information Visualizer software package by Xerox PARC [20] abandons the desktop metaphor and uses animated 3-D objects to visualize hierarchically structured data (Cone Tree Browser, for a current implementation in VRML, see [21]). In the Perspective Wall (Fig. 15.3), the central area of the screen allows users to examine the displayed information in greater detail, while context information is held at reduced magnification in the peripheral wings. This layout also uses the vertical dimension to codify layers of information. The use of either true or distorted monocular depth cues at the user interface has proven to be an elegant approach to cope with

[3] Perspective, interposition, contrast, shadow, texture gradient, motion, and motion parallax are termed monocular depth cues because they are visible with one eye and, hence, can be represented on a 2-D display. Stereoscopic or binocular depth cues require that two views of the same object from slightly different angles are presented separately to each eye, e.g. by means of a 3-D display.

[4] Spatial representations restricted to monocular depth cues are termed as $2^1/_2$-D visualization.

the limitations of screen space [22]. In this context, thumbnail images are particularly useful for conveying context information at a glance by showing simple pictures with fewer but more salient detail.

A recent approach using depth cues is a 3-D window manager developed by Microsoft Research (Task Gallery, Fig. 15.4) [23]. An appealing aspect of this approach is that existing Windows applications are hosted by the Task Gallery. A special redirection mechanism routes input and output between the 3-D environment and the unmodified 2-D Windows applications. User tasks appear as artwork hung on the walls of a virtual art gallery, with the selected task on a stage in the center of the screen. Multiple documents can be selected and displayed side-by-side using 3-D space to provide uniform and intuitive scaling. Currently unused windows are placed in stacks of overlapped windows to keep them organized. When windows are selected, the system moves them closer to the user for better legibility. The Task Gallery is composed of a sequence of virtual "Rooms" forming a linear hallway. A stylized body displayed in the corner of the screen indicates the user's current position and allows 3-D navigation. Arrows activated by a mouse click serve to glance in different directions and to jump one room back or forward. When moving through the hallway the user may carry commonly used tools and documents associated with the virtual body. User studies showed that the desktop metaphor has inadequate support for task switching, when users are engaged in a number of concurrent tasks (multitasking). In the Task Gallery, switching between tasks and viewing multiple windows simultaneously are simple actions. The gallery metaphor provides a strong spatial framework for encoding location information and front-to-back relationships, thereby engaging the user's spatial memory to help retrieve tasks and services.

In the examples mentioned above, monocular depth cues are used to represent three-dimensional data spaces on a 2-D display. On the other hand, a comparison of monocular and binocular depth cues has shown that spatial

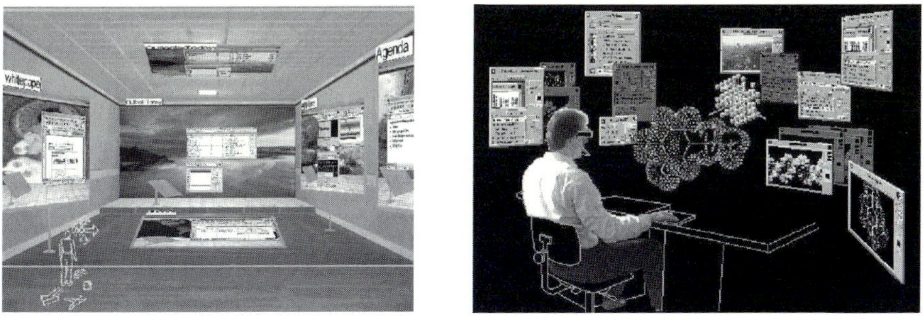

Fig. 15.4. The Task Gallery [23] (*left*, Courtesy George Robertson, Microsoft Corp., and ACM Publications) and the immersive 3-D user environment of InfoSpace [26] (*right*). From G. Robertson et al., *The Task Gallery: A 3-D Window Manager*, CHI'00, pp. 494-501 © (2000) ACM, Inc. Reprinted by permission

perception by viewing with both eyes is far superior to the spatial perception created by monocular depth cues. The stereo cues are particular useful at short distances, i.e. in the grasp area, which is also the common viewing distance to a computer screen [24]. Since human–computer interaction takes place in this area, it makes sense to implement full 3-D visualization (monocular plus binocular depth cues) at the user interface. Comparative experiments with $2^1/_2$-D vs. 3-D showed that cognitive orientation in complex information networks can be substantially supported by 3-D visualization [25]. A graphic user interface allowing complete 3-D visualization developed by the authors of this article will be presented in Sect. 15.3.

3-D interfaces have also been developed to allow users to fully immerse into the virtual data space. For example, the InfoSpace system shown in Fig. 15.4 (right) generates an immersive 3-D work environment with spatially arranged windows, dialog fields, status panels and 3-D models [26]. The user wears a head-mounted 3-D display (HMD). When the HMD is combined with a tracking system, full immersion in the data space allowing panoramic views is possible. Further immersive 3-D work environments were developed for special 3-D applications, e.g. in medical visualization or for product design using the Responsive Workbench (GMD) and the CAVE (University of Illinois).

15.2.2 Use of Auditory and Haptic Output at the Interface

The auditory and haptic output modalities are generally of particular value when the eyes are engaged in some other task. As sound events do not need the user to focus attention on the location of the source, sound has usually been applied to complement a standard visual interface in cases where the user has to be immediately informed about what is happening in the system (e.g. to alert or warn the user and for feedback). Haptic information is particularly useful as a feedback signal in teleoperation and direct manipulation of objects in a virtual world.

Sound can either be synthesized or digitally recorded and replayed. It can roughly be categorized into speech and nonspeech audio. Research on speech output shows different advantages for synthetic TTS (text-to-speech) and prerecorded voice systems. Synthesized speech is less comprehensible than natural speech, less pleasant to listen to, and perceived as qualitatively inferior [27–29]. Crucial characteristics of speech quality include overall speaking rate, interphrase pause length, and prosody [27].

The major advantage of TTS systems is their indefinite range of words, as they use phonemes as the building blocks of speech output [2] (phonemes are defined as the smallest units of sound affecting the meaning of a word). Hence, if auditory content is highly variable, TTS systems might be the choice. Research has also shown that users adapt to the particularities of synthetic voices [29]. On the other hand, for speech output with a limited vocabulary, it is recommendable to assemble the output from words or sentences that have been recorded by a human speaker. Mixing both styles is not advisable [28].

The following example describes how the authors of the Microsoft Task Gallery make extensive use of nonspeech sound to code functional information: "Buttons are spatialized according to left to right position and each button's timbre is subtly unique. There are rollover sounds that assist the user in locating specific tools. Objects are dragged on the pallet, objects are moved on the stage, and tasks are moved in the room – each with its own distinct dragging sound that is spatially and gesturally dynamic. Volume, timbre and stereo positioning are controlled in a way that simulates the sound of the type of object that is being moved (as an object is moved farther away from the user's viewpoint its sound becomes more distant). When the user moves a task there is a continuous sound proportional to the user's gesture and each surface is distinguishable: floor from ceiling, left wall from right. When the user moves a task from a wall to the floor an additional sound confirms the discontinuity caused by that interaction. This increases the user's sense of efficacy and is specific audio confirmation of that exact action. Run-time synthesis of pre-rendered basis sounds binds user interaction to audio. Simply playing back audio clips synchronized to user interaction does not have enough information to warrant the user's attention. We have tried to make each audio event convey useful information. User interaction is coupled to the audio, thus reducing the need for users to visually focus on the detail of every gesture." (Quotation of Microsoft Research [30])

Brewster [31] shows that using auditive feedback on a visual display can enhance pointing performance and reduce the subject's mental workload. This allows the reduction of button sizes on visually cluttered displays or when screen space is limited. Structured sounds indicating system states are often referred to as earcons (following the notion of icons) [32].

Haptic output has extensively been used in addition to graphics displays for tasks that involve perception of 3-D shape and force fields, e.g. to feel how molecules attract and lock-on or repel one another. Popular systems used in connection with desktop displays are the PHANToM force-feedback device from SensAble Technologies, a joystick or mouse with force feedback (e.g. the TouchSense Mouse from Immersion Corporation), and tactile gloves. Glove systems use force-reflecting exo-skeletal structures or pneumatic actuators (e.g. the NSF/Rutgers Master-II force-feedback glove) and small programmable vibrotactile stimulators attached to the fingers and the palm (e.g. the CyberTouch from Virtual Technologies Inc.).

Studies on pointing devices like the PHANToM or a Force Feedback Mouse reveal that providing haptic feedback enhances accuracy and time of pointing by 10 to 50% [33, 34]. Other questions concern the type of haptic output. Should haptically interacting surfaces have specified textures? Should they provide friction or be indicated by depressions? Or should interface elements exert gravitational forces attracting nearby pointers towards them?

Oakley et al. [35] investigated these issues and showed that haptically enhanced buttons and scrollbars work best when providing gravitational force feedback in opposition to texture, friction, or depressions. Subjects took less time completing tasks compared to other feedback conditions and preferred the gravitational version. Other challenges for research include haptic simulation of hand collision with objects and deformation of objects through the touch of the user. Due to the high sensibility of human haptic perception, feedback (with regard to the hand position) has to be displayed with sampling rates of up to 1000 times per second for inducing 'natural' sensations [36].

15.2.3 Multimodal Input

Numerous user studies revealed that multimodal input is desirable and generally preferred, because it may result in higher bandwidth, less communication errors, greater flexibility, less stress and higher naturalness of expression. The following input devices are candidates for a multimodal input system [37]: (1) the traditional keyboard for typing, (2) the 2-D/3-D mouse, trackball or joystick for pointing, selecting, and dragging, (3) speech input, (4) video-based lip-reading, particularly in order to improve speech recognition accuracy, (5) pen devices (tablets) to accommodate handwriting and pen-based gestures (pointing, drawing of signs, and symbols), (6) touch screens for pointing and selection with a finger, (7) 3-D hand gestures tracked by a data glove, (8) body movement, head movement, free hand gestures, facial expression, eye movement (eye fixations and saccades) measured by nonintrusive computer-vision methods, and (9) hand pressure sensed by a haptic device.

Speech Input. The most common way to communicate among humans is using speech. There has been tremendous research using speech input or automatic speech recognition (ASR) at the human–computer interface. Typical tasks in which ASR technology is deployed include word processing (dictation), giving commands to the interface, and formulating queries for information retrieval. Depending on the applications and techniques used, ASR can be divided into a number of categories: speaker-dependent vs. speaker-independent; small vocabulary (less than two hundred words) vs. large vocabulary (more than several thousand words); and isolated word recognition vs. continuous speech recognition. The choice of the proper ASR technique is dependent on the task to be solved, the user characteristics (gender, age, preferences, and expectations), and the inherent properties of the speech technology being used.

When using ASR for issuing commands, care must be taken of the vocabulary (easily memorable, meaningful, and distinguishable). To satisfy the needs of every potential user, synonyms for commands should be available. Studies looking at the overall time of dictation found that typing is often faster than using ASR. This is due to the fact that the correction of speech-input errors takes much more time than the correction of typing errors. Recognition rates for redictations are almost half the normal rate [38]. Altering voice characteristics can involuntarily contribute to further declining recognition rates [39].

The "holy grail" researchers try to achieve is a system capable of spoken language understanding (SLU) [40]. With such an interface spontaneous speech input on any topic would be possible. Progress on SLU has been made in limited domains such as air travel information services [41] including automatic language detection [42]. In SLU systems, the syntactic, semantic and pragmatic check of utterances and the interpretation of the spoken language are the most difficult and important tasks. The more a system approaches the notion of a "conversational interface" with initiative taken by both the user and the system, the more designers should take aspects of human–human communication into account. This includes the occurrence of hesitation sounds like "erm" and "uh", as well as turn-taking signals and other topics from dialog and discourse theory [40, 43]. However, building a natural language interface is far from trivial. [43]

Hand Gesture Input. Humans use their hands for grasping, moving, and manipulating objects as well as in order to signal and enhance information in human–human communication. Current research in human–computer interaction aims at developing interface devices that allow gesture input as well as direct object manipulation with the bare hands [44–47].

Gesture input can be categorized into deictic gestures (point at an object or to a direction), mimetic gestures (accept or refuse an action), iconic gestures (define an object or its feature), and manipulative gestures (change the size and form of an object). From a technical viewpoint, the human hand is a composite nonrigid object. In order to recognize hand gestures, the position, orientation, and motion of significant limbs, such as the fingers and the palm of the hand, need to be detected and identified using user-worn devices (data gloves) or video-based contact-free techniques. Several researchers have tried to model the human hand with a very high degree of freedom [48, 49] (about 26 degrees) or with Hidden Markov Models [50]. In this way, it is possible to track the position and movement of the hand and to recognize even very detailed gestures, such as the American Sign Language.

As humans are used to manipulate objects with both hands, e.g. when opening a bottle or playing a musical instrument, there is also increasing interest in two-handed human–computer interaction. According to the framework for bimanual action developed by Guiard [51], bimanual gestures are collaborative but generally asymmetric in the sense that (1) the nondominant hand serves as a reference for the dominant hand (e.g. one hand holds a sheet of paper while the other hand writes on it); (2) the nondominant hand acts first and is followed by the dominant hand; and (3) the dominant hand is capable of doing finer movements than the nondominant hand. Two-handed interaction has been explored using various conventional and experimental interface devices [46, 52, 53]. Overall, these studies have shown that poorly designed two-handed techniques can be very cumbersome to use [45].

Manipulating a real object differs in several aspects from interacting with a virtual object. One substantial drawback is the lack of force feedback when a user touches or moves a virtual object. Usually, visual feedback is the only information available to control the position of the hand in relation to the object. This makes detailed object manipulation an extremely difficult task. It has been suggested that the missing force-feedback signals could be mapped to other sensor modalities, e.g. by triggering an acoustic signal (auditive feedback) [2] or a local change of the object's color and/or form (enhanced visual feedback).

Unlike interaction with real objects, objects to be manipulated in a virtual environment may be located far out of reach of the user's hand. Objects shown on a 3-D display may appear behind the physical display screen. Therefore, techniques for remote object grabbing and manipulation are required. Various approaches including ray casting (grasping an object by casting a virtual ray of light on it) and arm extension (virtually extending the user's arm) have been developed and tested. [44, 54]

Gaze Input. Since people usually look at what they are working on, eye gaze in general contains a lot of information about the current task. Moreover, eye gaze indicates the potential intention of a user [55], because people tend to look at a target object prior to initiating an action [56]. A number of researchers have recognized the utility of using eye gaze for interacting with a computer, e.g. to select characters on a virtual keyboard [57] or to position a cursor [58, 59]. Performing eye movements can be done much more quickly and with less strain than moving a mouse [60].

On the other hand, there are some constraints for replacing the mouse pointer by eye gaze: eye movements are triggered automatically and unconsciously (eye jitter), and it takes much voluntary effort to control eye fixation. When a system interprets unintended gaze movements as input, it might start operations in an unforeseen and confusing manner ("Midas touch problem") [61]. The physiological function of eye movements is to pick up information for further processing and action control (e.g. eye-hand feedback loop). Gaze is not stable, even when users have the impression of staring at things [62]. The combined effect of gaze jitter and measurement errors limits the accuracy of gaze input to one or two degrees of visual angle [63].

Gaze indicates the focus of attention and can disambiguate input from other modalities. Hence, eye gaze has been found particularly useful, when used in conjunction with other input modalities [60–63]. In a perceptual user interface [59, 64, 65], gaze input should be evaluated by an intelligent interface agent and interpreted in the context of the current task, in order to derive the appropriate action (noncommand interaction [61]).

Head/Eye Movement Input. People continuously explore their environment by moving their head and eyes. Such movements are performed in order to inspect a single object from different perspectives or to find out the spatial relations of multiple objects in a three-dimensional scene. In a perceptual user interface, similar interactions could help users explore the virtual environment shown on the (3-D) display. Moreover, providing different perspective views in accordance with the user's head/eye position is a very natural means for virtually extending the usable screen space, which is physically limited by the size of the display (dynamic perspective, same head-motion parallax effect as when looking through a window from different viewpoints).

In order to present visual information adapted to the user's current viewpoint, a method for sensing the head/eye position and orientation is required. This information is also needed for optically addressing the user's eyes when autostereoscopic 3-D displays are used (see details in Sect. 15.3.3).

Combining Modalities. The crucial point in combining input modalities is to develop a semantic model which defines how to interpret the input [66] and how to integrate the streams of information at various levels – from early "signal" level to intermediate "feature" level and to late "decision" level [67]. Sharma et al. point out that it is helpful to know how the integrating modalities relate in a natural environment. Some modalities, like speech and lip movements, are more closely tied than others, such as speech and hand gestures.

Table 15.1 overviews various combinations implemented in existing systems. Speech input seems to be relevant in any multimodal system (probably because speech is the most powerful human output modality and research on speech recognition has a longer history than that on image analysis). Hand gesture, eye gaze, and lip reading are often combined with other modalities, depending on the characteristics of the application or task considered. Users generally prefer combined inputs to systems offering only a single modality used alone. Combining several modalities, when each modality is not performing perfectly, increases the overall system performance. The different input channels can complement and disambiguate each other thus making multimodal systems very flexible in terms of environmental conditions and user groups [68–70].

The various input modalities have quite different characteristics. For example, positioning a cursor by eye gaze interaction is quicker but less precise than positioning by hand movement [71]; natural-language speech input is perfectly suited for descriptive tasks, whereas hand gestures are the ideal solution for direct manipulation of objects [53]. Hence, system designers must decide which input or which combination of inputs to use for a specific task.

Human–computer interaction differs from interaction with natural objects in real environments and from communication among humans – and will probably never be the same. Challenges in designing an efficient user interface lie in how to minimize the time needed to switch among the input modalities

Table 15.1. Overview of advanced input modalities implemented in existing systems. The table is based on more detailed reviews found in [12, 37, 67]

Human-Action Modalities	Application	Results/Remarks
Speech and hand	Object manipulation Map manipulation Issue commands in a "Put-That-There" system Video presentation system control (Finger-Pointer)	Preferred by 71% of the subjects. Preferred by 95% of the subjects. Spoken words affirm gestural commands. Gestures disambiguate noisy speech. Gestures to specify positional information and simple object attributes, spoken commands to identify actions.
Speech, pen gestures and user positions	Military simulations (QuickSet)	Speed/efficiency advantages, 68% multimodal success rate including error correction; strongly preferred.
Speech, pen gestures and handwriting	Interacting with a calendar (Jeanie)	Performs with up to 80% accuracy.
Speech, eye gaze and space mouse	Manipulation of virtual objects (VisualMan)	Allows users to choose among modalities.
Speech and lip reading	Voice control	Improved recognition of speech in high noise environments.
Speech and touch	Robot control	Reduced number of operational errors.
Speech, mouse and keyboard	Drawing application	Reduced time (71%) and less commands needed (77%). Reduced amount of mouse pointer movements (53%).
Speech, hand and eye	VR-based CAD systems	Subjectively preferred by potential users.
Speech and handwriting	Error correction in text	Requires significantly less effort.
Speech, lip reading, gaze and touch	Medical applications; smart environments	Allows users at multiple locations to simultaneously interact with each other.

available and how to simplify the learning process needed to understand how interactions in the virtual world relate to corresponding processes in the real world used as a metaphor.

15.3 MUltimo3-D: a Testbed for a Multimodal 3-D PC

In the following sections we present in greater detail the mUltimo3-D system [64] developed as a demonstrator for a future general-purpose desktop computer and as a testbed for evaluations. We have implemented a 3-D graphic user interface adapted to a novel autostereoscopic 3-D display, computer-vision based devices for a perceptual user interface, and an operating system to accommodate trial applications in a multimodal human–computer environment.

15.3.1 Concept of the mUltimo3-D User Interface

The key idea behind the mUltimo3-D system is to offer users a clearly arranged virtual interaction space at the interface, where they find the tools and media objects for various desktop applications and the paths used, when browsing the Internet and other data bases. Instead of windows presented on a flat (2-D) display, the 3-D user interface provides a spatially extended field for interaction. A special interface agent helps the user to optimally place and scale all the visual elements (e.g. the documents, tools, and video communication windows) required for a specific task. In this paradigm, stereoscopic spatiality is used to gain an additional dimension in order to accommodate a larger number of visually well-distinguishable media objects on the display. In this context, advantage is taken of the human ability to focus attention on discrete distances or depth layers; the effect of perspective scaling causes a natural reduction in size of the displayed objects with increased distance.

Having available the third dimension, various novel approaches are conceivable in order to improve human–computer interaction. Discrete depth layers indicated, for example, by tags on a graphic z axis, could be linked with functional attributes, so that a particular action is executed when media objects are actively placed in a certain layer. Moreover, the third dimension could be used for encoding features associated with the displayed objects. For example, the apparent distance of a displayed object could indicate its importance or task relevance ("farther away" intuitively meaning "less important"). Similarly, in the case of telecooperation the state of authorization (whether other users have access to the media objects) could be encoded by the distance where the respective files appear ("far away" meaning "free access for others"). In video-communication, displaying a communication window in the foreground close to the user could indicate that a bidirectional video and audio connection exists ("I am being seen and heard by my tele-partners"). On the contrary, a distant video window appearing in the background would indicate: "The others can neither hear nor see me". In this manner, the depth cue would help users to control their personal domain.

The media objects could also be automatically fanned out in the display space according to typification defined by the user. For example, spatial structuring according to topics, work context, current validity, or presentation form of media objects (text vs. picture) is conceivable together with automatic rearrangement after the criterion has been changed (quick changing between "aspects" on the same data base). Functional grading of depth could be partially subjected to the control of an interface agent which magnifies certain objects by moving them to the foreground, reducing less important things to the size of thumbnails and sending them to the background, depending on where the user is focusing or as a response to external/internal events.

With today's input devices, the action space and perception space are separated: input devices like the keyboard or mouse are positioned some distance away from the display on which the result of an action (characters entered, cursor movement) is visible. The mUltimo3-D system foresees that the virtual interaction space is seamlessly integrated into the real work area using suitable visualization techniques. The merging of the display space and the real action space permits simultaneous visibility and handling of virtual and real objects (mixed reality). Above all, "natural" interaction with virtual objects in the user's grasping area should be possible. The user should be able to form (sculpture) or move objects which appear stereoscopically in the grasp area using just his/her hands. Since it is not possible to really touch the virtual objects[5], the absent tactile signals are simulated using other human sensor modalities in our system (by changing the color of the object at the point it is touched or by acoustic signals).

In the mUltimo3-D system, the viewing perspective is coupled to the user's head movements in order to provide head-motion parallax effects as in natural vision. Hence, objects in the foreground can voluntarily be shifted aside for looking at occluded objects (look-around function) without having to operate a device. By detecting the point of fixation (gaze), additional direct interaction at distances beyond the grasping area becomes possible. For example, looking at a media object can be used to display context information – automatically or on demand (via speech input). A synthetic depth of focus linked to the current point of focus helps the user to concentrate his/her attention on the selected media object. In this way, the advantages of a limited depth of focus inherent in natural vision can be exploited [72, 73]. Furthermore, eye gaze can be used to address or zoom objects on the screen (eye–mouse function [74]). Gaze is currently the only relatively reliable indication of the focus of visual attention [75].

The mUltimo3-D system foresees noncommand interaction [63, 76]. This means that an interface agent observes the user and registers the application context, and then, on the basis of this information, acts independently without any direct request from the user (e.g. tidies up the screen or starts an application). In order to make sensible decisions, the agent needs an internal model that is as comprehensive and up-to-date as possible, describing both the current aims of the user and the possibilities of the computer.

Such a concept relies on a perceptual user interface assuming that the agent has access to visual or other sensors in order to monitor the user (for example, cameras for measuring the viewing perspective and the point of focus [76, 77], microphones for capturing speech). After gathering the various kinds of sensor information, the agent has to conclude the likely intention of the user. The agent could, for example, offer assistance in cases of repetitive routines.

In general, noncommand interaction cannot be foreseen by the user and is hence contradictory to Norman's demand for a consistent conceptual user model of the system. This dilemma can be solved by restricting autonomy to a specific degree that is known to the user. An agent that handles within (known!) limits can be taken into account in the conceptual model and integrated into forecasting activities.[6] Hence, the user does not delegate control over the system to an agent, but frees himself from the direct monitoring of

[5] Force feedback produced by touching virtual objects would require the use of auxiliary systems such as the PHANToM.

[6] Automatic spell-checking and correction during input and online help systems in office applications are examples of established command-free interaction.

certain tasks. If the agent triggers undesired actions, there must be a simple way to undo it. This may be carried out directly by command or implicitly by observing the user's reactions (Affective Computing [78]). Due to a lack of empirical results, these concepts are, however, to a large degree still the subject of controversial discussion.

15.3.2 Implemented Operating System and Interface Agents

The central part of the mUltimo3-D user interface is an operating system that supports both 3-D visualization and noncommand interaction. The operating system is based on a C++ class library with which the system itself and the various applications were developed. The library is based on the virtual reality development environment dVS/DIVISION from PTC Ltd.

The dVS database manages all the information required in order to display the media objects in the interaction space. dVS uses special software components for processing user-generated control signals. These so-called actors primarily process signals from conventional virtual reality (VR) input devices such as the mouse, data glove, and head tracker. We have modified and extended the dVS actors for speech, gaze and gesture input. The actors calculate intersections and collisions of 3-D objects and enable access to the dVS database. The operating system uses additional actors which manage user data and permit access to the file system.

Interface agents position the objects in the interaction space, collect data on the user, and support interaction between the user and the computer. The dVS database acts as the interface between the agents at one end, and the system processes, the system kernel, and the applications at the other end. Within the agency, the collector agent plays a special role. It continuously collects and interprets data on the user and makes this information available to the other agents. It registers how often and in which preferred setting certain applications are used, in which context this happens, and when interactions last took place. These observations are interpreted in order to conclude the learning status (automatic aids can be offered accordingly on a different level) or in order to remove applications that have not been used for some time. All the data collected is stored in the form of a user model for later use.

The placer agent is responsible for arranging the media objects in the interaction space. The aim of the placing strategy is to avoid overlapping, so that all the applications are clearly visible to the user. For this purpose, all the objects are surrounded by an invisible bounding box. The dimensions of the bounding box also define the "catch space" for gaze-controlled addressing of a media object. Automatic removal of an application from the display happens gradually; the placer agent gradually moves an application that is little used into the background and its size is slowly reduced (reducer agent) until it is ultimately fully deleted. This protects the interaction space from becoming overfilled with media objects. Other forms of signaling are also possible for the reduction procedure, e.g. by iconification or by slowly fading an application (transparency) in order to indicate aging. A starter agent is responsible for the automatic starting of applications, for example, when a user logs in or when files are opened.

Critical actions, such as the deleting of data, must be acknowledged by the user. That means that the agent must hold a dialog with the user. For this purpose, the visualization agent appears as an avatar (Fig. 15.5) and requests, in the form of gestures, the user to make decision. The avatar acts almost like a spokesman for the agency. It establishes the direct contact between the user and the agents and permits the user to directly intervene in decisions by the agents. The avatar can also act as a tutor for help and learning systems.

If, for example, a yes/no decision is required, the visualization agent activates the avatar and places it in a preferential position. The user activates (gaze-controlled) either the green button (acceptance) or the red button (rejection), indicating to the avatar his/her decision. Other forms of status visualization are also possible by changing the body color of the avatar. When the avatar refers to a critical situation, its body color could change to bright red. Other useful status displays could be: "planning" (the agent is currently in the process of drawing up possible actions), "working" (an action is currently being carried out), "satisfied" (the last action carried out by the agent was given positive confirmation), "surprised" (the user made an unexpected decision), or "confused" (the user rejected the last action). At this point, the operating system supports a wide range of interaction possibilities which have yet to be researched.

15.3.3 3-D Display

Based on the assumption that future users of desktop computers will not tolerate encumbering interface devices attached to their head and body, both the devices for the perceptual user interface as well as the 3-D display should rely on contact-free nonintrusive technologies. Hence, the mUltimo3-D system uses autostereoscopic 3-D displays creating the 3-D effect without the aid of polarizing glasses or other head gear occluding the eyes. Various

Fig. 15.5. Communication states of the avatar: idle state, confirmation request, and inquiry

autostereoscopic techniques have been suggested, including electroholography [79], volumetric displays and super-multiview displays [80]. Pastoor [81] and Sexton [82] give comprehensive and in-depth reviews of 3-D display techniques.

Autostereoscopic displays based on lenticular lens techniques have been developed at our laboratory and used to implement the mUltimo3-D system. The latest development is the Face-to-Face display. Figure 15.6 shows the optical principle. A lenticular plate made of very thin, vertically oriented cylindrical lenses is placed in front of a liquid crystall (LC) panel. Behind each single lens, a thin column of the left and right stereo image is displayed in the width of a pixel, with the odd pixels presenting the image information for the right eye and the even pixels that for the left eye. The lens pitch and the distance between the lenticular plate and the LC panel are calculated in such a way that, within the range of a nominal viewing distance, two quadrilateral zones are formed, where the two constituent stereo images are completely separated (stereo zone).

A device for tracking the stereo zone makes sure that the user's eyes remain optically addressed after head movement. The display is mounted on a platform which turns around the vertical axis when the user moves to the left or right. Hence, the display continuously orients itself to the user's eyes and keeps itself "face-to-face" with the user. This perfectly avoids spatial and optical distortions frequently observed in other 3-D displays when looking from a nonorthoscopic viewpoint. The display allows a wide range of head movements, still guaranteeing an excellent image quality. The screen diagonal length is 20.1 inches and the spatial resolution is 1280 × 1024 pixels (RGB). The display provides an excellent stereo image separation (the crosstalk between the stereo images is less than 1% and visually not perceptible; for perceptual limits of crosstalk in stereo displays see [83]).

Lenticular lens 3-D displays along with most of the known 3-D display concepts (with the exception of holographic displays) have one serious disadvantage when it comes to supporting direct virtual object manipulation with

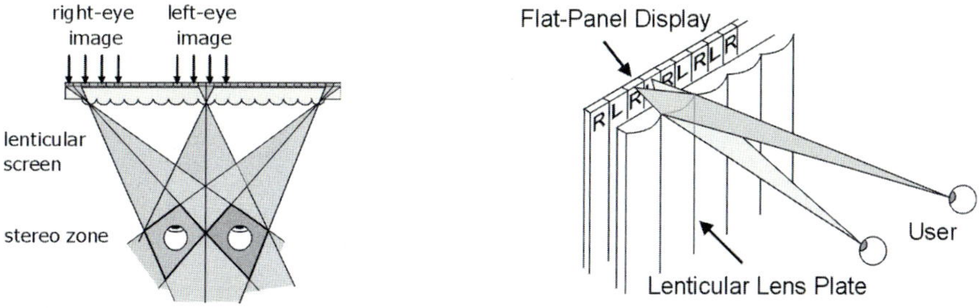

Fig. 15.6. 3-D display using an array of lenticular lenses to channel the left and right eye images to the respective eye (L and R refer to image stripes of the left and right images of a stereo pair)

the hands or other physical tools. This problem is due to the fact that the stereo images are displayed on the physically existing display screen, while the spatial position of the virtual media object may stereoscopically appear at a distance in front of the display screen (in the grasp area). The user has to accommodate on the distance of the screen, in order to see an unblurred, sharp stereo image, whereas his/her lines of sight converge at the apparent position of the virtual object. This unnatural condition is a potential cause of eye strain. Moreover, because of the limited depth of focus of the human eye, the user will see either the virtual object (accommodation on the screen) or the real hand or tool (when looking at it) in full focus. Natural interaction where both the virtual and real objects are seen in full focus is hence not possible (Fig. 15.7).

We have found a solution to this problem in the Accommodation 3-D Display [84]. As opposed to a fixed stereo image plane, the stereo image pair in this display is focused on an aerial plane floating in front of (or behind) a large convex lens (e.g. a Fresnel lens, Fig. 15.8). This aerial image plane constantly changes its location according to the actual fixation point of the user, so that the user perceives an aerial image of the virtual object being looked at in his/her accommodation distance. Hence, the link of accommodation and convergence is supported as in natural vision. The Accommodation 3-D Display enables a holography-like image with a large depth range. On the other hand, it requires on-line measuring of the user's gaze direction and adaptation of the optical components of the display.

A prototype of the Accommodation 3-D Display providing a stereo resolution of $2 \times (1024 \times 768)$ pixels (RGB) has recently been constructed. The photo sequence in Fig. 15.9 shows what the user sees with his/her left eye when approaching the left-eye viewing window of the display (top-left to bottom-right photos). The two viewing windows have a diameter of 65 mm each and are physically located at a distance of 70 cm from the display. The windows are formed by the real images of the exit pupils of the projection lenses at the designed nominal viewing distance. Looking through the win-

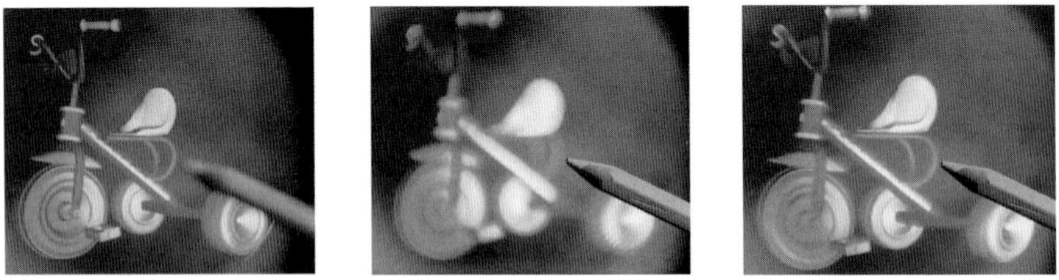

Fig. 15.7. With conventional 3-D displays, interaction between virtual and real objects in the grasp area is affected by blurred effects, either the virtual object is sharp (*left*) or the real object is sharp (*middle*), depending on what object the user is looking at. On the Accommodation 3-D Display both objects are seen in full focus (*right*)

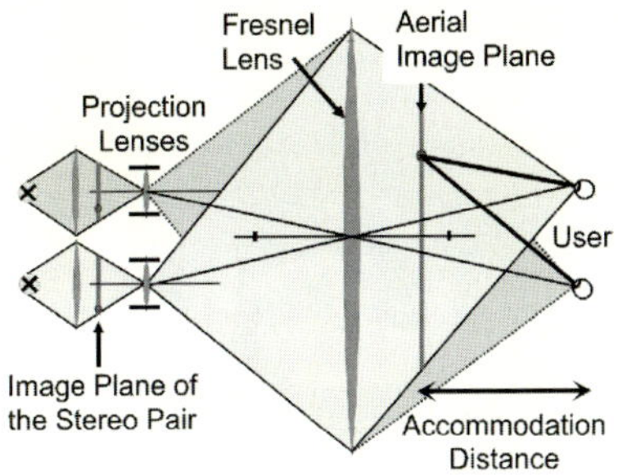

Fig. 15.8. Top view of the Accommodation 3-D Display principle. The user accommodates on the aerial image plane and perceives a floating 3-D object. The Fresnel lens projects the exit pupils of the stereo projector to the left and right eye, respectively (field-lens principle) and, hence, completely separates the constituent stereo images

dows, the left- and right-eye images are fused, forming a 3-D image which floats at a distance of 30 cm in front of the display. Hence, with this prototype the accommodation and convergence distances are fixed at about 40 cm, allowing comfortable and unhindered hand interaction with the displayed objects.

A 3-D display concept, where accommodation and convergence are linked as in natural vision has also been developed at ATR Labs (Kyoto). The ATR "display with accommodative compensation" is a head-mounted display with movable relay lenses interposed between the screen showing the left- and right-eye images and the eyepiece lenses. The convergence distance is sensed by a gaze tracker. The position of the relay lenses is constantly adjusted, so that the screen surface (image plane) appears at the same distance as the convergence point of the gaze lines. In a later version, the real (physical) distance of the display screen is adapted to the current point of fixation [85].

A "holo-haptic display" has been developed at MIT Media Lab [86]. The display creates stereograms using video-based holographic techniques. A PHANToM force-feedback device is combined with the display, in order to carve a virtual (holographic) object floating in front of the display. Because it is not yet possible to calculate complex holographic images in real time, a set of precomputed holograms is used to "assemble" the visual representation of the carved surface of the object. For updating the object geometry, the appropriate elements from the set of precomputed holographic images are mapped onto the object instead of recomputing the entire hologram. This display allows people to see, "feel", and interactively manipulate simple holographic objects.

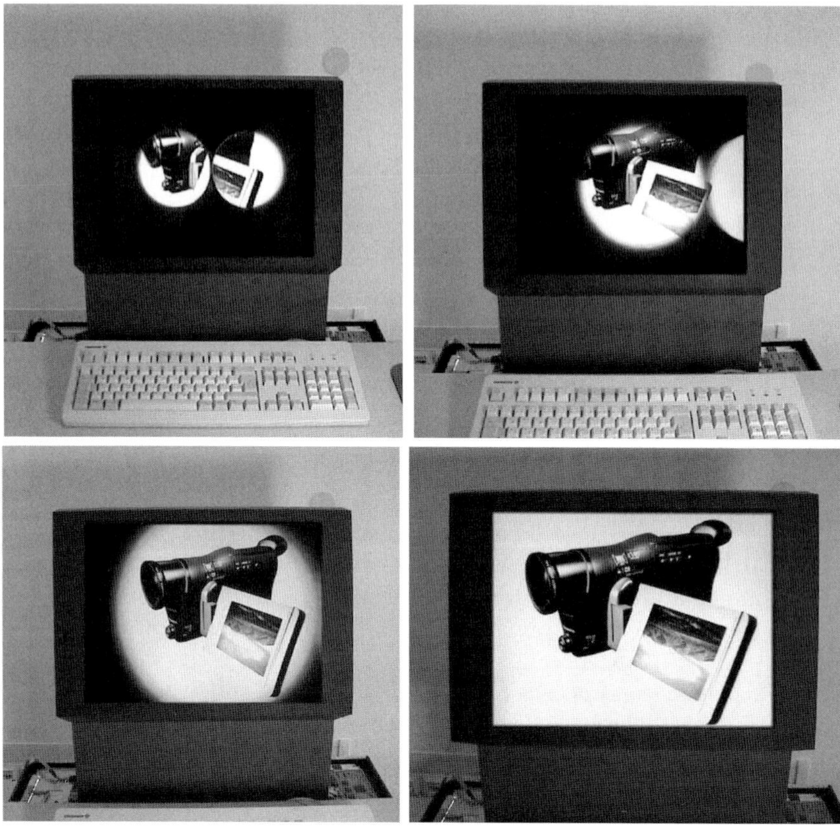

Fig. 15.9. Prototype Accommodation 3-D Display. The stereo viewing windows float in space at a distance of 70 cm in front of the display. The sequence shows what the user sees with the left eye, when approaching the left-eye window

15.3.4 Input Devices for the Perceptual User Interface

A perceptual user interface (PUI) requires detection, tracking, and recognition of input signals articulated by, for example, speech, facial features, hand gestures, and body motion, in order to interpret the user's intentions. As explained earlier, we suppose that future human–computer interfaces require unencumbered contact-free (e.g. video-based) techniques measuring the user's actions from a distance.

The video-based approaches for analyzing and interpreting the various user actions (hand, gaze, body, and facial gestures) have a lot in common. For example, face and hands have similar skin color. Hence, several approaches use skin-color information to track both the user's face and hand simultaneously (e.g. in Hongo et al. [87]). Based on the similarity of the articulated structures in whole-body and hand movements, Gavrila [88] developed 3-D pose estimation algorithms to analyze user movements. Several approaches use more than one camera (e.g. with different focal lengths) to sense and track different parts of the human body and to avoid occlusions. Boehme et al. [89] applied two pan-tilt cameras with zoom lenses to localize the user (using the general-view camera) and to recognize his/her gestures (using the gesture camera). Hongo et al. [87] used four cameras in order to track the user's head and hand. In their system, two fixed wide-angle cameras form a stereo pair in order to estimate the location of the face and hand. Guided by the known 3-D positions, the other two cameras are aimed at the hand and face to deliver zoomed images of appropriate sizes. Thus, the following recognition and classification steps were simplified (because of size invariance). Utsumi and Ohya [90] described a five-camera system for shape recognition of both hands. 3-D hand postures are estimated based on a small number of reliable image features. Using multiple-camera setups, the self-occlusion and hand-hand occlusion problems were significantly reduced.

Recent review papers [48, 91, 92] summarize the extensive research work in this area. The most critical aspects in image processing are the extraction of useful and meaningful features, and the classification and tracking of different feature sets. Although constantly growing efforts have been made in struggling against varied and quite different capturing conditions [93–97], the problems of recognizing and tracking complex patterns with arbitrary orientation, location, and scale under changing lighting conditions still remain unsolved.

The mUltimo3-D system uses four cameras for detecting and tracking the user and various gestures (Fig. 15.10). One camera is fixed on the top of

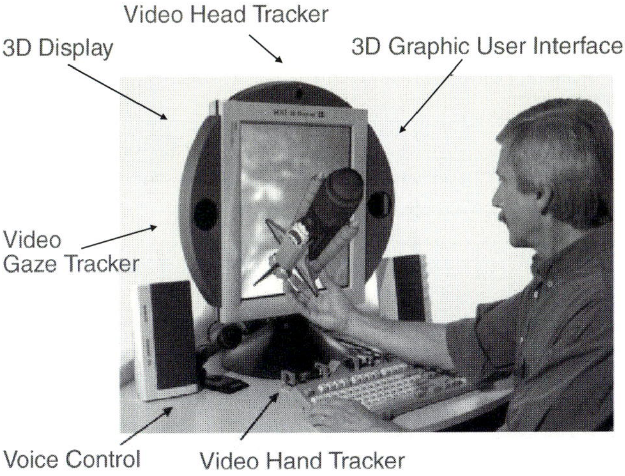

Fig. 15.10. Prototype of the mUltimo3-D system. Apart from conventional input devices, the system also provides a video-based perceptual user interface for multimodal interaction

the display in order to track the user's head and eye position (head tracker). Another camera with a larger focal length is mounted on a pan-tilt platform at one side of the display. Guided by the head tracker, this camera constantly aims at one of the user's eyes and is used to sense the user's gaze direction (gaze tracker). The other two cameras are close to the keyboard and form a stereo pair. They are used to estimate the 3-D hand position and to recognize simple hand gestures (hand tracker). For speech recognition, a commercial software solution (IBM Via Voice) has been integrated. For speech output we use the Logox3 speech synthesis software from G-DATA. In order to detect the user's actions in real-time and to keep the overall system delay small, the various signals are processed by separate PCs. The information about the 3-D position of both eyes, the 3-D hand position and the recognized hand gestures, the gaze direction, and the speech input is constantly forwarded to the interface agents, which combine the various information sources in order to guess the user's intentions. The overall system diagram in Fig. 15.11 shows the basic components which have so far been implemented in the mUltimo3-D system.

Video Head Tracker. In our system, the exact 3-D positions of both eyes must be known in order to (1) optically address the user's eyes by the 3-D display, (2) to adapt the 3-D perspective to the user's current viewpoint (for head-motion parallax), and (3) to aim the gaze camera at one eye for precise gaze measurements. Real-time head tracking is currently not the most serious problem. Various systems have been presented [98–100]. Most approaches use skin color as a tracking feature, because it is scale- and rotation-invariant and

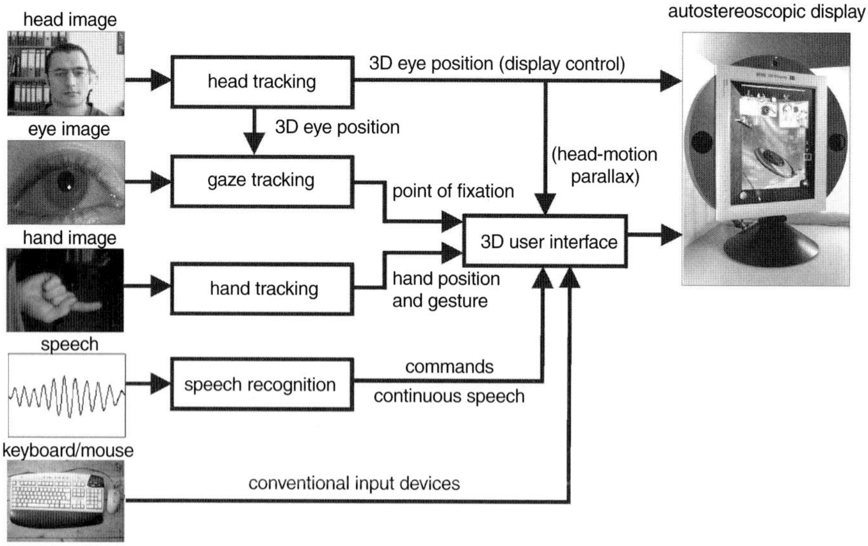

Fig. 15.11. The overall system diagram of the mUltimo3-D testbed

robust when a part of the head is occluded. On the other hand, color information is very susceptible to changes in the lighting conditions and camera characteristics. There is no optimal color space known for common use.

In the mUltimo3-D system, only luminance signals are exploited. During initialization we detect the user's blinking to find eyes in the video images (blinking is a natural and involuntary human action that keeps the eyes moist). The simultaneous closing and opening of the eye lids provide a very specific spatio-temporal pattern which can be used to distinguish blinking from other changes in successive video images. These patterns are used in the following tracking procedure as user-specific references. Detection of the eyes in the following video image sequences is based on a block-matching technique [98].

The critical points in the tracking process are to cope with changes in size, orientation and illumination. A luminance-adaptive block-matching technique has been developed, where the luminance distributions of both the stored reference pattern and the search region in the current image are equalized and expanded to the maximal luminance range before matching (nonlinear transformation with resulting luminance values ranging from 0 to 255; see Fig. 15.12). This makes the head tracker robust against changes in the lighting conditions. For tracking eyes in various sizes and orientations, a rotation factor and a scaling factor have to be taken into account. In order to speed up tracking, not the current image but the stored patterns are rotated and scaled at various degrees. Thus, the matching process has merely to compare the current image with the prestored patterns, starting with the "probably true" pattern, which is known from the previous tracking results.

After localizing both eyes in an image, their 3-D positions are derived by taking the human interocular distance and the camera parameters into account. This assumes that the user's eyes are in a plane parallel to the

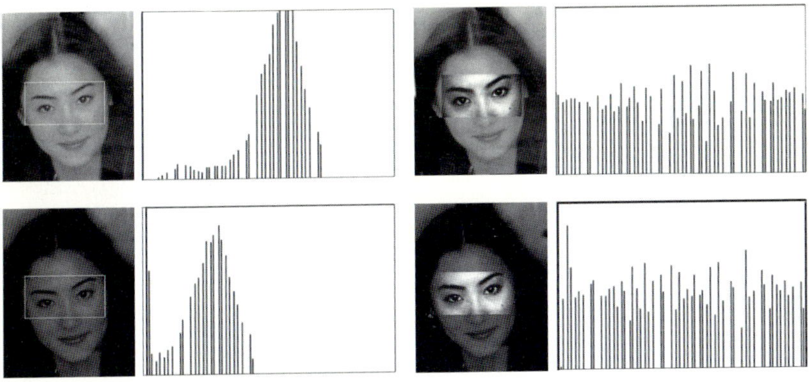

Fig. 15.12. Effect of luminance equalization for two different lighting conditions (search regions and luminance distributions before and after equalization)

image plane. Typically, the accuracy is 2 – 3 mm in the lateral and 1 cm in the frontal directions. The update rate is 50 Hz (PC, 933 MHz).

Video Hand Tracker. Hand tracking has a lot in common with head tracking, especially if the skin color is used as the primary feature. On the other hand, the shape and orientation of the hand is very flexible and nonuniform so that geometrical relationships are of little use in hand tracking. Experimental studies have shown that the highest frequency in hand movements is about 5 Hz. This suggests that tracking systems must operate at least at 10 Hz [101]. Several real-time systems can be found in references [47, 90, 102–105].

The mUltimo3-D hand tracker (Fig. 15.13) uses two infrared-sensitive cameras in a stereo setup and an infrared light source next to the cameras. The infrared light is reflected by the hand and provides a bright image, since the hand is close to the light source. The intensity of light reflected from the distant background is comparatively weak. This significantly simplifies the hand detection and segmentation process. During initialization, a reference image without the user's hand is captured. When the hand appears in the video image, it is detected by calculating the luminance difference between the current image and the stored reference image.

Luminance differences exceeding a predefined threshold indicate candidate regions of the hand in the image. The difference images undergo a temporal low-pass filtering to reduce noise effects. After a smoothing process using a morphological filter (for connecting small areas and further noise reduction), the largest connected area defines the hand region in the current image. Next, a circle is matched to the palm area and a set of vectors pointing from the center of the circle to the boundary of the hand region (hand vectors) are calculated. Fingers are indicated by long hand vectors. Simple hand gestures are recognized by analyzing the distribution of the hand vectors and comparing it with prestored distribution patterns (Fig. 15.14). The hand segmentation and gesture recognition processes are performed on one of the two stereo images only.

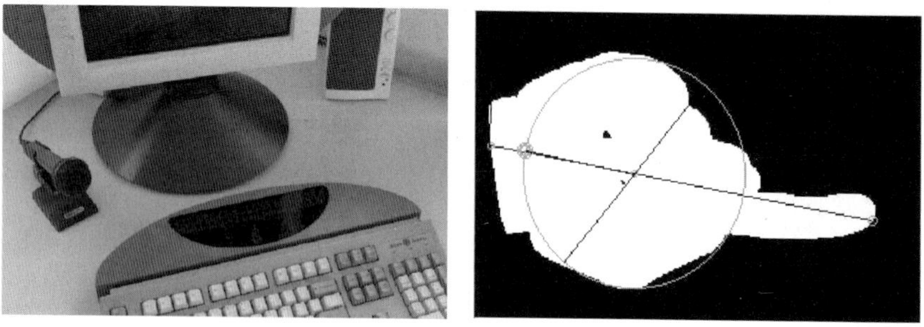

Fig. 15.13. Video hand tracker (the device attached to the keyboard) at the mUltimo3-D user interface and the segmented hand

A small set of feature points including the center of the palm and the finger tips are selected in the preprocessed image, and their correspondences are searched for in the other stereo image using a feature-point-based block-matching technique. Based on the geometrical parameters of the stereo camera setup, the 3-D position of the hand is calculated. Taking into account the position, gesture, and status of the hand (still or in motion) the interface agent decides whether the user is currently moving or grasping an object or pointing at a target. Currently, the 3-D position of the finger tip is measured at an average accuracy of about 2 – 3 mm in the horizontal plane and about 1 cm in the vertical distance with 13 Hz update rate (PC, 500 MHz).

Video Gaze Tracker. The user's gaze direction is sensed for two reasons as follows. First, the stereo images are rendered in such a way that only the currently fixated object appears in full focus; other objects are considered as less important and therefore shown out of focus. Second, the gaze direction indicates the user's focus of attention and is interpreted by the interface agents for noncommand interaction.

We have opted for the cornea-reflex method because of its high precision and stability and because it is nonintrusive [105]. The viewer's eye is illuminated with low-intensity infrared light. An array of LEDs is mounted at a distance from the optical axis of the gaze camera. As a result, the pupil appears as a black elliptical region. The center of the pupil and the reflection of the light from the cornea are found rapidly due to a new algorithm. There is a monotonic geometrical relationship between the vector pointing from the center of the pupil to the center of the light reflection (eye vector) and the gaze direction [106]. After calibration, the gaze direction can be derived from the eye vector.

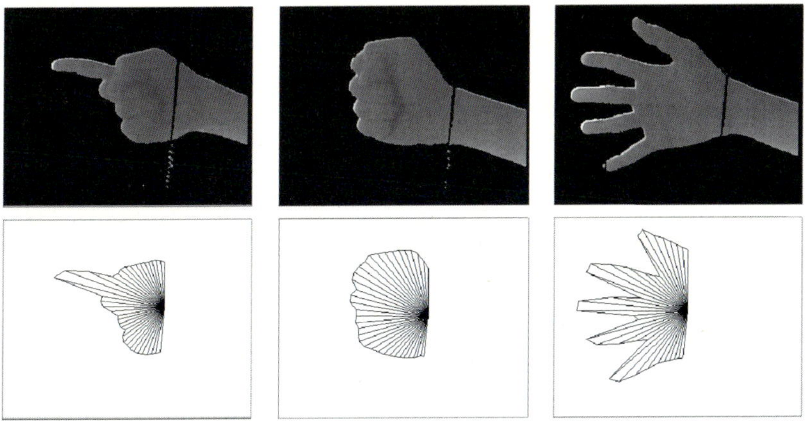

Fig. 15.14. Segmented hand regions and hand vectors indicating simple hand gestures

Recently, suggestions have been made for a self-calibrating system [107] and an auto-calibrating system [108]. Shih et al. [108] used two cameras and 3-D computer vision techniques to determine the gaze direction. Using two cameras and two light sources is a sufficient condition for auto-calibrated gaze tracking. The technique presented in Liu et al. [107] has been implemented in the mUltimo3-D system. This method uses one camera and more than one light source. The accuracy depends directly on the number of light sources used. On the other hand, the segmentation of the pupil and the highlights (reflections) becomes increasingly difficult with a larger number of light sources, because the pupil would be largely covered by the highlights and there would be more reflections on eye glasses worn by a user (Fig. 15.15). Three to five light sources seem to be optimal.

Our "self-calibrating" gaze tracker requires initialization before it can be used for the first time. In this phase, a "reference eye" is defined by calibrating the system with a "reference user". The reference user will have to fixate a set of calibration points shown on the display and the corresponding gaze vectors will be registered. Then, a mapping function (MF_r) is calculated which maps the gaze vectors of the reference eye onto the calibration points. In this way, both the relationship between the gaze vectors and the gaze direction and the geometric arrangement of the camera and the light sources are taken into account in the calibration process. When a new user looks at the display, he/she will be presented with an eye-catching visual object instinctively attracting his/her gaze. At that moment, four gaze vectors are determined simultaneously. A user mapping function (MF_u) is deduced which maps the set of gaze vectors of the actual user (GV_u) onto the set of gaze vectors of the reference user. The user's gaze direction is then determined by the function $MF_r(MF_u(GV_u))$.

In order to ensure exact measurements at any head position, we have introduced a head-fixed coordinate transformation that compensates for measurement deviations caused by head movements [109]. The gaze tracker runs on a PC (933 MHz) with an update rate of 50 Hz and has an accuracy of about 0.7° after precise individual calibration and of about 1.2° when using the "self-calibrating" method, respectively. Figure 15.15 shows that the algorithm is quite robust against distracting reflections on the user's spectacles. As long

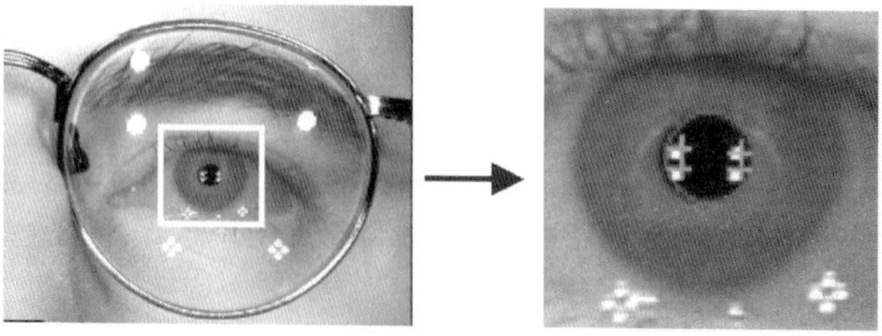

Fig. 15.15. The eye image captured by the gaze tracker (*left*) and an enlarged part of the eye showing the pupil and light reflections on the cornea (*right*)

as these reflections do not seriously cover the pupil, the pupil and the corneal reflections can be correctly detected.

15.3.5 Trial Applications

Several trial applications have been developed that accommodate for different task demands. For example, a 3-D CAD application and different versions of media browsers were implemented, tested, and revised in the course of development.

For example, the cubes in Fig. 15.16 (upper part of the screen) were implemented for the selection of tools and applications via head movement and gaze fixation. The cubes display icons of frequently used applications (media player, file/internet browser, calculator, etc.). By slightly moving his/her head, the user can make all sides visible. When the user looks at an icon, an animation is started for feedback, enlarging the icon. When the gaze remains on the icon (for 150 ms [63]), the agency starts the related program.

Further gaze-controlled interactions are available after an application has been launched. For example, after a "music box" has been instanced, the box is positioned by the placer agent in the center of the interaction space. When looked at, the box moves to the foreground. The user can select a title and play it by looking at the virtual play button. When no further interaction takes place, the application moves to a medium depth level and is automatically removed after playback has been completed.

The screenshot in Fig. 15.16 (left) shows a clock, the music box, and the Web browser as the active applications. The information displayed by the Web browser automatically rolls on when the user looking at the information comes to the edge of the document window. If the user gazes at a hyperlink,

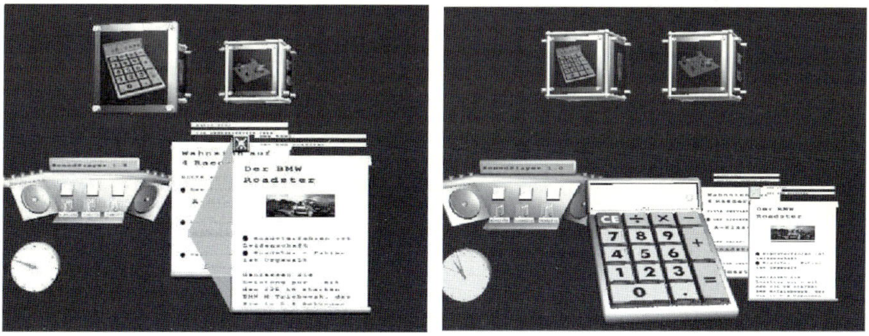

Fig. 15.16. The placer agent arranges the display of the media objects in the interaction space. The tool objects (cubes) show icons for gaze-controlled selection of applications. The cubes turn over-proportionally when the head moves, so that all sides become visible with slight head movements

the document linked to this is automatically downloaded. The documents previously read then move somewhat into the background so that the current search path and the position in the network are permanently visible. If the user wishes to return to an earlier document, he/she merely needs to look at it in order to bring it back to the foreground. By changing the viewing perspective, hidden documents can also be viewed. In Fig. 15.16 (left) a pocket calculator is being started. The placer agent positions the calculator close to the user (Fig. 15.16, right), so that the buttons can be pushed with the finger tip.

In a more recent version of the graphic user interface, the Internet browser presents itself as a 3-D web space (Fig. 15.17, top). The current page floats at full size in the middle of the space and can easily be read. Buttons on the left "wall" permit the user to store frequently used URLs by special categories and to instantly reactivate them by gaze control. Some "favorites" hang on the right "wall" in the form of preview images (c.f. the MS Task Gallery presented in Sect. 15.2.1). At the bottom left of the web space, additional thumbnails illustrate the way through the web with lines indicating the links. In order to return to a previous page the user has to look at it, and then call it by verbal confirmation. All buttons are sized and spaced in such a way that they can be selected by gaze. If required, information on the wall will appear at a more favorable angle, when the user slightly moves to the side (change in perspective controlled by the head tracker).

The "3-D CAD" program gives the user a realistic three-dimensional impression of the designed object (Fig. 15.17, bottom). Objects and operations can be selected by gaze, speech, touch, or mouse input, respectively. Basic geometric elements can be taken out of the "object box" and manipulated in the central work space. A set of different tools allows a change of object properties, like color and texture, as well as transforming, grouping, and moving of objects in three-dimensional space. Gazing at the object box, it will tip out and make the elements easily selectable by gaze, hand, and speech input. For example, by looking at an object and saying "select", the chosen object will be placed in the work space for further manipulation.

The current object will change its appearance (e.g. wire frame, shaded surface, lighting effects) when the corresponding options are selected. The system allows users to use speech input in a more natural way by not being tied to a particular syntax. The recognition of commands is based on keyword spotting, i.e. the input is checked for the occurrence of specified keywords. The test system supports video conferencing and cooperative manipulation of CAD objects.

15.4 System Evaluation with Users

It is now widely accepted that users should be involved right from the start in the process of developing and evaluating new computer hardware and

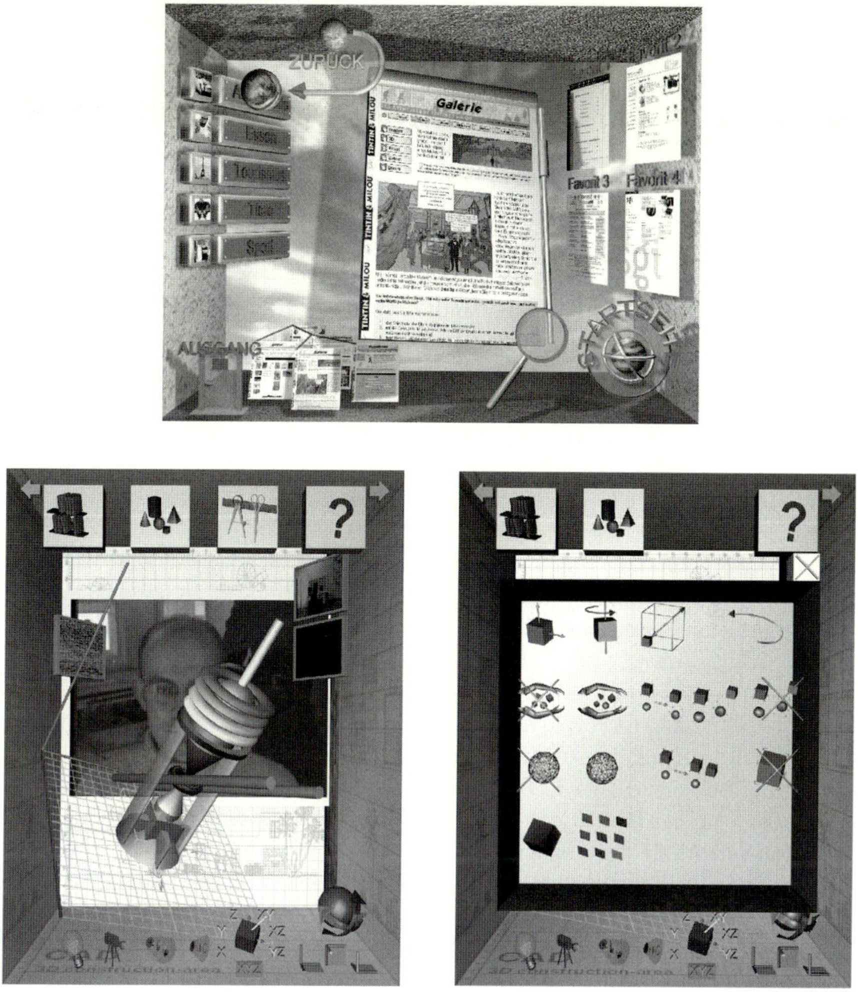

Fig. 15.17. Screenshots of the Internet browser (*top*) and the multimodal 3-D CAD application (*bottom*). By changing the viewing perspective (head movement), hidden objects can be viewed and addressed by eye gaze. In a telecooperation application, the user can select video windows showing the partner images. The menu boxes in the upper part of the CAD screen (system, objects, tools, and help boxes) are opened by gaze interaction. Subsequently, the user may optionally select sub-items by speech input, gaze input, or by pointing with the mouse or the finger tip (virtual 3-D touch screen)

software concepts (user-centered design) [2]. The evaluation process should be repeated at various stages to define subsequent development steps (iterative design). Moreover, an interdisciplinary approach with cooperation among experts in computer science, engineering, psychology, and design is indispensable. In the following we report some preliminary experiences with the mUltimo3-D system.

When presenting the mUltimo3-D system in our laboratory and at public exhibitions most of the users were highly impressed by the 3-D presentation and the possibility of communicating with a computer simply by looking at the object of interest and/or by changing the viewing position. Critical remarks concerned the design of the graphic elements (some icons were too small for easy gaze interaction in our initial setup and some were irritating because they were not consistent in style) and the lack of some quite basic and common functionalities (e.g. an "undo" function in case of a user's mistake or a misinterpretation by the system). Moreover, users frequently asked for a "park" function (which is not commonplace with today's windows-based user interfaces) in order to anchor an application at a fixed 3-D position on the display. The interface agent implemented in our system normally moves idle applications to the background in order to tidy up the display. For such autonomous forms of interaction, new functions must be considered in the system design.

Another point was the way how noncommand interactions had been implemented. In our first approach, noncommand actions were launched by an interface agent depending on how long the user was looking at a particular object. A fixed dwell time, however, does not seem to be the best alternative, since it appeared to be too long for certain applications and users while too short for others. Moreover, sensitive actions, like deleting a file, should be confirmed unmistakably by the user.

As pointed out earlier in this chapter, multimodal interfaces are supposed to enable a higher bandwidth, to provide the right modality for a given task, to reduce errors or disfluencies, and to be more natural, intuitive, enjoyable, and engaging [1, 11]. Systematic system evaluation should concentrate on testing such effects. Additionally, established criteria in software ergonomics, like suitability for the task, controllability, or error tolerance [110], have to be taken into account.

In order to determine the most salient problems we conducted heuristic evaluations [111] with the aid of usability experts. The experts rated prototypes of the applications shown in Figs. 15.16 and 15.17. We analyzed their comments on the functionality of the software, interface design, and interaction techniques.

Speech recognition rates were fairly low in our initial implementation. The experts therefore preferred to have feedback from the system to check whether their words had been correctly recognized. They rated speech input particularly useful when used as a shortcut. For example, the selection of design primitives like a cylinder in the CAD application took three steps in total when using the mouse (move the cursor, select the object, drag and drop it). Using speech it might take just one command, saying "cylinder". But how do we design appropriate commands for speech input? Users always tried to address screen objects directly. Hence, it should be possible to have alternatives, either to say the name of the screen object (e.g. "magnifying glass") or to say the name of the function this object is illustrating ("search"). Command vocabularies should allow for synonyms and should be easily memorizable by the user.

For data-retrieval purposes the use of natural language should be possible, e.g. "which cinemas are showing 'Titanic' tonight?" Careful selection of the tasks carried out with speech commands is necessary. Dragging and dropping objects, for example, is not well supported by speech input alone. Generally, the experts expected higher use of the keyboard than speech, at least for keyboard-experienced users. But will users be influenced by system affordances, e.g. will they speak more often when the computer talks to them as well? This should serve as a hypothesis for further empirical testing. A more general question raised here is how system output or feedback will influence the user's attitude, e.g. to chose among various input modalities available. A better understanding of such interactions would significantly simply and improve the design process.

Gaze interaction imposed several problems on users which were not well calibrated or changed their head position rapidly. They had difficulties with selecting narrowly spaced small buttons. Various options for gaze feedback were tested. Acoustic feedback was felt irritating and sometimes could not be related by the subjects to the gaze input. When feedback was given by highlighting the fixated object, the users' focus of attention was narrowed by controlling the highlight position – thus they often did not notice changes in other parts of the screen. Moreover, the rapidly jumping highlights between selected objects were felt to be annoying. The evaluators rather suggested a general status indicator showing whether the system received valid gaze data. With such an indicator, the workload scores measured were significantly lower.

Under experimentally controlled conditions with immediate gaze feedback (continuous display of the current gaze position by a cursor; discrete highlighting of an object when the system detected it as "being looked at") users took more time and made more errors than without such feedback. Two-thirds of the subjects tested felt that the visual feedback signals were irritating and distracting from the actual task. We conclude that neither continuous nor discrete feedback is recommended when using gaze input. Further user testing will concentrate on the effect of various combined modalities.

15.5 Conclusions and Future Work

The ongoing integration of IT systems is offering computer users a wide range of new (networked) services and the access to an explosively growing host of multimedia information. Consequently, apart from the currently established computer tasks, the access to and exchange of information is generally considered as the driving computer application of the next decade. Major advances are required in the development of the human–computer interface in order to enable end-users with the most varied backgrounds to master the added functionalities of their computers with ease, and to make the use of computers in general an enjoyable experience.

Research efforts in computer science are concentrated on user interfaces that support the highly evolved human perception and interaction capabilities better than today's 2-D graphic user interfaces with a mouse and keyboard. Thanks to the boosting computing power of general-purpose computers, highly interactive interfaces are becoming feasible, supporting a variety of input and output modalities. Multimodal interaction not only makes working with a computer more "natural" and "intuitive" but can substantially help to disambiguate the exchange of information in both directions between the user and the computer.

Recent approaches aim to create a (virtual) three-dimensional interaction environment where users find a clear arrangement of the media objects, paths, and links used, as well as highly responsive tools supporting continuous interaction. This chapter has overviewed the current concepts in general and described an approach developed by the authors in greater detail as a case study of a 3-D PC (the mUltimo3-D project).

This particular approach uses 3-D displays which do not require stereo glasses to present a 3-D graphic user interface. A newly developed 3-D display makes it possible to seamlessly integrate the virtual interaction space into the real working space. Video trackers remotely sensing the user offer new modalities for unencumbered interaction with the displayed objects. Interface agents watch and interpret the user's actions with regard to the task context, fuse the input from different interaction modalities, and organize the visual interaction space accordingly. The concept aims at the vision of a command-free, continuous human–computer interaction. A test bed has been implemented and used for system evaluation with users, and the first results have been reported.

Because of the high technical complexity of a multimodal interface and the many facets of interface design and evaluation, interdisciplinary research and cooperation among computer specialists, engineers, psychologists, and designers is required. Future work in the mUltimo3-D project will focus on the improvement of the overall system performance (in terms of the interface hardware and software) and on the optimal integration of various input/output channels. Apart from higher accuracy, reliability, and recognition rate of every single input mode, the important issue is to combine multimodal information in the best possible way. In order to achieve this, design issues have first to be clarified at the level of each separate modality, thus reducing the potential design space to meaningful alternatives. Simultaneously, investigations will be extended to the integrated modalities. The focus will be on test environments for tasks such as navigation in complex data structures and manipulation of complex objects in a 3-D CAD application. Using these tasks, the specific advantages of multimodality in connection with 3-D desktop computing will be tested. Following the principles of user-centered design, empirical evaluations with users will further accompany the technical design and implementation, in order to recognize and avoid potential human factor problems in the various stages of development of the mUltimo3-D system.

15.6 Acknowledgements

This work was supported by grants from the German Federal Department of Education and Research (BMBF) and the Senate of Berlin. The authors are exclusively responsible for the contents of this paper.

References

1. D.A. Norman: 'The Psychopathology of Every Day Things'. In: *Readings in Human–Computer Interaction: Towards the Year 2000*, ed. by Baecker, Grudin, Buxton, Greenberg (Morgan Kaufman, San Francisco 1995)
2. J. Preece, H. Sharp, D. Benyon, S. Holland, T. Carey: *Human–Computer Interaction* (Addison-Wesley, Harlow 1994)
3. B. Shneiderman: 'Interview with Ben Shneiderman'. In: *Human–Computer Interaction*, ed. by J. Preece et al. (Addison-Wesley, Harlow 1994)
4. J. Rosenberg, R.E. Kraut, L. Gomez, C.A. Buzzard: *IEEE Communications Magazine*, **30**, 20 (1992)
5. J. Conklin: Hypertext, IEEE Computer **20**, 17 (1987)
6. N.I. Benimoff, W.B. Whitten: 'Human Factors Issues in Multimedia Conferencing', Proc. 14th Int. Sympos. Human Factors in Telecommunications, 14th Int. Symposium on Human Factors in Telecommunication, Darmstadt, May 11–14, 1993; ed. by G. Schenck; R. v. Decker's Verlag, Heidelberg 1993; pp. 227–236
7. M. Weiser: Communications of the ACM **36**, 75 (1993)
8. A. Pentland: IEEE Trans. on PAMI **22**, 107 (2000)
9. B. Shneiderman: *Designing the User Interface – Strategies for Effective Human–Computer Interaction*, 2nd edn. (Addison-Wesley, Reading, MA 1992)
10. L. Wu, S.L. Oviatt, Ph.R. Cohen: IEEE Trans. on Multimedia **1**, 334 (1999)
11. G. Robertson: 'Leveraging Human Capabilities in Perceptual Interfaces', (invited talk, http://www.research.microsoft.com/PUIWorkshop/PUI98.htm), PUI (1998)
12. L. Schomaker, S. Münch, K. Hartung (eds.): *A Taxonomy of Multimodal Interaction in the Human Information Processing System*, A Report of the ESPRIT Project 8579 MIAMI (1995)
13. P. Maes: Commun. ACM **37**, 31 (1994)
14. S. Fertig, E. Freeman, D. Gelernter: SIGCHI Bull. **28**, 66 (1996)
15. E. Freeman, D. Gelernter: ACM SIGMOD Bull. **25**, 296 (1996)

16. B.B. Bederson, J.D. Hollan, J. Stewart, D. Rogers, D. Vick, L.T. Ring, E. Grose, C. Forsythe: 'A Zooming Web Browser'. In: *Human Factors in Web Development*, ed. by Ratner, Grose, Forsythe (Lawrence Erlbaum Assoc. 1998) pp. 255–266
17. T. Fox: 'Kognitiv ergonomische Benutzerschnittstellen – Entwicklung interaktiver 3-D-Visualisierungen und multimedialer Simulationen: das TUTOR-System COCARD zur Einführung in Ultraschall-Untersuchungen des Herzens', GMD-Studien Nr. 218, Sankt Augustin: GMD (1993)
18. B. York: Adv. Imaging **4**, 26 (1994)
19. M. Hemmje: 'Lyber-World - Eine 3-D-basierte Benutzerschnittstelle für die computer-unterstützte Informationssuche in Dokumentmengen', Der GMD-Spiegel 1/93 (1993)
20. S.K. Card, G.G. Robertson, J.D. Mackinley: 'The information visualizer, an information workspace', SIG CHI'91, April 27–May 2, New Orleans, LA, ACM Press, New York, p. 181 (1991)
21. R.W. Webster: 'VRML Conetree Generator', http://zansiii.millersv.edu/work2/conetree/ (Millersville University 1999)
22. R.A. Reaux, J.M. Carroll: 'Human factors in information access of distributed systems'. In: *Handbook of Human Factors and Ergonomics*, 2nd edn. ed. by G. Salvendy (Wiley, New York 1997)
23. G. Robertson, M. Dantzich, D. Robbins, M. Czerwinski, K. Hinckley, K. Risden, D. Thiel, V. Gorokhovsky: CHI 2000, The Hague, The Netherlands, 1–6 April 2000, ACM Press, New York (2000) pp. 494–501
24. S. Nagata: Proc. SID **25**, 239 (1984)
25. C. Ware, G. Franck: *Evaluating stereo and motion cues for visualizing information nets in three dimensions*, University of New Brunswick Technical Report TR94-082 (1994)
26. J. Leftwich (IDSA-Orbit Interaction): 'InfoSpace: A Conceptual Method for Interacting with Information in a Three-Dimensional Virtual Environment', 3rd Int. Conf. on Cyberspace, The University of Texas at Austin, May 1993 (http://www.well.com/user/jleft/orbit.infospace/)
27. K.H. Richardson: 'Prosody and User Acceptance of TTS', HCI '99, Edinburgh Conference Centre, Riccarton, Edinburgh, Scotland 30th August–3rd September 1999, pp. 1332–1336
28. A.D. Christian, B.L. Avery: 'Speak out and Annoy Someone: Experiences with Intelligent Kiosks', CHI 2000, The Hague, The Netherlands, 1–6 April 2000, pp. 313–320
29. J. Lai, D. Wood, M. Considine: 'The Effect of Task Conditions on the Comprehensibility of Synthetic Speech', CHI 2000, The Hague, The Netherlands, 1–6 April 2000, pp. 321–328
30. http://research.microsoft.com/ui/TaskGallery/pages/cues.htm; (2000)
31. S.A. Brewster: 'Sound in the Interface to a Mobile Computer'. HCI 1999, Pittsburgh, Pennsylvania, USA, May 15–20, pp. 43–47
32. M. Blattner, D. Sumikawa, R. Greenberg: 'Earcons and icons: Their structure and common design principles', Hum. Comput. Interact. **4**, 11 (1989)
33. R. Arsenault, C. Ware: 'Eye-Hand Co-ordination with Force Feedback', CHI 2000; The Hague, The Netherland, 1–6 April 2000, pp. 408–414
34. J.T. Dennerlein, D.B. Martin, C. Hasser: 'Force-Feedback Improves Performance for Steering and Combined Steering-Targeting Tasks', CHI 2000; The Hague, The Netherlands, 1–6 April 2000, pp. 423–429

35. I. Oakley, M.R. McGee, S. Brewster, P. Gray: 'Putting the Feel in 'Look and Feel' ', CHI 2000; The Hague, The Netherlands, 1–6 April 2000, pp. 415–422
36. G. Bishop, H. Fuchs: ACM Computer Graphics **26**, 153 (1992)
37. M.T. Vo: 'A framework and toolkit for the construction of multimodal learning interfaces', Ph.D. Thesis, School of Computer Science, Carnegie Mellon University, USA (1998)
38. C.M. Karat, C. Halverson, D. Horn, J. Karat: 'Patterns of Entry and Correction in Large Vocabulary Continuous Speech Recognition Systems', CHI 1999, Pittsburgh, Pennsylvania, USA, May 15–20, 1999, pp. 568–575
39. K. Fischer: 'Repeats, Reformulations, and Emotional Speech: Evidence for the Design of Human-Computer Speech Interfaces', HCI 1999, Edinburgh Conference Centre, Riccarton, Edinburgh, Scottland 30th August–3rd September 1999, pp. 560–565
40. C. Kamm, M. Helander: 'Design Issues for Interfaces Using Voice Input'. In: *Handbook of Human–Computer Interaction*, ed. by M. Helander, T.K. Landauer, P. Prabhu (Elsevier Science, Amsterdam 1997) p. 1043
41. J. Dowding, J.M. Gawron, D. Appelt, J. Bear, L. Cherny, R. Moore, D. Moran: 'GEMINI: A natural language system for spoken-language understanding', http://www.ai.sri.com/pubs/papers/Dowd93-54:GEMINI/document.ps.gz, (1993)
42. T. Schultz, A. Waibel: 'Das Projekt Globalphone: multilinguale Spracherkennung', 4th Conf. NLP – Konvens-98, Bonn, Germany, 5–7 Oct 1988, pp. 423–429 (1998)
43. W.C. Ogden, P. Bernick: 'Using Natural Language Interfaces'. In: *Handbook of Human–Computer Interaction*, ed. by M. Helander, T.K. Landauer, P. Prabhu (Elsevier Science, Amsterdam 1997) pp. 137–161
44. D.A. Bowman, L.F. Hodges: 'An Evaluation of Techniques for Grabbing and Manipulating Remote Objects in Immersive Virtual Environments', Proc. 1997 Sympos. on Interactive 3-D Graphics; Providence, Rhode Island, April 27–30, pp. 35–38
45. P. Kabbash, W. Buxton, A. Sellen: 'Two-Handed Input in a Compound Task', CHI 1994; Boston, MA, USA, April 24–28, pp. 417–423
46. K.-H. Jo, Y. Kuno, Y. Shirai: 'Manipulative Hand Gesture Recognition Using Task Knowledge for Human Computer Interaction', FG 1998, April 14–16, 1998 Nara, Japan, Nara-ken New Public Hall (Big Roof), pp. 468–473
47. D. Heckenberg, B.C. Lovell: 'MIME: A Gesture-Driven Computer Interface'. In: Visual Communications and Image Processing 2000, SPIE 4067, 261 (2000)
48. V.I. Pavlovic, R. Sharma, T.S. Huang: IEEE Trans. PAMI **19**, 677 (1997)
49. F. Lathuilière, J.Y. Hervé: 'Visual Tracking of Hand Posture with Occlusion Handling', Proc. 15th Int. Conf. On Pattern Recognition - ICPR 2000; Barcelona, Spain, 3–7 September 2000, 3, pp. 1129–1133
50. T. Starner, A. Pentland: 'Real-Time American Sign A Gesture-Driven Computer Interface', SPIE 4067, Perth, Australia, 20–23 June 2000, pp. 20–23
51. Y. Guiard: J. Motor Behavior **19**, 486 (1987)
52. W. Buxton, B. Myers: 'A Study in Two-Handed Input', CHI 1986; Boston, MA, USA, April 13–18, 1986, pp. 321–326
53. G.W. Fitzmaurice: 'Graspable User Interfaces', Ph.D. Thesis, University of Toronto (1996)

54. M. Wloka, E. Greenfield: 'The Virtual Tricorder: A Unified Interface for Virtual Reality'. ACM Sympos. on User Interface Software and Technology (UIST) 1995; Pittsburgh, Pennsylvania, November 15–17, 1995, pp. 39–40
55. J.H. Goldberg, J.C. Schryver: In: *Eye Movement Research*, ed. by J.M. Findlay, R. Walker, R.W. Kentridge (Elsevier Science, Amsterdam 1995)
56. W.F. Helsen, D. Elliott, J.L. Starkes, K.L. Ricker: J. Motor Behavior **30**, 249 (1998)
57. G. Edwards: 'A Tool for Creating Eye–aware Applications that Adapt to Changes in User Behavior', ASSETS '98, http://eyetracking.stanford.edu/assets/assets.html (1998)
58. Gips, P. Olivieri: 'Eagle Eyes: An Eye Control System for Persons with Disabilities', BCCS-96-01. (http://www.cs.bc.edu/techreps.html) (1996)
59. B.A. Smith, J. Ho, W. Ark, S. Zhai: 'Hand Eye Coordination Patterns in Target Selection', ACM Symp. on Eye Tracking Research and Applications, Palm Beach Gardens, FL, USA, Nov. 6–8 2000, pp. 117–122
60. E. Sibert, R.J.K. Jacob: 'Evaluation of Eye Gaze Interaction', CHI 2000; April 1–6, 2000, The Hague, The Netherlands, pp. 281–287
61. R.J.K. Jacob: 'Eye Tracking in Advanced Interace Design'. In: *Virtual Environments and Advanced Interface Design*, ed. by W. Barfield & T.A. Furness III (Oxford University Press 1995) p. 258–288
62. A. Duchowski, R. Vertegaal: In: Siggraph 2000 Course Notes (http://www.vr.clemson.edu/eyetracking/sigcourse)
63. R.J.K. Jacob: 'Eye movement-based human-computer interaction techniques: toward non-command interfaces', *Advances in Human–Computer Interaction*, Vol. 4 (Ablex Publishing Co., Norwood, N.J. 1993)
64. S. Pastoor, J. Liu, S. Renault: IEEE Trans. Multimedia **1**, 41 (1999)
65. D.D. Salvucci, J.R. Anderson: 'Intelligent Gaze-Added Interfaces'. CHI 2000, The Hague, The Netherlands, 1–6 April 2000, pp. 273–280
66. T. Sowa, M. Fröhlich, M.E. Latoschik: 'Temporal Symbolic Integration Applied to a Multimodal System Using Gestures and Speech'. In: Gesture-based Communicaton in Human–Computer Interaction – Proceedings GW 1999; Gif-sur-Yvette from 17 to 19 march 1999, Braffort, A., Gherbi, R., Gibet, S., Richardson, J., Teil, D. (Eds.) (Springer (LNAI 1739) 1999), pp. 291–302
67. R. Sharma, V.I. Pavlovic, Th. Huang: Proc. IEEE **86**, 853 (1998)
68. S.L. Oviatt: Commun. ACM **42**, 74 (1999)
69. S.L. Oviatt: 'Mutual disambiguation of recognition errors in a multimodal architecture', CHI 1999, Pittsburgh, Pennsylvania, USA, May 15–20, 1999, pp. 576–583
70. S.L. Oviatt, E. Olsen: 'Integration themes in multimodal human-computer interaction', ICSLP 1994, Yokohama, Japan, 18–22 Sep 1994, pp. 551–554
71. V. Tanriverdi, R.J.K. Jacob: 'Interacting with Eye Movements in Virtual Environments'. In: CHI 2000, The Hague, The Netherlands, 1–6 April 2000, pp. 265–272
72. M. Wöpking: J. SID **3**, 101 (1995)
73. W. Blohm, I.P. Beldie, K. Schenke, K. Fazel, P. Pastoor: J. SID **5**, 307 (1997)
74. B.M. Velichkovsky, J.P. Hansen: 'New Technological Windows into Mind: There is More in Eyes and Brains for Human-Computer Interaction', CHI 1996, Vancouver, British Columbia, Canada, April 13–18, 1996; pp. 496–503

75. P. Hansen, M. Pomplun, H. Ritter, A. Sprengler, B.M. Velichowski, H.J. Volke: 'Neue Anwendungen der Augenbewegungsforschung in der Mensch-Computer-Interaktion'. In: *Blickbewegungen in der Mensch-Maschine-Systemtechnik*, ed. by M. Rötting (Pro University Verlag, Sinzheim 1999) pp. 56–70
76. J. Nielsen: Commun. ACM **36**, 83 (1996)
77. A.J. Glenstrup, T. Engell-Nielsen: *Eye-Controlled Media: Present and Future State*, Ph.D. Thesis, University of Copenhagen, Denmark (1995)
78. T. Klein: *Computer Response to User Frustration.* MIT SM Thesis (1999)
79. M. Lucente, T.A. Galyean: 'Rendering interactive holographic images', ACM SIGGRAPH 1995, Los Angeles, USA, 6–11 Aug. 1995, pp. 387–394
80. Y. Kajiki, H. Yoshikawa, T. Honda: 'Hologram-like video images by 45-view stereoscopic display'. SPIE 1997, Stereoscopic Displays and Virtual Reality Systems IV, San Jose, USA, 11–14 Feb 1997, 3012, pp. 154–166
81. S. Pastoor, M. Wöpking: DISPLAYS **17**, 100 (1997)
82. I. Sexton, Ph. Surman: IEEE Signal Processing Magazine **16**, 85 (1999)
83. S. Pastoor: Human Factors of 3-D Imaging: 'Results of Recent Research at Heinrich-Hertz-Institut Berlin'. 2nd International Display Workshop, Hamamatsu, Japan, 1995; 3, pp. 69–72
84. S. Pastoor, G. Boerger: 'Autostereoscopic display providing a natural link of accommodation distance and convergence distance and adaptive depth-of-focus' (in German), DE 195 37 499, Patent Pending (1997)
85. S. Shiwa, T. Miyasato: 'A New HMD causing natural accommodation changes', SID 1997, Boston, MA, 13–15 May 1997, pp. 265–268
86. W. Plesniak, R. Pappu: 'Coincident Display Using Haptics and Holographic Video'. CHI 1998, Los Angeles, ACM, April 1998
87. H. Hongo, M. Ohya, M. Yasumoto, K. Yamamoto: ICPR 2000, 15th International Conference on Pattern Recognition, Hotel Barcelo Sants, Barcelona, Catalonia, Spain, 3–8 September 2000, pp. 925–928
88. M. Gavrila: Compu. Vis. Image Underst. **73**, 82 (1999)
89. H.-J. Boehme, U.-D. Braumann, A. Brakensiek: 'User Localisation for Visually-Based Human-Machine-Interaction' FG 1998, April 14–16, 1998, Nara, Japan; Nara-ken New Public Hall (Big Roof), pp. 486–491
90. A. Utsumi, J. Ohya: 'Multiple-Hand-Gesture Tracking using Multiple Cameras', CVPR 1999, Ft. Collins, CO 23–25 June 1999
91. R. Chellappa, C.L. Wilson, S. Sirohey: Proc. IEEE **83**, 105 (1995)
92. K. Jain, R.P.W. Duin, J. Mao: IEEE Trans. PAMI, **22**, 4 (2000)
93. G.D. Hager, P.N. Belhumeur: IEEE Trans. PAMI **20**, 1025 (1998)
94. D. Jacobs, P.N. Belhumeur, R. Basri: 'Comparing Images Under Variable Illumination', CVPR 1998, Santa Barbara, California, June 23–25, 1998, pp. 610–617
95. S. Georghiades, D.J. Kriegman, P.N. Belhumeur: 'Illumination Cones for Recognition Under Variable Lighting: Faces', CVPR 1998, Santa Barbara, California, June 23–25, 1998, pp. 52–58
96. M.L. Cascia, S. Sclaroff, V. Athitsos: IEEE Trans. PAMI **21**, 322 (1999)
97. R. Alferez, Y.-F. Wang: IEEE Trans. PAMI **21**, 505 (1999)
98. L.-P. Bala, K. Talmi, J. Liu: 'Automatic detection and tracking of faces and facial features in video sequences', *Picture Coding Symposium 1997*, Berlin, Germany, 10–12 Sep. 1997, pp. 251–256
99. C. Morimoto, D. Koons, A. Amir, M. Flickner: 'Real-Time Detection of Eyes and Faces', PUI 1998, San Francisco, CA, USA, 5–6 Nov. 1998

100. N. Oliver, A. Pentland, F. Bérard: 'LAFTER: Lips and Face Real Time Tracker', CVPR 1997, San Juan PR, USA, June 17–19, 1997, pp. 123–129
101. R. Kjeldsen, J. Kender: 'Interaction with On-Screen Objects using Visual Gesture Recognition', CVPR 1997, San Juan PR, USA, June 17–19, 1997, pp. 788–793
102. Ch. Maggioni: 'GestureComputer – New Ways of Operating a Computer'. In: Proc. of Intl. Workshop on Automatic Face and Gesture Recognition, Zurich, Switzerland, 26–28 June 1995, pp. 166–171
103. http://www.toshiba.com/news/980715.htm (2000)
104. H. Koike, Y. Sato, Y. Kobayashi, H. Tobita, M. Kobayashi: 'Interactive Textbook and Interactive Venn Diagram: Natural and Intuitive Interfaces on Augmented Desk System', CHI 2000, 1–6 April 2000, The Hague, The Netherlands, pp. 121–128
105. L. Young, D. Sheena: Behav. Res. Methods Instrum. **7**, 397 (1975)
106. K. Talmi, J. Liu: 'Eye and gaze tracking for visually controlled interactive stereoscopic displays'. Signal Processing: Image Communication **14**, 799 (1999)
107. J. Liu, S. Pastoor: 'A computer-supported method for contact-free, video-based determination of gaze direction for eye-controlled human-computer-interaction and a device for the implementation of the technique' (in German), Patent No. DE 199 53 835.2. (2000)
108. S.-W. Shih, Y.-T. Wu, J. Liu: 'A calibration-free gaze tracking technique', CVPR 2000, Barcelona, Spain, 3–7 Sep. 2000, 4, pp. 201–204
109. J. Liu: 'Determination of the point of fixation in a head-fixed coordinate system', ICPR 1998, Brisbane, Australia, 16–20 Aug. 1998, 1, pp. 501–504
110. ISO EN 9241: Ergonomische Anforderungen für Bürotätigkeiten mit Bildschirmgeräten, Teil 10: Grundsätze der Dialoggestaltung (1996)
111. J. Nielsen: *Usability Engineering* (Academic Press, Boston 1993)

Part III

Psychological Visual Science

16 The Influence of Optical Distortions and Transverse Chromatic Aberration on Motion Parallax and Stereopsis in Natural and Artificial Environments

J. Faubert

16.1 Introduction

The main purpose of this paper is to illustrate the importance of motion parallax and stereoscopy on judgments of relative depth and discuss factors that may impact on these judgments. I will first discuss some basic properties of motion parallax, its interaction with stereopsis, and follow with an analysis on how these cues can be disturbed or distorted in different viewing conditions. I will raise some practical and theoretical issues that have received little attention in the scientific literature but are of critical importance in our daily functions and when viewing artificial displays.

16.2 What is Motion Parallax?

When an observer is in motion, the visual scene surrounding the person is represented as a drifting image on the retinas of the observer's eyes. The drift speed on the retina depends on the relative distance of a given object in the image. If the object is close to the observer, the drift speed of this object on the retina will be faster than when the object is further away from the observer. It is well known that this relative motion of the visual image on the retinas, known as motion parallax, is used by the visual system to generate a sensation of depth [1, 2]. An illustration of how the distance of objects in a visual scene can generate different drift speeds on the retina of the corresponding points of the light rays is shown in Fig. 16.1.

This is one of the many cues used by the brain to estimate depth in a visual scene that does not require information from both eyes. For instance, superposition or occlusion (one object overlapping the other), texture gradients (the density of elements in the image changes as a function of distance), perspective (the shape of elements changes with distance), shadows and shading, and luminance (brighter objects appear closer), among others, are all examples of depth cues used by the human brain to estimate distance and depth that do not require binocular visual input. However, it is clear that among all the cues mentioned above, motion parallax is often considered as being a very efficient cue to generate a sensation of relative depth [1, 2].

Figure 16.1 demonstrates how two points that are aligned relative to the eye in a visual scene, but are distant from one another in depth, will generate

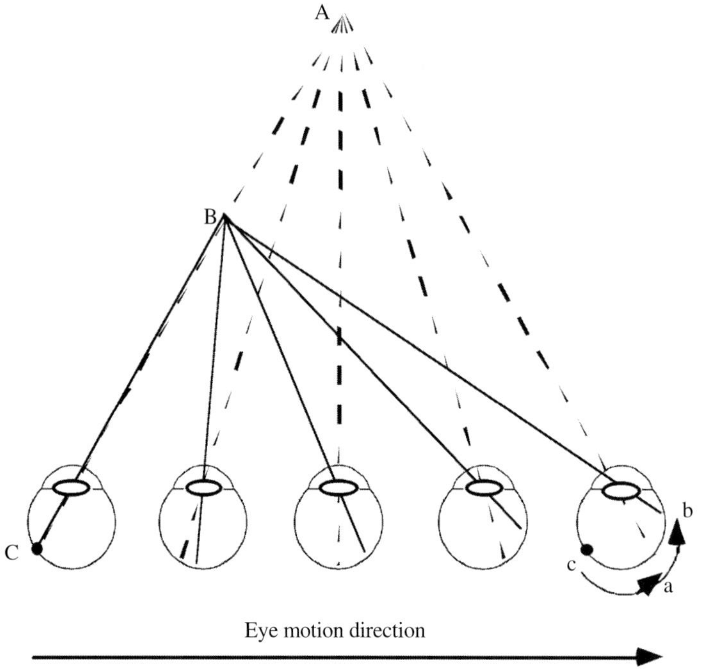

Fig. 16.1. Illustration of motion parallax generated by two points in depth. Point A is distant from the eye while point B is closer and on the same visual axis. As the eye moves from left to right, the projected points a and b on the retina are dissociated because of their respective geometry. The consequence is a longer travel distance on the retina of the projected ray B-b than ray A-a resulting in faster drift speeds for B-b

a different amount of displacement on the retina when movement is involved. Whether the scene itself is in movement and the eye stationary, or whether the eye position is changing and the scene is static (for example when looking out the window when traveling in a train) the relative displacements are the same on the retina. In this particular example the eye is in motion while the scene remains fixed. The figure illustrates two points in depth within the scene. Point B is close to the eye while point A is more distant. Both points A and B project an image on position C of the retina. As the eye moves (from left to right in this example) points A and B, given the geometrical constraints, travel at different relative distances on the retina and the end result can be seen at the projections a and b, where there is a substantial difference in their final positions. The different distances correspond to different perceived velocities, as both rays A-C and B-C traveled different distances within the same time span.

16.3 How Efficient is Motion Parallax?

As for most perceptual phenomena, the relative efficiency of motion parallax for producing estimates of relative depth depends on spatio-temporal factors and what adaptation state the visual system is in (luminance levels). However, one can illustrate the efficiency of motion parallax for generating a sensation of relative depth by directly comparing it with binocular disparity, which is the ultimate cue for generating small depth estimates. A simple experiment conducted in our laboratory demonstrates this point [3].

16.3.1 Motion Parallax Versus Binocular Disparity (Stereoscopy)

Let us suppose that we must judge the relative distance of an object that we are fixating in the central part of our visual field (foveal fixation). In this context, it is possible to assess how efficiently an observer judges the amount of separation in depth between two rods. This is a classic depth judgment based on the Howard method [4], where the smallest amount of separation that is necessary to see the difference between two rods in depth under binocular viewing conditions (stereoacuity thresholds) is generally determined. The same judgment task may be used when motion parallax is the only depth cue available.

Such thresholds were obtained with 38 young healthy observers who were enrolled in an optometry course. All had binocular vision as assessed by the randot test and 6/6 visual acuities. The apparatus and procedure are described in detail in Faubert [3]. Thresholds were determined for the binocular (stereoacuity) condition and for a motion parallax condition (monocular). In the motion parallax condition, the chin rest was mobile and was displaced at a frequency of 1 Hz for a distance of 6.7 cm. In other words, the head moved all the way to one side and back to the original side within one second. The viewing in this case was monocular.

The results are summarized in Fig. 16.2, where we can directly compare the steroacuity thresholds with parallax acuity thresholds. The mean log thresholds show that the stereo thresholds are better than the monocular parallax thresholds under these conditions. This confirms that binocular disparity is superior to monocular motion parallax for judging relative depth in the central visual field.

Can we conclude from this that the motion parallax cue is inefficient at generating relative depth judgments? Not necessarily. First, we must consider that the task that we evaluated is the optimal condition for binocular disparity because it involves the central visual field. It is clear that binocular fusion is limited to a certain area of the visual field, beyond which binocular fusion is rendered inefficient. This binocular fusion area is called Panum's area and has been well described in the literature [5]. Therefore, there are areas where we may be able to judge the relative depth of objects by motion parallax when binocular fusion is impossible. This would most certainly be

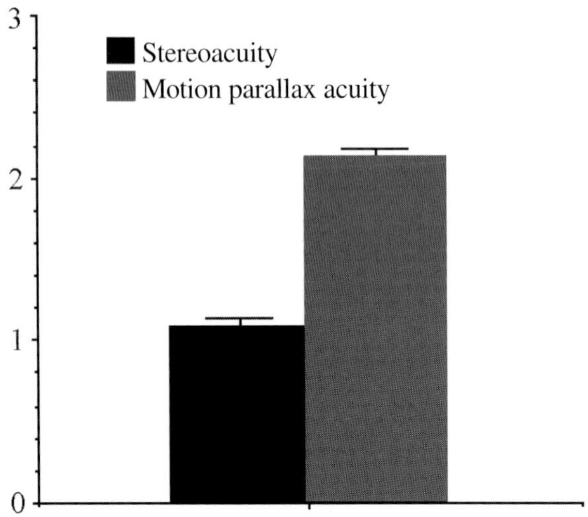

Fig. 16.2. Mean Log depth acuity thresholds in seconds of arc as a function of the testing conditions for the same subjects. The dark column represents the stereoacuity thresholds while the shaded column represents the motion parallax acuity thresholds. The error bars show the standard error of the mean for each condition

true under natural viewing or conditions with large viewing displays. If fact, the notion that stereopsis can only be efficient in a restricted portion of the visual field has been used to develop compression algorithms for transferring stereoscopic information contained in standard video signals [6].

Another reason why we must be careful about the interpretation of the present results for estimating the relative efficiency of motion parallax in depth judgments is the fact that our subject population consisted of healthy binocular viewers. They may not require the use of motion parallax to make judgments about relative depth in the central visual field under these viewing conditions. It is possible, therefore, that when parallax is the only source of information, it may also be a quite efficient cue for making such judgments.

16.3.2 Motion Parallax in Monocular Individuals Versus Stereoscopy in Binocular Individuals

To test this hypothesis, we assessed the motion parallax thresholds of three observers who had lost the use of one of their eyes due to accidents [3]. The three observers had lost the use of their eyes at least 2 years but no more than 5 years before participating in this study. All had 6/6 visual acuity in their functional eye. The results are presented in Fig. 16.3 and show that, although there is a difference between the motion parallax acuity thresholds of the monocular observers and the mean stereoacuity thresholds of our binocular observers, the difference is reduced by about half as compared to the control subjects' own motion parallax thresholds. Furthermore, the three observers' values fell well within the 95% confidence interval (mean + 2SD) established from the normative data. In other words, there were a number of normal observers who actually had equivalent or worse stereoacuities than the motion parallax acuity thresholds obtained for our three monocular observers. This demonstrates that, when binocular disparity is not available, motion parallax can be used to obtain relative depth judgments that can, at the very least, be considered within the normal range as far as stereoacuity is concerned.

16.4 Interactions Between Motion Parallax and Stereoscopy

The next question is whether both stereoscopic and motion parallax information can be used to improve our ability to make depth judgments. To illustrate this point we can refer to previous experiments performed in our laboratory that were designed to address practical questions of the visual requirements involved in telemanipulation environments [3, 7]. In the electric power industry there are frequent occasions where wires have to be mounted at high levels from the ground and where high currents are involved. To circumvent this problem, electric power companies use robots that are manned by trained

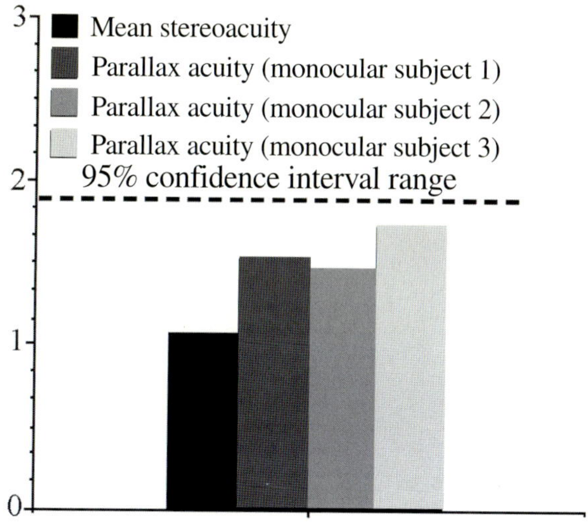

Fig. 16.3. Log depth acuity thresholds in seconds of arc as a function of the testing condition. The dark column represents the mean stereoacuity thresholds for the normal group while the shaded columns represent the motion parallax acuity thresholds for the three monocular observers. The dashed line corresponds to the upper limit of the 95% confidence range for the normal observers

personnel. Normally the operator is contained within a cabin that is raised in proximity to the wires and he or she then manipulates the robotic arms. An alternative approach is to operate the robotic arms at a distance via a virtual environment. The questions that were asked in this particular study were: what are the necessary and sufficient depth cues to perform these specific tasks, which involve positioning instruments in holes and assembling cables? Is stereopsis sufficient or should we also consider using motion parallax? The outcome of such questions may have a dramatic impact on the technology used and the production costs.

To answer this question we used a simple depth judgment task similar to the task used in the previous experiment. The experimental setup along with the view from the perspective of the cameras are presented in Fig. 16.4.

Two viewing distances were used. In Experiment 1 the cameras were positioned at 1 m viewing distance and in Experiment 2 the cameras were positioned at 2.6 m. The distances were based on the requirements of the operators' tasks in real-life situations when mounting cables for the electric power company. In one task the operators had to place and pick up instruments in holes when they were positioned at a distance of 1 m, and in the other task they assembled the cables at a 2.6 m viewing distance. As illustrated in Fig. 16.4, the experimental condition consisted of cameras that were facing the target. The camera movements were made possible by placing them on a sliding platform and the platform was yoked to the observers helmet via a steel rod. In this way the lateral displacements of the observer's head produced a perfect correspondence with the camera movements in front of the visual stimulus.

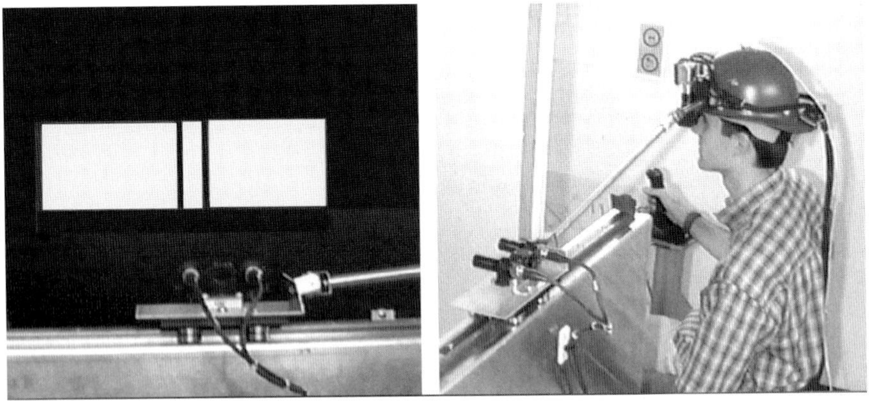

Fig. 16.4. Right image shows the subject with virtual reality helmet and the cameras used for viewing the image. The cameras were yoked to the head movements by being placed on a sliding platform. A steel rod connected the sliding platform and the helmet. The left image shows the view from the perspective of the cameras. See text for details

The visual information was transmitted to the viewer by a virtual reality helmet equipped with two monochromatic CCD video displays separated by 64 cm from each other. These displays were each linked with the corresponding left and right cameras, permitting a stereoscopic image. We were able to switch from a stereoscopic to a monocular view by allowing the information of a single camera to supply both visual displays. The focal distance of the screens was set at 7.5 mm and the visual field generated by the CCDs subtended 46° of visual angle in width and 35° in height with 100% overlap. The screens consisted of a 500-line horizontal resolution.

The subjects were required to align a mobile rod positioned to the right to match the position of the left rod with the use of a joystick (see Fig. 16.4). The four conditions tested consisted of (1) same image to both eyes with no motion parallax, (2) stereoscopic views with no motion parallax, (3) same image to both eyes with motion parallax, and (4) stereoscopic views with motion parallax.

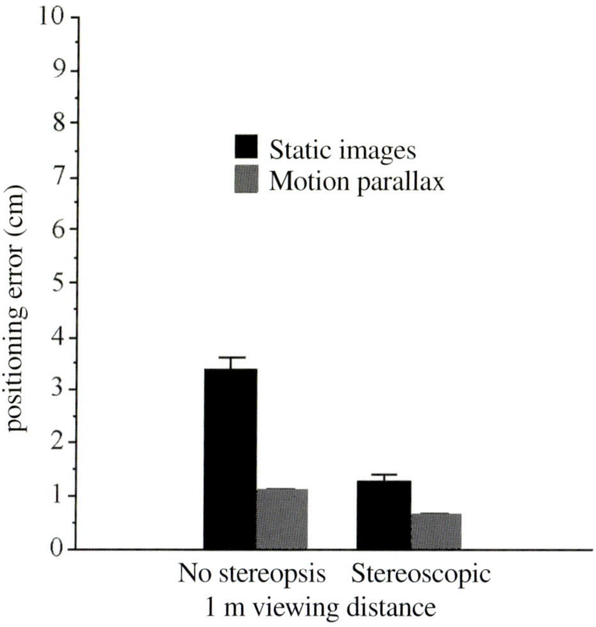

Fig. 16.5. Relative positioning error in cm as a function of experimental conditions for the 1 m viewing condition. The dark columns consist of the data obtained when the images contained no motion, while the shaded columns correspond to the conditions containing motion parallax. The two left-hand columns show the data for the monocular viewing conditions, while the right-hand columns represent the data for the stereoscopic conditions. The error bars show the standard error of the mean for each condition

The results for both experiments (1 m and 2.6 m viewing distances) are shown in Figs. 16.5 and 16.6 respectively. The results are quite interesting as they reveal several aspects of stereoscopic and motion parallax cue interactions. Considering the 1 m viewing distance, we can observe that in both the conditions where only stereoscopic or motion parallax cues are presented, there is a dramatic improvement in performance. This is demonstrated by the statistically significant reduction in the magnitude of the positioning error of the test rod relative to the reference rod. Furthermore, these cues in isolation appear equally efficient for judging relative depth in this type of task. The results for the interaction condition when both stereoscopic and motion parallax cues are available demonstrate that there is an additive effect resulting in a statistically significant improvement in the depth judgment task as compared to when the cues were presented in isolation.

The results of the second experiment (2.6 m viewing distance) show a similar trend, with the exception that the improvement produced by presenting both cues was about half the magnitude of the improvement obtained with

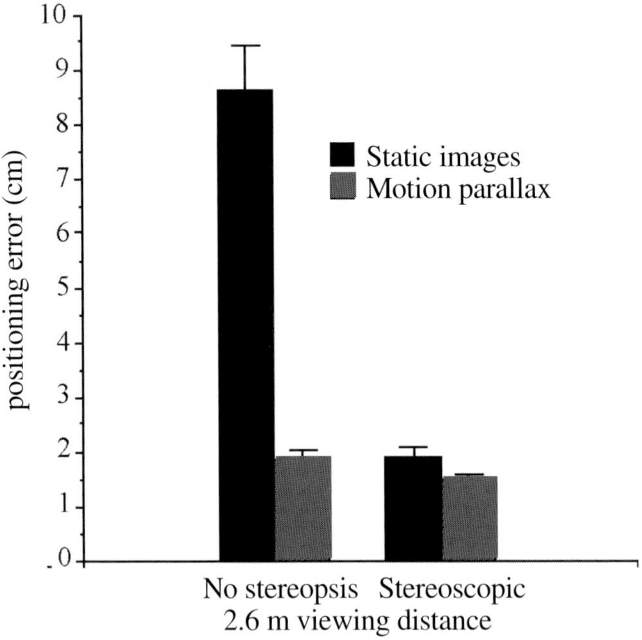

Fig. 16.6. Relative positioning error in cm as a function of experimental conditions for the 2.6 m viewing condition. The dark columns consist of the data obtained when the images contained no motion, while the shaded columns correspond to the conditions containing motion parallax. The two left-hand columns show the data for the monocular viewing conditions, while the right-hand columns represent the data for the stereoscopic conditions. The error bars show the standard error of the mean for each condition

the 1 m viewing condition and did not reach statistical significance. Generally the observers were not as efficient in performing this task as when the 1 m condition was used.

What is striking in these results, and in apparent contrast with the first study, is that motion parallax is equally efficient as the stereoscopic cue for making depth judgments. However, this can be explained by the fact that in this particular study both eyes receive visual input with motion parallax. In the first study only one eye was used in the motion parallax condition. Binocular summation was therefore available in the second study and not in the first. It is well known that visual performance will improve under dichoptic viewing conditions as opposed to when only one eye is used [8], although the reverse effect can happen if there is asymmetrical input between the eyes [9]. In a sense, the present study is a better estimate of the relative efficiency of motion parallax for generating depth judgments because we can directly compare stereoscopy and motion parallax when both eyes receive visual information.

Another interesting result is that there is an improvement in judging relative depth when both the stereoscopic and motion parallax cues are presented to the observer as opposed to when they are presented in isolation. This implies that these two depth cues share common mechanisms.

Finally, we observed that that the facilitation, obtained when both depth cues are present, will vary depending on where the object is positioned in depth relative to the observer. We saw that the relative improvement obtained when both cues were present was much smaller in the 2.6 m viewing condition than in the 1 m viewing condition. We can explain this difference by referring back to Fig. 16.1 and analyzing the simple geometrical effects of viewing distance on motion parallax. As we saw in Fig. 16.1, the closer an object is to the eyes in depth, the more velocity will be generated by the same lateral displacement of the eyes. Therefore, in the 1 m condition there is a larger relative displacement of the visual image on the retina than in the 2.6 m condition, resulting in greater motion parallax. Other studies have clearly shown that motion parallax is most efficient at close viewing distances [10].

16.5 Optical Distortions and Motion Parallax

What we have established with the studies above is that motion parallax, under certain circumstances, can be as efficient a depth cue as stereoscopic vision for measuring relative depth, and this is so even in foveal vision, where stereoscopic function is optimal. As was mentioned above, there is a large zone of the visual field where stereoscopic vision is not available because it is beyond the fusional capacity of the visual system (Panum's area). It is logical that the visual system uses motion parallax under these circumstances to estimate depth. If we believe this assumption, then we must consider factors that may affect the perception of motion parallax in the peripheral visual field. One such case is when we use optics in visual displays or simply when observers wear optical corrections (ophthalmic lenses). An optical distortion effect affecting motion parallax would be particularly important in the case of progressive lenses used to correct for presbyopia.

To illustrate this point we can start with simple examples of optical distortions produced by positive (hyperopic correction) or negative (myopic correction) lenses. The model chosen to make this point is based on the Le Grand–Fry approach [11, 12]. These optical models have the advantage of establishing possible distortions produced by ophthalmic lenses from a wearer's perspective [13].

Figure 16.7 demonstrates a simple viewing perspective when a viewer is placed at a given distance from a target. There are no distortions present in the target because no lens is placed in front of the eye. If, on the other hand, we place a positive lens and we maintain foveal fixation (as shown in Fig. 16.8), then the geometrical outcome of the image changes. What is shown in the image is that point B in the object image is in fact displaced in the periphery by a certain amount depending on the power of the lens (magnification). This displacement is obtained by tracing back from the angle of incidence of the ray from point B on the image, which has a changed angle relative to the eye once it has been refracted by the spherical lens. Point B from the viewer's perspective can be projected back and is now displaced to point B′. The distortion produced by such lenses can be demonstrated by tracing the line AB as viewed by the observer through the lens generating the curve B′A′. If there was no distortion produced by the lens, then the edge B′A′ would consist of a straight line. This is shown by the straight dotted line perpendicular to axis X projecting to point B′. However, because of the optical properties of the lens the actual projection of the ray is shown by the trace B′A′. The original line BA seen through the lens is actually seen as the curve B′A′.

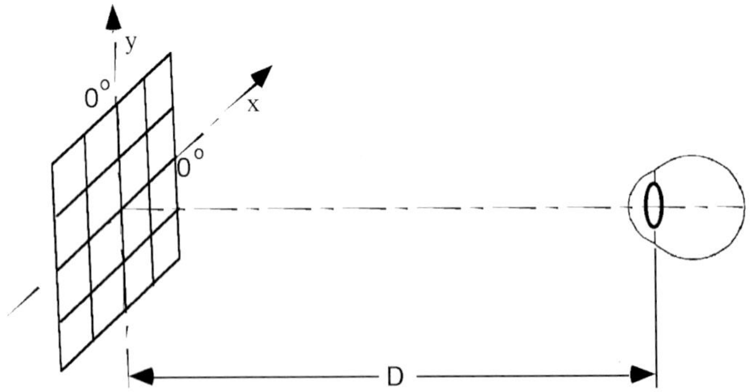

Fig. 16.7. Standard viewing condition and image grid used to generate the optical model projections of distorted images as a function of lens type and correction

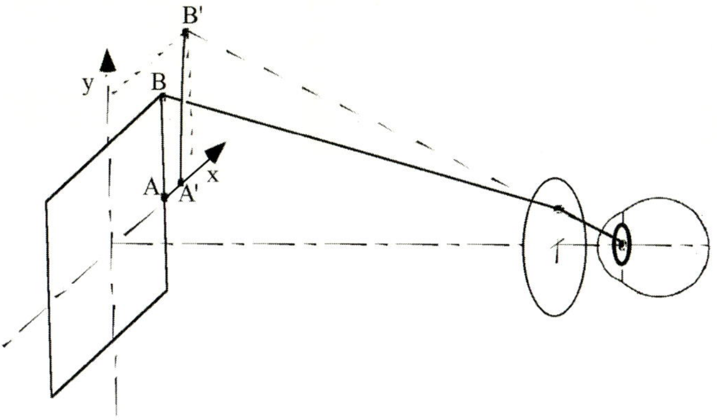

Fig. 16.8. Example of the distortion produced by a positive spherical lens from a viewer's perspective. Edge B-A is now projected as the curve B′-A′. It is assumed that fixation is perfectly lined up with the reference optical axis of the lens, which in turn is perfectly aligned with the center of the image grid

Figure 16.9 represents the new projection of the entire image relative to the viewer's perspective. What can be seen from this is that the corners have been displaced outward and the image has a pincushion distortion at the edges. We can simulate the distortion produced by a plus or minus lens by producing a frontal view of the object grid based on this model. Such projections are shown in Figs. 16.10 and 16.11 for plus and minus lenses respectively. The lens specifications are presented in the figures. The amount of distortion will of course depend on the power and the type of the lens. Figure 16.11 demonstrates that a myopic correction will in fact compress the image and produce barrel distortions at the edges from a viewer's perspective.

What becomes immediately obvious from such geometrical projections of ophthalmic lens distortions is that the motion parallax obtained under these conditions will differ from what would normally be obtained without an ophthalmic lens correction. This situation is even more critical if progressive lenses to correct for presbyopia are used. Presbyopia results from changes in the optical properties of the human eye, in particular changes related to the crystalline lens, due to the normal aging process. Almost every human in their 40s will undergo these changes and require some form of positive addition lens correction to see objects up close. The exceptions are myopic observers, who can compensate by simply removing their myopic correction given that they already have a natural positive correction from the optics of their own eyes.

An example of the object distortion obtained from a progressive addition lens is given in Fig. 16.12. It is obvious from such an illustration that the bulk of the distortion is in the bottom half of the image relative to the fixation point (identifiable by a dot in the top-center part of the image).

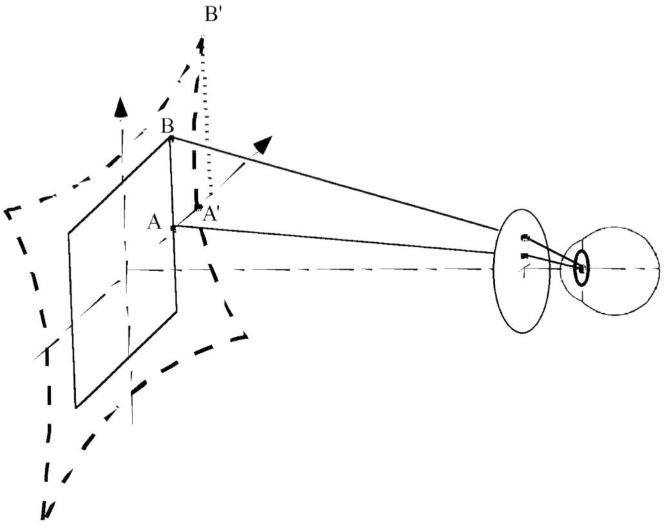

Fig. 16.9. Example of the model projection of all four edges when seen through a positive spherical lens. See text for details

The potential complications arising from the use of a progressive lens come from the interaction of two factors that vary simultaneously. As the image moves away from the central axis of the lens, there is both a magnification effect resulting from the increased power gradient of the progressive lens and, as a consequence, there is also an increase in distortion of the perceived image. If the head is in motion while the eyes remained fixated on the same point, one can perceive strong changes in the spatio-temporal components of the image. In other words, the motion-parallax component that is perceived under natural circumstances is dramatically changed by the refractive optics. Observers report a sensation of sway when moving their heads and fixating a single point, or when the eyes move behind the lens while the head is maintained in one position [14]. This is the single most important problem to consider in the process of ophthalmic lens design for progressive addition lenses [15].

To illustrate the effect of progressive lens distortions on a natural image, we have applied the optical distortion model above to a photograph. The image in Fig. 16.13 shows a view from our laboratory. Figure 16.14 is the result of a possible distortion of this image when viewed from a progressive addition lens if the image is placed in the bottom half of the standard grid. Because there are as many outcomes as there are different corrections, we will not discuss the specific lens characteristics or viewing conditions. What we show is a coarse approximation of the distortions that would be perceived under real conditions. However, it is obvious from the simulations that the world appears distorted and that the images appear projected forward in the

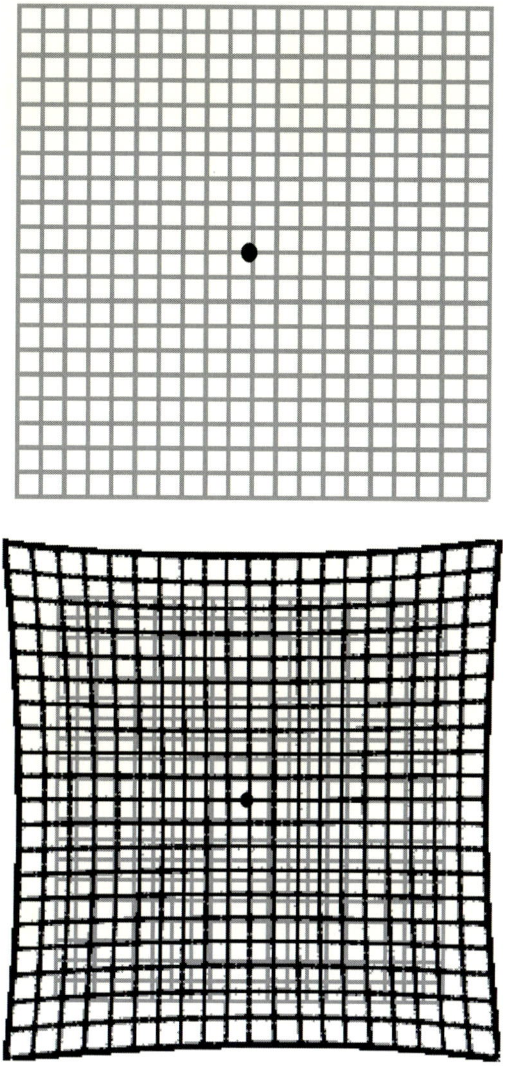

Fig. 16.10. The *top image* shows a front-view image of the image grid without lens distortion (light gray). The *bottom image* reproduces the nondistorted grid superimposed by a front-view projection of the distorted image from a positive spherical lens. The black dot in the center of the image shows the fixation point

lower visual field. The effect would be even more striking with an image in motion or with the observer in motion. We can also illustrate what happens when a viewer looks at oblique angles through the lens or when an observer maintains fixation while moving the head. Figure 16.15a shows a simulation of the effect of maintaining fixation on an image while moving the head sideways

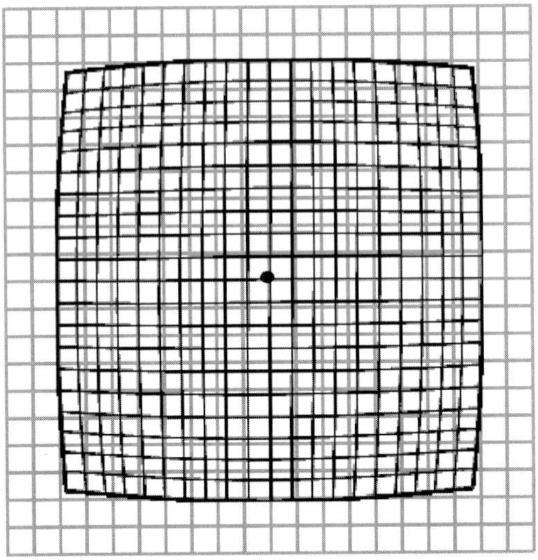

Fig. 16.11. The nondistorted grid presented in Fig. 16.10 (*top image*) superimposed by a model projection of a minus spherical lens

when progressive lenses are used. An asymmetrical distortion of the image results. The effect would be the opposite if the head was turned in the other direction from the fixation point (Fig. 16.15b). The reader can imagine the swaying effect when moving the head back and forth in these conditions.

16.6 Binocular Interactions With Progressive Lenses

We saw from Studies 1 and 2 above that binocular input to motion parallax is used and is important for the visual system. There is also recent evidence that velocity differences between the eyes can generate a perception of depth [16]. This creates an additional problem for the design of progressive lenses. In fact, Essilor International has recently identified this as the single most important cause of discomfort from progressive lenses. To minimize the interocular difference in motion parallax generated by progressive addition lenses, they now calculate a velocity difference gradient to improve their lens design [15]. To illustrate this point, let us consider what the two eyes are viewing through progressive lenses while fixation is maintained. Figure 16.16a shows that when the head is moved in one direction, the incidence angles of the image rays are different relative to each eye. This situation will produce differences in the geometry of the projected image to each eye. An example of the superimposed images of the two eyes that are out of phase is shown in Fig. 16.16b. The problem here is that the phase difference is not constant, as can be seen from Fig. 16.16a. The results can be dramatic differences in perceived velocity

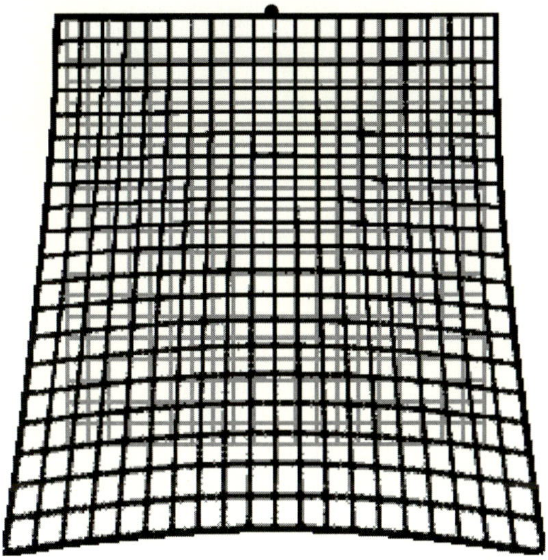

Fig. 16.12. Standard grid superimposed with a model projection calculated for a typical progression addition lens. The *black dot* at the top of the image grid shows the fixation point and the reference axis of the lens. Note that the distortion from a viewer's perspective is in the lower visual field area and is nonuniform as a function of eccentricity

Fig. 16.13. Image and grid used for the simulations of the optical distortion model

between corresponding points of the visual image. This question has not been studied and must be addressed to understand the observer's perceptual experience of motion parallax under such viewing conditions.

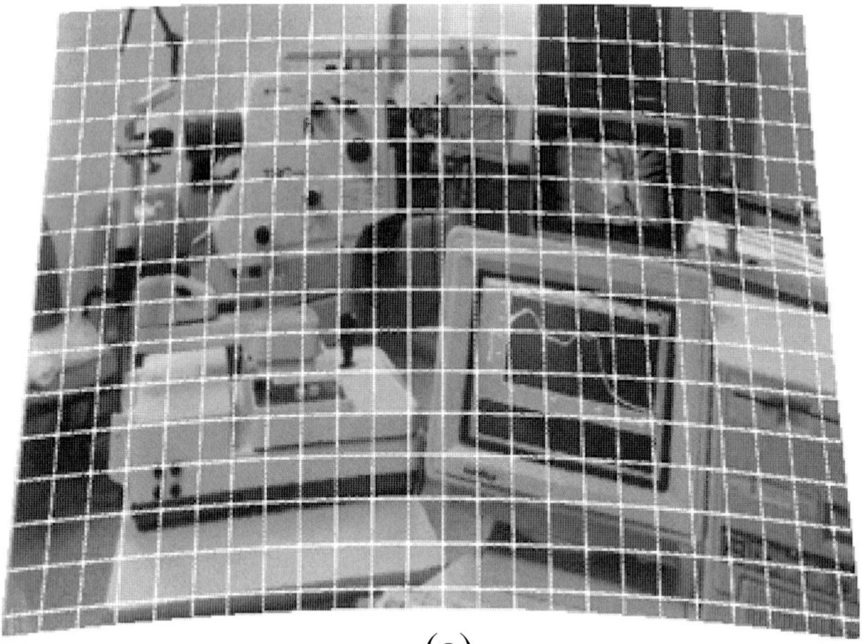

(a)

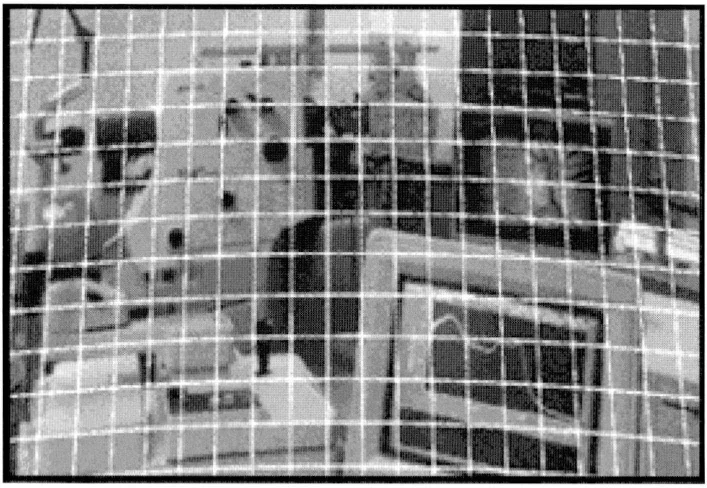

(b)

Fig. 16.14. (**a**) Simulation of the optical distortions produced by a progressive addition lens. The entire grid image is distorted by the lens. (**b**) Simulation of the optical distortions produced by a progressive addition lens. What is now seen after distortion in the same visual field area that was covered by the original image before distortion (Fig. 16.13)

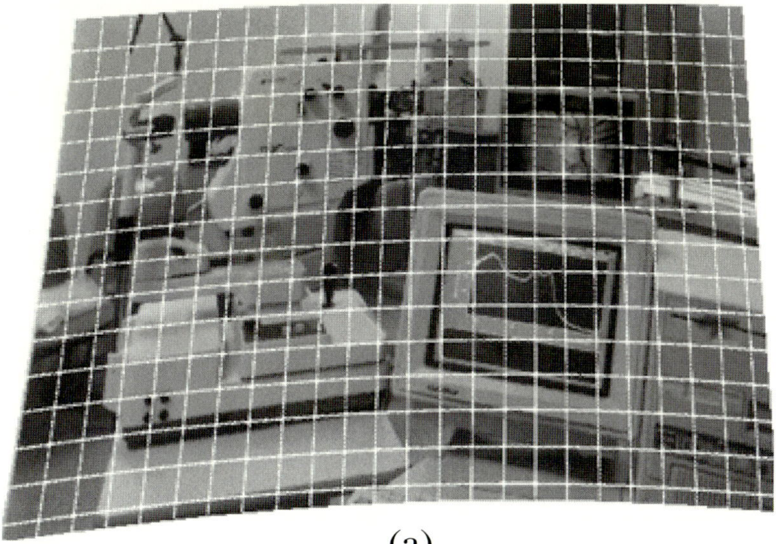

(a)

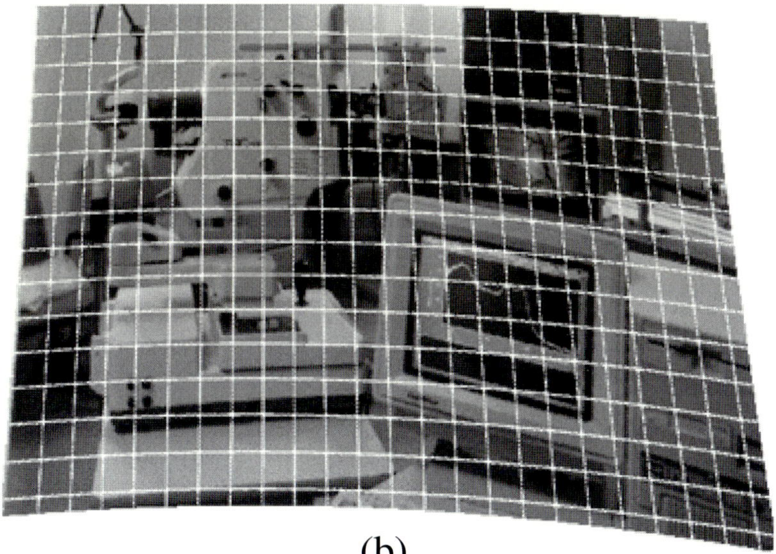

(b)

Fig. 16.15. (**a**) Simulation results of the distortion model when the observer maintains fixation on the same point of the image but moves the head away from the image (See Fig. 16.16a). Simulation results are for when head is moved rightward. (**b**) Simulation results of the distortion model when the observer maintains fixation on the same point of the image but moves the head away from the image (See Fig. 16.16a). Simulation results are for when head is moved leftward from the fixation target

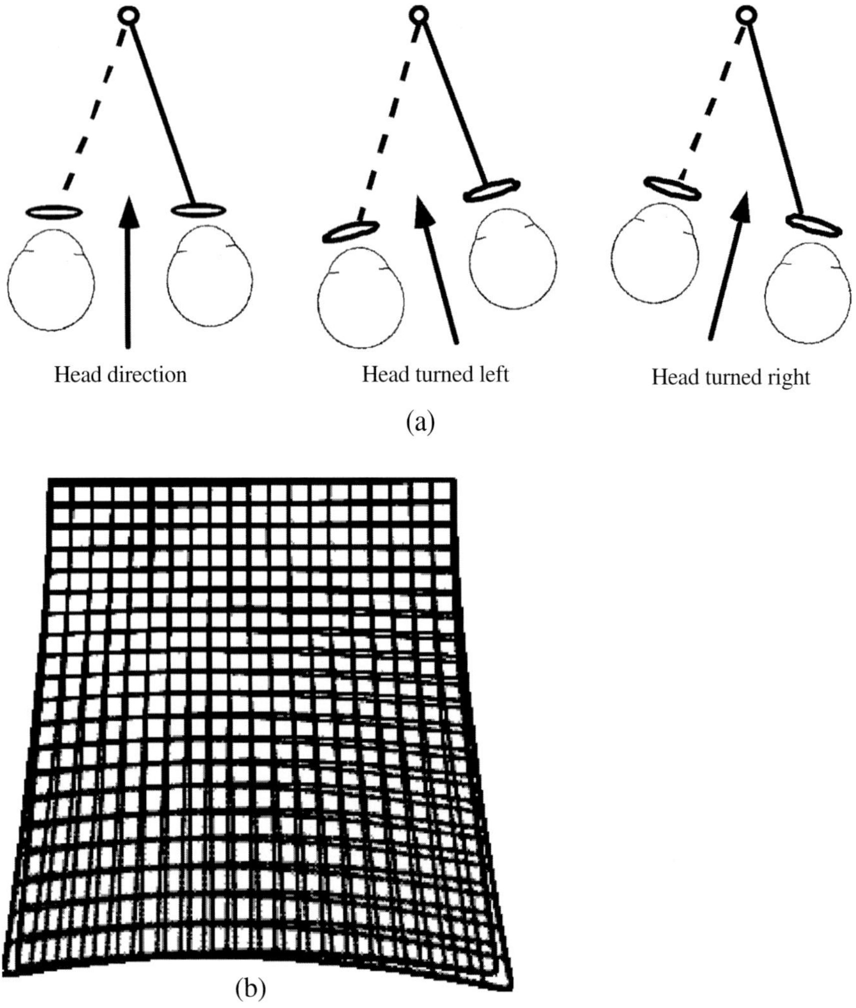

Fig. 16.16. (**a**) Example of how maintaining fixation while moving the head will generate difference incidence angles relative to the lens for each eyes. (**b**) Model results showing phase differences for a given position of binocular oblique viewing. These phase differences will vary with the angle of the head relative to the fixation point, resulting in differences in perceived velocities between the two eyes

16.7 Natural and Lens-Induced Transverse Chromatic Aberration, Stereopsis, and Motion Parallax

16.7.1 What is Transverse Chromatic Aberration?

It is well known that the optics of the eye produce chromatic aberrations [17–19]. Chromatic aberration results from the fact that light refracts

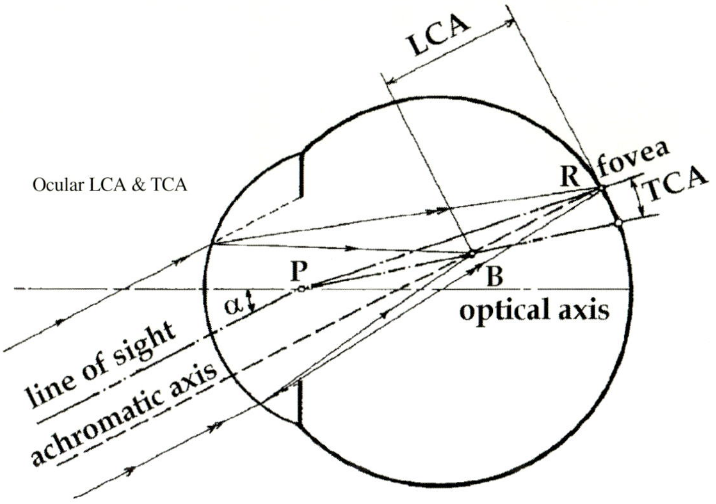

Fig. 16.17. The two components of chromatic aberration at the fovea. Longitudinal chromatic aberration (LCA) corresponds to the difference of focus with wavelength between the red image (R) and the blue image (B). The red image has been arbitrarily positioned at the fovea. Natural transverse chromatic aberration (TCA) corresponds to the difference of position of the retinal images with wavelength. In the figure, natural TCA is shown as the angle between the principle rays of the two retinal images. For the sake of clarity the drawing is not to scale, and it has been assumed that the whole refractive system of the eye is situated at the cornea and that the entrance and exit pupils are superimposed (adapted from [22])

differently depending on its wavelength. There are two major components of *ocular* chromatic aberration at the fovea. Longitudinal chromatic aberration (LCA) corresponds to the difference of focus with wavelength while transverse chromatic aberration (TCA) corresponds to a difference of position of the retinal images with wavelength. Figure 16.17 illustrates the two major components of *ocular* chromatic aberration at the fovea. LCA corresponds to the difference of focus with wavelength between, in this example, the red image (R) and the blue image (B). The red image has been arbitrarily positioned at the fovea. In the top figure, *natural* TCA is shown as the angle between the principle rays of the two retinal images. For clarity the drawing is not to scale, and it has been assumed that the whole refractive system of the eye is situated at the cornea and that the entrance and exit pupils are superimposed.

Figures 16.18a to 16.18d are schematic diagrams representing examples of possible interactions between *natural* and *induced* TCA at the fovea by a prismatic lens. For the sake of clarity, a reduced eye is used and the distances and angles are exaggerated in the graph.

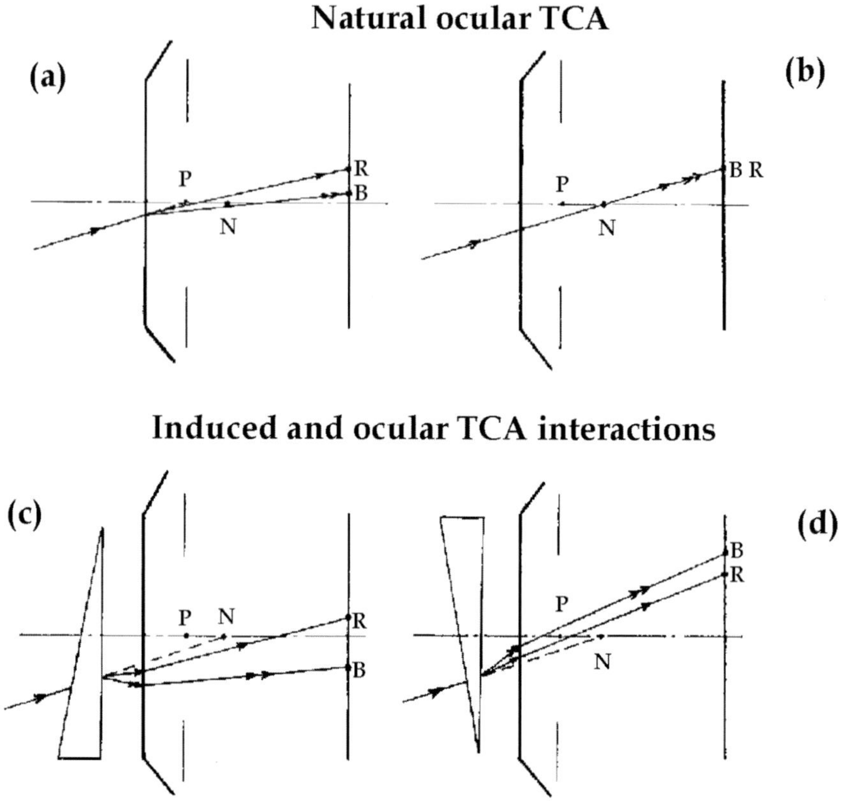

Fig. 16.18. Ocular TCA at the center of the entrance pupil (**a**). No ocular TCA when the incident ray is superimposed with the achromatic axis (nodal axis) (**b**). TCA induced by a prism (**c**). This induced TCA is additive with the ocular TCA. TCA induced by an inverted prism (d). The effects of induced and ocular TCA are opposed (adapted from [53])

Figure 16.18a illustrates a condition where there is a naturally occurring TCA as opposed to when the incident ray is co-incident on the achromatic axis (see Fig. 16.18b), which results in the absence of TCA. Figure 16.18c gives an example of induced TCA from a prismatic lens. Note that in this case the direction of the natural TCA and the induced TCA are in the same direction and are therefore additive. If the prism is inverted as shown in Fig. 16.18d, the effect of induced and natural TCA are opposed to one another and, as a consequence, the total TCA is diminished. Although ocular TCA and induced TCA are distinguished below, it must be understood that the consequences of both types of TCAs on visual performance should be exactly

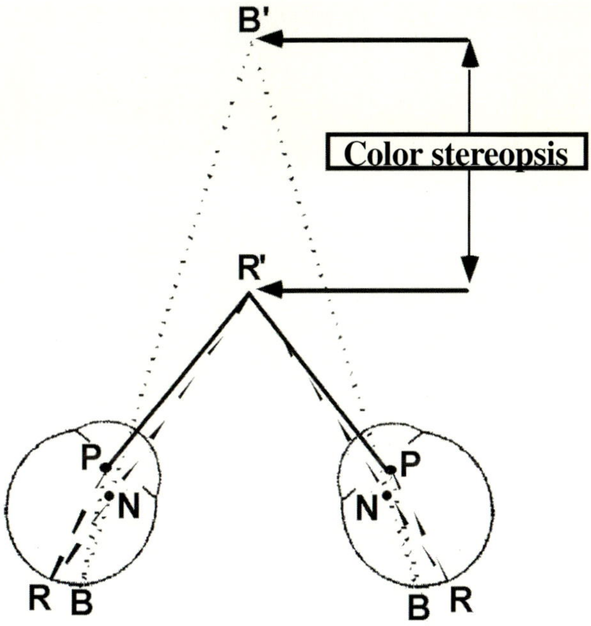

Fig. 16.19. Illustration of how chromatic parallax (TCA) can induce color stereopsis. The image is focused at point R′. The rays of the red and blue portions of the image are displaced at different angles on the retina producing retinal disparity between the eyes (adapted from [32])

the same because both produce selective color phase shifts at the retinal level [20].

16.7.2 Consequences of TCA on Visual Performance

TCA can decrease achromatic spatial contrast sensitivity [21–25] dramatically if the spatial-frequency component of the stimulus is greater than 1 cpd. It has been shown that correcting TCA with achromatizing lenses can produce more problems than it solves if the lens is not perfectly aligned with the achromatic axis [20].

16.7.3 TCA and Stereopsis

The illusion of depth obtained from viewing two-dimensional colored targets has been reported for many years and has generally been attributed to some form of chromatic aberration [26]. The most compelling example of seeing depth with colored stimuli is evident in conditions of chromostereopsis. Chromostereopsis is when two colors (generally red and blue or red and green) presented in the same depth plane are binocularly perceived as residing in

Fig. 16.20. Stimulus example where some of these effects were first observed on a computer screen. Some effectiveness of the stimulus is diminished by presenting it on paper due to the loss of saturation and luminance control. See [17, 32, 33] for other examples

separate depth planes [18, 27–31]. The usual stimulus for the study of chromostereopsis consists of a target with red and blue regions where the red portion is either perceived in front of the blue, called positive chromostereopsis, or behind the blue, representing negative chromostereopsis [28]. Over the years several models have been proposed to explain this effect, which include longitudinal and transverse chromatic aberrations [26]. However, recent work attributes most of the effects to transverse chromatic aberrations with possible interactions resulting from the Styles–Crawford effect and pupil decentralization [17, 18, 31–33]. The work from other laboratories have focused primarily on optical factors to explain chromostereopsis while I have demonstrated that depth relations in chromostereopsis involve more than optical effects alone by presenting novel color depth illusions [32, 33]. Such effects can explain why we sometimes perceive depth from colored stimuli and sometimes not.

Figure 16.19 demonstrates in a simplified manner how the TCA from the eyes can generate stereopsis and Fig. 16.20 shows an example of where the effects described by Faubert [32] were first observed.

16.7.4 Why Are Such Illustrations Important to Understand When We Actually See or Do Not See Color Depth Effects?

Color information is present all around us. We know from previous work that the optics of the eye will refract the different colors on different corresponding points of the retina (disparity) due to TCA. There are many circumstances where two given colors (e.g. red and green) are seen on different depth planes while the same colors will be seen as having no depth information in a different context. What generates these different percepts given that the optics should produce exactly the same disparity between the colors on the retina? The studies reported by Faubert [17, 32, 33] address this very issue. Based on observations and measures, the following factors are particularly important in color depth effects.

Same Color on Different Planes and Different Colors on the Same Plane. The first noticeable difference between the usual conception of chromostereopsis and the present phenomenon becomes obvious when carefully observing the pattern in Fig. 16.1. For clarifying purposes, I will refer the small inner square of the stimulus as the center square, the larger background square as the surround, and the remaining as the background. In a given stimulus, one can observe the same color in a different depth plane and different colors in the same depth plane. In Fig. 16.1, observers generally see the entire inner square composed of both black and green dots as being on the same depth plane and in turn are on a different depth plane than the red and green surround. The surrounding red and green dots are generally perceived as being on the same depth plane. All the green contained in the image, whether it is from the inner square, the surround, and the background, is exactly the same green (i.e. identical in chromaticity and luminance). The red and green regions perceived on the same depth plane represent areas where disparity from chromatic aberrations must be present, yet cannot be used by the visual system. At the same time, the same green is perceived on two separate depth planes, which must be the result of other factors than disparity induced by chromatic aberrations.

A Notion of Reference Plane? In stimuli such as Fig. 16.20, the removal and replacing of the black dots with the surrounding color eliminates the perception of depth [32]. Yet, the colors responsible for chromatic aberrations are still present. The black is a reference without which the disparity caused by chromatic aberration cannot be used with natural pupils. However, replacing the black dots with bright white dots produces similar depth perception as when using black dots, but the images are generally less distinct because of the polychromatic components of white. Therefore, in these stimuli the black and white portions (luminance information) must be used as a reference depth plane by the brain. Later manipulations demonstrated that the color of the reference plane mattered little as long as it was distinctly brighter or darker than the red and green. The dots normally perceived as being on a reference plane will be referred to as "reference dots". When the reference dots were removed from the stimuli, none of the six observers reported seeing illusory contours; therefore, illusory contours do not appear to be responsible for this effect.

Spatial Configuration Effects: Dot Density and Dot Size Play a Role. The effect was still visible when the dot density was diminished from 50% to 15%, although some integration time was required by the observers in initial trials. Most observers could not see the depth when first presented, but after staring at the stimulus for many seconds (in some cases several minutes) the depth effect came back, implying some form of mental construction.

There appears to be an optimal dot size to maximize depth. At 1.22 m, the optimal size of the random dots was two to three pixels (about 3 min) for all observers. When the dots were smaller (1 pixel) they became difficult to resolve and the effect was not visible. Increasing the size of the dots quickly decreased the impression of depth. Thus, this phenomenon appears to favor high spatial frequency components (fine details).

Another interesting observation was made in relation to spatial configuration when reference dots (black dots in this case) were placed individually in different places of the surround of Fig. 16.20. In this case the random dots did not merge with the surround but either appeared in front or behind the surround congruent with the perceived depth of the center portion of the stimulus also containing reference dots. The reference dots and the red/green dots do not have to be adjacent to produce the depth effect.

Effect Still Present With a Stereoscope and at Infinity Viewing Distances. In an attempt to segregate convergence with the depth effect, identical pairs of slides were made of the display and presented dichoptically with a standard stereoscopic viewer. The horizontal placement of the slides was adjusted for each observer so that the slides were parallel to each eyes and the image was fused. Under these conditions, all observers saw depth in the way described originally when observing the screen. In addition, the two observers who initially did not perceive depth when observing the computer screen, reported clearly seeing depth in the stereoscope condition. Observers were also asked to view the image when closing one eye and all reported not seeing depth monocularly.

Another way to segregate convergence is to increase viewing distance. A slide of the stimulus was projected on a wall in a large room and shown to five of the observers. The green background was $1.03\,\mathrm{m} \times 0.8\,\mathrm{m}$ in size. The surround was 0.66 m and the center 0.38 m in size. All observers reported seeing binocular depth. The observers were asked to move backwards until the depth was no longer visible. All observers except one moved to the very end of the room where the depth was still visible at 12.8 m. The last observer stopped seeing the depth about 0.5 m before reaching the end of the room.

He reported that the dots of the stimulus were no longer visible; thus he could not see the depth. The observers, without exception, reported that the perceived depth increased while increasing the distance. The observers were also asked to move forward to determine whether the depth was still visible at all distances. Depth was visible until the 4 m mark, where shorter distances produced no depth.

Effect of Artificial Pupils (Size and Displacement). Images were viewed through 3 mm and 1 mm artificial pupils by three observers for the computer screen, stereoscope, and wall conditions. The artificial pupils were placed on trial frames positioned very close to the eyes. The horizontal positions were varied from side to side to change the transverse chromatic aberration. Adjusting the horizontal position of the 3 mm pupil changed the perceived depth, demonstrating that transverse chromatic aberration has an effect on this perceived depth illusion.

Two of the three observers reported seeing the depth with a 1 mm pupil for all three conditions (screen, stereoscope, and wall). One observer could not see the depth when the stimulus was projected on the wall because the luminance was too low for him to detect the stimulus properly, but he saw the depth in the stereoscope and on the screen through 1 mm pupils. Moving the artificial pupils did have an effect on the perceived depth, implying that TCA affects perception under this condition.

The Influence of TCA. To determine the influence of TCA on these illusions, the effect of displacement of artificial pupils relative to the achromatic axis on the perceived depth was determined. If, as expected from the pilot studies, the perceived depth in this stimulus context is based on TCA, then we should observe a relationship between the amounts of displacement of two artificial pupils from the achromatic axis (the achromatic axis represents the optical axis that generates no differences in chromatic parallax between colors). However, if for the same chromatic conditions no depth is perceived when the reference dots are absent, we can conclude that TCA is present but cannot be used by the visual system under this context.

The results clearly supported the notion that the extent of the perceived depth for these stimuli, when reference dots are present, is based on the amount of TCA. However, the fact that under the same pupil and chromatic conditions, but without reference dots, all observers reported that *no depth was perceived* even if the TCA was still induced, clearly demonstrates that factors other than TCA alone are essential to explain the perceived depth.

16.8 Seeing Depth with Black and White and Only One Other Color

Further studies [33] demonstrated that it is not necessary to have two colors of the opposite ends of the spectrum (red and green or red and blue) to perceive color depth effects.

Using images such as those proposed by Faubert [32] to facilitate and optimize the detection of chromostereopsis, it was hypothesized that depth should be visible in conditions where only achromatic information and one other color are present in an image. This was done by making assumptions about where the relative position of chromatic and achromatic components in an image should be located on the retinas and, therefore, where this informa-

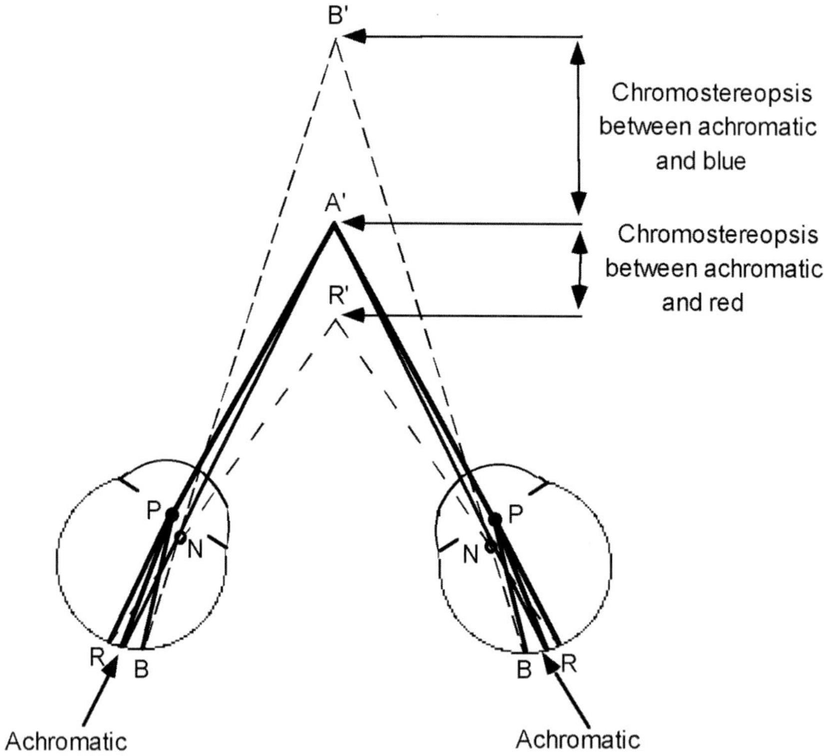

Fig. 16.21. Model of the lateral displacement of colored images on the respective retinae and the resulting luminance profile for a positive chromostereopsis condition, tracing the light rays through the eyes and the relative positions of the red and blue lights on the retinae and the achromatic information. R represents the red light position on the retina and B the blue light position. P represents the pupil entry and N the nodal point. R′ and B′ represent the perceived depth of the red and blue portions of the image

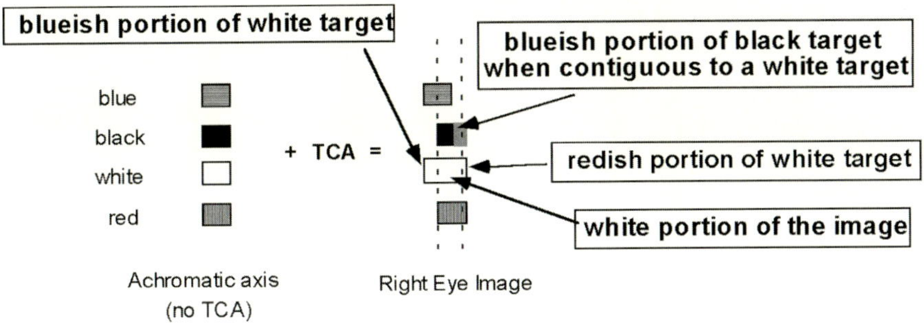

Fig. 16.22. The resulting luminance profiles on the retina for the right eye based on the model presented in Fig. 16.21

tion should be located in relative depth planes produced by chromostereopsis. Another important assumption presently made is the way the visual system constructs the different depth surfaces. This assumption is based on two properties demonstrated by the visual system when perceiving color depth effects, which have been previously discussed by Faubert [32]. Firstly it was proposed that achromatic (luminance) texture information was often used by the visual system as a reference plane upon which the depth perception is constructed. Furthermore, when the achromatic information was removed from the image, and the other colors were left in, the color-induced depth went from very visible to difficult (if not impossible) to perceive. Secondly, it was proposed that the image information that was embedded in texture information was often perceptually dragged into the same depth plane (capture) even if the optical model predicts that they have different disparity information.

Based on the proposition that gray and black contiguous segments and the gray and blue (or red) contiguous segments in an image should be used as unique surfaces by the visual system, regardless of whether the pupil or other optical sources produce a misalignment relative to the achromatic axis, three predictions were made relative to the depth perceived in images with achromatic light and one other color: (1) depth should be perceivable with blue/achromatic and red/achromatic images and not require a red and blue portion simultaneously in the image, (2) The depth in the blue/achromatic images should be perceived in the opposite direction than in the red/achromatic images with identical TCA levels and identical pupillary positions relative to the achromatic axis, and (3) The perceived depth should be greater for the blue/achromatic images than for the red/achromatic images.

The optical predictions are based on the model presented in Fig. 16.21 and the outcome on the retina of the right eye is illustrated in Fig. 16.22.

As demonstrated in Fig. 16.22, the action of TCA on a white light target is generally to spread this target. The optical model predicts that the contiguous gray and black portions of the stimulus are positioned between unique blue

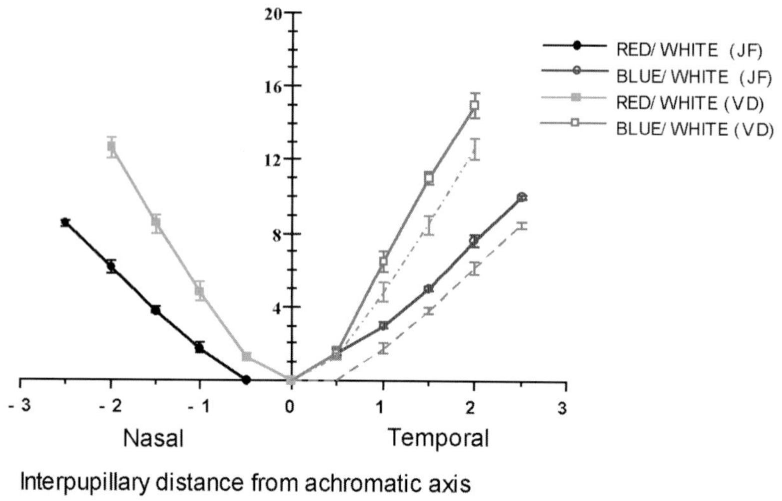

Fig. 16.23. Results of TCA manipulation on the perceived depth for two observers. Note that to obtain depth in the same direction for the two image conditions, we used nasal displacements relative to the achromatic axis for the red/achromatic condition and temporal displacements for the blue/achromatic position. The red/achromatic results are reproduced by dotted lines on the positive side of the graph for comparative reasons

and red light rays. As argued above, it was hypothesized that both the gray and black portions together would be used as a unique reference frame to construct the image.

On the other hand, the blue and red elements of the image are shifted in opposite directions relative to the achromatic elements and, consequently, should create opposite disparities for the same TCA conditions if the achromatic portion is used as the reference plane. Moreover, because the blue light rays in the presence of TCA are situated further from the reference plane than red light rays, the magnitude of the perceived relative depth should be greater for a blue/achromatic than for a red/achromatic stimulus.

The results for the two observers are shown in Fig. 16.23. The horizontal axis represents the amount of displacement from the foveal achromatic axes where negative values are nasal displacements and positive values are temporal displacements. The vertical axis represents the forward perceived depth in mm between the inner square and the surround square as measured by the position of the reference rod. The results clearly support the three predictions made earlier. Firstly, that chromostereopsis can be perceived with achromatic light and a single other color. Secondly, that the perceived depth is in opposite directions when using blue/achromatic versus red/achromatic for the same TCA levels. Thirdly, that the perceived depth is greater for the blue/achromatic stimuli.

Figure 16.23 clearly demonstrates that, as the pupillary distance from the foveal achromatic axis is increased, the perceived depth also increases, demonstrating the role of TCA in this effect. It was also observed that the inner square of the image was perceived as moving forward with nasal displacements in the case of the red/achromatic images, and was also perceived as projected forward when temporal displacements for the blue/achromatic images were used. When nasal displacements and blue/achromatic images or temporal displacements with red/achromatic images were used, the opposite depth effect was produced. That is, the inner square was perceived as being behind the colored surround square on the screen. Figure 16.23 demonstrates clearly that the perceived depth for the blue/achromatic condition is generally greater than the red/achromatic condition.

The blue/gray and red/gray images (see [33]) have been shown to many observers under natural viewing conditions and many of these reported seeing depth. Furthermore, when depth was perceived in both images, the observers reported that the relative depth in the images was in opposite directions for the blue/achromatic images and the red/achromatic images and many reported that the perceived depth was greater for the blue/achromatic than the red/achromatic images. Another important point is that the black and gray contiguous elements are always perceived as being on the same depth plane, supporting the notion that the brain constructs these elements into only one surface even if the optical model predicts two different disparity values.

The colors generated in the present study were broadband, thus containing other chromatic information, and the achromatic information was composed of the same blue and red elements in addition with a green element. The results obtained under these conditions suggest that the visual system uses the matching dominant wavelengths in broadband color images to establish stereoscopic correspondence.

It is important to realize that what was proposed in the Faubert [32, 33] studies was that optical and perceptual factors are closely intertwined in these phenomena. Neither the optical factors nor the perceptual factors can be taken in isolation to completely explain the color depth effects described. In these studies, a number of concepts have been raised such as the notion of reference plane, perceptual integration time, capture, texture element characteristics, etc. These notions attest to the fact that perceiving depth in color images is a multifaceted problem, which must be viewed in a global context in order to fully appreciate these phenomena.

An interesting extension of the optical model predictions is that the gray and black contiguous segments have different disparity information, as demonstrated in Fig. 16.1. The fact that this kind of achromatic texture information is never seen as having depth information leads us to two speculations. The first possible explanation is that, because the disparity difference between the gray and black elements is very small, the visual system cannot resolve the difference or that, even if the visual system is theoretically capable of detecting the difference, it nevertheless will perceive only one surface, a phenomenon known a pyknostereopsis [34]. Although this explanation is quite plausible, it is not totally satisfactory because, as demonstrated previously, red and green contiguous elements are generally perceived on the same depth plane in natural viewing conditions even if they are quite efficient in producing stereopsis in other parts of the same image. The answer to these questions can possibly lie on how the visual system constructs the different surfaces. One possibility is that the brain generally attempts to construct solid surfaces to make sense of an image with depth information. The visual system may bias against perceiving transparent static images because such images are not frequent in a natural environment. However, under extreme TCA conditions when the disparity between colors is very large, such as when prisms are used, transparent surfaces can be perceived in color images. Perhaps there is a constant struggle between the visual system's natural tendency to construct solid surfaces in depth and the amount of disparity information available in an image.

In the studies by the present author relative to color depth effects, the question of how we perceive such effects in natural environments is always an underlying concern. An extension of this concern involves the following questions. What, if any, is the use of this type of effect in nature? Can animals perceive color depth effects? If so, do predators use this perception to identify preys or do the preys themselves have color characteristics that can help their disguise? It is clear that many animals have some of the basic requirements in order to perceive such effects, that is, they have chromatic dispersion [35] and binocular vision [36]. What is not clear, however, is how much TCA is present in different animal types. In any case, it strikes me as quite plausible that animals can perceive color depth effects and that this can in some way be incorporated into the different behavioral schemes which have evolved over the years. An example of a very strong effect that I have observed involves several types of butterflies that have very distinct circle marks on the bottom and sides of the wings, which may appear as two eyes, and can be used as some form of camouflage. What is striking is that these spots can actually appear as being forward or receding in depth, giving a remarkable effect of protruding or receding eyes. The color depth effect is clearly involved because the perceived depth from these markings can be reversed or eliminated by displacing artificial pupils nasally or temporally in front of the eyes. I have observed this with the *Precis* type of butterflies (Precis *rhadama*, Precis *orythia*, and Precis *clelia*) and with the *Megistanes japetous* butterfly. Here is a possible example where natural selection may have developed these color and texture schemes because it was particularly efficient at producing the illusion of protruding or receding eyes of a much larger organism than the actual butterfly and, as a consequence, keep potential predators at bay.

In conclusion, more research is needed in both optical and perceptual aspects to understand the phenomenon in its entirety and to elucidate the whole range of visual experiences produced in color depth effects under natural viewing conditions. Future research will also focus on the role of color and texture patterns that have evolved in the environment, and what may be the role of such arrangements in the behavioural schemes of the different organisms involved.

16.9 TCA and Motion Parallax

If we imagine the scenario presented in Fig. 16.16a, we can deduce that TCA will differ between the colors as we view the lens through different angles. The result will be a difference in perceived velocity between the eyes, which, as we stated earlier, can generate a sense of depth for portions of the image of different colors.

The motion parallax generated by the color of objects in an image can be easily demonstrated. What we have done in the laboratory is to use apparent motion to illustrate such effects. Figure 16.24a shows a simple experimental protocol where the color characteristics of the objects are changed frame by frame (chromatic flicker). The two objects are rectangles vertically aligned with one another. The perceptual experience obtained when no TCA is present is that of up-and-down apparent motion. However, when enough TCA is present, the perception is lateral motion (see Fig. 16.24b). This effect can be easily reproduced in the laboratory if one looks at the flickering targets through the periphery of the lens or when the stimuli are fixated eccentrically.

16.10 Effect of TCA on Color-Defined Motion

The role of the chromatic pathway in motion processing has been the subject of some debate [37–51]. Evidence for chromatic motion sensitivity using the motion nulling paradigm has been reported by a number of researchers [37, 38, 41, 52]. For this reason, we decided to directly assess the effect of TCA on chromatic motion thresholds [53].

The operational descriptions of the stimuli are presented in detail in [53]. Essentially, the task consists in opposing an isoluminant (equivalent luminance input for the visual system) red–green (R/G) grating drifting in one direction with an isochromatic (just luminance information) grating drifting in the opposite direction. This generates a perceptual tug-of-war where the winner (isoluminant versus isochromatic) will dictate the direction of the motion perceived. When the two sinusoidal drifting gratings are perceptually equivalent, the perception is that of a counterphase flicker (no clear direction of motion). This null point is assessed and used to measure the relative strength of the color system to motion perception. The dependent measure

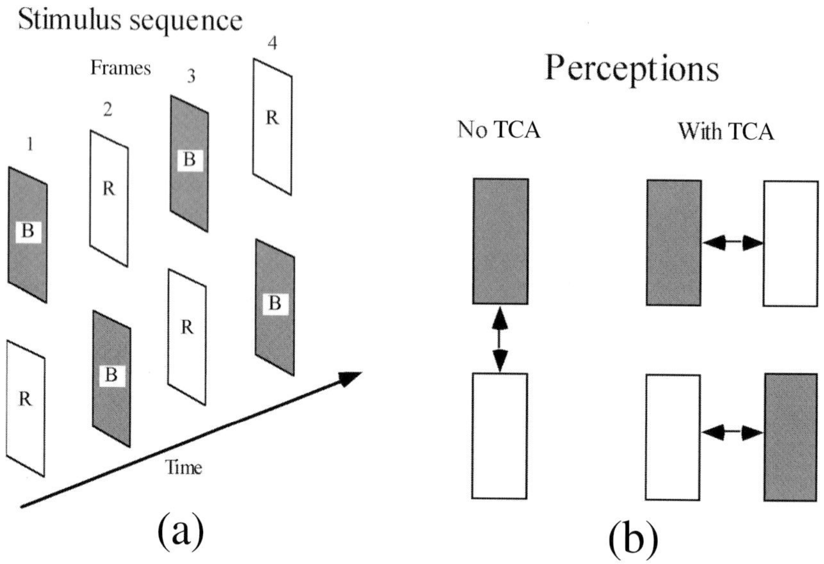

Fig. 16.24. (**a**) A simple way of illustrating the chromatic parallax effects. The red and blue rectangles are vertically aligned and the colors are alternated with subsequent frames. (**b**) The perceptual experience when viewing stimuli in (**a**) differ as a consequence of the amount of TCA. When no TCA is present, the rectangles are perceived as moving up and down. When TCA is present, the percept is that of lateral movement. This can be observed with eccentric viewing or by looking at angles through ophthalmic lenses

is the luminance contrast of the isochromatic grating that is necessary to null the chromatic motion sensation (percentage equivalent contrast). We manipulated the spatial frequency content of the stimulus (cycles per degree of visual angle or cpd) and the amount of TCA (prismatic diopters or Δ).

The R/G motion thresholds are shown in Figs. 16.25a to 16.25c for prism base out and 25d for prism base in. The data are plotted as the luminance contrast necessary to null isoluminant gratings (equivalent contrast) on the y axis and the spatial frequency component of the target on the x axis. The control data for subject JF showed greater chromatic motion sensitivity than the other two subjects, but his values are similar to values previously reported for normal observers [41].

All three subjects show increases in equivalent contrast for R/G motion when the gratings have spatial frequencies of 1 cpd or higher. Subject PS had very low sensitivity, as a result of his deuteranomalous color defect, and then showed dramatic increases in sensitivity with the addition of TCA levels. The data were consistent with our predictions based on the optical effects of TCA in that increasing levels of TCA should produce increases in chromatic motion-nulling equivalent contrasts as a result of luminance inhomogeneities produced by phase shifts. Furthermore, the data also show that for a given TCA value the effect is greater with increasing spatial frequency. All the data for a given TCA increase dramatically with increasing spatial frequency above 1 cpd. The only two exceptions are the values for JF and LB (base in condition) for the highest spatial frequency in the 0.40Δ prism condition. However, this is an artificial ceiling effect, as the maximum equivalent contrast available in the computer program was 30% for the parameters that we used. An interesting result is the fact that for two of the observers (JF and LB) the 0.11Δ condition produced lower sensitivities at high spatial frequencies than the control condition. It is possible that this effect is due to the naturally occurring TCA of the eye being offset by induced TCA in the opposite direction, thereby producing lower sensitivity values. To test this hypothesis, we retested subject LB who, based on the results, appeared to have the highest level of natural TCA (as demonstrated by the highest increase in equivalent contrast values between conditions 1 cpd and 4 cpd) with the base of the prism facing in the opposite direction. According to the hypothesis mentioned above, the sensitivity should now be increased for the 0.11Δ and 0.22Δ conditions relative to the control condition. The results are shown in Fig. 16.25d and the data are just as predicted where the 0.11Δ and 0.22Δ conditions now produced chromatic motion sensitivities which were greater than the control condition, demonstrating some form of additivity between the naturally occurring TCA and the induced TCA. This supports the notion that naturally occurring TCA may produce decreases in chromatic motion thresholds when using the motion-nulling paradigm.

16.10.1 Higher Spatial Frequency Effects

The reason that higher spatial frequencies are particularly affected by TCA is illustrated in Fig. 16.26, where we simulated the effect of TCA on sine waves composed of different spatial frequencies. Figures 16.26a and 16.26b simulate the presumed isoluminant conditions, with the remnant luminance profiles (dotted lines), for the 1 cpd and 0.25 cpd spatial frequency condition, respectively. Note that under perfect isoluminant conditions (when no TCA is present) the remnant luminance should theoretically remain at zero over space and time. For simulation purposes and for the sake of simplicity the isoluminant gratings are presented as having the same physical luminance values. In reality these values would differ between individuals, but the relative luminance artifact produced by a given level of TCA would be of the same magnitude as that presented in the model for each observer.

Figure 16.26c demonstrates that for high spatial frequencies, the relative shift of 0.1 degrees (which is roughly equivalent to the phase shift produced by the 0.22Δ condition) of the R sine wave as compared to the G sine wave in an isoluminant R/G grating produces a large luminance contrast artifact, as shown by the luminance profile generated from the sum of the two sine

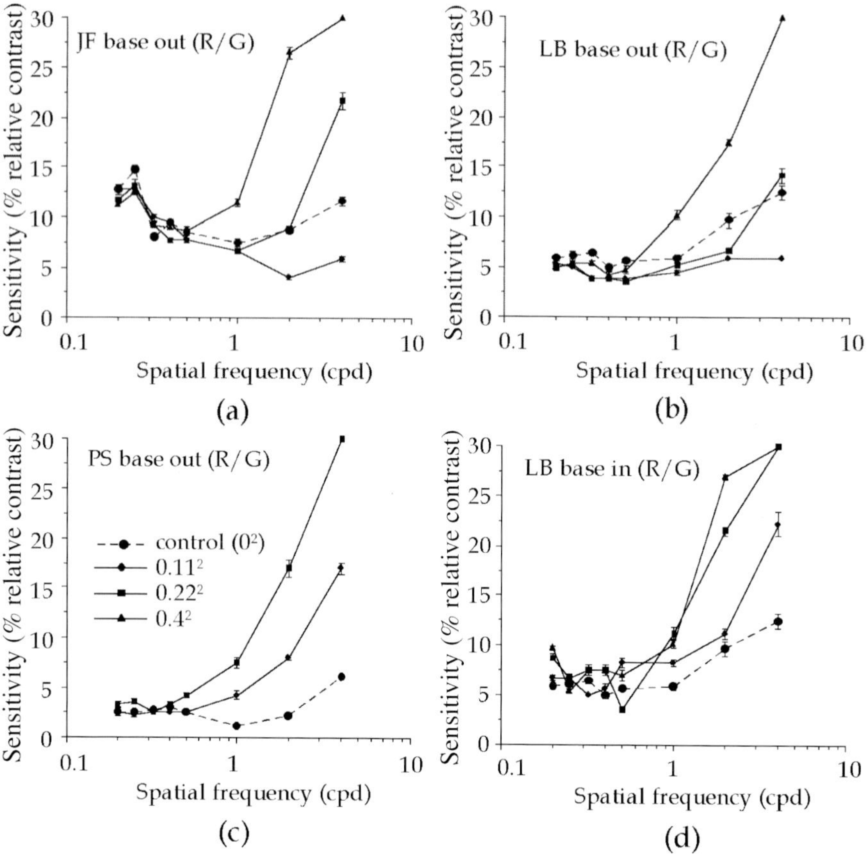

Fig. 16.25. Relative sensitivity in percentage contrast for isoluminance R/G gratings as a function of spatial frequency and amount of TCA: (**a**) for subject JF; (**b**) for subject LB; (**c**) for subject PS for prism base out conditions; and (**d**) LB for prism base in

waves. Conversely, Fig. 16.26d shows why the same TCA value could produce relatively small effects on lower spatial frequency isoluminant gratings. In this case the luminance contrast necessary to oppose the chromatic grating would be very small if the motion signal came only from the luminance artifact. In the case of high spatial frequencies, however, the luminance contrast of an isochromatic grating would require a relatively high luminance contrast just to null the motion sensation generated from the luminance-contrast artifacts produced by TCA. Figure 16.26e directly compares the luminance-contrast artifacts which would be produced by a 0.1 degree phase shift in the two spatial frequency conditions simulated. It is clear from this illustration that the result on sensitivity values when using motion nulling to measure chromatic motion sensitivity would differ dramatically for the same phase shift conditions but with stimuli of different spatial frequencies.

The results show that a careful control of induced and naturally occurring TCA is necessary when measuring chromatic motion sensitivity at isoluminance for spatial frequencies above 0.5 cpd for R/G stimuli. We can further conclude that chromatic motion sensitivity throughout the visual field can be measured with motion nulling providing that the spatial-frequency components of the stimuli are low enough and are scaled with eccentricity [37, 38]. Finally, the chromatic motion-nulling paradigm may provide a very sensitive functional measure of natural and induced TCA. Measuring TCA is a difficult problem, particularly in the visual periphery. Perhaps one can exploit the use of motion sensitivity in the peripheral vision to determine the amount of TCA that is present with increasing eccentricity.

16.11 Conclusions

In summary, we have described in general terms what is meant by motion parallax and we have contrasted its performance with stereopsis by illustrating studies performed in our laboratory. We have also raised issues on how optical conditions may influence our perception of depth with motion parallax and, finally, we have discussed conflicting situations that may arise between the motion-parallax information perceived from the two eyes.

The attempt was to present some issues which have received little attention in the motion-parallax literature and that have practical and theoretical applications. From the issues raised above, we can conclude that several future research orientations are imperative for us to better understand how motion-parallax cues to depth can be affected in our day-to-day situations.

First, more research is required relative to the role of motion parallax throughout the visual field, in particular areas where stereoscopic cues become inefficient. Several studies by Regan's group have assessed the processing of stereomotion throughout the visual field [54]. They found that stereomotion visual fields are in fact quite restricted to the central portion of the visual field (20° or less). We still need to know how well motion parallax can be used to perceive depth at different locations of the visual field, as there are many known asymmetries for different perceptual tasks [37, 38, 55–58].

We also need to understand better how optical corrections such as progressive addition lenses and other optical corrections influence our motion-parallax judgments. As demonstrated above, the majority of us will require ophthalmic lens corrections as we go beyond 40 years of age. All of us will be faced with distortions and aberrations that will directly influence how we perceive motion parallax in our daily operations.

Finally, we need to study further the binocular integration of motion parallax, particularly in the circumstances where the motion parallax between the eyes may give rise to conflicting cues as illustrated above.

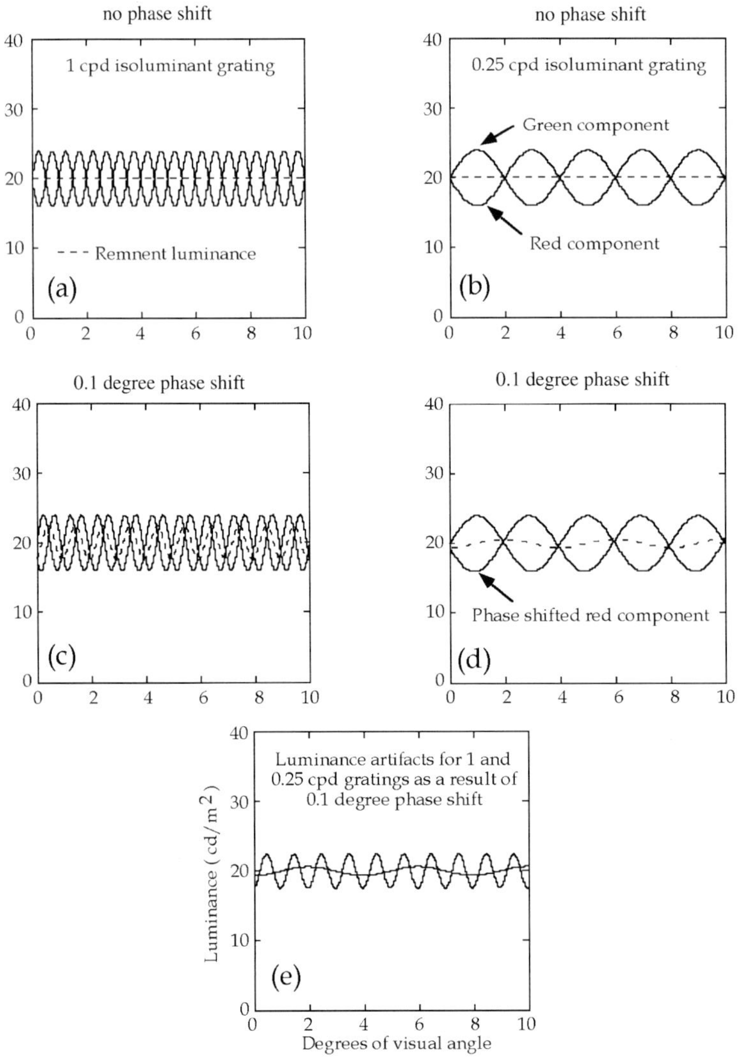

Fig. 16.26. Simulation of color and luminance profiles for sine wave gratings under experimental isoluminant conditions. (**a**) The color gratings (*solid lines*) and the remnant luminance (*dotted line*) for a 1 cpd stimulus when no TCA is present. Note that the remnant luminance is flat. (**b**) The color gratings and the remnant luminance for a 0.25 cpd stimulus when no TCA is present. (c) The color gratings and the remnant luminance for a 1 cpd stimulus when a 0.1° TCA-induced phase shift is present. Notice that the remnant luminance function has a sinusoidal waveform. (**d**) The color gratings and remnant luminance for a 0.25 cpd stimulus and a 0.1° TCA-induced phase shift is present. (**e**) A direct comparison of the remnant luminance artifacts produced as a result of the 0.1° phase shift for both simulated spatial frequency conditions. Note that the luminance contrast artifact for the 1 cpd condition is much larger

References

1. H. von Helmholtz: *Handbuch der Physiologischen Optik*, Hamburg, Voss, 12866. Trans. from 3rd German ed. by J.P.C. Southall (Optical Society of America, 1924/1925. New York: Dover 1962)
2. J.J. Gibson: *The Perception of the Visual World* (Houghton Mifflin, New York 1950)
3. J. Faubert: Proc. SPIE **CR76**, 168 (2001)
4. H.J. Howard: Am J. Ophthal **2**, 656 (1919)
5. C.W. Tyler: In: *Vision and Visual Dysfunction*, Vol. 9, ed. by D. Regan, J.R. Cronly-Dillon (MacMillian, New York 1991) pp. 19–37
6. F. Labonte, C.T. Le Dinh, J. Faubert, P. Cohen: IEEE Trans Circuits Syst. Video Tech. **9**, 144 (1999)
7. P. Rondot, J. Lessard, J.M. Robert, J. Faubert: "Étude de la parallaxe de mouvement avec un système de visualisation de type casque utilisé en téléopération," *Compte rendu du XXXe congrès de la Société d'Ergonomie de Langue Française*, Biarritz, 27–29 septembre, pp. 429–436, 1995
8. A. Arditi: In: *Handbook of Perception and Human Performance. Vol. I. Sensory Processes and Perception*, ed. by K.R. Boff, L. Kaufman, J.P. Thomas (John Wiley, New York 1986)
9. J. Faubert, O. Overbury: J. Am. Geriatrics Soc. **48**, 375 (2000)
10. M.E. Ono, J. Rivest, H. Ono: J. Exp. Psychol. Hum. Percept. Perform **12**, 331 (1986)
11. Y. Le Grand: Ann. Opt. Oculaire **4**, 1 (1956)
12. G.A. Fry: Am. J. Optom. Physiol. Opt. **54**, 282 (1977)
13. P. Simonet, B. Bourdoncle, C. Miège, J. Gresset, J. Faubert: Vision Res. **35**, S245 (1995)
14. J. Gresset, C. Fauquier, B. Frenette, M. Lamarre, B. Bourdoncle, P. Simonet, P. Forcier, J. Faubert: In: *Vision Science and its Applications*, OSA Technical Digest Series (Optical Society of America, Washington, D.C. 2000) pp. 218–221
15. B. Bourdoncle: Points de Vue **42**, 1 2000
16. S. Shioiri, H. Saisho, H. Yaguchi: Vision Res. **40**, 2565 (2000)
17. J. Faubert: *Proc. 3rd Int. Display Workshop: 3-D Display Technologies and Human Factors*, Kobe, Japan, November 27–29, SID Vol. **2**, pp. 481–484, 1996
18. P. Simonet, M.C.W. Campbell: Ophthalmic and Physiological Optics **10**, 271 (1990)
19. L.N. Thibos, A. Bradley, D.L. Still, X. Zhang, P.A. Howarth: Vision Res. **30**, 33 (1990)
20. X. Zhang, A. Bradley, L.N. Thibos: J. Opt. Soc. Am. A **8**, 686 (1991)
21. A. El-Kadouri, W.N. Charman: *Trans. 1st Int. Congress in Optometry*, Vol. **2** (British College of Ophthalmic Opticians, London, UK 1984) pp. 154–160
22. J. Faubert, P. Simonet, J. Gresset: Ophthalmic and Physiological Optics **19**, 336 (1999)
23. P. Simonet, J. Faubert, J. Gresset: Ophthalmic and Visual Optics Technical Digest Series **3**, pd21 (1993)
24. C.Y. Tang, W.N. Charman: Ophthalmic and Physiological Optics **12**, 340 (1992)
25. G.C. Woo, F.W. Campbell, B. Ing: Ophthalmic and Physiological Optics **6**, 415 (1986)

26. J.M. Sundet: Scand. J. Psychology **19**, 133 (1978)
27. W. Einthoven: *Stereoscopic durch Farbendifferenz.* Albrecht von Graefes Archiv für Ophthalmologie, **31**, 211–238 (1885)
28. H. Hartridge: Philos. Trans. R. Soc. Lond. **232**, 519–671 (1947)
29. B.N. Kishto: Vision Res. **5**, 313 (1965)
30. P. Simonet, M.C.W. Campbell: Vision Res. **30**, 187 (1990)
31. M. Ye, A. Bradley, L.N. Thibos, X. Zhang: Vision Res. **31**, 1787 (1991)
32. J. Faubert: Vision Res. **34**, 1165 (1994)
33. J. Faubert: Vision Res. **35**, 3161 (1995)
34. C.W. Tyler: In: *Vision and Visual Dysfunction*, Vol. 9, Binocular Vision, ed. by D. Regan (MacMillan, New York 1991) pp. 38–74
35. R.O. Kreuzer, J.G. Sivak: Ophthalmic and Physiological Optics **5**, 33 (1985)
36. R.H. Smythe: *Vision in the Animal World.* (St. Martin's Press, New York 1975)
37. L. Bilodeau, J. Faubert: Vision Res. **37**, 2073 (1997)
38. L. Bilodeau, J. Faubert: Vision Res. **39**, 757 (1999)
39. L. Bilodeau, J. Faubert: J. Opt. Soc. Am. A **16**, 1 (1999b)
40. P. Cavanagh: In: *From Pigments to Perception: Advances in Understanding Visual Processes*, ed. by A. Valberg, B.B. Lee (Plenum Press, New York 1991) pp. 151–164
41. P. Cavanagh, S.M. Anstis: Vision Res. **31**, 2109 (1991)
42. P. Cavanagh, J. Boeglin, O.E. Favreau: Perception **14**, 151 (1985)
43. P. Cavanagh, O.E. Favreau: Vision Res. **25**, 1595 (1985)
44. P. Cavanagh, D.I.A. MacLeod, S.M. Anstis: J. Opt. Soc. Am. A **4**, 1428 (1987)
45. P. Cavanagh, C.W. Tyler, & O.E. Favreau: J. Opt. Soc. Am. A **1**, 893 (1984)
46. S.J. Cropper, A.M. Derrington: Nature **379**, 72 (1996)
47. A. Gorea, T.V. Papathomas: J. Opt. Soc. Am. A **6**, 590 (1989)
48. M.S. Livingstone, D.H. Hubel: Science **240**, 740 (1988)
49. A.B. Metha, A.J. Vingrys, D.R. Badcock: J. Opt. Soc. Am. A **11**, 1697 (1994)
50. T.V. Papathomas, A. Gorea, B. Julesz: Vision Res. **31**, 1883 (1991)
51. V.S. Ramachandran, R.L. Gregory: Nature **275**, 55 (1978)
52. M.A. Webster, J.D. Mollon: Vision Res. **37**, 1479 (1997)
53. J. Faubert, L. Bilodeau, P. Simonet: Ophthalmic and Physiological Optics **20**, 274 (2000)
54. D. Regan: In: *Vision and Visual Dysfunction Vol. 9*, Binocular Vision, ed. by D. Regan, J.R. Cronly-Dillon (MacMillan, New York 1991)
55. D. Regan, J. Faubert, M. Pinard, P. Simonet, J. Gresset: In: *Vision Science and its Applications*, OSA Technical Digest Series (Optical Society of America, Washington, D.C. 2000) pp. 35–38
56. S.S. Fukusima, J. Faubert: Vision Res, **41**, 2119 (2001)
57. J. Faubert: *Proc. Int. Sympos. on Computer Graphics, Image Processing and Vision* (Rio de Janeiro, Brazil, IEEE Computer Society Press 1998) pp. 2–9
58. J. Faubert: *Perimetry Update 1990/1991* (Kugler, Amsterdam/New York 1991)

17 Depth Perception of Moving Objects

S. Shioiri, A. Morinaga, and H. Yaguchi

17.1 Introduction

Vision reconstructs the three-dimensional (3-D) world in the brain from the pair of retinal images. Although the images on the retinas are two-dimensional (2-D), we see, or perceive, 3-D objects in a 3-D world. The visual system succeeds in recovering the third dimension from a pair of retinal images. This contrasts with computer vision, which shows difficulties in reconstructing a 3-D object from a 2-D image particularly in natural scenes. Vision scientists have investigated how the visual system solves the problem, and yet, much effort is still needed to understand how the vision system reconstructs the 3-D world.

One of the important factors that are not fully understood is the depth perception of objects in motion. Compared to studies with static images, a much smaller number of studies have been reported with dynamic images [1, 2]. Temporal factors are important since retinal images change with time due to the motion of objects and of the viewer. The investigations of the dynamic characteristics of human 3-D perception should provide crucial features of visual processing. They are important for designing 3-D displays, efficient presentation of 3-D information, efficient coding of 3-D images and other applications, as well as for understanding our visual functions.

In this paper, we review studies of the dynamic features of depth perception to summarize our knowledge of the depth perception of moving objects. We focus on three topics: the temporal characteristics of stereopsis, the effect of motion on depth perception, and the cues for motion in depth.

17.2 Spatio-Temporal Characteristics of Stereopsis

17.2.1 Depth Modulation in Space and in Time

The spatial and temporal frequency characteristics of visual functions is evaluated from the sensitivity for detecting sinusoidal gratings. The sensitivity is derived from measurements of the minimum stimulus modulation with which the observers can detect the stimulus, i.e. the modulation threshold. The sensitivity defined as a reciprocal of the threshold as a function of spatial or temporal frequency represents the basic spatio-temporal characteristics.

For depth processing, the modulation of depth in the stimulus is varied to determine the sensitivity. A sinusoidally corrugated surface in a random-dot stereogram is used to vary the spatial frequency. Figure 17.1 shows the spatial frequency tunings measured by several researchers [3–8] with a schematic view of the stimulus. The amplitude of the depth modulation is adjusted by the observer so that the depth modulation is just detectable to determine the threshold. Figure 17.1b shows the sensitivity, the reciprocal of threshold, expressed in minutes of arc on a logarithmic scale.

The spatial-frequency characteristics are similar across studies. They show a bandpass function with a peak of about 0.5 c/deg. The absolute threshold varies depending on the experiment, perhaps due to the differences of the

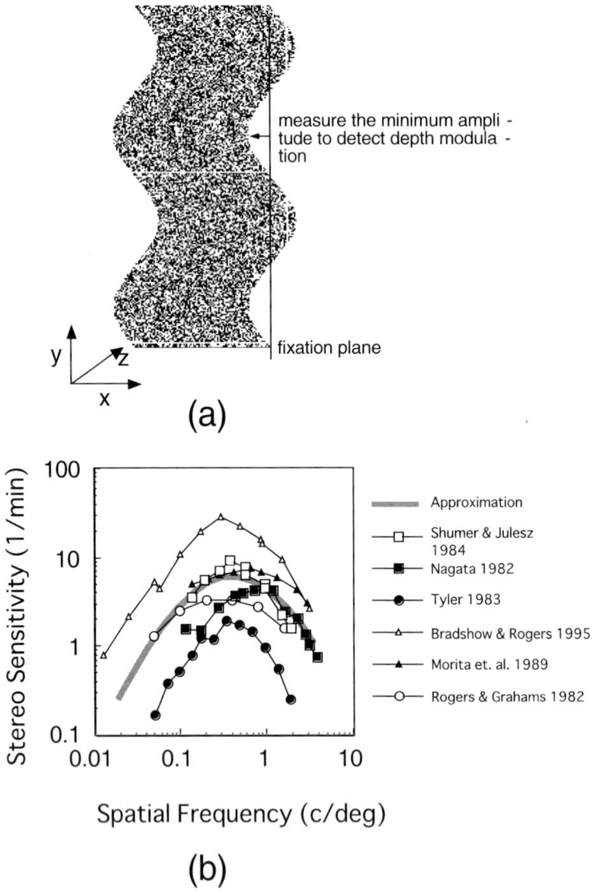

Fig. 17.1. (**a**) Schematic view of depth modulation in a random-dot stereograms. (**b**) Disparity threshold as a function of spatial frequency of depth modulation from several studies. The gray line indicates a function fitted to the data as an approximation

experimental conditions, and the highest is about 2 sec of arc. The shape of the sensitivity function is similar to that obtained for detecting luminance modulation, although the peak frequency is much lower than that for luminance (about 3 c/deg).

To show the general trend of the function across the studies, we used an equation to approximate the experimental data. The gray line in Fig. 17.1b depicts the hyperbolic curve in the log–log coordinate fitted to the data by a least-square method. The formula of the function is

$$\log m = A(B + (\log f - \log f_0)^2)^{0.5} + C ,$$

where, f represents the spatial frequency and m represents the modulation threshold. The parameters A and B determine the shape of the function and the parameters f_0 and C determine the location of the peak. We used the same values of A and B for all the data sets while variing f_0 and C across the data sets individually for each data set. The function shown in Fig. 17.1b used the average values of f_0 and C across data sets, with A and B optimized for the best fit to all. The values of A, B, f_0 and C are 2.46, 1.69, 0.44 and 3.95, respectively. The peak sensitivity is at 0.44 c/deg on average. Figure 17.1b shows that the hyperbolic function approximates the experimental spatial-frequency characteristic functions well.

Similarly, the temporal frequency characteristics have been measured for detecting depth using stereograms. Unfortunately, the number of measurements of the temporal-frequency function is smaller. We found only three measurements of stereoacuity reported as a function of the temporal frequency of depth modulation. One of them used a square-shaped random-dot stereogram [5]. The other two used line stereograms [9, 10]. Figure 17.2 compares the sensitivity functions of the three experiments. The shapes of the functions are similar among the experiments. As for the spatial char-

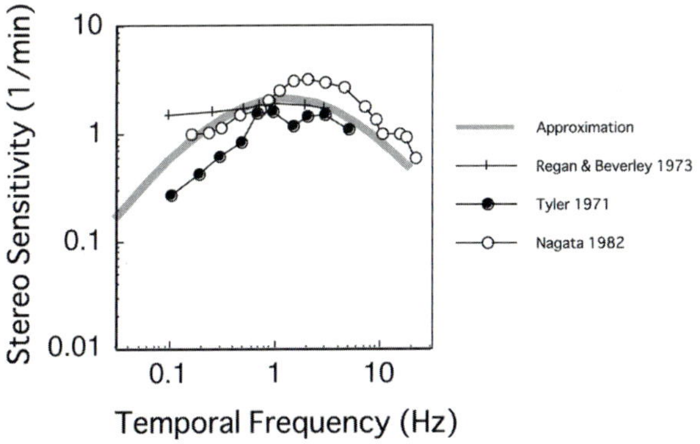

Fig. 17.2. Stereosensitivity as a function of temporal frequency of depth modulation from three studies. The gray line indicates a function fitted to the data as an approximation

acteristics, the sensitivity peaks at a temporal frequency, showing a band-pass characteristic. The same hyperbolic function is used to approximate the threshold data. The fitting result by the same least-squares method is shown by the gray line. The values of the fitting parameters A, B, f_0, and C are 1.23, 2.12, 1.52, and 2.30. The peak sensitivity is at 1.5 Hz on average. As for spatial frequency, the shape of the sensitivity function is similar to that for the luminance modulation, with the peak frequency lower than that for luminance (about 8 Hz).

The effect of stimulus presentation duration on stereoacuity is consistent with the temporal frequency tuning in terms of the differences from sensitivity to luminance modulation. A number of measurements have been reported for the dependence of the disparity threshold for depth detection (stereoacuity) on the presentation duration [9, 11–16]. All the data show that the threshold for depth detection becomes lower with the increase of the presentation duration. The effect of duration is stronger for shorter duration (slope is about −1 in a log–log plot) and decreases with the increase of duration (the slope is close to 0 for longer duration). The limit of temporal summation may be estimated from the duration beyond which there is no effect of duration. However, such a critical duration is not often clearly seen and it varies, depending on the experimental conditions. We can estimate the temporal summation limit from the slope change of the threshold-duration function. In most of the data, the change of the slope is seen between 100 and 1000 ms. This gives a rough idea of the limit of temporal summation for binocular depth perception. This estimate of the temporal summation limit is longer than that for detecting luminance increments, which is less than 100 ms when the stimulus size is not too small [17].

Contrast sensitivity for luminance modulation has been measured as a function of combinations of spatial and temporal frequencies [18]. It is important to know the sensitivity in the spatio-temporal frequency domain to understand the dynamic properties of a system. No experiment has been

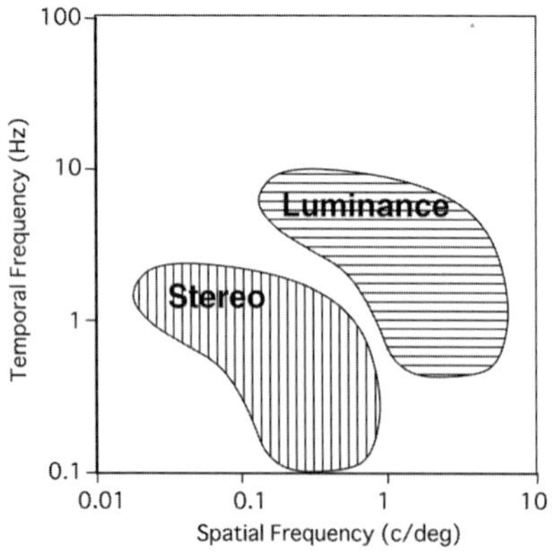

Fig. 17.3. A predicted sensitivity function in the spatio-temporal frequency domain for depth modulation with a sensitivity function measured for luminance modulation. The hatched areas represent the regions with sensitivity higher than the half of the peak

reported for the disparity modulation. A prediction might be possible for the spatio-temporal frequency tuning for depth modulation from that for luminance modulation. The bandpass characteristics for detecting depth modulation in space and in time are similar to those for detecting luminance modulation, although the peak frequencies are lower. If we assume that the tuning function is the same for depth and luminance except the peak frequency, the spatio-temporal frequency tuning for depth perception can be predicted from that for luminance. Figure 17.3 shows a schematic view of the spatio-temporal frequency tuning function for depth modulation with that for luminance modulation. The hatched area indicates the area with sensitivity higher than the half of the peak sensitivity. The area for luminance modulation was depicted based on an actual measurements and the location is shifted to the estimated sensitivity peak for depth modulation. This prediction shows that the sensitivity is highest either when the spatial frequency is about 0.4 c/deg or when the temporal frequency is about 1.5 Hz in the region of spatial frequency between 0.02 and 1 c/deg and of temporal frequency between 0.1 and 2 Hz. To know the actual tuning, of course, we need threshold measurements with various combinations of spatial and temporal frequencies for depth modulation.

17.2.2 Luminance Modulation in Space and in Time

Spatio-temporal characteristics in the luminance domain are as important as those in the depth domain. A detection process of features in the luminance distributions on each retina should precede the depth process. Whatever feature is used to process binocular disparity, there should be some influence of the sensitivity to luminance modulation. One possibility is that the sensitivity tuning in the luminance domain is the same for detecting luminance modulation and for detecting depth modulation, although it may be different. For spatial frequency, several studies have shown that contrast sensitivity to luminance modulation for detecting depth is different from that for the detection of luminance modulation itself [19–21].

The differences of contrast sensitivity between the two tasks are both in absolute sensitivity and in the dependency on spatial frequency (i.e. the shape of the sensitivity function). First, the sensitivity for depth is usually less than that for detection of luminance patterns. The contrast sensitivity for depth is lower by the factor of 2 or more. Second, the peak of the sensitivity function varied with the disparity of the stimulus. Figure 17.4 shows contrast sensitivity functions for detecting depth difference measured in our laboratory [22] with an example of stimulus. In this experiment as in others, spatially filtered version of random-dot stereograms were used (Fig. 17.4a). This was to restrict the spatial-frequency contents in the square stimulus. The observer's task was to identify the direction of the depth difference between the upper and lower halves of the stimulus (Fig. 17.4b). The temporal frequency was also controlled by counter-phasing the stimulus contrast at 1 Hz. To remove the

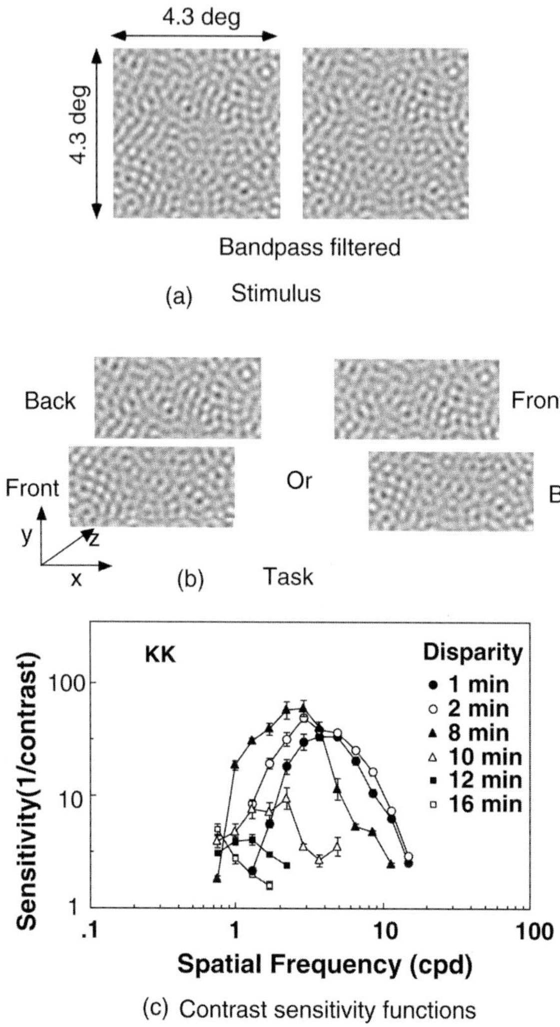

Fig. 17.4. (**a**) A bandpass-filtered version of a pair of random-dot stereograms. (**b**) The observer's response: whether top half appeared to be closer or further. (**c**) Contrast sensitivity functions against spatial frequency for different disparities

temporal edges at the start and the end of the stimulus presentation, gradual on and off of the stimulus was realized by multiplying a sinusoidal function with the contrast during the first and the last third of total duration of 1.9 s.

Figure 17.4c shows contrast sensitivity functions with several disparities, or several depth differences. Clearly, the disparity of the stereograms influences contrast sensitivity. The peak of the sensitivity shifts to lower values with the increase of disparity (size–disparity correlation). Although individual variations for the relative sensitivity among different disparity conditions were not small, four observers showed clear size–disparity correlation (the fifth observer had measurable sensitivity only with small disparities). The sensitivity to luminance modulation cannot predict the sensitivity to depth modulation. It should be noted, however, that size disparity correlation may not be found and a similar shape of the contrast sensitivity function may be obtained when longer presentation duration is used due to vergence eye movements [19, 21]

There are only a limited number of reports on the effect of temporal frequency on the stereopsis, as for depth modulation. We reported that the increase of temporal frequency decreased the sensitivity [22]. Figure 17.5 shows contrast sensitivity functions with spatial frequency in three different temporal frequencies. The stimulus and task were the same as those for the data in Fig. 17.4 except that the temporal frequency was varied. The functions shown are similar in shape for all temporal frequency conditions (left-hand panels). The spatial frequency tuning changed only slightly for a given disparity. The right-hand panels show the threshold relative to that at 1 Hz. There was a slight decrease of the ratio with spatial frequency in the results of temporal frequencies of 8 and 16 Hz with the 1- and 2-minutes disparities. With the 8-minute disparity, no systematic effect of temporal frequency is seen. The results for small disparities are consistent with the report by Patterson [14]. He found that the stereoacuity increases with the increase of temporal frequency for higher spatial frequency stimuli and changed little for lower spatial frequency stimuli.

The results with small disparities suggest that the detection of disparities in higher spatial frequency stimuli is less sensitive to higher temporal frequencies. This tendency agrees with the fact that the mechanisms sensitive to high spatial frequencies show low sensitivity at higher temporal frequencies for detection of luminance modulation. Assuming more than one channel with different spatio-temporal frequency tunings explains the shift of the peak spatial frequency with temporal frequency. If there is a channel sensitive to high spatial and low temporal frequencies as well as a channel sensitive to low spatial and high temporal frequencies, the peak spatial frequency should change with temporal frequency. The story, however, cannot be simply extended to the depth process. The effect is slight with small disparities and small with larger disparities. This is perhaps because disparity determines the spatio-temporal frequency channel to detect it, and only one or two channels with close spatio-temporal tunings contribute to detect a given disparity. The reduction of sensitivity of a single channel would cause a simple reduction of sensitivity for all spatial frequencies.

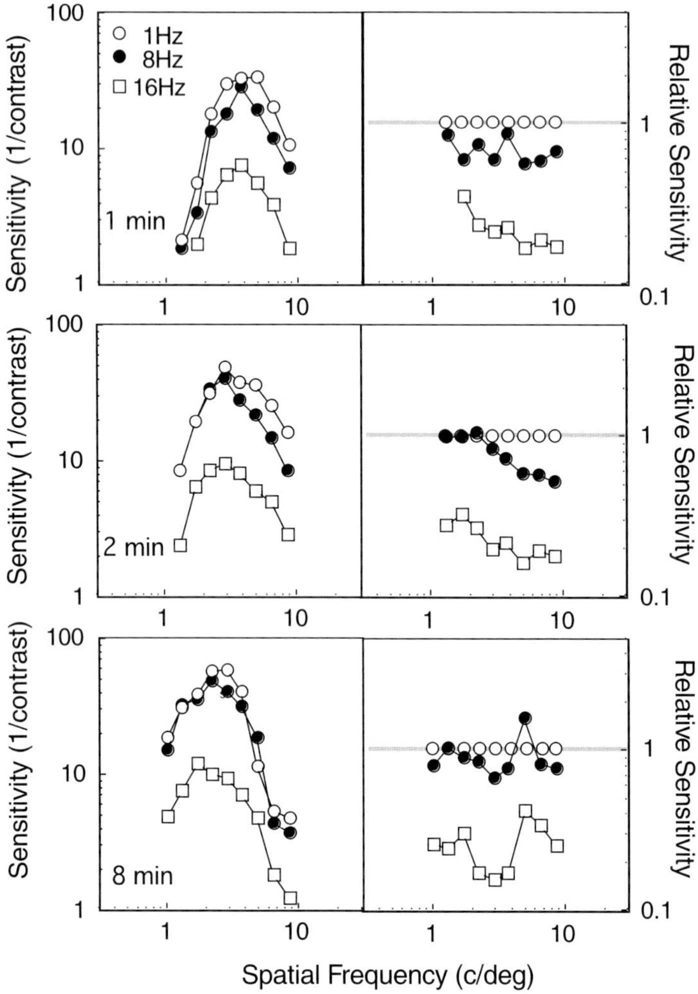

Fig. 17.5. Contrast sensitivity functions versus spatial frequency for different temporal frequencies and disparities (*left*) and the ratio to the sensitivity at 1 Hz for each condition (*right*)

17.3 Motion Influence on Depth Perception

The previous section has shown that the temporal modulation of the luminance in stereograms influences the sensitivity of depth perception. This indicates that stimulus motion also influences stereopsis, since the speed and temporal frequency are linearly related for a given spatial frequency. This section focuses on the influence of motion on depth perception, because it is useful to consider the temporal characteristics in terms of motion to deal with moving objects in natural scenes. Besides the simple effect predicted from temporal characteristics, several studies have shown close relationships between motion and depth processing. We first review the studies investigating the interaction between depth and motion processing, and then describe our experiments that showed a new type of interaction between them.

17.3.1 Disparity Contingent Motion Aftereffect

There is a report of motion aftereffect specific to a depth plane. Motion aftereffect is a phenomenon where motion is seen in a stationary image after prolonged presentation of moving stimulus (Fig. 17.6a). The perceived direction of motion aftereffect is in the opposite direction to that of the moving stimulus. This indicates that motion aftereffect is one of negative aftereffects, which are usually explained by the fatigue of cells that respond to the preceding stimulus. Motion aftereffect is not seen when the motion direction of the preceding stimulus alternates between the opposite directions or when the stimulus contains elements moving in opposite directions (Fig. 17.6b). However, when different disparities are given to stimuli moving in different directions of motion, the aftereffect comes back [23]. This indicates that there is an interaction between the motion-sensitive mechanisms and disparity-sensitive mechanisms. This is supported by the independent motion aftereffect seen simultaneously in different depth planes [24] (Fig. 17.6c). Verstraten et al. used the preceding or adapting stimulus moving, say, leftward with crossed disparity and that moving rightward with uncrossed disparity. After the adaptation period, motion aftereffect was seen in the rightward direction on the cross disparity plane while motion aftereffect was seen in the leftward direction on the uncross disparity plane in about half of all the trials.

17.3.2 Cells in Monkey MT

Physiological studies have revealed that many cells in monkey middle temporal area (MT), which are believed to specialize in motion analysis, are sensitive to both motion and disparity [25]. Some of these cells even show that a stimulus moving in the opposite direction inhibits the response to motion in the preferred direction only when it is in or near the preferred depth plane [26]. Disparity tuning of the cell is shown in Fig. 17.7a. Addition of the dots moving in the opposite direction reduces the response to the random-dots with the preferred disparity and the preferred motion direction. However, the amount of the response reduction becomes smaller with the increase of the relative disparity between the dots with preferred motion and the dots moving in the opposite (Fig. 17.7b). Similarly, moving dots in the opposite direction in different parts of the stimulus (motion-border stimulus) inhibits the response, and the amount of the response reduction becomes smaller with the increase of the relative disparity (Fig. 17.7c). These motion-sensitive cells segregate motion signals on different disparities. The role of

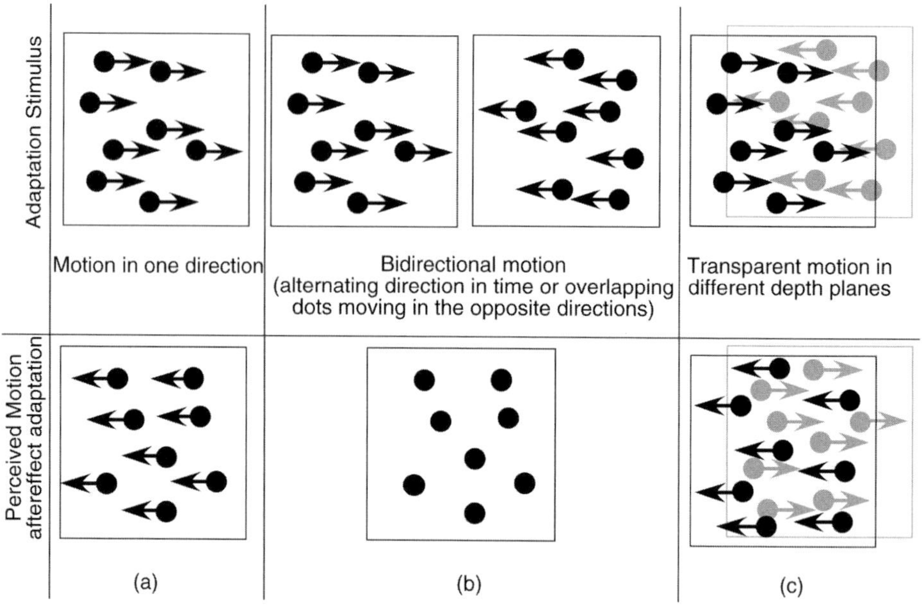

Fig. 17.6. Experimental paradigm for disparity contingent motion aftereffect [25]. (**a**) Motion aftereffect is seen after the exposure to unidirectional motion. (**b**) Motion aftereffect is not seen when bidirectional motion is used as the adaptation stimulus. (**c**) Depth contingent motion aftereffect is seen after the exposure of bidirectional motion with depth difference

the disparity tuning of motion-sensitive cells in MT may be to distinguish motion signals from different surfaces such as distinguishing the motion of an animal from that of leaves in front [27]. These kinds of cell are, perhaps, useful detecting transparency or depth boundaries.

17.3.3 Motion Processing Independently From Disparity

If a motion-detecting mechanism or a cell in the visual system has selectivity to disparity, we expect little interaction between stimuli moving in different depth planes as far as such a mechanism is concerned. The sensitivity to a moving stimulus on the preferred depth plane should be the highest and the suppression by a stimulus moving in the opposite direction to the preferred direction should decrease with the increase of the disparity from the preferred disparity. These are direct predictions from MT cells' performance described. However, several psychophysical studies have reported the absence of such a effect [29–32]. Hibbard et al. measured the minimum number of dots moving in one direction (signal dots) to identify the direction among dots moving in random directions (noise dots) [28]. The number of the dots at threshold did not vary with the disparity range of the signal dots. That is, the sensitivity

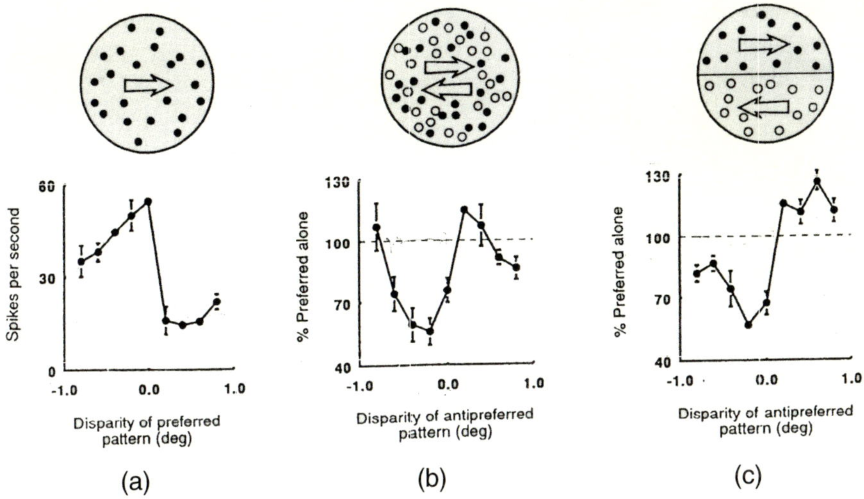

Fig. 17.7. Responses of a MT cell [27]: (**a**) disparity tuning of a cell; (**b**) disparity specific inhibition of noise stimulus moving in the opposite direction to the preferred direction of the cell; and (**c**), disparity specific inhibition of moving dots in the opposite direction in different parts of the stimulus (motion-border stimulus). From D.C. Bradley, N. Qian, R.A. Andersen: Nature **373**, 609 (1995). Reprinted by permission

for motion detection was similar even when the number of depth planes with signal dots varied (Fig. 17.8a). No disparity tuning is necessary to explain the results, disagreeing with the prediction from the existence of a motion-sensitive mechanism with disparity tuning. Similar results were reported by McKee et al. [29]. They measured the detectability of the trajectory of a moving dot through random-dot noise. The observer was asked to detect the signal dot that moved without direction change while noise dots kept moving in direction. They compared the detectability of the signal dot in a 3-D condition and that in a 2-D condition. In the 3-D condition, the signal dot moved in three dimensions through a 3-D random-dot pattern while all dots moved in two dimensions in the 2-D condition (Fig. 17.8b). They found no better sensitivity in the 3-D condition than that in the 2-D condition. Differentiation of the target dot from noise dots by disparity did not help in detection of the motion trajectory.

In these experiments, a rather small number of signal dots carries motion information. This contrasts to the stimulus display used in the experiments of motion aftereffect and of MT cell response measurements. One possible interpretation of the discrepancy among experiments described so far might be that disparity influences motion perception only when it provides a clear surface segmentation cue [30]. The signal-to-noise ratio at threshold may be too low for providing such a cue. This is consistent with a psychophysical study that showed facilitation of transparent motion by disparity signals with clear depth differences [30] as with a psychophysical study that showed facilitation of depth processing by common motion signals using 50% random-dots patterns [31].

17.3.4 Effect of Motion on Depth Discrimination

The previous three sections described the influences of disparity on motion detection. This section focuses on the effect of motion on depth perception. We investigated the effect of motion on depth perception and found that common motion signals facilitated depth perception [32]. In the experiments, contrast sensitivity was measured with sinusoidal gratings for depth discrimination to measure the basic characteristics of depth perception in moving stimuli. The study considered two aspects of motion influence on depth perception. First, motion should influence the sensitivity of depth detection, at least as predicted from the temporal-frequency characteristics. Measuring the sensitivity for depth perception with different stimulus speeds reveals how the sensitivity depends on motion. Second, we compare the sensitivity for depth perception between stimuli with and without motion differences. If a mechanism sensitive to both motion and disparity is responsible for detecting a depth boundary (Fig. 17.7c), the border defined by motion differences would facilitate the detection of disparity differences. In other words, the detection of depth difference between stimuli moving in the opposite directions might be easier

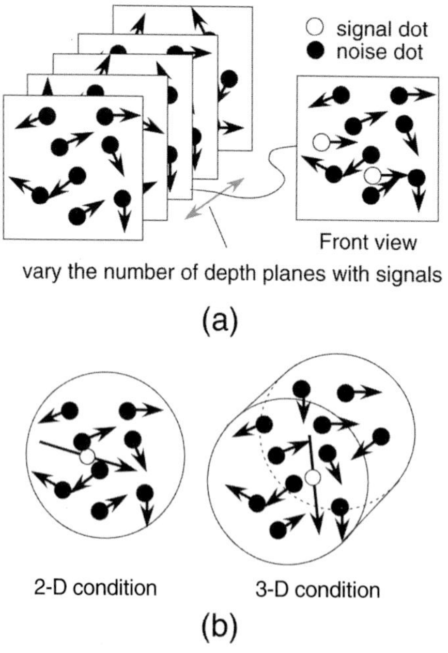

Fig. 17.8. (**a**) Stimulus to vary disparity distribution of stimulus dots. (**b**) Stimulus to measure detectability of the trajectory of motion target in 2-D and 3-D spaces [30]

than that of the same depth difference between stimuli moving in the same directions.

The stimulus was four patches of gratings arranged in a 2×2 array. The observer was asked to distinguish between the two types of stimuli that were different in depth relationship, as shown in Fig. 17.9a. Each diagonal pair of the grating patches had the same disparity, and the direction of disparity differentiated one pair from the other. All the gratings moved in the same direction in one condition (the uniform motion condition), while the gratings of one diagonal pair moved in the opposite direction to that of the other in the other condition (the relative motion condition). The relative phase of the gratings was fixed in the conditions we report here. Similar results were obtained by randomizing relative phase from trial to trial. Two spatial frequencies, 0.23 and 0.92 c/deg, were used with several stimulus speeds. The contrast threshold was determined by a conventional staircase method. Only the contrast was varied in an experimental run and the observers ran each condition three times on different days. Three observers with normal or corrected-to-normal acuity participated in the experiment.

Figure 17.9b shows a typical result of contrast sensitivity for depth discrimination in the uniform motion condition. The contrast sensitivity in the coordinates is the reciprocal of the contrast threshold, the minimum contrast to discriminate the direction of the relative disparity. The contrast sensitivity peaks at about 5 Hz independently of the spatial frequency. Since the speed of the grating motion with a constant temporal frequency is inversely proportional to the spatial frequency, the constant peak of the temporal frequency indicates that the peak in terms of speed is higher for lower spatial frequency. The results from the other two observers are similar. This peak temporal frequency is slightly lower than the sensitivity peak for detecting luminance modulation (about 8 Hz). The temporal frequency characteristics for stereopsis reflect, perhaps, those for detecting luminance modulation. The results of processing luminance signals at a stage prior to the depth processing should input to the process of depth perception. The dependency of the depth perception on speed, therefore, is not an indication of interaction between the mechanisms for motion and depth.

The second point of the experiment was to investigate the effect of motion differences between the grating pairs to detect depth differences. If motion difference helps to see the difference of depth, we expect higher sensitivity in the relative motion condition. If, in contrast, motion does not influence the depth perception, except for temporal modulation of stimulus luminance, we expect no differences between the two motion conditions. The experimental results were not consistent with either prediction. Figure 17.10 compares the sensitivity between the conditions with and without velocity differences. Contrast sensitivity was lower with relative motion stimulus than with uniform motion stimulus when the spatial frequency was 0.92 c/deg. The difference is about a factor of 3 on average across the speeds. Since the depth relationship

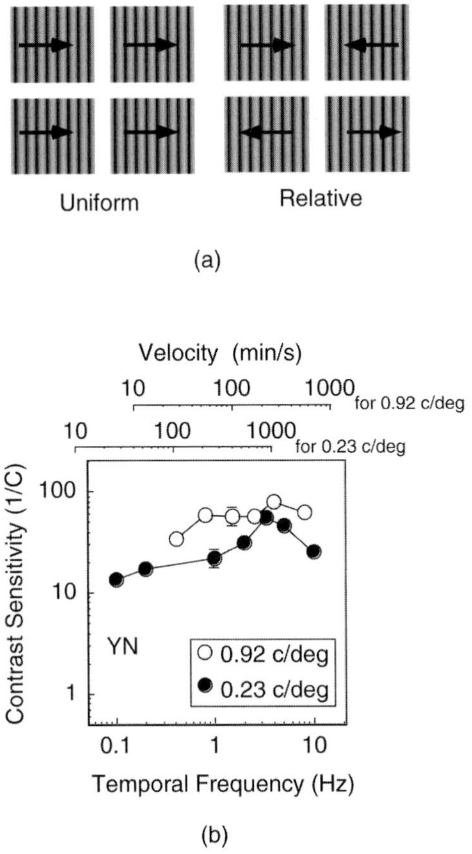

Fig. 17.9. (**a**) The stimulus configuration and task for the contrast sensitivity measurements of the relative depth discrimination. (**b**) The contrast sensitivity as a function of temporal frequency. The velocity is also shown for each spatial frequency

was identical in the two conditions, the results suggest that the difference of motion directions reduces the sensitivity. When the spatial frequency was 0.23 c/deg, the sensitivity was similar between the two conditions. No sensitivity reduction was found in the relative motion condition with 0.23 c/deg gratings (not shown).

It is not easy to explain the results from the known interactions between motion and disparity. A simple prediction from mechanisms with motion and disparity selectivity is the higher sensitivity to relative motion stimulus than that to uniform motion stimulus. Having the differences of both motion and disparity is expected to improve the discrimination of the two depth planes compared to having only disparity difference. This is opposite to our experimental results. Lower sensitivity was found in the relative motion condition. However, it is not necessary that our results contradict the existence of the mechanisms with motion and disparity selectivity. The results for relative motion stimulus in our experiment indicate that motion difference between planes with different disparities reduces the sensitivity to distinguishing the

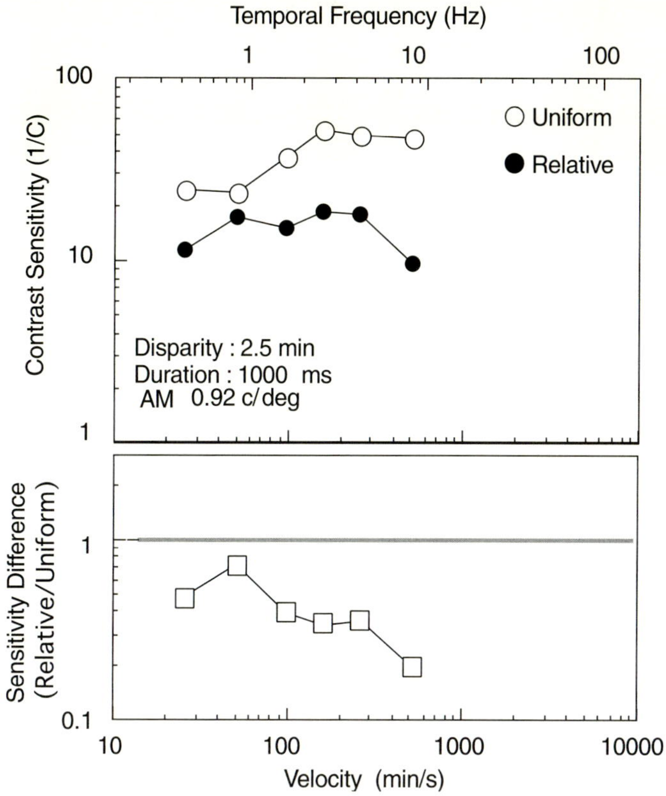

Fig. 17.10. Contrast sensitivity functions for relative and uniform motion stimuli (*top*) and the sensitivity ratio of the two (*bottom*)

planes. Interactions between motion and depth considered in the previous reports do not provide any direct prediction about ability to detect relative depth.

Eye movement is one factor that might cause the differences between the uniform and relative motion conditions. Possible differences of eye movements between the two motion conditions may explain the difference between the two motion conditions. Tracking eye movements of the moving grating should reduce the speed of the retinal image in the uniform motion condition, while it should not affect the relative velocity in the relative-motion condition. This difference in retinal motion may cause different results. To examine whether eye movements are crucial for the different sensitivity for depth between the motion conditions, we repeated the measurements with shorter presentations using 0.92 c/deg gratings. Since there is a latency of 100 ms or more for eyes to start tracking a stimulus, a smaller effect of eye movements was expected for shorter durations. Figure 17.11 shows the sensitivity at three speeds as a function of presentation duration. The sensitivity is lower in

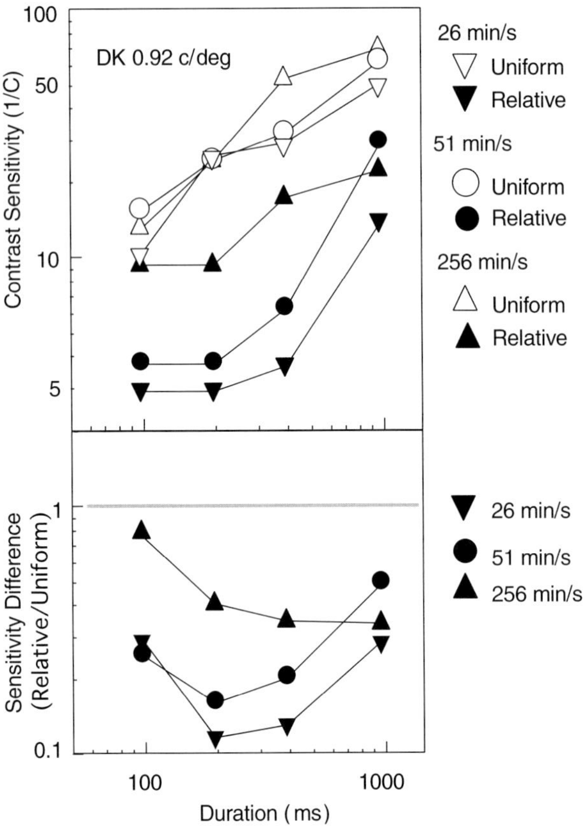

Fig. 17.11. Contrast sensitivity as functions of presentation duration for three speeds (*top*). The sensitivity ratio between the relative and uniform conditions (*bottom*)

the relative motion condition even with the shortest duration used (100 ms). It is also important to note that the effect of duration is not monotonic. The difference is the largest at about 200 ms duration. This is not consistent with the prediction for eye movements because the reduction of retinal speed is expected to decrease monotonically with the decrease of duration. These results suggest that the sensitivity difference between the relative and uniform motion conditions is caused by some interaction between motion and depth processes.

There may be a functional significance for recognizing object shapes to have better sensitivity for depth between surfaces moving in the same direction. Relative disparity is useful information for identifying the 3-D shape of objects. Since the surfaces of an object move together, it can be useful to be sensitive to the relative depth between the surfaces moving in the same direction. In contrast, surfaces moving in different directions usually belong to different objects. When motion difference segregates an object from others, the detection of the disparity differences between surfaces of different objects may be less important. This is particularly so for small disparities, since depth differences are expected to be large in this situation.

Our results together with those of other researchers indicate that there is interaction between depth and motion processing. However, the interaction may not be as simple as the prediction from the responses of MT cells. There are at least two aspects that should be considered. The first is the spatial integration of local motion and/or disparity signals, which is related to the experiments of motion aftereffect and of MT cell response measurements. The second is the detection of motion and/or disparity signal differences across space, which is related to the effect of relative motion for detection of relative disparity. These aspects can be related to different underlying mechanisms that would show different depth–motion interactions.

17.4 Perception of Motion in Depth

The final topic of this chapter is related to the question how the visual system detects motion in depth from 2-D retinal images. There is usually a list of depth cues in a textbook of visual perception. Most cues in such a list are conceived as depth cues for the static depth. They may or may not work as cues for motion in depth, although all cues are candidates for cues for motion in depth. Examination of the use of the cues to see motion in depth is limited to a few cues. In this section, we first describe the cues that may be useful to see motion in depth. Second, we discuss the roles of the two binocular cues for motion in depth. This is to investigate how multiple depth cues are used to see motion in depth. Different cues may have different roles while they may simply increase reliability.

17.4.1 Cues for Motion in Depth

We list some depth cues that are suggested to be used by experimental studies or by our preliminary observation. This list includes pictorial cues, motion cues, and binocular cues.

Overlap. If object A hides part of object B from the sight of the viewer, object A is perceived as being in front of object B (Fig. 17.12). Hiding or overlapping provides relative depth information. The sphere appears to be closer than the cube. Overlaps seems to accumulate the relative depth between objects. The distance between the sphere and the cube appears to be greater in Fig. 17.12b than in Fig. 17.12a. The objects inserted possibly increase the relative depth between the two by adding more overlap cues.

Continuous overlapping of objects at one place provides a strong sensation of motion in depth (Fig. 17.13). An object overlapping the pre-existing

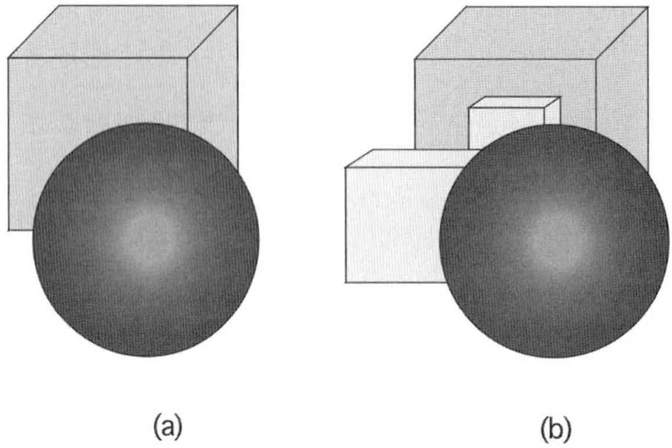

Fig. 17.12. Examples of overlaps as depth cues. Increase of the number of the overlapping objects seems to increase the relative distance in depth

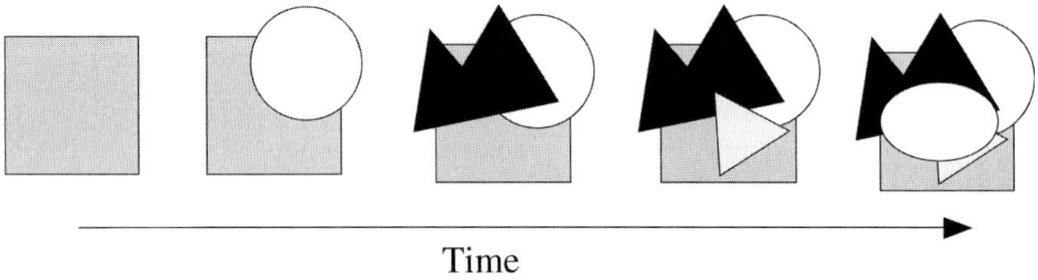

Fig. 17.13. An example of animation sequences of continuously overlaps

objects appears to be the closest, and something appears to be approaching toward the observer by continuous overlaps. This observation indicates that continuous overlaps can be a cue of motion-in-depth. This is an interesting effect since there is no real signal corresponding to this type of stimulation. The effect of size changes, which is a strong cue for motion in depth, cannot explain the phenomenon. The sensation of motion-in-depth remains even when the increase and decrease of the size across the sequential presentations are balanced. Transient signals by the presentation are the possible cause of this motion-in-depth sensation.

Texture Gradient. Visual texture is one of the strong depth cues. Textures provide 3-D information of surfaces of objects as well as those of fields or grounds. There are three independent factors of texture gradient as depth cues: the change of aspect ratio of texture elements, the change of spacing between the elements, and the change of element size [33]. These factors can be considered as different depth cues and their roles in the depth perception may be different. It has been suggested that the aspect ratio change is the dominant cue to surface curvature, and changes of the element size and spacing are the dominant cues to the slant of the planes [33–35].

Considering the dynamic change of each factor, the change of the element size is difficult to isolate from another cue for motion in depth. The changing the size of the image is a strong cue for motion in depth and changing the size of the texture elements may be regarded as changing size. Beverley and Regan tried to compare the effect of changing the element size in a texture and that of changing the size of an image. They used the size change of a uniform surface and that of a textured surface to find that the texture coarseness influenced the perception of motion-in-depth from the changing size. This suggests that the size changes of texture elements provide motion-in-depth signals independently from size change cues, although there may still have been the size change cues of the elements of the texture.

For the other two texture cues, our preliminary observation showed that both spacing changes and the aspect ratio changes provided some impression of motion in depth. Figure 17.14 shows the first and last frames of the animation sequences used. To remove motion cues, new sets of random dots were generated for each frame with a gradual change of either cue. Presentation of the sequence back and forth at about 1 Hz gave some impression of the moving surface in depth. The importance of the cues for motion in depth, however, is not clear. The impression may come from the change of the static impression of a surface in time.

Shadings, Shadows, and Highlights. Gradient changes of luminance in paintings provide a good impression of the 3-D structure of the objects. Similarly, good depth can be seen with cast shadows, occluding shadows, and highlights. Although perception of depth from these cues is often easily influenced by other depth cues such as contours [36, 37], their importance for depth perception is clear from the comparison between line drawings and drawings with tones.

Since the position change of an object relative to the light source produces the positional changes of its cast shadow, the movement of the cast shadow has information of motion in depth. However, the movement of the light source also causes changes of the cast shadow. The visual system has to decide which of the object or the light source has moved when it faces the movements of cast shadow. There is empirical evidence that an object can be perceived moving in a 3-D world when the cast shadow of the object moves even when the object itself is stationary [38]. This indicates that motion of the cast shadow is an effective cue for motion in depth. Shadings, occluding shadows, and highlights may show similar effects, although experimental studies are needed to verify this.

Size Change. When the image of an object grows larger on the retina, this produces a sensation of motion towards the observer. The change of the image size is known to be one of the strongest cues for motion in depth. This cue may not be directly related to the size change of objects because a recent report showed that spatial frequency changes of the stimulus can provide information

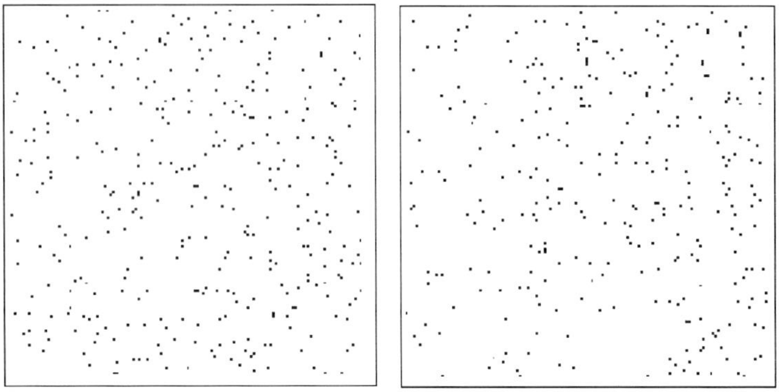

Uniform Gradual change

(a) Density change

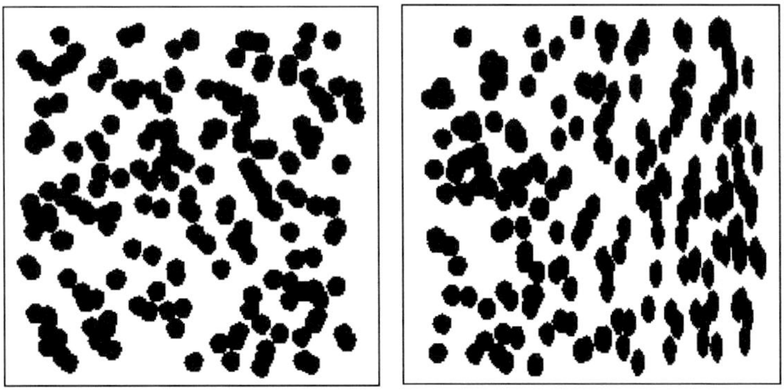

Uniform Gradual change

(b) Aspect ratio change

Fig. 17.14. The first and last images of the animation sequence for dynamic changes of texture gradient cues. (**a**) Density changes and (**b**) aspect ratio changes of elements. The right image is perceived to be slanted while the left image is perceived to be front parallel

of expansion rates without visible changes of stimulus shapes [39]. Size change cue has been investigated and compared with the binocular cues [40]. In particular, Regan and his colleagues have been extensively working on this issue. One of the important findings from their investigation in terms of the relative importance of cues is the effect of the object size. The size change cue is weak when the object size is small because the changing rate of the retinal image becomes low. The relative importance of this cue depends on the size of the objects.

Structure From Motion. Motion in images provides a good impression of the depth or 3-D structures as shown by the phenomena named structure from motion. We can recover 3-D shapes when only motion cues are available. An example is seeing the projection of random dots on the surface of a rotating cylinder, with which no structure can be seen when the cylinder is stationary [41–43]. Three-dimensional shapes can be perceived in appropriate velocity fields.

We simulated the object motion in depth using a structure from motion display, presenting a cylinder moving in depth that was recovered in a velocity field of random dots. The motion in depth of the object was clearly seen in the display. However, this display contained a size change of the 2-D image of the dot field. When the size of the dot area was fixed by putting a window in front of the stimulus, the impression of motion in depth became weak. Although motion signals of dots or some change in the structure recovered form motion may provide a cue for motion in depth, the size change of the image seems more important to see motion in depth.

Binocular Cues. There are two known binocular cues for motion in depth. One is disparity change in time. Retinal images of a moving object in depth change its disparity with time, based on which motion signals in depth can be calculated. The other is interocular velocity differences. The retinal velocities for a moving object in depth are different between the two eyes, based on which motion signals in depth can be calculated. The use of these cues is discussed in detail in the next section.

17.4.2 Binocular Motion Cues for Motion in Depth

There is no doubt that binocular difference is one of the most important cues for depth perception. The prime issue of the technologies for 3-D display is how to project different images onto the two retinas. There are three lines of evidence that the human visual system has mechanisms that specialize in detecting motion in depth [9, 11, 40, 44, 45]. First, the detection threshold of motion in depth is elevated by selective adaptation of motion in depth stimuli [41]. This implies that there are independent mechanisms for detecting static depth and motion in depth since the same adaptation stimulus produces little threshold change for detecting static disparity. Second, threshold and suprathreshold responses to a pulse of disparity are different for approaching and receding pulses [9]. Such a difference between the motion directions should be attributed to a motion-sensitive mechanism. Third, the binocular visual field of normal observers contains areas with no sensation of motion in depth by disparity oscillation of stimulus. In some of these stereomotion-blind areas, a binocular-fused bar appears to remain stationary as the amplitude of the disparity oscillation increases up to the limit of binocular fusion [45].

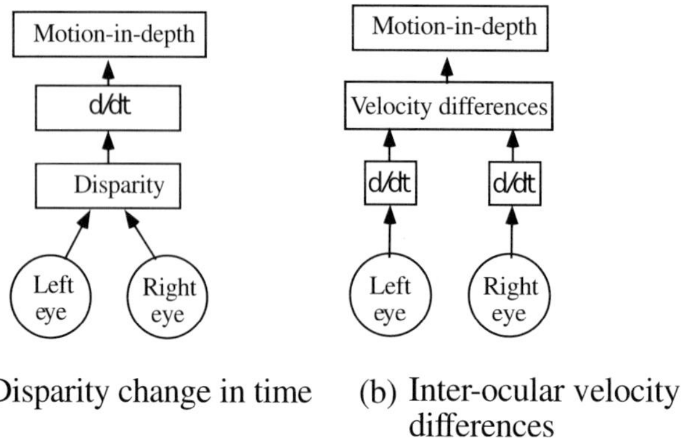

Fig. 17.15. Models of the mechanisms sensitive to motion in depth. Detection mechanism for disparity change in time (**a**), and that for interocular velocity differences (**b**). Reprinted from Vision Research, 40/19, Shiori, Saisho and Yaguchi, Motion in depth based on inter-ocular velocity differences, 2565–2572, © (2000), with permission from Elsevier Science

Two Cues for Motion in Depth. There are two binocular depth cues for motion in depth [46–48]. One is that based on disparity change with time and the other is that based on interocular velocity differences. To use the disparity cue, disparity is detected first, and then its change over time is calculated (Fig. 17.15a). To use the velocity cue, monocular velocities are calculated first, and then compared (Fig. 17.15b). The existence of cues and the use of them are different questions.

The disparity-change cue and the velocity-difference cue exist together in the retinas when we see moving objects in depth. To examine whether the visual system uses either of the cues, isolating one from the other is a required condition. Several researchers have used stimuli, called dynamic random-dot stereograms. Dynamic random-dot stereograms consist of a series of stereograms where left and right images are always correlated whereas new random-dots are used for each frame of the sequence. The monocular view of dynamic random-dot stereograms shows a twinkling of dots without structure. When the disparity changes in the dynamic random-dot stereograms, the observer sees motion in depth in the stimulus [49]. Since there is no coherent monocular motion, seeing motion in depth in the stimulus indicates the processing of a temporal change of disparity.

To isolate the velocity-difference cue from the disparity cue, we can use a similar technique to the use of dynamic random-dot stereograms [48, 50]. Shioiri et al., for example, used binocularly uncorrelated random-dot kinematograms [48]. Their stimulus consisted of a pair of random-dot kinematograms. Each of the kinematograms was a simple displacement of dots in a direction by a given amount, which produced a robust impression of motion in the direction when presented alone. By making the motion direction of the stimulus in one eye different from that in the other eye, they added interocular velocity differences in the stimulus. There was no correlation between the left and right images, and therefore the stimulus provided interocular velocity differences without binocular disparity. They reported that the observers were able to identify the direction of motion in depth in this stimulation. This suggested that interocular velocity differences play a role for seeing motion in depth.

There are two potential problems in the measurement of interocular velocity differences using binocularly uncorrelated kinematograms. The first problem is the effect of the detection of monocular motion. The experiment assumes that the observers response is based on the perception of motion in depth rather than any other clues from motion direction in the stimulus. Theoretically, the observers can guess correct answers if they are is able to detect the motion direction on either retina with the knowledge of which eye the motion signal is from. It is unlikely that there is information of "eye-of-origin" that is consciously available in general. However, there are cues that may be used to find the "eye-of-origin" of information in some conditions [51]. Stimulus rotating in depth can remove the monocular cues from the display. The velocities of the left and right random dots were determined to simulate the rotation in depth. For one cycle of clockwise rotation (Fig. 17.16b), the motion sequence of the left image may be rightward-stationary-leftward-stationary with that of the right image being stationary-leftward-stationary-rightward. For counter-clockwise rotation, the motion sequence of the left image may be stationary-rightward-stationary-leftward with that of the right image being leftward-stationary-rightward-stationary. When the starting phase of the sequence is randomly varied from trial to trail, it is impossible to identify the direction of rotation only with signals from either retina (see Fig. 17.16a). Using the stimulus, Shioiri et al. showed the success in identifying the rotation direction is better than a chance response [48]. This confirms the perception of motion in depth based on interocular velocity differences.

The second problem is the effect of the residual disparity cues in the binocularly uncorrelated random-dot kinematograms. Although there is no correlation between the left and right images, a certain number of dots could be paired with randomly varying disparity. These pairings should be considered as disparity cues. If they are strong enough to provide the sensation of motion in depth, the disparity cues, not velocity differences, may cause an impression of motion in depth even in binocularly uncorrelated random-dot kinematograms. To avoid random pairing of dots in binocularly uncorrelated random-dot patterns, Shioiri et al. replaced parts of the random-dot display in each eye with uniform gray horizontal bands that alternated in vertical position in the left and right images (Fig. 17.17a). Each image consisted of a square-wave horizontal grating of random-dot bands with gray bands. Since the bands with dots occupied alternate positions in the left and right images, there was no overlap with appropriate binocular fusion. Figure 17.17b shows

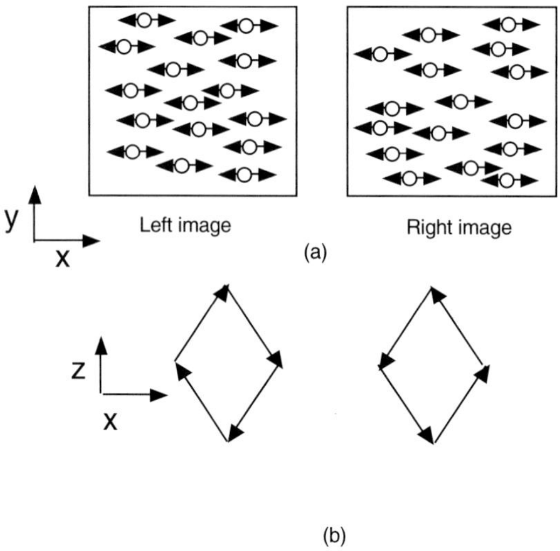

Fig. 17.16. Motion-in-depth stimulus for interocular velocity differences without monocular cues

the percentage of correct responses in this condition as a function of the band size. The results show that the direction of motion in depth can be identified even without direct binocular overlaps of dots. Direct overlaps of left and right dots are not necessary for identifying the direction of motion in depth. We also found that the performance was worse for the stimulus with disparity change in time in dynamic random-dot stereograms than that in binocularly uncorrelated random-dot kinematograms when there was no direct overlap between the images. To explain the results, we should assume that the visual system has a mechanism to detect the signal of motion in depth from interocular velocity differences.

Similar problems in the experiments of dynamic random-dot stereograms may arize. First, static depth cue, instead of motion cue, usually provides sufficient information to respond to the direction of motion in depth correctly. The dissociation of motion and displacements has been known to be a difficult problem in the experiments on 2-D lateral motion. One of the methods for the dissociation is to investigate the motion aftereffects. Since motion aftereffect is explained by assuming fatigue of motion-sensitive cortical cells in the visual system, seeing motion aftereffect indicates the existence of a motion-sensitive mechanism. Regan et al. showed motion aftereffect of motion in depth by dynamic random-dot stereograms [52]. Mechanisms sensitive only to static disparity cannot explain motion aftereffect, because it is a direction-specific phenomenon. To explain the phenomenon, we should assume that the visual system has a mechanism sensitive to disparity change in time for processing motion in depth. This is also supported by the fact that the effect of mean

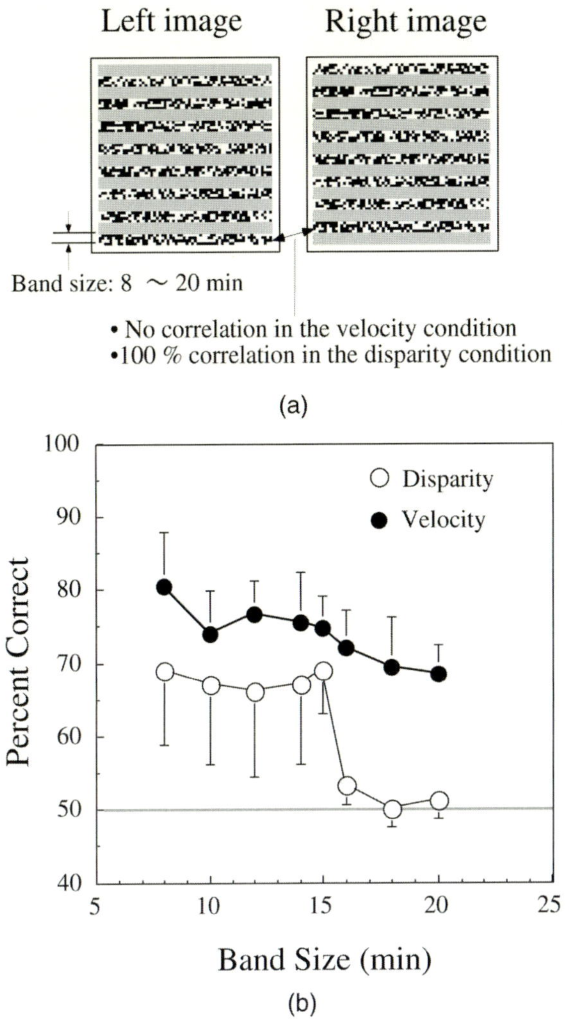

Fig. 17.17. (**a**) Motion-in-depth stimulus for interocular velocity differences without binocular overlap of dots. (**b**) Percentages of correct identification of motion direction in depth as a function of band size in the velocity and disparity cue conditions. The error bars represent the standard error across observers. Reprinted from Vision Research, 40/19, Shiori, Saisho and Yaguchi, Motion in depth based on inter-ocular velocity differences, 2565–2572, © (2000), with permission from Elsevier Science

disparity of the stimulus was different between detection of motion in depth and detection of static depth [53].

Second, there may be effects of the residual velocity cues in dynamic random-dot stereograms. Although there is no temporal correlation in the images, a certain number of dots could be paired with the randomly varying velocity in the horizontal directions. These pairings should be considered as velocity cues for motion in depth. If they are strong enough to provide the sensation of motion in depth, the velocity cues, not disparity changes, can cause the impression of motion in depth in dynamic random-dot stereograms. This question has not been examined seriously, perhaps because no coherent motion is seen monocularly in dynamic random-dot stereograms. However, seeing no monocular coherent motion is not sufficient to rule out the possible influence of velocity cues in dynamic random-dot stereograms. The motion direction of a dot in the left image may happen to match the monocular motion determined by the direction of the motion in depth. In this case, the corresponding dot in the right image also moves in the same manner, keeping the binocular correlation. This should provide appropriated interocular velocity differences. Combination of the binocular information could increase the motion-in-depth signal even when either of monocular signals does not. One answer to the question comes from an experiment that compared the performance for detecting motion in depth with and without velocity information (i.e. temporally correlated random-dot stereograms versus dynamic random-dot stereograms). If the velocity cue is crucial to see motion in depth, we expect a large reduction of sensitivity by removing coherent monocular motion and higher sensitivity with temporally correlated random-dot stereograms than that with dynamic random-dot stereograms. Cumming and Parker showed somewhat higher sensitivity for dynamic random-dot stereograms [46]. This disagrees with the prediction from the use of the residual velocity cues in dynamic random-dot stereograms. Thier results suggest that there is little effect of velocity cues in dynamic random-dot stereograms, at least in the condition they used.

The above arguments derive the conclusion that there are two cues used by the visual system to see motion in depth: disparity change with time and interocular velocity differences. The use of velocity cues for motion-in-depth processing agrees with the possible use of velocity information for judging the speed of motion in depth [54], although this is still in dispute [55].

Comparison of the Two Cues. The issue we address in this section is the differences in the use of the two binocular cues for motion in depth. The detector of disparity changes and that of interocular velocity differences, perhaps, have different tunings for spatio-temporal frequencies, velocities of motion in depth, stimulus sizes, and so on. If they did not have any differences, one with less sensitivity should be superfluous. In the experiment that compared the performance for detecting motion in depth with and without velocity information, Cumming and Parker reported a variety of experimental conditions showing little influence of coherent motion [46].

However, the results in Fig. 17.17b show that the velocity cue can be more significant than the disparity cue in the condition where larger spatial summation is required [48]. The addition of vertical disparity (as shown in

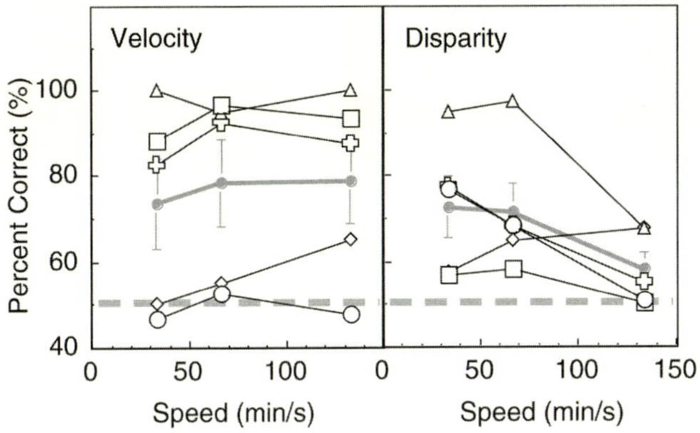

Fig. 17.18. Percentages of correct identification of motion direction in depth as a function of the speed of the stimulus in the velocity and disparity cue conditions. The gray line indicates the average of five observers with standard errors across observers

Fig. 17.17b) made the performance of motion in depth in dynamic random-dot stereograms seriously worse, whereas the effect of vertical disparity was slight in binocularly uncorrelated random-dot kinematograms. The performance was better in the velocity condition and the difference was larger with band sizes (or vertical disparity) larger than 15 min. The result for dynamic random-dot stereograms showed a chance-level performance for a bandwidth of 15 min or larger. The result indicates that the velocity-sensitive mechanism integrates retinal information in a larger area than the disparity-sensitive mechanism.

We investigated the differences between the two motion-in-depth processes in the effects of velocity and contrast, using similar displays as in the experiments of Shioiri et al. The stimuli were dynamic random-dot stereogram (the disparity condition) and binocularly uncorrelated random-dot kinematograms (the velocity condition). The motion in depth simulated was the linear motion along the axis for depth. The upper and lower halves of the squared field ($4.3° \times 4.3°$) moved in the opposite directions and the observer responded to the direction of motion of the upper half. The effect of velocity was measured with three different speeds, 33.3, 66.7 and 133.3 min/s, which was expressed by the monocular speed in the velocity condition and its equivalent of disparity change in the disparity condition (33.3 min/s corresponded to 1 pixel (1 min) moved every 2 refresh frames (2/67 sec)). The Weber contrast of the dots, which is the luminance ratio of the difference between the dots and background against the background, was 0.65 with the background of 46 cd/m [2]. This experiment used the dot size of $2' \times 2'$ and the dot number of 400 with presentation duration of 0.96 s. Figure 17.18 shows the average performance with the individual data shown separately for each condition.

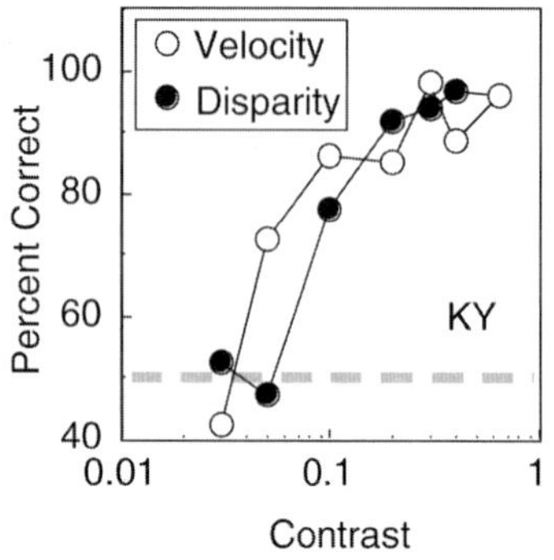

Fig. 17.19. Percentages of correct identification of motion direction in depth as a function of stimulus contrast in the velocity- and disparity-cue conditions for one observer

With the increase of stimulus speed, the performance increased slightly in the velocity condition while it decreased in the disparity condition. This tendency is consistent with the individual results, although there are large individual differences in the performance level. These results suggest different velocity tunings between the use of the disparity and velocity cues and that interocular velocity differences are more important than disparity change in time for moving stimuli at high speeds. This was also supported by an experiment using rotation in depth, where no monocular cue was available to respond correctly [56].

The effect of contrast was measured with the speed of 33.3 min/s. This experiment used the dot size of $6' \times 6'$, the dot number of 200, and the presentation duration of 0.73 s. The larger pixel of dots was used to produce the conditions that would provide appropriate percentages of correct responses throughout the range of stimulus contrasts. Figure 17.19 shows the percentages of correct responses against contrast in the two conditions. The performance was higher in the velocity condition than in the disparity condition at contrasts of 0.1 or lower, whereas it was about the same or slightly lower in the velocity condition at high contrasts. This was true for the other two observers who participated in this experiment. The results suggest that the relative importance of the two cues varies with contrast. Interocular velocity differences are more important than disparity change in time for low-contrast stimuli.

Interestingly, the results in the velocity condition are consistent with the general properties for motion processing. Motion processes are known to have higher sensitivity at low contrasts and high speeds as well as larger integration area, which corresponds to tuning to lower spatial frequencies, in comparison with the position- or pattern-based processes. The results of the analysis of monocular motion are, perhaps, used to process motion-in-depth information at a later stage, as the model in Fig. 17.15 shows.

Our results described so far apparently contradicts Cumming and Parker's result, which showed little effect of coherent monocular motion for most of the temporal frequencies they used (ranging from 0.5 Hz to 4 Hz). A close look at their results, however, revealed some contribution of the motion signals at high temporal frequencies. They showed that the sensitivity for motion in depth in the dynamic random-dot stereogram (which had no coherent monocular motion) was highest at 2 Hz. The sensitivity with the temporally coherent random-dot stereogram, however, peaks at 4 Hz or higher. This is consistent with the notion that the velocity cue is more important at high speeds. There is another factor that potentially interprets the differences between the experiments. Our experiments showed large interobserver differences in the relative sensitivity for the detection of motion in depth between the two conditions, as shown in Fig. 17.18. Three of five observers showed better performance in the velocity condition, whereas the other two showed better performance in the disparity condition. The relative importance of the two cues possibly depends on the individuals as well as the stimulus conditions. Different observers in different studies possibly provide different relative sensitivities to the velocity and disparity cues.

17.5 Conclusions

We reviewed three aspects of the dynamic features of depth perception, which are the temporal characteristics, the effect of motion, and the cues for motion in depth. The temporal frequency characteristics indicate that the depth process is a slow process compared with the detection process of luminance modulation. One basic effect of the influence of motion on depth perception is the effect of temporal frequency in stimulation. The sensitivity to depth of a moving stimulus is high with the speed that is equivalent to the temporal frequency with the highest sensitivity. However, several studies showed that there are other types of interactions between the depth and motion processes. First, both physiological and psychophysical studies have suggested the existence of the mechanism sensitive both to motion and to disparity. Second, we described the experiments that suggest that relative motion reduces the sensitivity of the depth perception. Adding relative motion reduces the contrast sensitivity to the detection of relative depth.

For perception of motion in depth, we list possible cues. Many of them have not been seriously studied and future studies are necessary to understand their functions. Two binocular cues for motion in depth have been identified. The visual system uses both disparity change in time and interocular velocity differences to see motion in depth. The difference between the uses of the cues may be the differences of sensitivity to stimulus variations, such as the spatio-temporal frequency content.

17.6 Acknowledgments

The preparation of this paper was supported in part by a grant from Hoso Bunka Kikin to SS. The authors thank Taiki Shionaga, Ayako Nakajima, and Daisuke Kakehi for their help for gathering data and preparing the manuscript.

References

1. I.P. Howard, B.J. Rogers: *Binocular Vision And Stereopsis* (Oxford University Press, Oxford 1995)
2. S. Shioiri: 'Basis Characteristics of Steropsis', In: *Science of 3D image and human*, ed. by H. Harrashima, N. Motoki S. Yano (Ohmsha, Tokyo 2000)
3. M.F. Bradshaw, B.J. Rogers: Vision Res. **39**, 3049 (1999)
4. M. Morita, H. Isono, M. Yasuda: *Effects of Luminance and Contrast on the Spatial Frequency Characteristics for Depth Modulation*, presented at the Annual meeting of ITE, 1989
   M. Yasuda: *The mechanism of depth perception in Foundation of 3D images*, ed. by T. Izumi (Ohmsha, Tokyo 1995)
5. S. Nagata: J. Jpn. Soc. Medical Electronics and Biological Engineering **20**, 154 (1982)
6. B. Rogers, M. Graham: Vision Res. **22**, 261 (1982)
7. R.A. Schumer, B. Julesz: Vision Res. **24**, 533 (1984)
8. C.W. Tyler: In: *Vergence Eye Movements: Basic and Clinical Aspects*, ed. by K.J. Ciuffreda (Butterworth, Boston 1983) pp. 199–295
9. D. Regan, K.I. Beverley: Vision Res. **13**, 2369 (1973)
10. C.W. Tyler: Science **174**, 958 (1971)
11. K.I. Beverley, D. Regan: Vision Res. **14**, 357 (1974)
12. J.M. Foley, C.W. Tyler: Perception and Psychophysics **20**, 125 (1976)
13. K.N. Ogle, M.P. Weil: A.M.A. Arch. of Ophthalmology **59**, 4 (1958)
14. R. Patterson: Optometry and Vision Science **67**, 123 (1990)
15. C.W. Tyler: In: *Binocular Vision*, ed. by D. Regan (Macmillan, London 1991) pp. 38–74
16. R.J. Watt: J. Opt. Soc. Am. A **4**, 2006 (1987)
17. H.B. Barlow: J. Physiol. **141**, 337 (1958)
18. D.H. Kelly: J. Opt. Soc. Am. **69**, 1340 (1979)
19. J.P. Frisby, J.E. Mayhew: Perception **7**, 423 (1978)
20. M. Minematsu, S. Shioiri, H. Yaguchi: Invest. Ophthalmol. Vis. Sci. (ARVO abstract) **37**, S284 (1996)
21. H.S. Smallman, D.I. MacLeod: J. Opt. Soc. Am. A **11**, 2169 (1994)
22. K. Kimura, S. Shioiri, H. Yaguchi, S. Kubo: Vision (Winter meeting of Vision Society of Japan abstract) **6**, 46 (1994)
23. S.M. Anstis, J.M. Harris: Perception **3**, 153 (1974)
24. F.A. Verstraten, R. Verlinde, R.E. Fredericksen, W.A. van de Grind: Perception **23**, 1181 (1994)
25. J.H. Maunsell, D.C. Van Essen: J. Neurophysiol. **49**, 1148 (1983)
26. D.C. Brandley, N. Qian, R.A. Andersen: Nature **373**, 609 (1995)

27. C. Jennings: Nature **373**, 563 (1995)
28. P.B. Hibbard, M.F. Bradshaw, B. DeBruyn: Vision Res. **39**, 961 (1999)
29. S.P. McKee, S.N. Watamaniuk, J.M. Harris, H.S. Smallman, D.G. Taylor: Vision Res. **37**, 2047 (1997)
30. P.B. Hibbard, M.F. Bradshaw: Perception **28**, 183 (1999)
31. M.F. Bradshaw, B.G. Cumming: Proc. R. Soc. Lond. B, Biol. Sci. **264**, 1421 (1997)
32. S. Shioiri, A. Morinaga, H. Yaguchi: Invest. Ophthalmol. Vis. Sci. (ARVO abstract) **41**, S236 (2000)
33. B.G. Cumming, E.B. Johnston, A.J. Parker: Vision Res. **33**, 827 (1993)
34. J.E. Cutting, R.T. Millard: J. Exp. Psychol. Gen. **113**, 198 (1984)
35. J.T. Todd, R.A. Akerstrom: J. Exp. Psychol. Hum. Percept. Perform. **13**, 242 (1987)
36. B. Mapperson, W. Lovegrove: Perception **12**, 545 (1983)
37. V.S. Ramachandran: Sci. Am. **259**, 76 (1988)
38. D. Kersten, D.C. Knill, P. Mamassian, I. Bulthoff: Nature **379**, 31 (1996)
39. P.R. Schrater, D.C. Knill, E.P. Simoncelli: Nature **410**, 816 (2001)
40. D. Regan: In: *Binocular Vision*, ed. by D. Regan (MacMillan, London 1991) pp. 137–160
41. M.L. Braunstein: In: *Visual Detection of Motion*, ed. by R. Snowden (Academic Press, London 1994) pp. 367–393
42. G. Sperling, M.S. Landy: J. Exp. Psychol. Hum. Percept. Perform. **15**, 826 (1989)
43. H. Wallach and D. N. O'Conell: J. Exp. Psychol. **45**, 205 (1953)
44. K.I. Beverley, D. Regan: J. Physiol. **235**, 17 (1973)
45. X. Hong, D. Regan: Vision Res. **29**, 809 (1989)
46. B.G. Cumming, A.J. Parker: Vision Res. **34**, 483 (1994)
47. D. Regan: Vision Res. **33**, 2359 (1993)
48. S. Shioiri, H. Saisho, H. Yaguchi: Vision Res. **40**, 2565 (2000)
49. B. Julesz: *Foundations of Cyclopean Perception* (University of Chicago Press, Chicago 1971) pp. xiv, 406
50. M. Maeda, M. Sato, T. Ohmura, Y. Miyazaki, A.H. Wang, S. Awaya: Invest. Ophthalmol. Vis. Sci. **40**, 3031 (1999)
51. H. Ono, R. Barbeito: Vision Res. **25**, 289 (1985)
52. D. Regan, C.V. Portfors, X. Hong: Invest. Ophthalmol. Vis. Sci. (ARVO abstract) **38**, S1168 (1997)
53. B.G. Cumming: Perception **24**, 105 (1995)
54. J.M. Harris, S.N. Watamaniuk: Vision Res. **35**, 885 (1995)
55. C.V. Portfors-Yeomans, D. Regan: Vision Res. **36**, 3265 (1996)
56. D. Kakehi, S. Shioiri, H. Yaguchi: Vision **13**, 209 (2001)

18 Visual Perception of Stereoscopic Images at the Point of Binocular Fusion

S. Yano

18.1 Introduction

In this chapter, we first roughly describe the characteristics of binocular fusion and the basic function of convergence eye movement and accommodation based on established knowledge and our experimental results. Next, the experimental results for the examination of visual fatigue caused by stereoscopic images are described. Finally, we summarize this knowledge and the results, and highlight future issues for the production of strain-free stereoscopic images.

We shall, in order, explain the basic visual perception of stereoscopic images and describe disparity, which serves as an important cue in the depth perception of stereoscopic images. Disparity generally has a close relationship with convergence eye movement, and the relationship between convergence eye movement and the accommodation function is important in the account of stereoscopic vision. This relationship is shown by Donders' line in ophthalmology, and we describe its characteristics, briefly. When taking account of stereoscopic vision, we have to clarify more of the characteristics of the human visual system at the Donders' in the design of stereoscopic TV systems. We therefore describe some results from our related experiments. We also describe the perception of stereoscopic images based on the accommodation function. We show our evaluations for the stereoscopic scenes in terms of visual comfort and fatigue based on the above-mentioned human visual perception characteristics. Finally, we point out future research themes for the design of stereoscopic TV systems.

18.2 Binocular Stereoscopic Images

18.2.1 Perception of Binocular Stereoscopic Images

Stereoscopic images are displayed using binocular disparity, as shown in Fig. 18.1. In Fig. 18.1, D is the distance between the viewer and the screen, and t is the distance between the right and left pupils of the viewer. The x, y and z axes are set in the horizontal, vertical, and depth directions from the frontal parallel surface, respectively. The positions of the left and right eyes of the viewer are located on $EL(XL, YL, ZL)$ and $ER(XR, YR, ZR)$.

When the points of the images for the left and right eyes are $Sl(Xl, Yl, Zl)$ and $Sr(Xr, Yr, Xr)$ respectively, one point $P(Xp, Yp, Zp)$ of the stereoscopic image is perceived by the viewer by geometrical calculation in the condition of binocular fusion.

When the distance ΔX between Sl and Sr on the screen is shown by the pupil distance t, ΔX is the difference between Sl and Sr on the x axis:

$$\Delta X = Xl - Xr \,, \tag{18.1}$$

Since,

$$\frac{Xl - Xr}{Zr - Zp} = \frac{t}{ZR - Zp} \,.$$

Therefore,

$$\Delta X = \frac{t}{ZR - Zp}(Zr - Zp) \tag{18.2}$$

and

$$p = \frac{Zr - ZR}{Zp - ZR} \,.$$

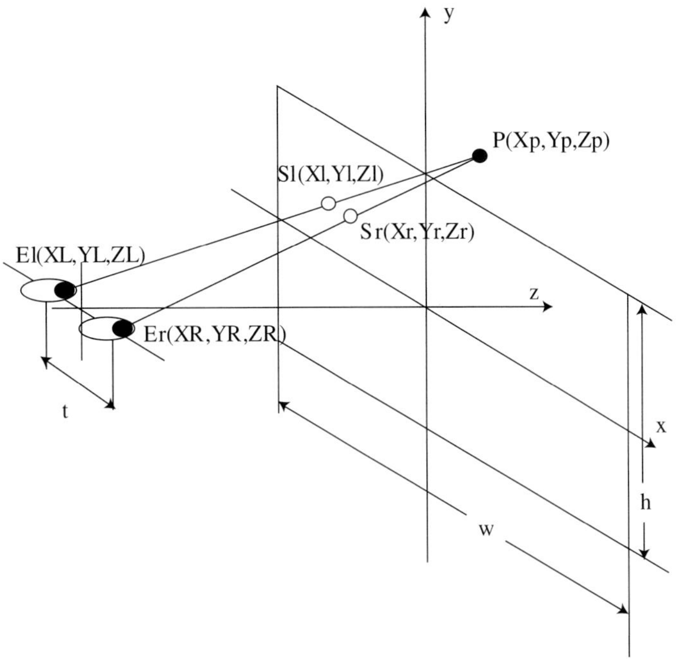

Fig. 18.1. Principle of stereoscopic imaging

So,

$$\Delta X = (1 - p)t\,. \tag{18.3}$$

Therefore, we conclude as follows. In the case where $p = 1$, $Zr - ZR = Zp - ZR$, and $\Delta X = 0$, the perceived image lies on the screen. When $p > 1$, $Zr - ZR > Zp - ZR$, and $\Delta X < 0$, the perceived image is located in front of the screen. When $p < 1$, $Zr - ZR < Zp - ZR$, and $\Delta X > 0$, the perceived image is in the rear of the screen.

The displayed stereoscopic image is limited to within the space of the quadrangular pyramid formed by connecting from the viewer's position to the four corners of screen. When the displayed stereoscopic image is behind the screen, the images are within the enlarged area of the quadrangular pyramid space.

The binocular stereoscopic images are displayed according to the above principles, but the following issues must be investigated in order to secure sufficient picture quality and natural viewing comfort in the watching of stereoscopic images.

- *Perception of distance and size.*
  It may be said that the absolute distance cannot be perceived from the information provided by binocular disparity. The major problems in the case of viewing stereoscopic images are the size and depth perception. These problems only arise with stereoscopic images, but not with two-dimensional images. Whenever the convergence eye movement of the viewer is considered, the perception of distance must be also examined. Moreover, the Puppet Theater and the Cardboard effects are also well-known visual phenomena in stereoscopic images, and these phenomena are particular to binocular stereoscopic images.
- *Tolerance range of binocular fusion for the characteristic differences of right and left images.*
  It is readily supposed that binocular rivalry arises from the large differences of geometrical or physical characteristics as perceived in the left and right images. The tolerance of many parameters, including geometrical gap, luminance difference and chrominance difference, must be examined at the point of binocular fusion.
- *Limitations of binocular fusion.*
  Basically, the visual functions of binocular fusion must be examined using only two points, which correspond to the right and left eyes. The first problem is that depth perception depends on the distance between two points. The second is that the binocular fusion is limited to the distance between two points.
  In the stereoscopic image system, if the distance between two points is zero, no depth perception is caused, and only a two-dimensional image is perceived. The human visual characteristics in the limitations of binocular fusion are related to the depth range of the displayed image and also to visual fatigue.

- *Perception of visual information from only right and left images.*
  The point of convergence eye movement and the focused point due to the accommodation function are principally located by the differences of depth direction when viewing stereoscopic images. Moreover, the perceived images are distorted by the movement of the viewer, because the stereoscopic images are composed of only left and right images. Visual fatigue may be caused by the former because convergence eye movement and accommodation function tend toward the same point, but the eye movement and accommodation function in the viewing of stereoscopic images differ from the natural condition.

In this chapter, we describe (3) the limitation of the binocular fusion range, and (4) the perception of visual information from only right and left images.

18.2.2 Visual Perception in the Depth Direction

The convergence eye movement and accommodation function perform an important role in gazing and the pursuit of the visual target in the depth direction. The accommodation function is controlled by the changing thickness of the crystalline lens. The refractive power exists not only in the lens, but also the cornea, and so on through the optical path. The accommodation function plays a role in giving a clear gaze image to detectors on the retina. The convergence eye movement carries out the role of setting the gazed-at-target on the fovea for cognition of the object.

In the human visual system, the accommodation function or the convergence eye movement can detect the distance of a visual target. Although an object in front or behind of the point of gaze is reflected to different positions on the retina of the right and left eyes, only one object is perceived near the gazed-at-point. This phenomenon is underpinned by the existence of binocular disparity detection cells in the visual cortex. Single vision is caused in the periphery of the point of gazing, and this range is called Panum's fusional area.

Donders' line shows the relationships between the accommodation function and convergence eye movement in the depth direction without consideration of Panum's fusional area [1]. Figure 18.2 shows these relationships.

In Figure 18.2, the x axis shows the convergence point for eye movement (unit: meter-angle), and the y axis the focal point by the accommodation function (unit: diopter). From this figure, 1 meter-angle of the convergence point corresponds to the position of 1 diopter. The area enclosed by the solid line shows the available range of single vision with binocular vision. When the gazed-at-point is located at a 3 meter-angle and the accommodation function fixes the same point, PRC shows the direction of convergence eye movement and NRC the direction of divergence eye movement. When the gazed-at-point is located at a 3 meter-angle but the accommodation function changes, the accommodation function focus moves closer in the case of PRA, and further

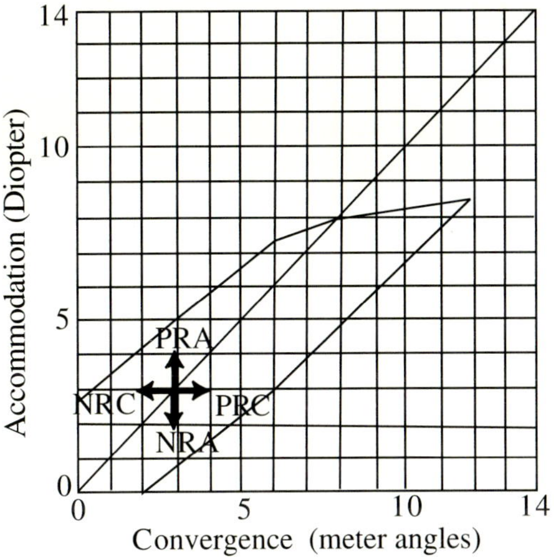

Fig. 18.2. Donders' line

away in the case of NRA. Therefore, the position of the conventional two-dimensional image is located on the line in Figure 18.2. A parallel short line to the x axis shows the binocular stereoscopic images.

In the range of single vision, visual perception gives the same response for the gazed-at-object. However, the degree of visual fatigue is different and depends on the position of convergence and accommodation behavior in the range. It is hypothesized that the degree of visual fatigue increases in the periphery of this range. Perceval's visual comfort area is proposed as the range of visual non-fatigue. This range of visual non-fatigue is shown as the central 1/3 range of the total and is limited from 0 to 3 meter-angle. In distance, this visual non-fatigue range is limited to the zone from 33 cm to the finite point. This range is well known as the visual non-fatigue area.

However, visual perception for stereoscopic images is not simply applied for Donders' line. Donders' line derives from research in ophthalmology, so the range of distance in this figure is mainly set within hand-reachable distances. Moreover, both the x and y axes are reciprocal to distance.

The viewing distance of TV images is quite different. 6H or 7H (H: display screen height) and 3H are the typical viewing distances for the conventional NTSC and HDTV display, respectively. As for the absolute distance, it is presumed from the TV display size that the viewing distance is from 1.5 m to 2.5 m. For this difference, the range of single vision, namely binocular fusion, must be investigated for the binocular stereoscopic images. It is important to examine the characteristics of the binocular fusion in the spatial frequency domain for images. It is also necessary to elucidate the characteristics of binocular fusion in the time-frequency domain for moving images.

18.3 The Limits of Binocular Fusion

When visual perception for stereoscopic images is examined, the characteristics of the limits of binocular fusion must be studied in terms of spatial frequency, motion velocity, temporal frequency, and so on [2]. These characteristics must also be examined for display size, frequency components and so on, which are fundamental requirements for displaying stereoscopic images. We outline the results obtained from experiments.

18.3.1 The Limits of Binocular Fusion in Terms of Spatial Frequency and Velocity of Movement

(a) Fusion Characteristics With Respect to a Sinusoidal Wave Grating

Experimental Method. Images were displayed on a stereoscopic image display using a polarized filter and binocular fusion. We used two frame memories, 1024 pixels (horizontal) $\times$ 512 lines (vertical), one for the right visual field and other for the left. A static or moving target was displayed as an RDS (Random-Dot Stereogram, hereafter RDS) in a 512 pixel $\times$ 64 line area in the total memory area, with the remaining area left at the black level.

We generated an RDS with software (generation rate: 50%), provided its necessary portions using parallax, and displayed it as right-eye and left-eye images. For transfer to the memory, one of these images was exposed to phase modulation by the means of the subject's mouse control in proportion to the parallax disparity, so that the RDS had a specified sine-wave spatial frequency.

Then, in order to obtain a target that moves horizontally at a constant speed, we used the frame memories to provide an RDS with an interrupt mode and I/O control by utilizing a vertical synchronization signal as an interrupt signal so that we could control the reading start address of the memory in accordance with the target's velocity. To assess the stationary image target using this method, the reading start address may be kept set at the initial value during interrupts.

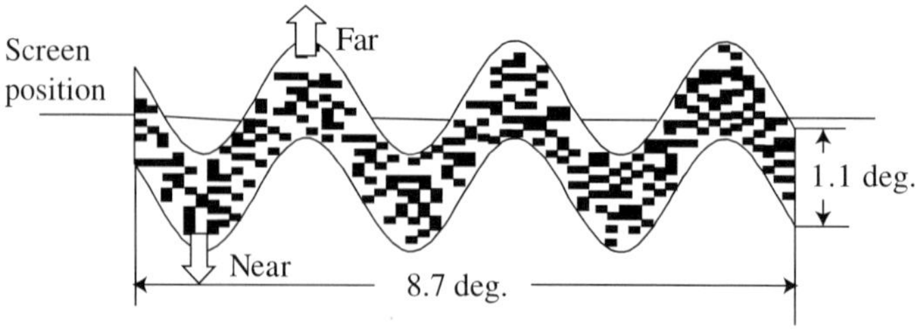

Fig. 18.3. Sinusoidal wave grating for the experiment

First, we used a horizontally moving sinusoidal wave grating, as shown Fig. 18.3. The sinusoidal wave grating used as the target was set to achieve a portion without parallax on the CRT screen. On this sinusoidal wave grating, which gives a feeling of depth, we defined the maximal spatial frequency for binocular fusion by using parallax disparities and horizontal motion velocities as parameters. The target's speeds were set stationary and 1, 5, 10, and 15 dps (degree per second, hereafter "dps"). To identify the maximum fusible spatial frequency, each subject was told to manipulate the presented pattern with a mouse so as to pinpoint the maximum extent that it can be seen as continuously waving strips.

This test for determining the binocularly fusible area was conducted with parallax and velocity of movement as the parameters under two different conditions: free viewing and with eye fixation. The parallax D of the target used in this experiment may be formulated as follows:

$$D = A\sin(wx - vt) \tag{18.4}$$

where A is the parallax disparity, w the spatial frequency, and v the velocity.

The spatial frequency here was calculated by taking account of the size (305 mm × 39 mm) of the RDS grating pattern, the number of pixels (512 pixels × 64 line), and the viewing distance (2 m) employed in the experiment.

Experimental Results. Figures 18.4a,b and 18.5a,b plot the results of the free viewing and visual fixation tests, respectively. In each of these graphs, the solid lines indicate the vertical spatial-frequency limit for binocular fusion viewing that is reproducible on the experimental equipment and the horizontal spatial-frequency limit for two-dimensional images, while the broken line represents the spatial frequency limit (3 cpd) for depth perception. The inclined line shows the limitation of a parallactic cycle for the formation of a sinusoidal or rectangular wave.

It is seen from Fig. 18.4 and 18.5 that the maximum fusible spatial frequency for static patterns on parallactic disparities has a peak value of some 2.5 cpd over a disparity range of surround 120 to 180 seconds. This is the maximum fusible spatial frequency that can be perceived with this experimental equipment.

The above-mentioned value obtained from the experimental results was somewhat lower than the fusible limit frequency of 3 cpd conventionally estimated from depth direction discrimination characteristics. This may be due to signal distortion incurred as a result of the quantization of binocular parallax information in the form of pixels. In other words, it is due to the deficiencies of the conventional TV system.

It was also shown that when the sinusoidal wave grating moves horizontally at a constant velocity, the maximum binocularly fusible spatial frequency for free viewing gradually decreases from the level for a static grating as the grating velocity increases, and drops sharply when the grating velocity reaches the range of 10 – 15 dps. The maximum spatial frequency

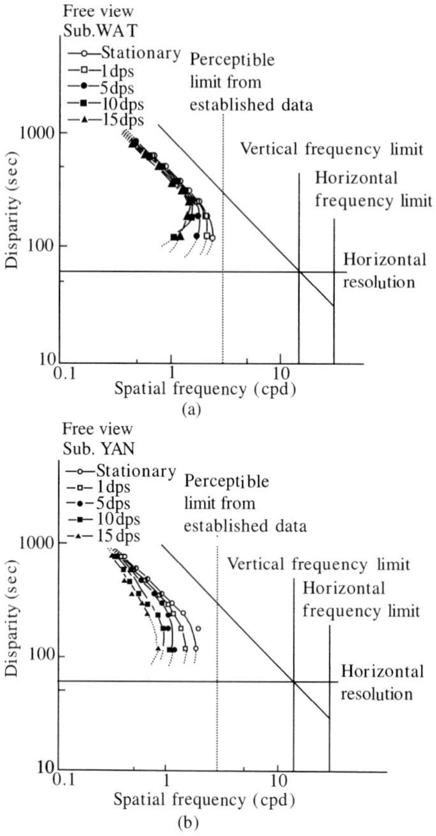

Fig. 18.4. Limit of binocular fusion for the moving target in the case of free viewing

shows a particularly sharp drop in areas where the parallactic disparity is small and a very slight drop in areas of large parallactic disparity. This does not mean, however, that no fusion occurs at a grating velocity of 15 dps or above. In the case of visual fixation, it was shown that the maximum fusible grating velocity is 10 dps or greater and that when the velocity exceeds this level, no depth perception is available through binocular fusion. In this case, the fusible spatial frequency drops far more sharply than in the case of free viewing.

(b) Fusion Characteristics With Respect to a Rectangular Wave Grating

Experimental Method. The experimental method and conditions and presentation method for the evaluation of fusion characteristics in relation to a rectangular wave grating were the same as those for the sinusoidal wave grating test. In this experiment, the presented patterns came in two types: the longitudinal and transverse wave. Figure 18.6 shows the test target of the

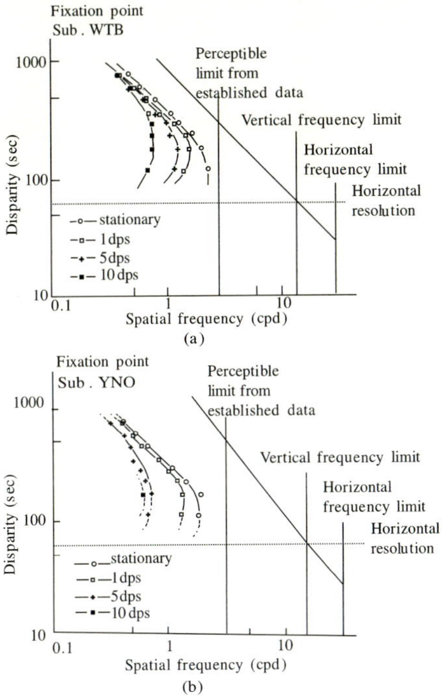

Fig. 18.5. Limit of binocular fusion for the moving target in the case of setting fixation

transverse rectangular wave grating. The maximal fusible spatial frequency was identified for different parallactic disparities for each static pattern.

Experimental Results. The experimental results are shown in Fig. 18.7a,b. It was observed that the maximum fusible spatial frequencies at different parallactic disparities were within a range of 2 – 3 cpd and that these frequencies dropped by very little even at high disparities, as opposed to the sinusoidal grating evaluation described in the preceding section. It is generally known that with respect to the spatial frequency characteristics obtained at the contrast sensitivity threshold of two-dimensional images, rectangular wave patterns have a higher sensitivity than sinusoidal wave patterns. One may consider, therefore, that this supports the so-called multichannel theory that, even in the case of binocular parallax stereoscopy, the spatial frequency characteristics for the entire visual system are composed of many narrow-width channels instead of being made up of a single wide channel. Our test, however, showed that the maximum fusible spatial frequencies did not drop

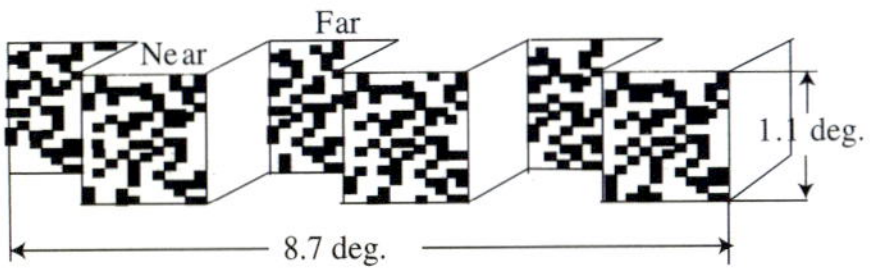

Fig. 18.6. Transverse rectangular wave grating for the experimental

as much as they did in the case of the sinusoidal wave grating evaluation, despite the increase in parallactic disparities. It can be said from these results that the rectangular wave grating not only offers better visibility than the sinusoidal wave grating but also draws attention to the most easily fusible frequency components for a feeling of viewer presence in stereoscopy.

The graphs in Figure 18.7 also indicate that, as the parallactic disparities become higher, there is a greater difference between the maximum spatial frequency of a longitudinal wave and that of a transverse wave. This comes from the fact that binocular parallax information is treated as horizontal dislocation, which conforms to the results of the another experiment conducted on sinusoidal wave gratings. Such longitudinal transverse wave differences are also produced by the fact that determining the vertical spatial frequency automatically defines the maximum parallactic disparity and restricts the fusion area.

18.3.2 Temporal Frequency Characteristics of Stereoscopic Vision

(a) Experimental Equipment and Method

Experimental Equipment. The display for stereo images uses the polarized filter method, where two projectors are equipped with polarizing filters and the subject wears polarizing glasses. The 120-inch display screen has an aspect ratio of 4:3. During display, the size of a random dot is 2 pixels by 2 lines. A simple linear code sequence generator (number of steps: 18; code length: 262,143; maximum length line tap: [18, 7]) is used to generate pseudorandom dots. This sequence of dots is sent to 1-bit frame memories of two pages of 720 pixels × 480 lines (360 × 240 random dots) each. Next, a 1-bit frame memory synchronized with these memories is set up to act as a key signal generator. Based on this signal, the magnitude and shape of possible stereo regions is determined.

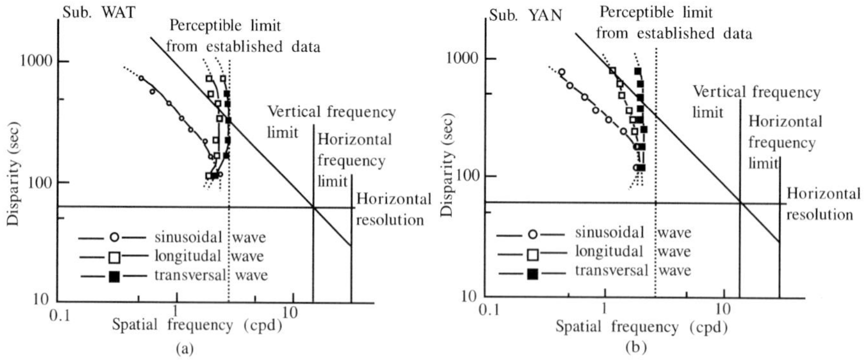

Fig. 18.7. Limit of binocular fusion for rectangular and sinusoidal wave gratings

The random dots shown to one eye are output from the memory unchanged for display on the projector. The random dots shown to the other eye are delayed by using a programmable delay line. A personal computer detects the vertical driving signal, and the delay, i.e. the parallax, is given to the programmable delay line by polling the I/O port of the PC. The gate then selects the delayed random dots and nondelayed dots and a DRDS (Dynamic Random-Dot Stereogram, hereafter DRDS) is displayed.

Experimental Method. The forward and backward motions of the parts of the target that are perceived to have depth are modulated by a sinusoidal wave and a rectangular wave. The repetition frequency of the target motion and the maximum fusible display are determined experimentally [3] both where the center of motion of the target that moves in depth is on the plate of the screen, as in Fig. 18.8a, and where the center of the depth motion is in front of or behind the screen, as in Fig. 18.8b. By shifting and setting the center of motion off the screen, as in Fig. 18.8b, we can observe the difference in the fusion characteristics in the depth direction of vision.

The subjects were instructed that the defining condition of binocular fusion is the state where the plane of forward and backward motion is clearly perceived. Depth motion perceived in only one direction is not, therefore, judged to constitute binocular fusion.

(b) Temporal Frequency Characteristics of Binocular Fusion

When Depth Motion is Centered on the Screen Plane. With a DRDS as the target and the screen as the center of motion, Figs. 18.9a–d illustrate the results for depth motion, given equal disparity. The horizontal axis is the repetition frequency of the target; the vertical axis is the maximum fusible parallax for binocular fusion. Figures 18.9a,b are the results during free viewing, and Figs. 18.9c,d the results when there is a fixation point. Similarly, the white circles are the results when the target is moved back and forth sinusoidally, and the black circles correspond to rectangular wave motion.

Based on the results of free viewing shown in Figs. 18.9a,b, the temporal frequency where fusion is possible is 7.5 Hz when the target is modulated by a sinusoidal or rectangular wave. When the fixation point is set in the retinal fusion, Figs. 18.9c,d show that the fusible parallax decreases. Frequencies where fusion is possible range from 3.75 Hz to 7.5 Hz for sinusoidal waves. For rectangular waves, the frequencies are from 1.88 Hz to 8.75 Hz.

This result agrees for the most part with the idea that, in general, eye convergence motion is possible up to about 3 Hz. However, as explained in this experiment during convergence motion, fusion is possible even over 3 Hz.

When the Center of Depth Motion is Shifted Off the Screen Plane. The experimental results are shown in Figs. 18.10a–d. No fixation point was

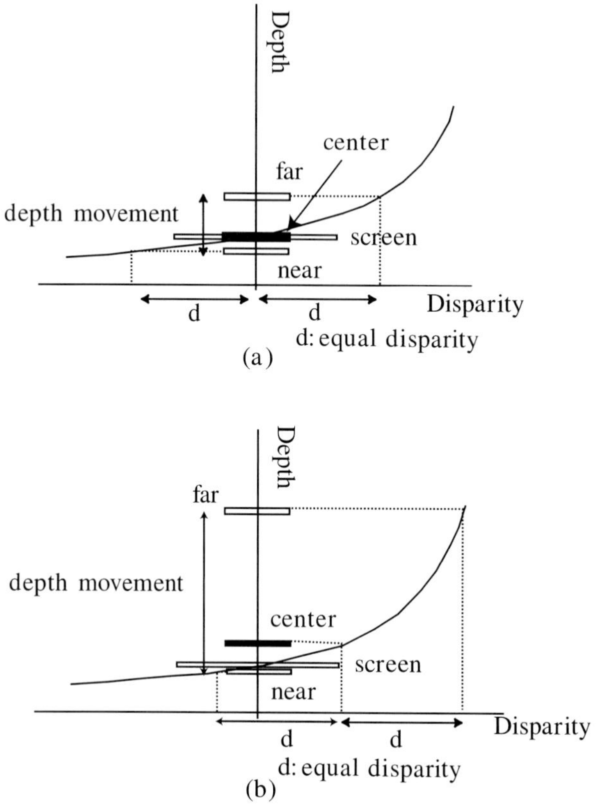

Fig. 18.8. Display method of depth motion in the temporal frequency domain

set in any of the cases. Also, the depth motion was modulated by rectangular waves. Figures 18.10a,b show the cases where the center-of-depth motion is shifted in front of the screen plane; Figs. 18.10c,d are for a shift backward in depth.

In the figures, zero designates the screen plane at the center of motion, and the shifted values of the center of motion are represented by visual angles. In other words, the initial value is indicated by the visual angle for the stimuli as seen by the left and right eyes of the subject. Next, only the positions on the left and right are changed temporally. The experimental results for values at and above 169.0 min., 84.5 min., and 211.3 min. are not shown in Figs. 18.10a,c,d. This is because binocular fusion was not established above these values. Values at low frequencies, as in Figs. 18.10a,c, are plotted and the high frequencies are not plotted. This demonstrates that binocular fusion was not established at the higher frequencies.

Based on the experimental results in Figs. 18.10a–d, regardless of a backward or forward shift in the center of depth motion for a DRDS, the parallax

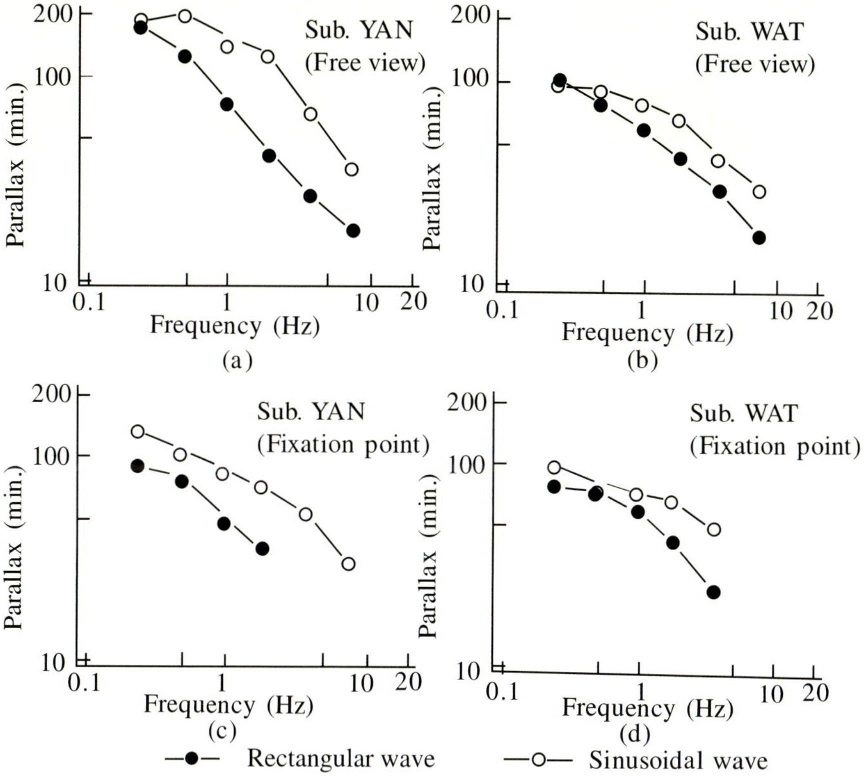

Fig. 18.9. Limit of binocular fusion in the temporal domain (depth motion on center)

values where binocular fusion is possible decrease at each frequency, or binocular fusion tends no longer to be possible at high frequencies. As the shift of the center of motion increases for the DRDS, the parallax values decreases somewhat, as shown in Figs. 18.9b,d, or fusion is no longer possible at high frequencies, as in Figs. 18.10a,c. The tendency illustrated in Figs. 18.10a,c was also observed for the other subjects. Further examination is required, since this is either the result of the positional relationship of the dots at zero parallax, which become background, or of the characteristics due to the basic viewing distance. If we consider the maximum fusible parallax at the macroscopic level to be uninfluenced by the size of the random dots in the background, these results illustrate the temporal-frequency characteristics in the visual field.

18.3.3 Relationship Between Display Size and Binocular Fusion

Experimental Equipment and Method. It was found that the limits of binocular fusion also depended on the size of the display screen. This relation-

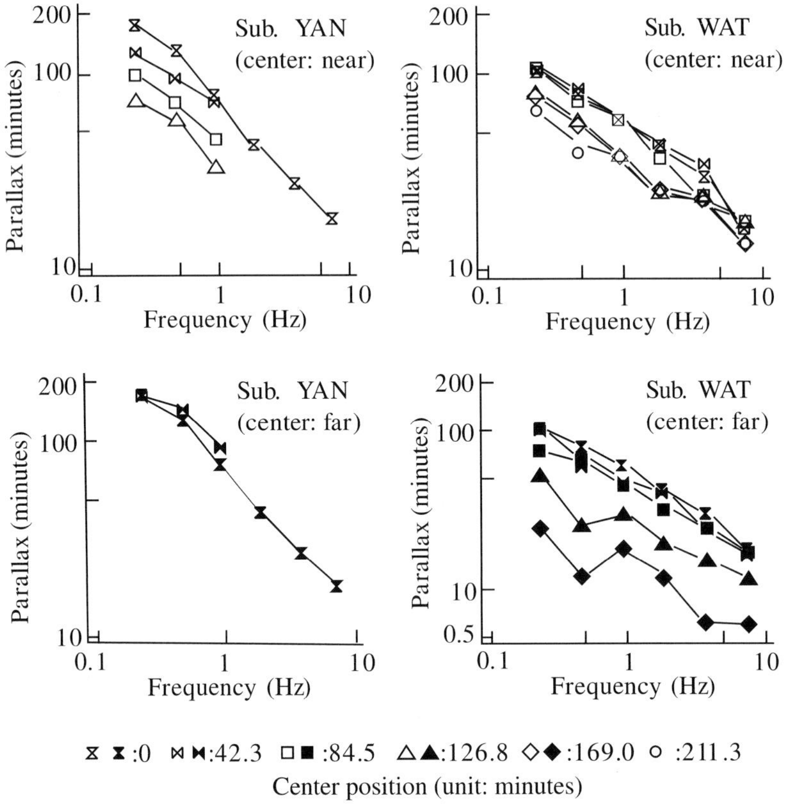

Fig. 18.10. Limit of binocular fusion in the temporal domain (depth motion off center)

ship was investigated using two projectors and a wide-visual-field screen, on which stereoscopic images were displayed using polarizing filters and glasses. The target in the experiment is shown in Fig. 18.11. In Fig. 18.11, a parallax is only given for the central quadrilateral, and a parallax in the periphery is zero. The area of the central quadrilateral was gradually enlarged, and the limit of binocular fusion was measured at each size.

The other side, the target in Fig. 18.11 has a parallax at the central quadrilateral and zero parallax in the peripheral area. The limit of binocular fusion was measured using this target, which was gradually enlarged in the peripheral area. The method of limits was applied for the measurement of the ultimate binocular limit.

Experimental Results Regarding Parallax Area. Figures 18.12 and 18.13 show the experimental results. Figures 18.12a, b show the results in the case of enlargement of a central area that has a parallax. Figures 18.13a,b show the results for enlargement of an area of nonparallax. Figure 18.12 shows

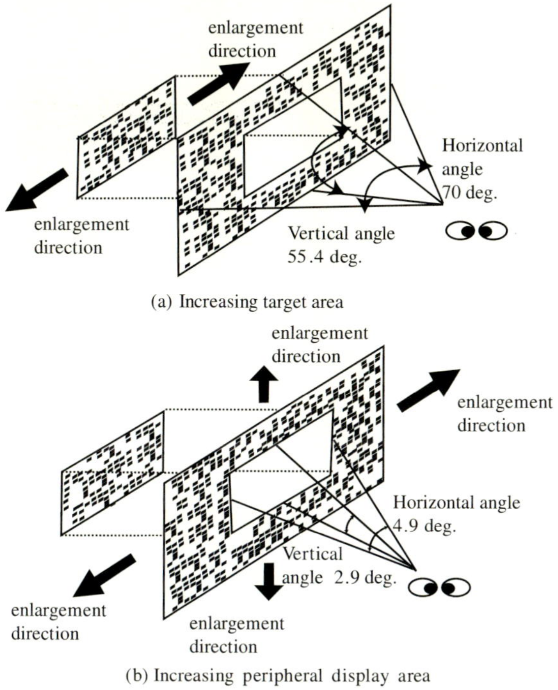

Fig. 18.11. Experimental target for the relationship between size and binocular fusion

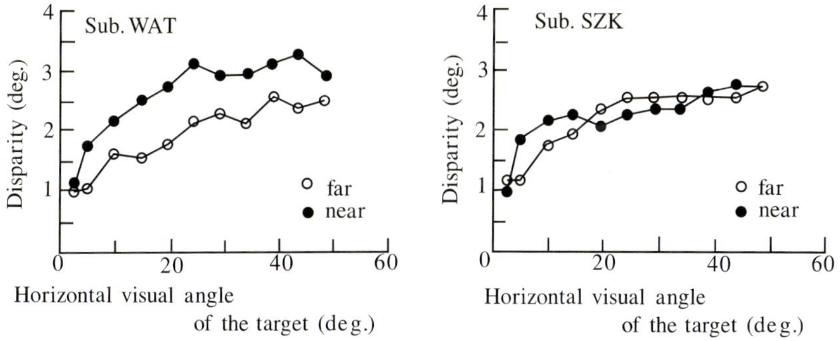

Fig. 18.12. Binocular fusion in the case of increasing target area

that when the area that has a parallax is large, the limit of binocular fusion is also extensive. The results shown in Fig. 18.13 suggest that the peripheral nonparallax area does not affect the limit of binocular fusion. Consequently, the limit of binocular fusion depends not only on the size of the display screen, but also on the display size that has a parallax.

Experimental Results Regarding Target Size. Figure 18.14 shows the results when enlarging the area in the horizontal direction with each constant vertical size, using the target shown in the upper part of Fig. 18.14. These

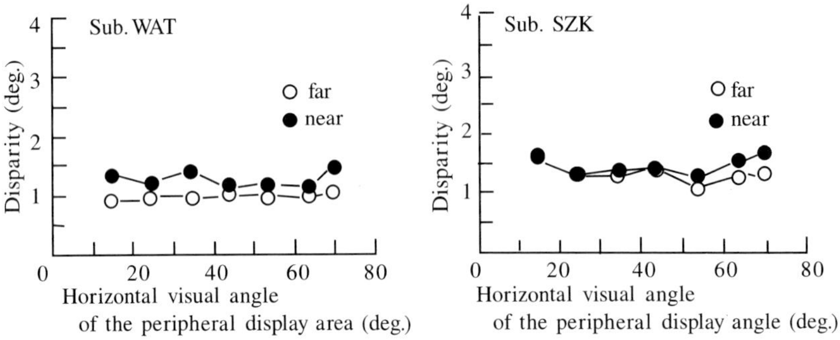

Fig. 18.13. Binocular fusion in the case of increasing peripheral display area

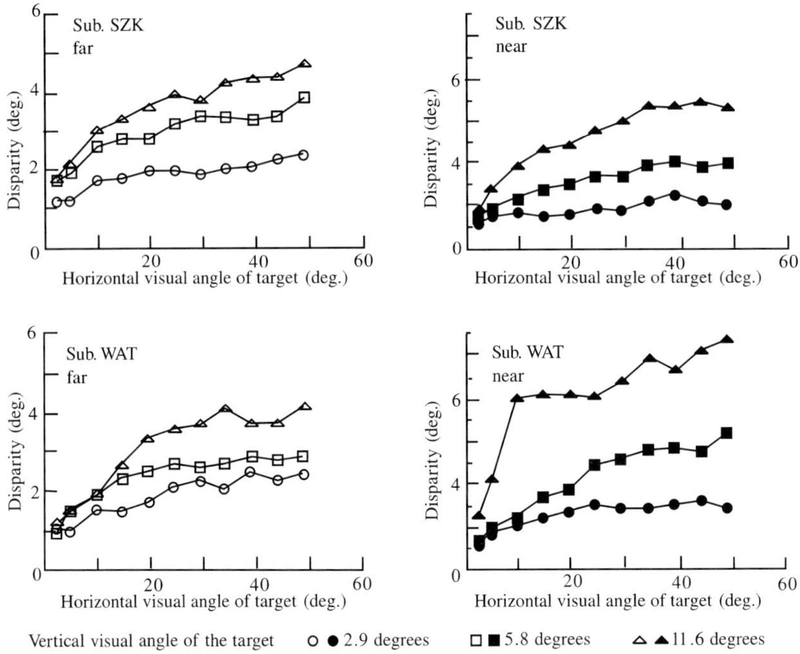

Fig. 18.14. Binocular fusion in the case of changing horizontal and vertical visual angle of the target

results reveal that the saturation tendency of the limit of the binocular fusion appears at about 30 degrees of the display target size.

Experimental Results Regarding Dot Size and Density on RDS. We also studied the effects of the dot size and density with the random dot stereogram (RDS) regarding the limit of binocular fusion. When the dot size of the RDS was large, the sampling interval in the RDS was long. As a result, the frequency components are restricted to the low-frequency range.

Conversely, when the dot size of the RDS is small, the RDS includes high-frequency components. When the dot size is constant and the dot density is changed in RDS, the frequency components are unchanged and only the power spectrum is altered.

Figure 18.15 shows the results for the limit of the binocular fusion when changing the dot size in the RDS. The limit of binocular fusion is small in an RDS consisting of low-frequency components. When the target sizes are enlarged to the peripheral vision range, the limit of binocular fusion is large in spite of the frequency components at each RDS. These results suggest that the characteristics of peripheral vision have a tendency towards a large limit of binocular fusion.

Figure 18.16 show the results for the limit of binocular fusion when changing the dot density in RDS. These results show that the limit of binocular fusion is not affected by the alternation of the dot density in the RDS.

18.4 Accommodation Function and Depth of Focus

18.4.1 Depth of Focus and the Effect of Pupil Size

The accommodation function and the function of convergence eye movement have a cross-link, as described in the explanation of Donders' line. Figure 18.17 shows a simplified model of this relationship between the accommodation and eye movement functions.

When a gazed-at-target is in the depth of focus for viewers, the accommodation function works in the condition of an open loop for the lack of feedback signals in the accommodation-function system. Consequently, the accommodation function is operated by the both the self-controlled signal and the convergence eye movement in the open-loop condition. Figure 18.18 shows the condition of operation for the relationship between the accommodation function and the convergence eye movement in that state [4]. The accommodation position and the gazed-at-position by eye movement are the same from $-0.2\,\mathrm{D}$ to $0.2\,\mathrm{D}$.

The depth of focus is generally affected by the size of the pupil. Figure 18.19 shows the relationship between the size of the pupil and the brightness of the target [5]. When the brightness of the displayed target is $10\,\mathrm{cd/m^2}$, the size of the pupil is 3 mm in diameter. The depth of focus is 0.5 diopter, using the Westheimer formula. This result seems larger than that obtained experimentally.

18.4.2 Display Area of Stereoscopic Images

These results derived from the characteristics of binocular fusion, the relationship from accommodation, and the function of convergence eye movement suggest the possible range of displaying stereoscopic images to be as shown

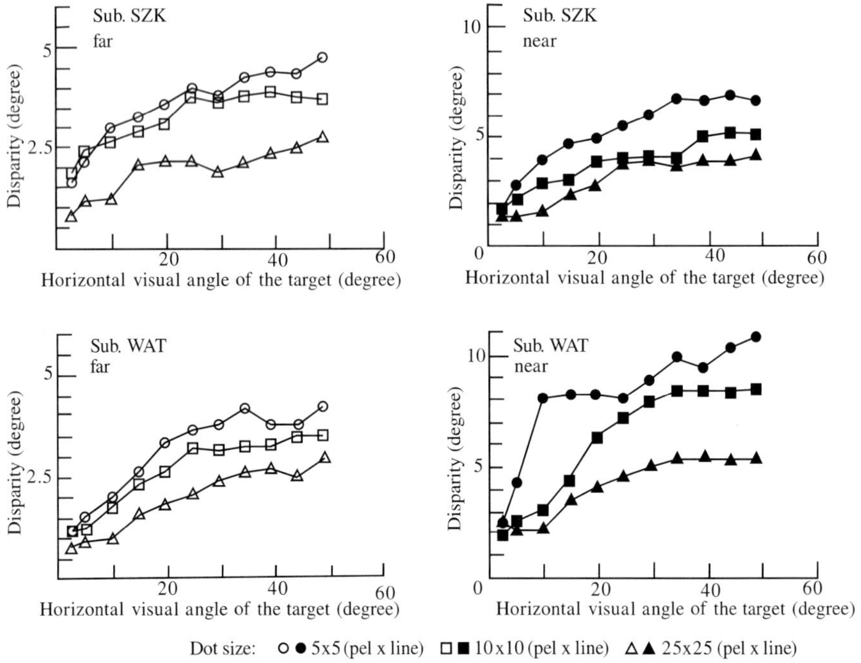

Fig. 18.15. Binocular fusion in the case of changing the dot size of the RDS

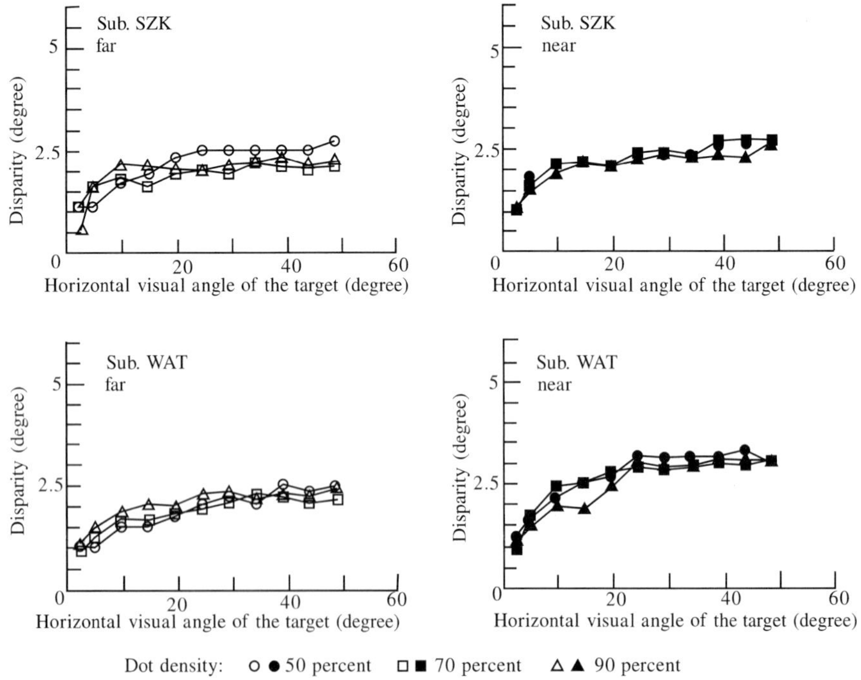

Fig. 18.16. Binocular fusion in the case of changing the dot density of the RDS

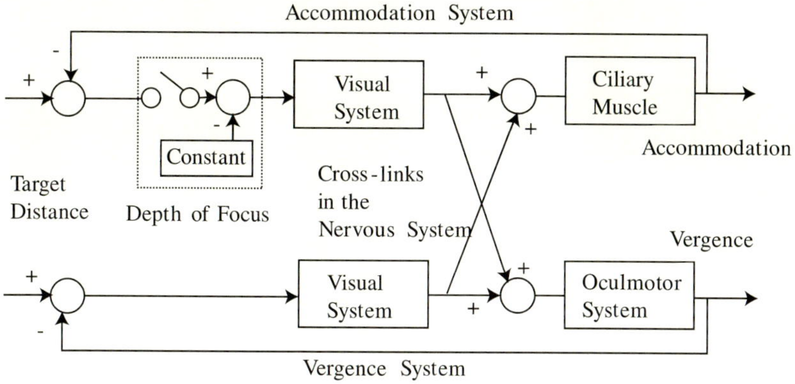

Fig. 18.17. Behavior of the accommodation and convergence control system

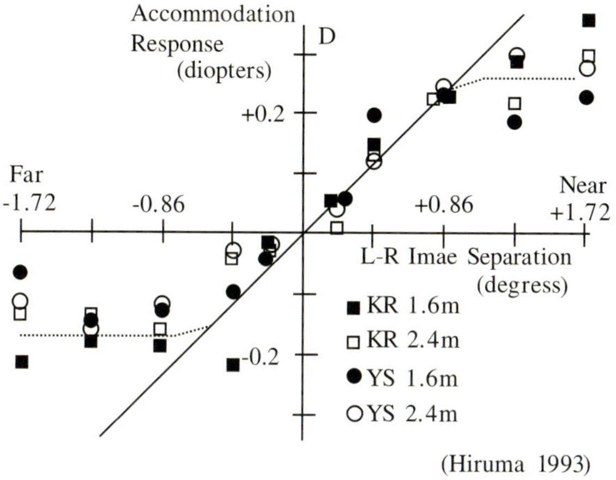

Fig. 18.18. Relationship between the L-R image distance and the accommodation response

in Fig 18.20. First, the stereoscopic images cannot be displayed extremely close to the viewer and far away from the screen due to the limit of binocular fusion. Second, the stereoscopic images can be displayed within the range of binocular fusion. This range depends on the characteristics of the viewer's visual perception, the screen size, and so on. As far as screen size is concerned, the size does not strongly affect the limit of binocular fusion for visual angles of over 30°. The binocular parallax for the stereoscopic images is effective in the range of less than 3 cpd (cycles per degree) for spatial frequency and less than 7.5 Hz for temporal frequency, when considered separately. The limit of binocular fusion decreases for a fast-moving target at higher levels. The results suggest that visual fatigue is not perceived strongly for the range shown in Fig. 18.20.

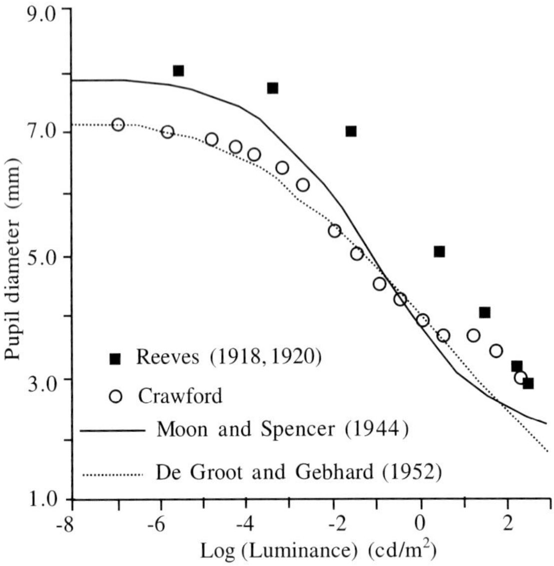

Fig. 18.19. Pupil diameter as a function of light level. After David A. Atchison, George Smith: *Optics of the Human Eye* (Butterworth, Heinemann, 2000) p. 24

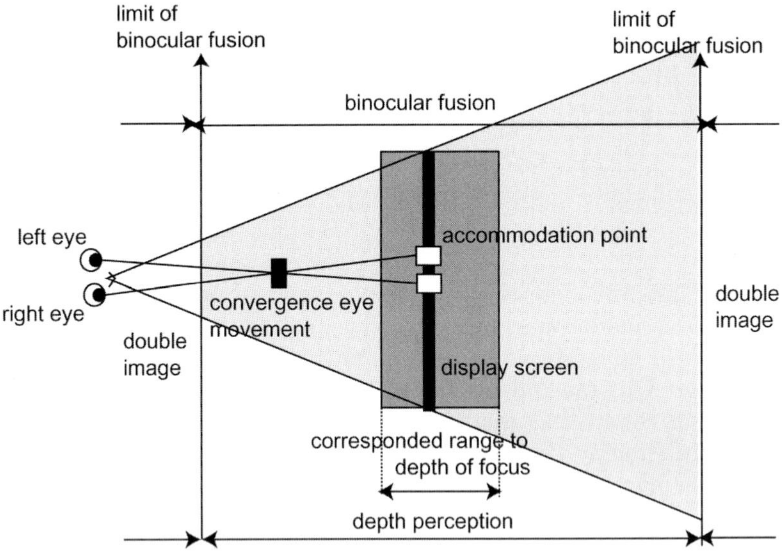

Fig. 18.20. Display range of stereoscopic images

18.5 Evaluation of Stereoscopic Images

18.5.1 Approach to the Evaluation Method for Stereoscopic Images

There are many causes of major visual fatigue in a stereoscopic image system, as explained below. In a stereoscopic image, the viewer watches two images corresponding to the right and left eyes with convergence and divergence eye movement, allowing the two pieces of the image to fuse. As a result, conflict occurs between convergence or divergence eye movement and the accommodation function thus causes visual fatigue.

An objective evaluation method that measures "accommodation response" is sometimes used in addition to subjective estimation of this visual fatigue. However, experimental results have shown that visual fatigue does not always occur. Furthermore, it is thought that even if visual fatigue is caused by viewing stereoscopic images, the visual function returns to the normal condition in about 30 minutes after viewing the images.

This chapter describes the results for evaluation of visual fatigue when viewing 3-D HDTV images [6]. Our experiment took a somewhat different approach compared to the former experimental method in two aspects. It is considered that screen brightness must be increased and viewing distance extended to reduce visual fatigue, because the conflict between convergence eye movement and accommodation function is a cause of visual fatigue. We considered this hypothesis and designed a long viewing distance of 4.5 m for this experiment.

Furthermore, to ensure that the subject looked at the image during the experiment, we imposed a continuous evaluation of the visual comfort of the pictorial image on the subjects. We examined the relationship between this subjective evaluation and the featured characteristics of stereoscopic test images.

18.5.2 Experimental Equipment and Method

Experimental Equipment. Figure 18.21 shows the condition of the experimental room. The stereoscopic 3-D HDTV image was presented using the binocular fusion method employing a polarized filter for separation of the right and left eye images. Accordingly we attached a polarizing filter to an HDTV projector and prepared a polarized filter (glasses) for the test subjects. The stereoscopic images were displayed using two 1/2-inch digital HDTV VCRs in synchrony. A 120-inch diagonal (horizontal 261 cm $\times$ vertical 147 cm) screen was used. Subjects watched the stereoscopic image, which we projected by this system from a viewing distance of 4.5 m.

The images that we displayed were of two kinds. One was a stereoscopic image sequence produced experimentally about 10 years ago called "Waffen". The other was a stereoscopic image recorded in Africa in 1998, entitled

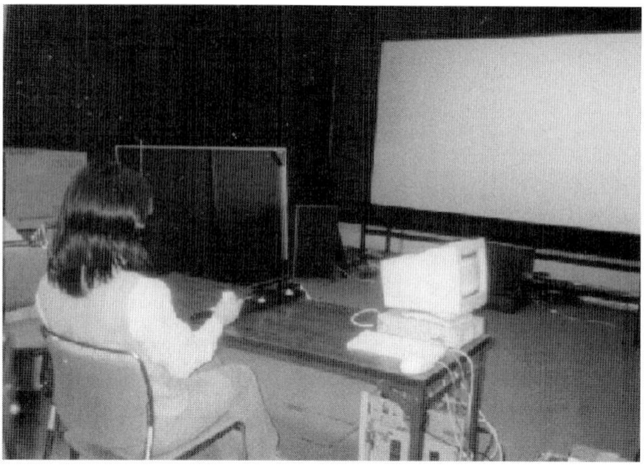

Fig. 18.21. 120-inch screen for 3-D HDTV images

"Africa". The difference between the two lay in the setup of the camera, that recorded the scene.

"Waffen" was recorded using a stereoscopic camera toe-in system. In this method, the distance between the two cameras and the convergence point of the camera were different in each scene. For "Africa", the two cameras were always set parallel to each other with a distance of 65 mm between the two, and the scene was recorded by the parallel method. Under these conditions, a zoom lens was also used for many shots as compared with when using the stereoscopic camera toe-in system. In the former setup, we can display an arbitrary position in the depth direction on the screen by setting the parallax at zero. With the second recording method, an infinite point in a recording becomes an infinite point in the presentation when the camera setup is parallel.

Experimental Method. Two-dimensional images were displayed by supplying a left eye picture from the stereoscopic pair to the two projectors. The subjects wore polarized filter glasses when viewing a two-dimensional image and the stereoscopic image. "Waffen" was displayed three times for 15 minutes each time with a break between presentations of 5 minutes. Accordingly, the total presentation time was 55 min. "Africa" was displayed 5 times for 10 minutes each time with a break between presentations of 5 minutes, giving the total presentation time of 70 minutes.

In the experiment, we measured the accommodation response in the before and after conditions for viewing the stereoscopic and two-dimensional images. The accommodation response was measured using the AA-2000 eye-mark (Nidekku Co.). Subjective estimation was carried out for the degree of visual fatigue with a five-grade evaluation scale in addition to the above when the presentation of each image was finished. The evaluation scale was: 5, my eyes are not tired; 4, my eyes are almost fresh; 3, I cannot say if my eyes are tired or not; 2, my eyes are a little tired; 1, my eyes are tired.

In addition, we evaluated the "viewing comfort" of the stereoscopic image by the single-stimulation continuous-evaluation method while displaying the stereoscopic sequences. In the case of the subjective estimation, the evaluation value of the subject's response was measured via the voltage change of a potentiometer, converted by an AD converter, and recorded on a PC. Five subjects were used.

18.5.3 Experimental Results and Discussion on the Change of the Accommodation Response

Figures 18.22a, b show example results for the accommodation response. Figure 18.22a shows a waveform of the accommodation response of two subjects before viewing the "Waffen" image. This figure shows the accommodation response of the four repeats. Figure 18.22b shows the response waveform after viewing the stereoscopic image sequence, "Waffen", for 55 min. The pulse waveform in the figure indicates the movement of the inner target when measuring the accommodation response. The target is located at $-0.16\,\mathrm{D}$ (Diopter, hereafter "D") as a criterion and, for ten seconds, moves in the position of $-5.10\,\mathrm{D}$ with a step response. For ten seconds, it then stands still, and then moves at the position of $-0.16\,\mathrm{D}$ again. Figure 18.22 shows the accommodation response of the two subjects who showed the most extreme results, in other words, the two subjects showing the biggest and the smallest change of accommodation response before and after viewing the stereoscopic images.

Figures 18.23a, b show the result when the subjects watched the stereoscopic picture image sequence "Africa". The results for the same two subjects were similar to those for the results of "Waffen".

The change of waveform of the accommodation response in Figs. 18.22b and 18.23b is related to the estimation of visual fatigue [7]. If the change of accommodation response is considered as an index of visual fatigue, these results suggest that one subject may perceive visual fatigue while another does not, because the results for one subject in Figs. 18.22b and 18.23b show visual fatigue, while the others in Figs. 18.22b and 18.23b for another subject do not. In other words, when viewers watched stereoscopic images for about one hour, visual fatigue was occasionally found.

Figures 18.24a, b show the results for five subjects regarding the change of the accommodation response in the cases of watching stereoscopic and two-dimensional images of "Waffen" and "Africa". The subjects' visual fatigue was evaluated from the change of the accommodation response. However, no standard analysis technique for evaluating the visual fatigue from the accommodation response has yet been established. To get the correct result from the waveform of the accommodation response at the point of visual fatigue, it is necessary to examine in detail what kind of physical parameter of

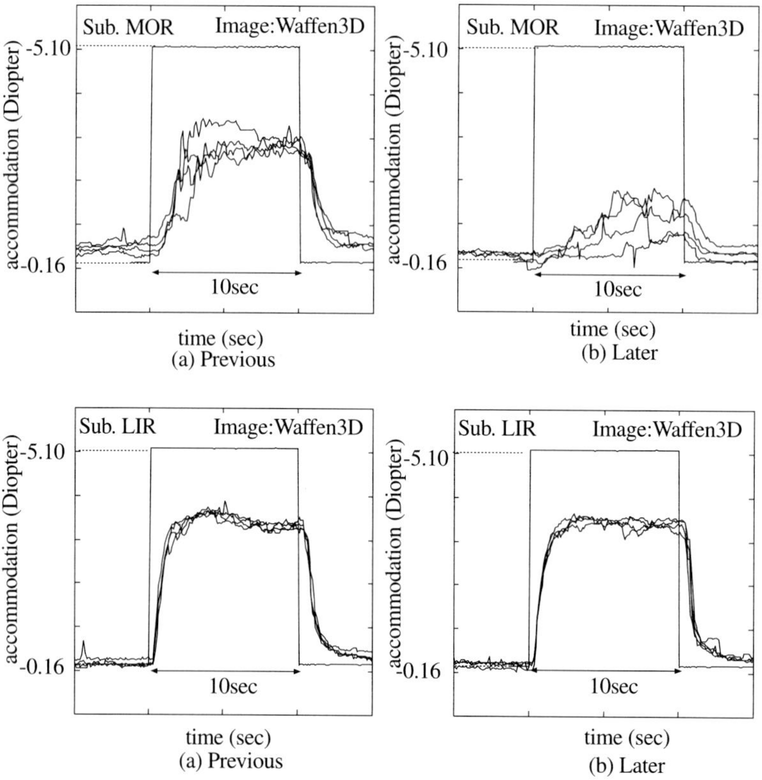

Fig. 18.22. Accommodation response: before and after viewing "Waffen"

the waveform of accommodation response should be analyzed. In this paper, the change of amplitude of the accommodation response was measured to evaluate visual fatigue.

Figure 18.24a shows the change of amplitude of the accommodation response before and after viewing the stereoscopic HDTV image, "Waffen". The white circles in the figure show the amplitude of the accommodation response before viewing the stereoscopic image, and the black circles show the amplitude after viewing the image. For the same subjects, the results of two repeated experiments are shown. Figure 18.24b shows the results in the case of the two-dimensional HDTV image, "Waffen". As in Figs. 18.24a, b, the white circles indicate the results before viewing, and the black circles those after viewing.

Figures 18.25a,b are the results for the stereoscopic HDTV and two-dimensional HDTV image, "Africa", using the same method as in Figs. 18.24a,b.

An asterisk (*) beside a black circle in these figures indicates that the change of amplitude is bigger than 0.5 D after viewing the images. We could

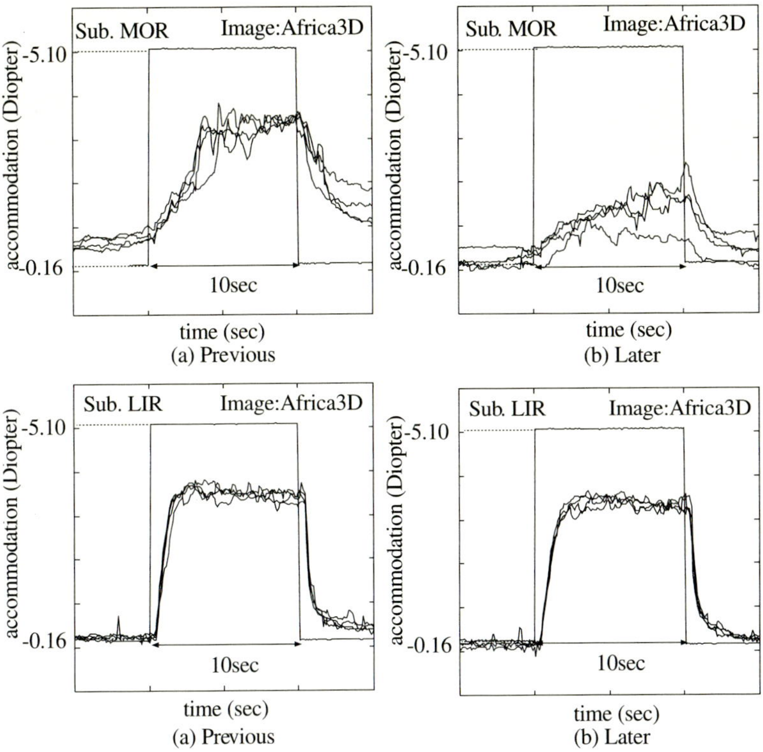

Fig. 18.23. Accommodation response: before and after viewing "Africa"

not find any change of amplitude of the accommodation response exceeding 0.5 D before and after viewing in the case of viewing of two-dimensional HDTV images. However, the responses of four out of five subjects deteriorated by more than 0.5 D for the stereoscopic HDTV image, "Waffen", at the point of accommodation response. In addition, three of the subjects deteriorated more than 0.5 D for the stereoscopic images, "Africa". The results of 0.5 D deterioration as shown in Figs. 18.24a and 18.25a occurred only one of the two times for one subject in each result.

In these results, when viewers watched stereoscopic HDTV images for about one hour, some felt visual fatigue, and the degree of fatigue depended on the display and pickup system of the images.

The subjective estimation of visual fatigue was 2.35 for the stereoscopic image, "Waffen", and 3 for the two-dimensional image, "Waffen", as a simple average of the five subjects. Similarly, the value was 2.5 for the stereoscopic image, "Africa", and 3.3 for the two-dimensional image, "Africa". These results of the subjective test resemble the tendency in the change of the accommodation response.

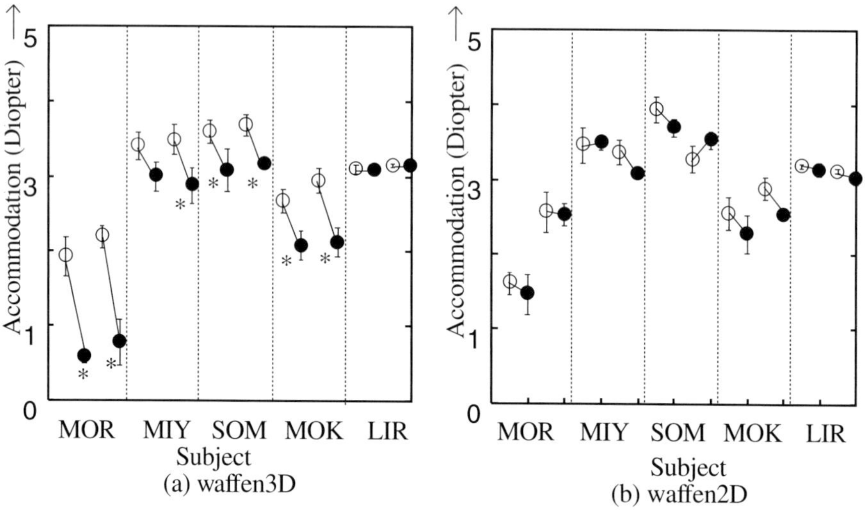

Fig. 18.24. Change of accommodation response: before and after for 3-D HDTV/HDTV "Waffen"

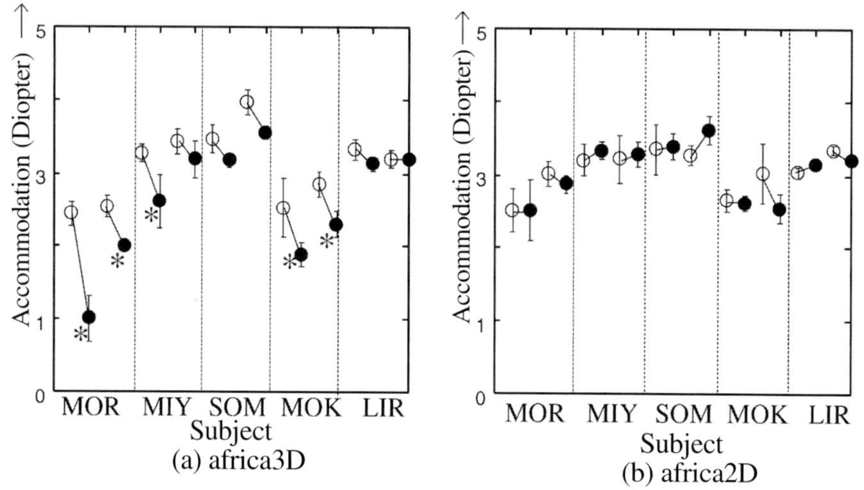

Fig. 18.25. Change of accommodation response: before and after for 3-D HDTV/HDTV "Africa"

18.5.4 Results of Subjective Estimation Experiments

The visual fatigue caused by viewing stereoscopic images was examined by the change of the accommodation response before and after watching stereoscopic and two-dimensional images. However, we cannot establish which image scenes affected viewers in terms of visual discomfort from only evaluating the accommodation response before and after watching test the images.

In this experiment, we required subjects to evaluate test images using a single stimulus quality evaluation (hereafter "SSCQE") regarding visual comfort and discomfort [8]. It is not clear whether scenes which cause visual discomfort are directly related to the causes of visual fatigue. However, it is necessary to grasp the degree of visual comfort and discomfort in the case of watching two-dimensional and stereoscopic images, to enable the production of stereoscopic images that do not cause visual discomfort. Also, this experimental condition, which requires viewers to continuously evaluate the image while watching test images, demands that the subjects watch the images carefully throughout the experiment.

In recent years, the same evaluation technique has been used for estimating digital encoding images. Studies on estimation and analysis methods are currently in progress for evaluating picture quality for digital encoding images. In this experiment, we evaluated the visual comfort and discomfort of two-dimensional and stereoscopic images instead of evaluating image degradation using the conventional method, and calculated the simple averages of the results for all subjects individually.

Each subjectivly evaluated value is sampled every 0.5 seconds. That is, the evaluated point for "Waffen" is composed of 1800 points, because it lasts for 15 minutes. The total data for "Waffen" is composed of five subjects' results, namely two trials with three showings in each trial. Accordingly, the results are the simple average of thirty evaluations.

The total data of "Africa" is composed of the same subjects' results, namely two trials of five showings each. In "Africa", there are 1200 points in the one trial for the 10-minute program. The results are calculated as a simple average of fifty evaluations.

Figures 18.26 and 18.27 show the results of subjective evaluation using SSCQE. Figure 18.26 shows an evaluation result using SSCQE for the stereoscopic and two-dimensional images, "Waffen". The horizontal axis shows the display time, where the total display time is 900 s. The vertical axis shows the subjective estimation value. The vertical dashed line shows a change in the timing of the scene. "Waffen" is composed of 57 scenes. A high-value curve shows the results for two-dimensional images, and a low-value curve those for stereoscopic images (Fig. 18.26). Figure 18.27, for "Africa", shows the results of SSCQE by a similar method to that used in Figure 18.26. There are 36 cuts in "Africa". In Fig. 18.27, the evaluation value of two-dimensional images is also higher than that of stereo picture images, based on the average value.

We carried out a qualitative analysis of these results in Figs. 18.26 and 18.27. In Fig. 18.26, the scenes having low scores are scenes 15, 22, 26–33, 42–48, and 57–58 in the case of stereoscopic images of "Waffen". The double-headed arrows in Fig. 18.26 show these scenes. The results suggest that the

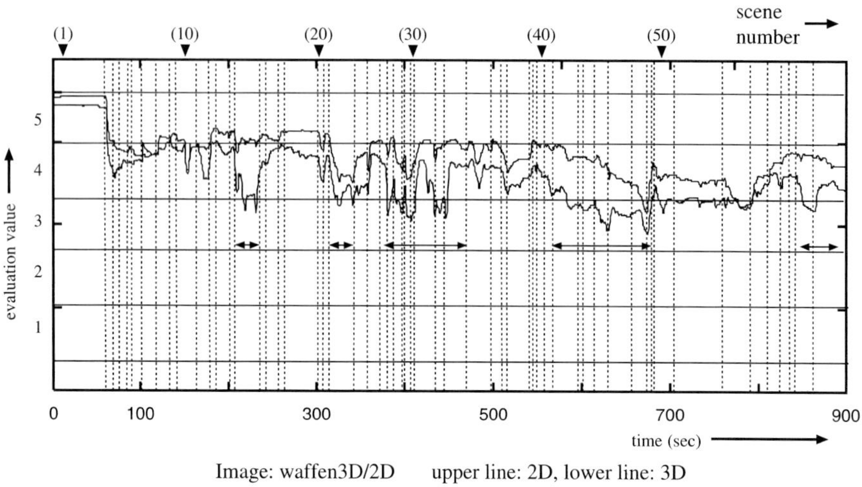

Fig. 18.26. Results of SSCQE for 3-D HDTV/HDTV "Waffen"

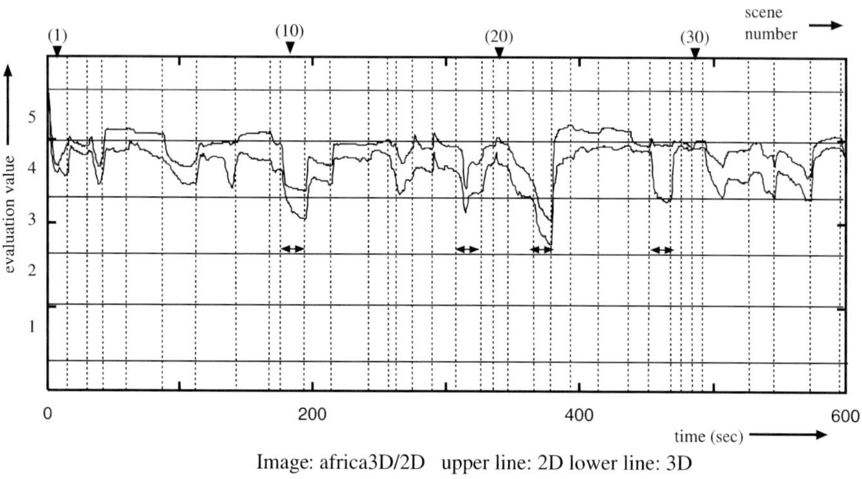

Fig. 18.27. Results of SSCQE for 3-D HDTV/HDTV "Africa"

scenes causing visual discomfort are composed of a main object located in the front of the screen that moves rapidly.

Figure 18.27 shows the result in the case of "Africa". In Fig. 18.27, scene 10 in the stereoscopic images of "Africa" had some vibration caused when shooting the original scene due to an unstable camera setting. As a result, the subjects judged the scenes as causing visual discomfort. The low evaluations for scenes 18, 22, and 27 of "Africa" are thought to be due to the large parallax and high speed of movement in each scene.

It was shown that the limit of binocular fusion is reduced if the target moves fast and that if there is a gazed-at-point besides the moving objects in the stereoscopic image, the limit of binocular fusion is further reduced. Reference to the human visual function suggested that the subjects perceive double images for moving objects in these stereoscopic images, and therefore give these scenes a low score.

18.5.5 Relationship Between Viewing Distance and Depth of Focus

This experiment was conducted under a viewing distance of 4.5 m. If the depth of focus of the human visual system ranges from -0.2 D to $+0.2$ D, as shown by previous research, then Fig. 18.28 shows the range, which is calculated using the depth of focus in the case of viewing distances of 1 m and 4.5 m, respectively. In these areas, the accommodation function and convergence eye movement have a linear relationship with the gazed-at-object in the depth direction. The depth of focus ranges from 125 cm to 83.3 cm at the viewing distance of 1 m, and from 50 m to 2.38 m at the viewing distance of 4.5 m, according to arithmetical computation. Outside of these areas, conflict is caused by the difference between the point of accommodation and the gazed-at-point. From these computational results, it is thought that the accommodation function and convergence eye movement do not generally conflict if the objects in the test stereoscopic images are displayed between a near point of 2.38 m and a far point of 50 m for the projection screen under the conditions of this experiment [9].

As far as the stereoscopic image of "Waffen" is concerned, there are scenes which jump out of screen, but such scenes are few. For "Africa", even fewer scenes jump out of screen. It is thought that almost all scenes in "Waffen" and "Africa" were displayed within the depth of focus. However, the accommodation response and subjective estimation results show visual fatigue among some subjects. Accordingly, visual fatigue is not caused only by the conflict of accommodation response and convergence eye movement in this experiment.

On the other hand, the "keystone distortion" was not zero in the experimental equipment because two HDTV projectors set in parallel were used, as shown in Fig. 1. Moreover, the electrical properties of the two HDTV projectors must be taken into account when considering visual fatigue. These problems concerning experimental equipment will have to be examined in the future. However, we assume that the image display method and electric properties in this experiment did not affect the evaluation of visual fatigue.

18.6 Conclusions

This chapter has outlined the evaluation of stereoscopic images based on binocular fusion. We have described the basic visual characteristics, called

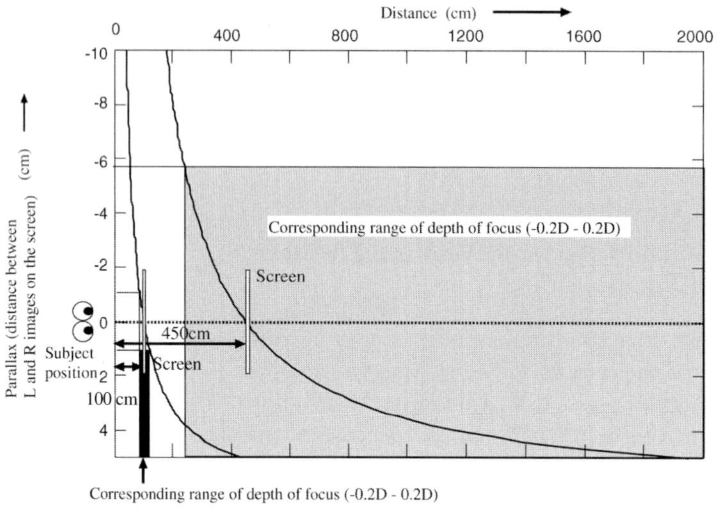

Fig. 18.28. Relationship between corresponded range to depth of focus and viewing distance

Donders' line, for viewing in the depth direction. We have also pointed out other study topics for stereoscopic TV images. Regarding these topics, we have shown the limit of binocular fusion in the spatial-frequency and temporal-frequency domains, and regarding display size. We have also shown the experimental results for visual comfort and visual fatigue for stereoscopic images based on these visual perceptions.

We measured the degree of visual fatigue from the change of the accommodation response before and after viewing images for about one hour, and evaluated these using a subjective test. We identified scenes causing visual discomfort in the test programs using the single-stimulus continuous-quality method for the viewed images.

The mechanism between the convergence eye movement and the accommodation function in the depth of focus appears to affect visual fatigue. This is in addition to conflict between the convergence eye movement and the accommodation function. The relationship between the results of SSCQE and the featured characteristics of test stereoscopic images were also examined. A local low subjective evaluation appeared for both large parallax and great motion in the test stereoscopic images. These results will be used for the study of visually comfortable stereoscopic images and investigation of the causes of visual fatigue when viewing stereoscopic images.

These results were considered as essential problems regarding the characteristics of binocular fusion. We found that visual fatigue was caused by watching stereoscopic images at an absolute viewing distance of 4.5 m. This means that we have to investigate not only a static function of convergence eye movement and accommodation, but also a dynamic function in the case of stereoscopic images.

The study of stereoscopic images has a long history. However, it is clear that stereoscopic systems are not yet accepted in the image utilization market. For practical use, the development of a stereoscopic display is one of the most important problems, and we must also identify the fundamental requirements for stereoscopic images based on the human visual system.

References

1. I.M. Borish: *Clinical Refraction* (Chicago Professional Press 1970) pp. 875–879
2. S. Yano: 'Spatio-Temporal Frequency Characteristics of Binocular Vision for Stereoscopic Images', Proc. TAO (Tokyo Japan, Dec 6 1993) ed. by J. Tsujiuchi, J. Hamasaki, M. Wada (Telecommunications Advancement Organization of Japan) pp. S-2-31–S-2-3-11
3. S. Yano, Tetsuo Mitsuhashi: Electronics and Communications in Japan **76**, Part-3 (1993) pp. 99–111
4. N. Hiruma, Tadahiko Fukuda: SMPTE J. **102**, 1137 (1993)
5. D.A. Atchison, G. Smith: *Optics of the Human Eye* (Butterworth Heinemann 2000) p. 24
6. S. Yano, S. Ide, H. Thwaites: J. ITEJ **55**, 711 (2001) [Japanese]
7. F. Okuyama: Proc. IDW'97 (Nagoya Japan, Nov 19–21) 839 (1997)
8. W. IJsselsteijn, H. de Ridder, R. Hamberg, D. Bouwhuis, J. Freeman: Displays **18**, 207 (1998)
9. W. Blohm, I.P. Beldie, K. Schenke, K. Fazel, S. Pastoor: J. SID **5**, 307 (1997)

19 A Display System for Natural Viewing of 3-D Images

T. Honda, Y. Kajiki, K. Susami, T. Hamaguchi, T. Endo, T. Hatada, and T. Fujii

19.1 Introduction

The human eye can usually accommodate to an object when the viewing distance is less than about 2 m. This means that a three-dimensional (3-D) image display that produces only binocular disparity, i.e. a stereoscopic display, is inadequate for natural viewing.

A new concept called the "super-multiview condition" has been proposed for displaying 3-D images [1, 2]. In a 3-D display system that satisfies this condition (light from at least two images from slightly different viewpoints enter the pupil). We have developed a 3-D display system (image capture, image processing, and display) that satisfies this condition.

In the next section, we explain the super-multiview concept and describe the characteristics of a system that satisfies the condition. In Sect. 19.3 we present the display method and system we developed to satisfy the condition: the focused light-source array (FLA) method and the cylindrical display system. In Sect. 19.4 we discuss the driving of the display devices, the positioning of the video cameras, and the development of the signal-processing equipment used to generate the video signals of the intermediate images. In Sect. 19.5 we discuss our evaluation of the functional response of the human eye when observing the super-multiview display. We conclude, in Sect. 19.6, with a few remarks.

19.2 Super-Multiview Concept

19.2.1 Conventional 3-D Display Systems

Various types of 3-D video display systems have been introduced for entertainment and communications. Although conventional 3-D display systems provide significant visual impact, viewers must tolerate certain constraints. Some systems require the viewer to wear 3-D glasses. Others require the viewer to wear a head-mounted display. Systems referred to as "autostereoscopic" do not require the viewer to wear anything, but the viewer's gaze must be fixed on a narrow viewing area on the display. An example of such a system is the combination of a parallax barrier and a liquid-crystal panel. Some systems do not impose a constraint on the viewer. Instead, they detect the eye position of the viewer and change the viewing area accordingly; however, such systems cannot track the different motions of multiple viewers.

In addition to these constraints, viewers are usually shown unnatural 3-D images. Systems that display images taken by a stereoscopic-camera do not enable the viewer to see the scene from a different perspective when the viewer moves side to side or forward and backward. To eliminate this unnaturalness, some systems detect the viewer's eye position and re-render the images to match the eye position; however, such systems still cannot track the different motions of multiple viewers. Moreover, although objects appear to project from the display screen, the stereo-pair images are actually displayed on the screen, so the stimulus for eye accommodation stays on the screen. In the real world, the stimulus is on the object itself, so the images shown by conventional 3-D display systems are physiologically unnatural.

To implement a 3-D video system on a 3-D-TV system and to bring 3-D TV into our daily lives, the constraints on viewers and the unnaturalness of images must be reduced or eliminated.

19.2.2 Multiview Stereo Display System

Conventional 3-D display systems produce two images, one for each eye, but some systems capable of producing more than two views have been developed. The former are classified as binocular stereoscopic display systems, and the latter are classified as multiview stereoscopic display systems, or "multiview systems" for short.

An example configuration of a multiview display system is shown in Fig. 19.1. Five projectors produce five different views at one time, and each view can be seen only through a particular virtual window. Viewer 1 sees the view of projector 1 through virtual window 1 with the left eye and sees the view of projector 2 through virtual window 2 with the right eye, without using 3-D glasses. Consequently, the viewer perceives the image depth from the binocular disparity between the views. At the same time, viewer 2 sees a 3-D image from the views of projectors 4 and 5. If viewer 1 moves to the position of viewer 2, viewer 1 sees the scene from the perspective of the new position. Therefore, with a multiview system, multiple viewers can see 3-D images of the same scene simultaneously and the scene is viewed from perspectives corresponding to the positions of the viewers.

The multiview system thus eliminates the constraints described above almost completely. However, if the viewer moves sideways within the virtual window, the perspective will not change. If the viewer's eye moves sideways across a boundary between virtual windows, the perspective will change discretely at the boundary. Therefore, the viewer may perceive some discreteness of views. This phenomenon is usually called "flipping" and produces a somewhat unnatural image. Moreover, the unnaturalness of the accommodative stimulus, is inherent in conventional multiview stereoscopic systems because the viewer sees a stereo-pair image on the screen and nothing more.

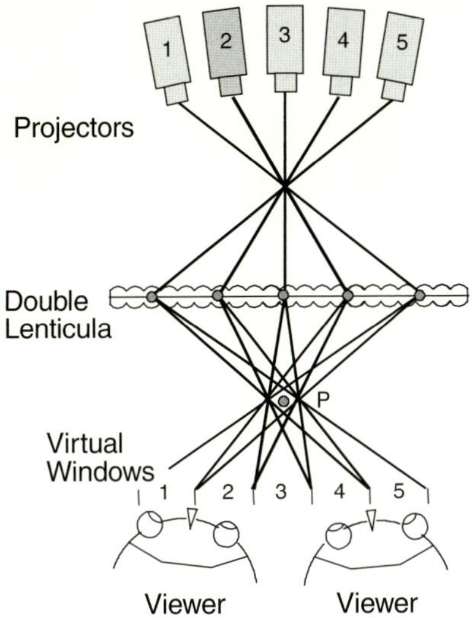

Fig. 19.1. Example configuration of a multiview display system

19.2.3 Super-Multiview Region of Stereogram

In a multiview system, the continuous parallax of the original scene is sampled, so the parallax of the produced images is discrete. If the sampling interval is narrower than the diameter of the viewer's pupil, multiple views will always pass through the pupil, as shown in Fig. 19.2.

The views are projected by the eye lens on the retina, where they overlap. The pupil masks the brightness of each view. When the eye moves a little,

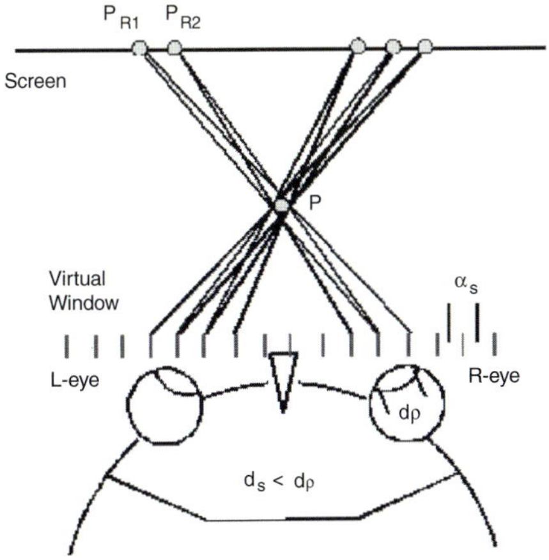

Fig. 19.2. Schematic of super multiview region

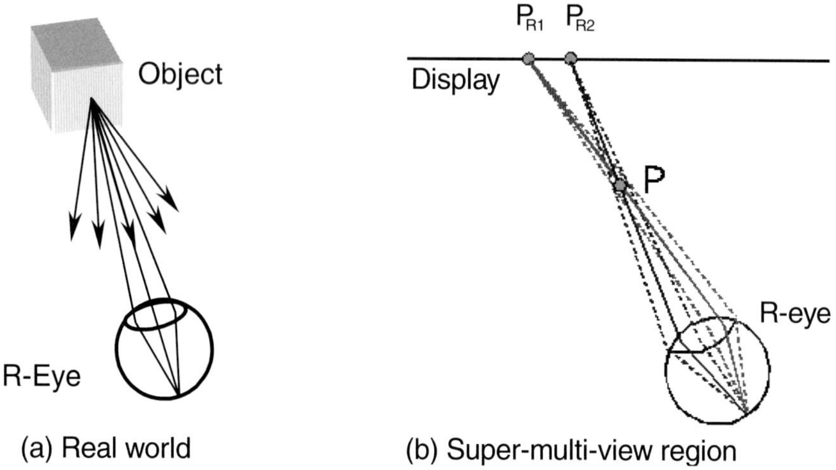

Fig. 19.3. Stimulus for eye accommodation

the brightness ratio of the views on the retina changes a little. When the eye moves a little more, another view comes into view, and the original view disappears. In other words, if the viewer moves sideways, the perspective of the scene changes smoothly as discrete views fade in and out. As a result, the viewer may barely perceive parallax discretely. We call the region where the parallax-sampling interval is narrower than the pupil-diameter the "super multiview region of stereogram."

In the real world, illumination light is usually reflected and diffuses over an object's surface. The diffused light becomes the stimulus for eye accommodation, as shown in Fig. 19.3a. For a multiview stereogram, the light diffused on an object's surface is sampled discretely by multiple views. However, more than at least two views always pass through the pupil of each eye in the case of the super-multiview region. That means that if a viewer focuses on the object's surface, multiple views passing through the pupils are projected on the same position on the retina, as shown in Fig. 19.3b. This suggests that the stimulus for accommodation can be reproduced so as to appear natural to human eyes by using the super-multiview region of a stereogram.

We thus propose using the super-multiview region of a stereogram to reduce or eliminate the constraints on viewers when viewing 3-D images and the unnaturalness of the images.

19.3 Satisfying the Super-Multiview Condition

19.3.1 Focused Light-Source Array Display Method

Background. As described above, our goal is to develop a 3-D image display system that eliminates the constraints placed on viewers by conventional systems and that presents images so that they appear natural. We intend to do this by using the "super-multiview region" of a multiview stereogram. To achieve the conditions necessary for the super-multiview region, we must produce multiple views with a parallax interval narrower than the diameter of the viewer's pupils. The diameters of a person's pupils range from about 3 to 8 mm. For example, if the diameter is 5 mm and the viewer can move over a range 200 mm wide, we would have to produce 40 or more views because the parallax interval is 5 mm or less.

While several types of multiview display systems have been developed, it is difficult to produce the number of views required for a narrow parallax interval. A common approach is to combine a lenticular sheet with a liquid-crystal panel. This technique is useful for binocular stereoscopic systems, but it is difficult to increase the number of views by using this technique. The total number of pixels in a liquid-crystal panel is limited; so if tens of views are displayed on the panel at once, the resolution of each view decreases accordingly. To overcome this problem, some display systems use multiple projectors instead of a single panel. This technique is useful for displaying multiple views without decreasing the resolution of each one, but it is difficult to display dozens or hundreds of views with a narrow parallax interval by using this technique because the projectors are not so small.

Display Technique. In 1995, our group proposed a technique for realizing a super-multiview display [1]. In this technique, the beam of a semiconductor light source is focused on a far point by beam-shaping optics, as shown in Fig. 19.14a. Next, multiple light sources are placed in an arc and their beams are focused on the same point, as shown in Fig. 19.4b. When a viewer looks at the focal point from direction A, the brightness of the focal point depends only on the intensity of light source A′; it is independent of that of the other sources. When a viewer looks at the point from direction B, the brightness depends only on the intensity of light source B′. Because the intensity of each source can be controlled independently, the focal point can be regarded as a multiview pixel. Conventional semiconductor light sources are very small, and so dozens, or even hundreds, of light sources can be positioned to have a narrow angular interval, giving the "pixel" dozens, or hundreds, of views with a narrow parallax interval. We call this array of light beams a "focused light-source array" (FLA).

The focal point of the FLA is scanned rapidly, as shown in Fig. 19.5. If the intensity of each light source is modulated by the image signal of each view, multiview images can be produced by a multiview pixel with raster scanning

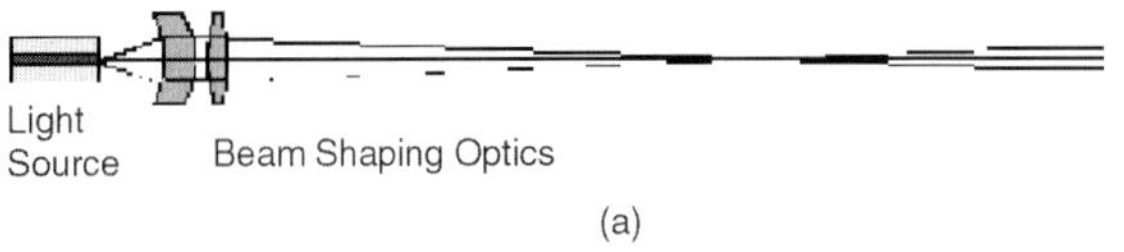

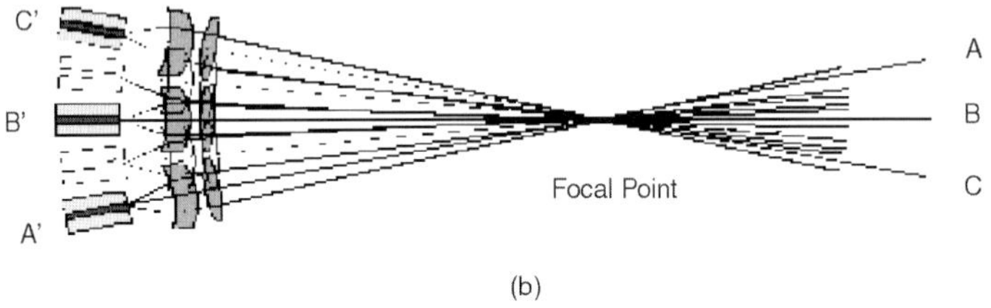

Fig. 19.4. Schematic diagram of focused light-source array: (**a**) beam from semiconductor light source is focused on a far point by beam-shaping optics; (**b**) multiple beams are arranged in an arc and focused on the same far point

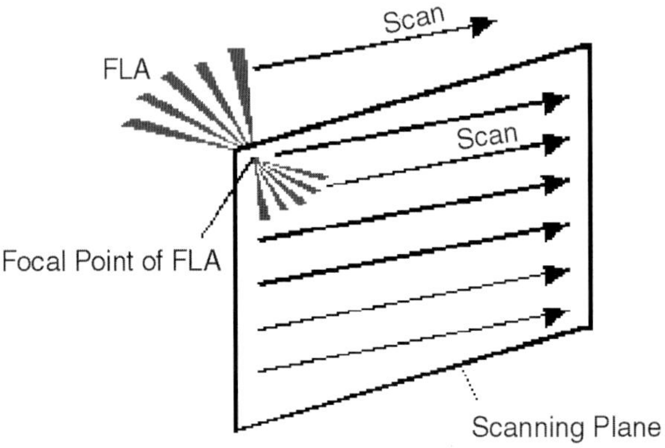

Fig. 19.5. Schematic diagram of FLA with two-dimensional raster scanning

and intensity modulation. This process is similar to that of a cathode-ray tube, but the image has multiple views in our arrangement. If the pixel has dozens of views at narrow parallax intervals, then the multiview images also have dozens of views at narrow parallax intervals. Moreover, the intensities of conventional semiconductor light sources can easily be modulated at hundreds of megahertz, which is sufficiently fast to draw high-resolution images with a high refresh rate. Therefore, this arrangement can be used to produce numerous perspective views with narrow parallax intervals at high resolution with a high refresh rate.

In practice, an FLA cannot be scanned as rapidly as the electron beam of a CRT because of inertia. Therefore, optical projection with scanning is used to implement the FLA technique, as illustrated in Fig. 19.6. An objective

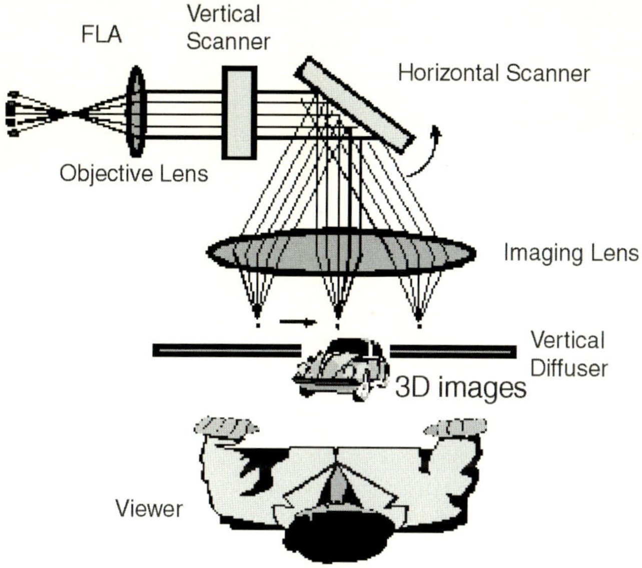

Fig. 19.6. Schematic diagram of the optics used

lens and an imaging lens project the focal point of the FLA. Horizontal and vertical mirror scanners between the lenses scan the image of the focal point horizontally and vertically.

If the light sources of the FLA are placed two-dimensionally on a spherical surface, the images produced have full parallax; however, if they are placed horizontally on an arc, the images have only horizontal parallax. In practice, horizontal-parallax-only (HPO) images are often used for technical and economic reasons. When they are, a vertical diffuser is used to widen the vertical viewing zone by diffusion.

Experimental Display. Our group has been developing experimental display systems by using the FLA technique [2, 3]. As shown in Table 19.1, our latest display system can produce 45 perspective video images with a parallax interval of 0.5 degrees at a refresh rate of 30 frames per second. A photograph of the system is shown in Fig. 19.7. Because red laser diodes (LDs) are used for the light sources, the images produced are only red (monochrome).

Two photographs of 3-D images produced by the experimental display are shown in Fig. 19.8. Several viewers could see the same 3-D image at the same time from any position in the viewing zone without wearing special glasses. The images had strong depth cues produced by natural binocular disparity. When a viewer moved to another position in the viewing zone, the viewer saw the same scene, but from the perspective of the new position. When a viewer moved around at about 2 m from the image, the viewer perceived a boundary of views because the brightness of the image changed at the boundaries, i.e.

Table 19.1. Specifications of latest experimental display system

FLA sources (Wavelength)	LDs (630 nm)
Number of Views	45 (Horizontal Parallax Only)
Angular Parallax Interval	0.5 degrees
Visual Field	185 mm (W) × 125 mm (H)
Angular Viewing Zone	22.5 degrees
Number of Pixels and Lines	400 pixels (V) × 400 lines (H)
Refresh Rate	30 frames/sec
Dimensions of Display	500 mm (W) × 530 mm (H) × 750 mm (D)

Fig. 19.7. Photograph of latest experimental display system

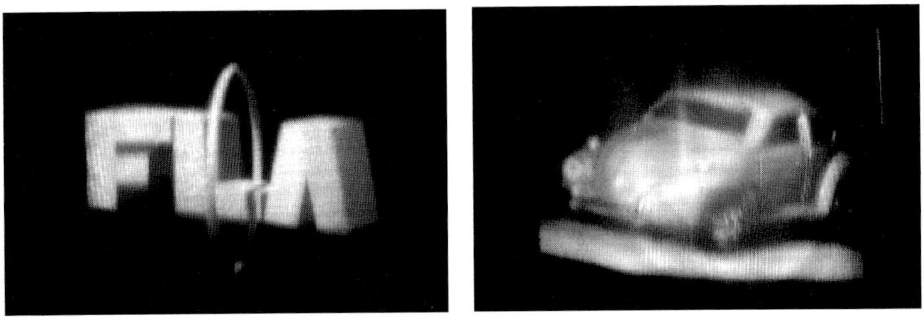

(a) (b)

Fig. 19.8. Photographs of static 3-D images produced by the experimental display: (**a**) generated by 3-D computer graphics; (**b**) taken of actual object using a TV camera from 45 directions

there was parallax flipping. However, when a viewer moved around at about 300 mm from the image, the viewer barely perceived the boundaries because they were partly smoothed. In addition to static images, moving 3-D images were produced using a VCR stack. The images were refreshed at 30 Hz, and so viewers saw natural 3-D video images that moved as smoothly as those on a conventional television screen.

We are now using the FLA technique to develop a color display with RGB laser-light sources.

Summary. We have described in this section a 3-D display method using a focused light-source array with scanning. This method can be used to display dozens or hundreds of perspective views with an angular interval narrower than 1 degree at high resolution. We used this method to develop an experimental 3-D display system that can produce 45-view video images with an interval of 0.5 degrees. Its resolution and refresh speed are close to those of a conventional television. We captured the images of actual objects with a TV camera, recorded the images onto a VCR stack, and displayed them on our experimental display. The FLA method was designed to realize the super-multiview concept, and it achieved this goal comparatively easily.

19.3.2 Cylindrical Display System [4]

Introduction. While 3-D image display systems are expected to be applied in many areas, a system satisfying the super-multiview condition works most effectively when the display is close to the viewer. This enables the viewer to observe more clearly the details in the image. The merit of a system satisfying the super-multiview condition is that it produces proper stimuli for eye accommodation, and these stimuli are most effective when the display is close to the viewer. An example of such usage is the multiplex hologram.

The multiplex hologram, a cylindrical holographic stereogram, is widely used for art, advertisements, and other applications because several viewers can see a finely detailed 3-D image from any direction, front, rear, or side. However, the image is static. Therefore, we first propose a basic multiview display system that enables several viewers to see a 3-D image from anywhere in a 360-degree horizontal direction. This system is based on the parallax panoramagram and uses a cylindrical parallax barrier and a one-dimensional light-source array that rotates along the inside of the cylindrical parallax barrier. We next improved this system so that the parallax barrier can move. We did this to satisfy the super-multiview condition.

Basic System. The principle of the proposed system is illustrated in Fig. 19.9. We consider a parallax barrier and a minute light source that moves along the parallax barrier while keeping a fixed distance. If the aperture width of the barrier is sufficiently narrow, the light passing through it is a thin flux; its direction is scanned by the movement of the light source. We call this kind of scanning "angular scanning".

When a viewer sees the aperture from a particular direction, the light reaches the viewer's eyes only at the moment when the light source is at a certain position. Therefore, a pixel whose luminosity differs with the viewing

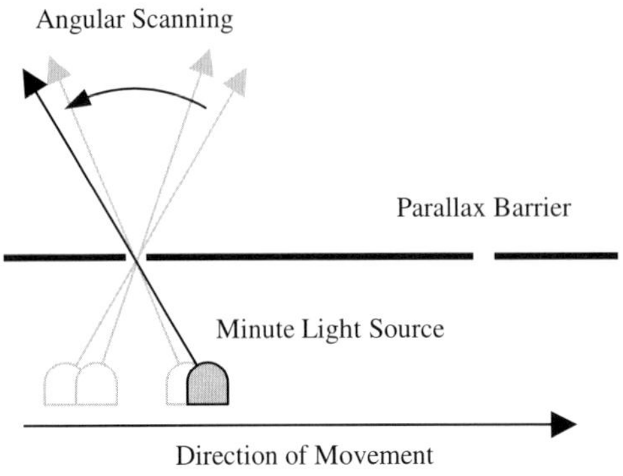

Fig. 19.9. Light passing through the aperture is scanned while moving the light source

direction can be displayed by synchronously changing the intensity of the light source as it moves. Furthermore, one light source can be used to display multiple pixels by successively passing it behind multiple apertures. A cylindrical multiview display can thus be constructed from a cylindrical parallax barrier and a one-dimensional light-source array consisting of semiconductor light sources, such as LEDs, aligned vertically. The array rotates along the inside of the cylindrical parallax barrier, as shown in Fig. 19.10.

In this system, the resolution of the images is determined by the aperture pitch of the parallax barrier, and the parallax number is determined by the ratio of the aperture width to the pitch. Therefore, narrow apertures are needed to produce high-quality images, but narrow apertures cause diffraction, which reduces the parallax number. This problem is not peculiar to this method; it is a fundamental limitation of the parallax panoramagram. However, it can be overcome by refining the scanning technique, and the ability to do this is one of the advantages of this system.

Improved System. We improved our basic system so that not only does the light source move but so also does the parallax barrier. This improves the resolution in spite of the limitation described above. As shown in Fig. 19.11, when the parallax barrier moves in the direction opposite to the movement of the light source, the light passing through the aperture is scanned as if there was a virtual parallax barrier finer than the actual one.

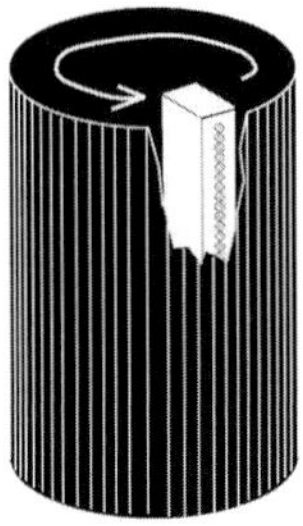

Fig. 19.10. Cylindrical multiview display constructed of a cylindrical parallax barrier and a one-dimensional light-source array. The display rotates along the inside of the barrier with synchronous intensity modulation

The aperture pitch p_ν of the virtual parallax barrier is expressed as

$$p_\nu = \frac{V_1}{V_1 - V_2} p_0 ,$$

where p_0 is the aperture pitch of the actual parallax barrier, $V_1 (> 0)$ is the velocity of the light source, and $V_2 (< 0)$ is the velocity of the parallax barrier.

This equation means that the movement of the parallax barrier reduces the pixel pitch to $V_1/(V_1 - V_2)$ times that of a system with a fixed parallax barrier. The range of the angular scanning and the divergence angle of the light passing through the aperture depend only on the distance from the light source to the actual barrier and on the shape of the barrier. This means that the image size and parallax number are independent of the movement of the barrier. As mentioned above, this increases the resolution independent of the image size and parallax number.

Experimental Display Model. We have constructed an experimental display model using the parallax barrier movement method described above.

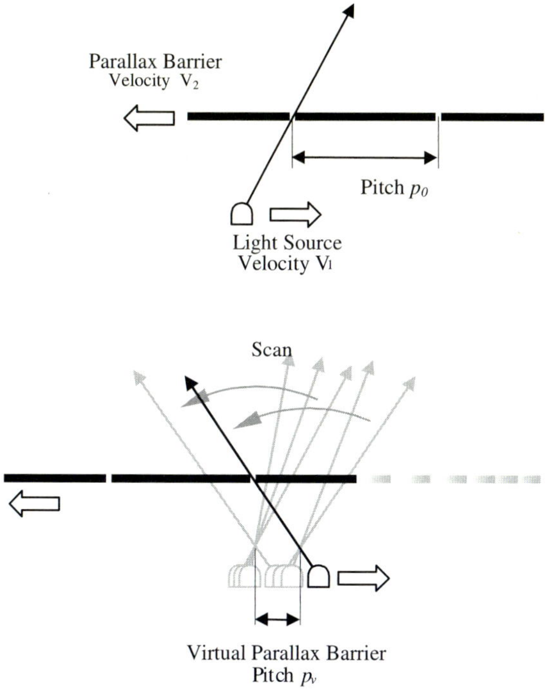

Fig. 19.11. Light source moves to right at velocity, and parallax barrier (aperture interval) moves to left at velocity. The light passing through the apertures is scanned as if there was a virtual parallax barrier with an aperture pitch narrower than that of the real one

Although the refresh frequency is 30 Hz, the rotation rate has been reduced to 56.25 rpm by using 32 LED arrays. The cylindrical parallax barrier has 38 apertures, and its diameter is 230 mm. The barrier rotates at 1800 rpm, which is 32 times the speed of the LED arrays, in the opposite direction; the number of pixels per rotation is $38 \times (1 - (-1800/56.25)) = 1254$. A photograph of this model is shown in Fig. 19.12.

Images produced with the model taken from various directions are shown in Fig. 19.13 and Fig. 19.14. The 3-D images could be naturally seen by both eyes. They had strong depth cues due to natural binocular disparity. When a viewer moved around the display, the viewer saw images corresponding to the viewing position, so that the objects appeared to be floating in air.

Discussion. Given its ability to display an image observable from any direction, we believe that the cylindrical display is suitable for applications that require detailed observation, such as for medical imaging and artwork display. In such applications, viewers would likely move their head actively to utilize the motion parallax. Accordingly, a very narrow parallax interval is needed so that the discontinuity in parallax is not conspicuous.

In our improved system, a very narrow parallax interval is achieved by moving the parallax barrier. We produced 3-D images at a parallax interval of less than 1 degree (which corresponds to a 5-mm interval at a 300-mm distance) with our experimental display model. The parallax interval could be narrowed even more by increasing the number of light-source arrays or by using other techniques. In addition, an accommodation stimulus for the viewer's eyes may be produced by the effect of the monocular parallax in the super-multiview region.

Fig. 19.12. Experimental Display Model

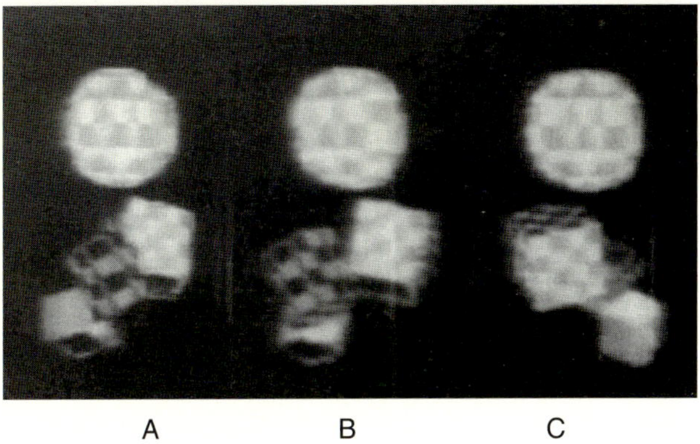

Fig. 19.13. Images produced with experimental display model

Fig. 19.14. Photographs shown in Fig. 19.13 were taken from directions A, B and C

Summary. To realize a multiview 3-D display that is observable from any direction, we proposed a technique in which a one-dimensional light source array is rotated inside a cylindrical parallax barrier. This technique is based on the parallax panoramagram, which was improved to display in all horizontal directions by introducing mechanical scanning. The parallax panoramagram imposes limitations on the resolution of images and the parallax number, which are caused by the effect of diffraction at its apertures. To overcome these limitations, we proposed the parallax barrier movement method and developed an experimental display model. The 3-D images produced with this model are observable from 360-degrees with smooth motion parallax.

19.4 Video Image Capture and Processing for the 3-D Display

19.4.1 View Interpolation Technology

Summary. Although research on view interpolation technology has been undertaken at numerous research institutes, the majority of this research has used a method whereby the three-dimensional shape, surface texture, and other physical data of an object are faithfully reproduced, after which images are interpolated at intermediate viewpoints by CG (Computer Graphics) techniques. In this research, emphasis is placed on the accurate reproduction of data of actual space rather than on high-speed processing and stability.

In the case of view interpolation technology for FLA and other super-multiview displays, interpolation images must be generated with high stability at all times, regardless of the shape or arrangement of the viewed objects or the type of light source and so forth. In addition, this technology must be able to rapidly generate interpolation images at the video rate. At present, block matching and filter interpolation are two promising techniques for satisfying these requirements.

Block-Matching Method. The block-matching method involves determining the parallax between stereo images for each block by using the block-matching technique, converting that parallax to the amount of parallax at the location of the intermediate viewpoint to be generated, and then using that amount of parallax to generate an interpolation image.

This technique is characterized by its simple algorithm and accurate determination of parallax for each block. However, it requires a large processing volume since an entire search must be performed, and it is unable to determine parallax for areas smaller than the blocks. It can be easily applied to multiview images.

Filter Interpolation Method. A characteristic pattern appears when multiview images are layered on top of each other to form a 3-D rectangular parallelepiped followed by cutting it into cross-sections. Filter interpolation is the technique whereby image processing is performed on this pattern to obtain a high resolution followed by the generation of images of intermediate viewpoints. More specifically, this method involves generating intermediate images by increasing the resolution of the ray-space data in the direction of the parallax axis by using a filter. One approach is to generate interpolation images by enhancing the resolution of the epipolar plane images (EPIs). Processing that increases the resolution of these images in the longitudinal direction is equivalent to processing that generates intermediate images.

Although filter interpolation is less susceptible to errors because there are no problems with error accommodation and it lends itself to high-speed processing because it does not require search processing, it generates less-accurate interpolation images compared with those of block matching and is unable to perform processing corresponding to local characteristics since only a single interpolation function is used.

This method is promising in terms of stability and interpolation speed. Furthermore, it will enable processing that includes noise removal and other types of preprocessing and postprocessing.

View Interpolation Processing Using Adaptive Filters. The "adaptive filter interpolation method" combines the characteristics of block matching and filter interpolation. It was developed at Nagoya University and involves preparing multiple filter sets in advance and switching filters corresponding to the local characteristics of the EPIs.

Figure 19.14 shows a block diagram of this technique. First, interpolation filters are prepared so as to be able to reflect the local characteristics of the EPIs. Next, each EPI is up-sampled in the longitudinal direction to obtain the input image. The number of times each EPI is up-sampled is the same as the number of interpolation images needed. Next, the pixels around the one to be interpolated are analyzed. The optimum filter is selected based on the results of this analysis. Convolution processing is then performed on all pixels to be interpolated to obtain an output image.

The results of conventional and adaptive filter interpolation for a computer-generated EPI are shown in Fig. 19.15. With nearest-neighbor interpolation (Fig. 19.16b), a stepwise error in the diagonal line sections is particularly conspicuous. With bilinear interpolation (Fig. 19.16c), although favorable results were obtained when looking at the interpolation image only, the image is obscured by large low-frequency components. With interpolation using

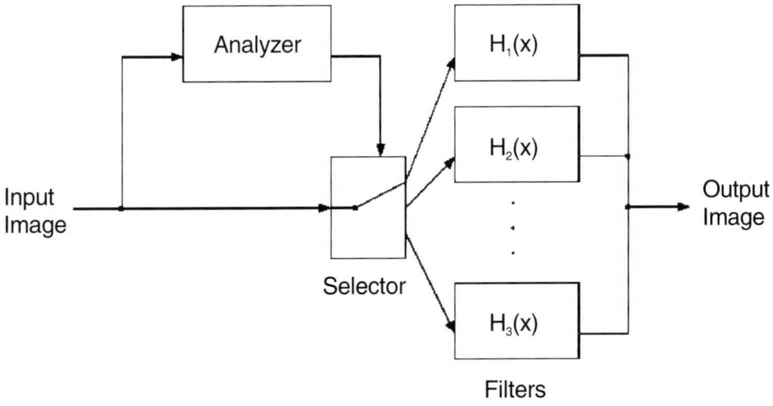

Fig. 19.15. Block diagram of interpolation using adaptive filters

adaptive filters (Fig. 19.16d), the highest quality results were obtained. In addition, the PSNR (Peak Signal to Noise Ratio) between the interpolation and parallax images actually captured at the intermediate viewpoints was 35 dB, which is better than with the other techniques by 1 – 6 dB.

19.4.2 3-D Video Capturing and Data Processing System

(a) Multiview Video Image Capturing

Image Capture Using Multiview Camera System

Multiview Camera System for Super-multiview Displays. The 3-D TV system we are researching displays the total images at viewpoints on a super-multiview display after increasing images by a factor of roughly 5–10 times by capturing the object images with a multiview camera system and generating images at viewpoints between the cameras by interpolation processing.

If a super-multiview display has an observation distance of 0.5 m and a viewing zone angle of 30 degrees, the parallax of the observer's pupils when entering the super-multiview area is about 0.3 degrees. Therefore, a super-multiview display having roughly 100 views is required to produce a 3-D display having a viewing zone angle of 30 degrees.

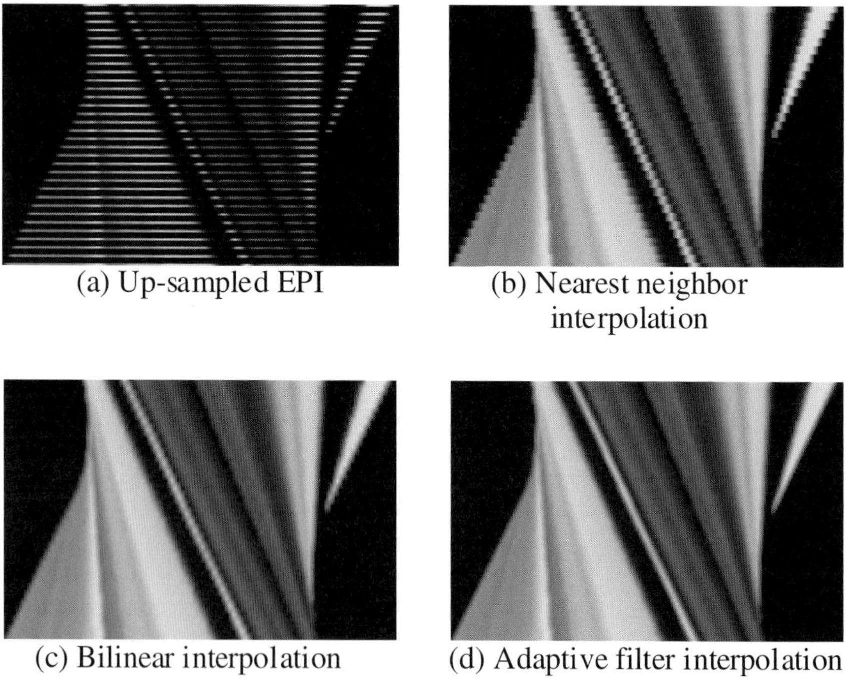

Fig. 19.16. Interpolation computer-generated EPI

In contrast, in a multiview camera system, the distance between the cameras and the object is set at 1 m, and the viewing zone angle is set at 30 degrees, equal to that of the display. Each camera is installed on an X–θ stage, the interval between cameras is variable, and convergence capturing is possible. When a value of 10 is used for the number of interpolation processings for generating the multiview images for a super-multiview display from the images captured by a multiview camera system, the number of cameras required is 10. Conceptual diagrams of the super-multiview display and multiview camera system are shown in Fig. 19.17.

Basic Evaluation System. The system we used to evaluate the 3-D TV system described above consisted of a multiview camera system, a multiview camera image incorporation block, and a multiview image-data processing block.

Although the cameras are arranged in a straight line with respect to the optical axis, it is best to set the camera interval and angles so as to allow both parallel and convergence capturing. We thus used a four-view camera system in which each camera had two degrees of freedom of X–θ. Each camera stage had a linear motor and a pulse motor. The minimum camera interval was set to 90 mm in consideration of the feasibility. We used color CCD cameras using the NTSC format. We used cameras that allowed external synchronization so that interpolation processing could be performed on the same frame for each camera image using the corresponding image frames of the cameras.

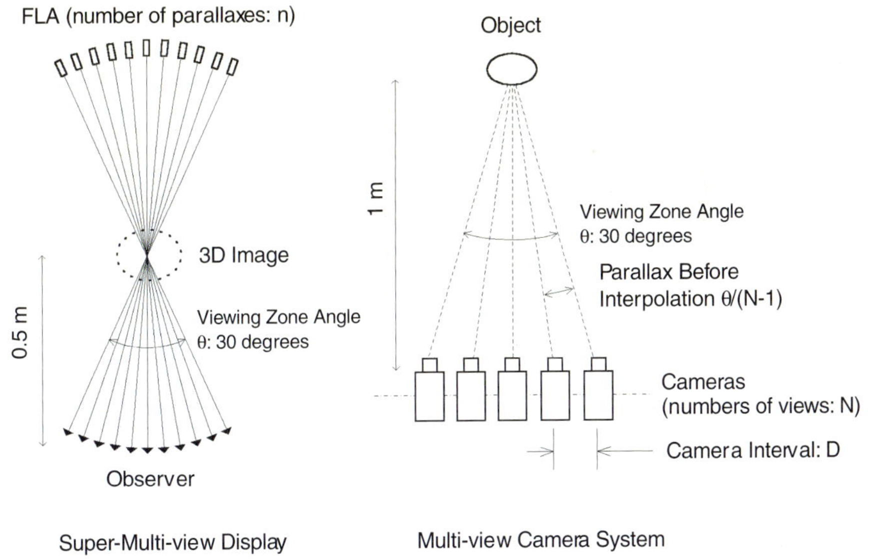

Fig. 19.17. Conceptual diagrams of the super-multiview display and the multiview camera system

Each camera could capture about 20 MB/sec (for VGA, YUV 4:2:2, and 30 fps) of video image data, a tremendous amount of data for a multiview camera system. We thus incorporated a processing block to incorporate the video image data from the cameras on a real-time basis and transfer that data to a data processing block. Although this system is currently compatible with 4 views, it was designed to enable compatibility with up to 20 views.

The data processing block performs a series of processings in real time: calibrating the cameras and interpolating the images. Systems using dedicated hardware and systems using digital signal processor (DSP) have been used for arithmetic processing of 3-D images. Although the former are advantageous in terms of speed, we used a system with a DSP board to facilitate verification of the view interpolation processing algorithm. An overall block diagram of the basic evaluation system is shown in Fig. 19.18.

Calibration of Multiview Camera

Need for Camera Calibration. The individual CCD cameras composing the multiview camera system have unique variations. Even when capturing an image of the same object from the same location and same direction, the images they capture differ. In addition, if the location of each camera and the orientation of each camera's optical axis cannot be determined precisely, views cannot be interpolated accurately. Camera calibration is thus performed to correct for these factors.

Geometric calibration is performed to determine the geometric relationship between three-dimensional space and the CCD plane, and the geometric conversion properties inside the camera. At present, when the coordinates of one point on a world coordinate system are designated as X, and the coordinates of the corresponding point on the image coordinate system of

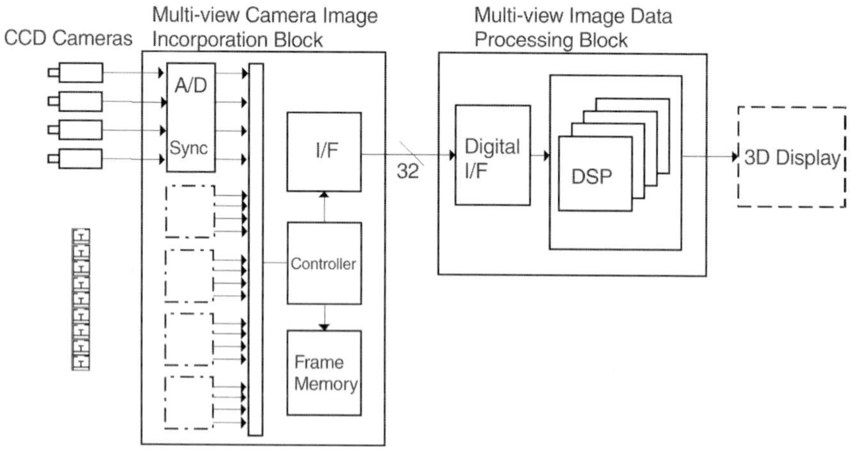

Fig. 19.18. Overall block diagram of the basic evaluation system

a CCD for that point are designated as m, the relationship between m and X is as shown below (s is a constant):

$$s\tilde{m} = P\tilde{X} = A[Rt]\tilde{X} \; . \tag{19.1}$$

Here, $P(= A[Rt])$ is referred to as the perspective projection matrix; it can be represented as the product of an internal parameter and an external parameter. The objective of geometric calibration is to determine the value of P.

Camera calibration is normally performed by capturing the image of a reference object having known geometrical properties, matching its characteristics (world coordinates, characteristic values) and the characteristics of the CCD (image coordinates, characteristic values), and then determining the camera's parameters based on a camera model.

Need for Rectification Processing. The view-interpolation method we are studying has the properties of an EPI in that corresponding points are arranged along a straight line. When multiview images are captured with a multiview camera system in which the optical axes of the cameras are parallel, the epipolar lines are parallel to the x axis, which is advantageous for facilitation of view interpolation. If the baseline is enlarged to widen the viewing zone for all the cameras, a wide-angle lens must be used. The result is inadequate resolution of the object images.

Therefore, we limit the depth of the object to a finite range. Each camera performs convergence capturing, and images are captured so that the object being captured is roughly positioned in the center of the viewing zone of each camera. Next, the image of each camera is converted into an image captured by parallel capturing, and interpolation images are generated that have adequate resolution. The conversion from images captured by convergence capturing to images captured by parallel capturing is referred to as rectification.

We consider correction of individual camera differences, correction of camera absolute positions, correction of optical-axis directions, and rectification processing to be camera calibration in the broad sense.

(b) Multiview Video Image Data Processing

Object Depth and Number of Filter Sets. The number of filter sets used in the adaptive filter interpolation method corresponds to the distribution of parallax vectors between cameras. For an object having large parallax, i.e. one having considerable depth, filter sets must be provided when performing view interpolation processing on that object. The number of sets needed corresponds to the depth of the object. Figure 19.19 shows the relationship between the depth and the required number of sets assuming, for example, a CCD camera in which the pixel pitch is 8.6 µm, the number of horizontal pixels is 640, and the lens focal distance is 25 mm, and using the number of interpolations as the parameter.

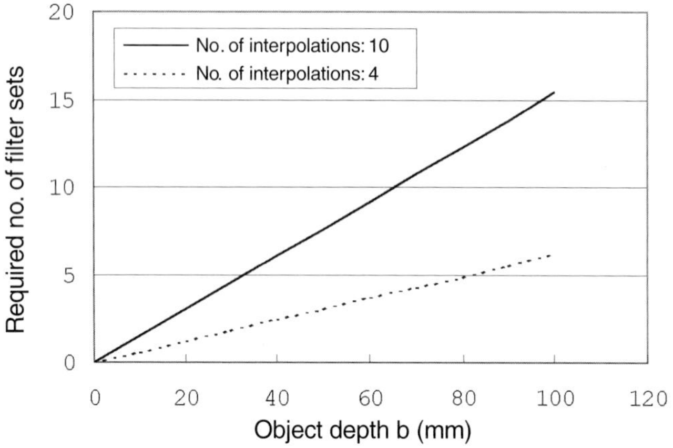

Fig. 19.19. Object depth and required number of filter sets

View Interpolation Processing Evaluation. The images of a reference object is captured in advance for each CCD camera to acquire the characteristic point data needed to calculate and accumulate the camera parameters, rectification matrix, and other data. During this image capturing, the correction and rectification processing, i.e. calibration, are performed in real time. The calibrated data is then subjected to view interpolation processing.

This processing is performed by the multiview image data processing system in the basic evaluation system in Fig. 19.19. We are now conducting experiments, including ones on view interpolation processing, based on the results of the research on the view interpolation method.

19.5 Visual Condition of Human Eyes in 3-D Image Observation

In our research and development of the super-multiview display, research into the functional responses (vergence, accommodation, and pupil response) of the eyes has become important. The main issue is whether the responses when observing a super-multiview image are closer to those of natural vision than those when observing a conventional stereoscopic display. In this section we discuss the functional response when observing the super-multiview display.

19.5.1 Ocular Vergence and Accommodation to Super-multiview Stereoscopic Images

In conventional stereoscopic displays, the information inducing vergence is located at the position where integrated binocular images appear to form, whereas the accommodative stimuli are located on the surface of the display (Fig. 19.20). Therefore, in the binocular fused images that appear to be embossed on a large binocular disparity, there may be a difference between the vergence and accommodation. On the other hand, in the super-multiview region display, the accommodative stimuli are presented at the position of the stereoscopic images, which may resolve the difference between the vergence and accommodation (Fig. 19.21). The super-multiview region allows the entry of images with multiple parallaxes into the right and left pupils. This is made possible by the focused light array (FLA) display method described in Sect. 19.3.1.

Simulation using cameras has shown that accommodation is possibly induced by the super-multiview images. In a recently reported study, the accommodation was measured using a super-multiview display [5]. Compared to monocular vision, binocular vision more readily induces an appropriate accommodation corresponding to the distance to the target, due to the vergence-induced effect on accommodation (vergence accommodation) [6]. However, since the vergence was not measured in that study, it is difficult to assess whether the change in the binocular accommodation is attributable to accommodative vergence or to the effect of the super-multiview display. Therefore, we investigated vergence and accommodation to the super-multiview image by measuring the vergence and the accommodative state simultaneously.

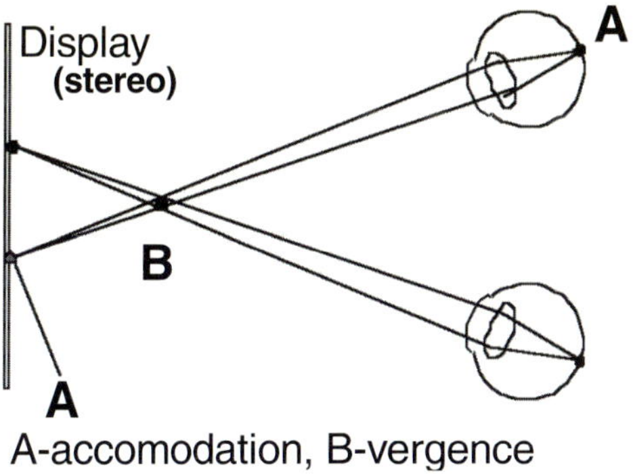

Fig. 19.20. Accommodation and vergence to conventional stereoscopic images

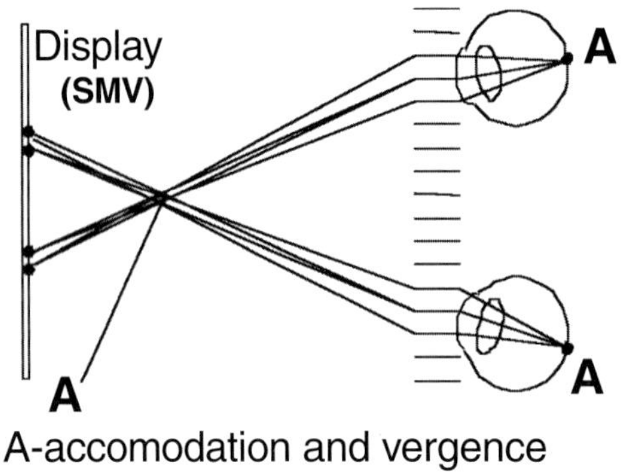

Fig. 19.21. Accommodation and vergence to super-multiview stereoscopic images

19.5.2 Methods

Stimuli and Conditions. The stimuli were obtained from an image displayed by an FLA red laser. To produce the super-multiview conditions, an image of 42 parallaxes with parallax intervals of 0.22 degrees was used. With an FLA image of 1.75 D (57 cm), 2 or 3 parallax images were expected to enter a pupil 5 mm in diameter.

The FLA images were displayed three conditions: monocular, stereoscopic, and super-multiview (SMV). Under the monocular conditions, the images were all the same, with no disparities among them. Therefore, the stimuli were given at 1.75 D on the images. Under the stereoscopic conditions, the images were divided in half, down the center. The left- and right-half images each had 21 parallaxes with the same disparity. The difference between the left and right images was set so that the binocular fused image from the stimuli of the upper and lower parts were positioned at 1.75 D (57 cm) and 2.13 D (47 cm), respectively (Fig. 19.22). The binocular disparity in the distance of the upper and lower parts was about 40′. Under the SMV conditions, the binocular disparity was set for each of the 42 parallaxes, and the patterns in the upper and lower parts were presented at the same viewing distances as those under the stereoscopic conditions. The stimuli were presented as two patterns: a vertical line and a radial pattern (Fig. 19.23).

Measurements of Vergence and Accommodative State. Eye movement and the accommodative state were measured using an Eye Tracker (Takei Scientific Instruments Co Ltd.) and an auto-refractometer (AA-2000; NIDEK). The convergence and accommodative state were measured alternately with these two devices using the high-speed time-division program for the infrared light of the Eye Tracker. In this measurement system, the pupillary area was calculated using the ocular infrared light image for measurement of the accommodative state. The data-sampling time was set at 190 ms, and the sampling duration for the vergence and accommodative state of each pupil was set at 5 ms for both.

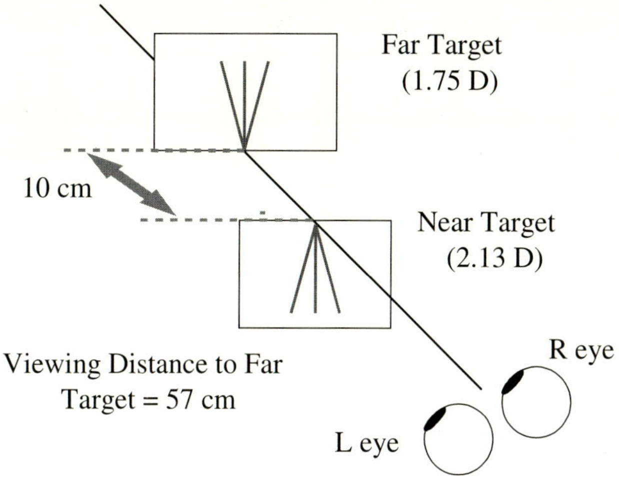

Fig. 19.22. Schematic configuration of targets

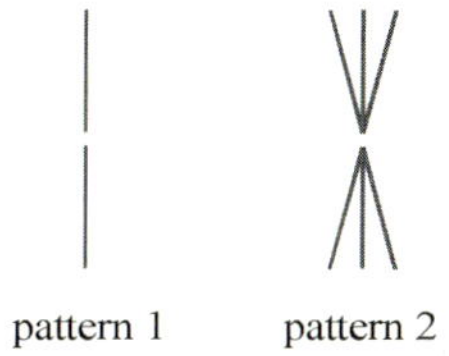

Fig. 19.23. Stimuli used for vergence and accommodation experiment

Procedure. The subjects were seated (one at a time) in a chair, and their heads were secured using a jaw rest and head belt. They observed stimuli presented on the screen of a super-multiview display in a dark room. They underwent two sets of trials. Each trial consisted of six sessions set by the combination of two stimulation patterns and three display patterns. In each trial, these six conditions were examined randomly. Immediately before the trial, the eye movement was calibrated. The vergence distance was calculated from the displacement angles of the eyes, setting the pupillary distance to 64 mm. The subjects alternately observed the stimuli of the upper and lower parts in the order lower-upper-lower-upper; the signals were computer generated every 3.8 seconds.

Subjects. Three subjects participated in this study: KM (23 years old); DN (23); KS (40). KM's vision was corrected with contact lenses, while that of the other two was not.

19.5.3 Results and Discussion

The results showed that there were changes in accommodation and vergence corresponding to the viewing distance to the far and near stimuli under the stereoscopic and SMV conditions. The diameters of the subjects' pupils were relatively large (about 8 mm) due to the dark room conditions, but there was no prominent change in pupil size in response to the viewing of the far and near targets. Consistent data on the accommodative condition were not obtained for DN and KS. DN's eyelashes were found to disturb the measured infrared light. KS has hyperopia (farsightedness), and although there were changes in vergence and the accommodative state with changes in the timing of the stimuli, no accommodation was observed at around 1.75 to 2.13 D, where the stimuli were presented. Consequently, only the data obtained for KM were evaluated quantitatively.

Example results obtained for KM under SMV conditions are shown in Fig. 19.24. The abscissa shows the time of stimulus presentation (sec), the left ordinate shows the positions of vergence and accommodation (diopter), and the right ordinate shows the pupil area (mm^2). The following abbreviations are used in the figure: vergence (VG), accommodation (ACC), pupil area (IRIS), and timing of stimulation-switching (STML). The F and N bars show the far and near viewing distance of the stimulus; the distance to the stimulus is shown by the dashed line. The changes in vergence and accommodation tended to appear shortly after the stimulus was switched, but can be considered to occur almost synchronously with the stimulus. The delay in accommodation change at the switching point was more evident than the delay in the change in vergence. The observed values for vergence and accommodation corresponded to the distance to the stimuli. No change in pupillary size with the change in distance to the stimuli was observed. These results indicate that the changes in vergence and accommodation in the present study were sufficiently detectable in the visual range of 47 to 57 cm.

We analyzed viewing-distance data obtained for KM by analyzing the variance for each set of trials. There was a statistically significant relationship between the viewing distance and display conditions, except for the first session of Pattern 1. Between the stereoscopic and SMV conditions, there was a difference in the accommodation to the far–near stimuli, suggesting that the accommodation-inducing effect varies with the method of display. Therefore, focusing on the stability of the vergence and accommodation, we performed a t-test to analyze the values obtained for the first and second fixations of the upper and lower parts of the stimuli within one trial under each of the display conditions. We used the data obtained 0.95 seconds before the viewing signal (5 values), taking into consideration the changes in vergence and accommodation following the stimulation-switching.

Table 19.2 and 19.3 show the results of the t-test on the vergence and accommodation, respectively. When the t-value is greater than 3.36, the value is considered to be statistically significant, with the level of significance being 1%. In other words, in the first and second viewing within one trial, the

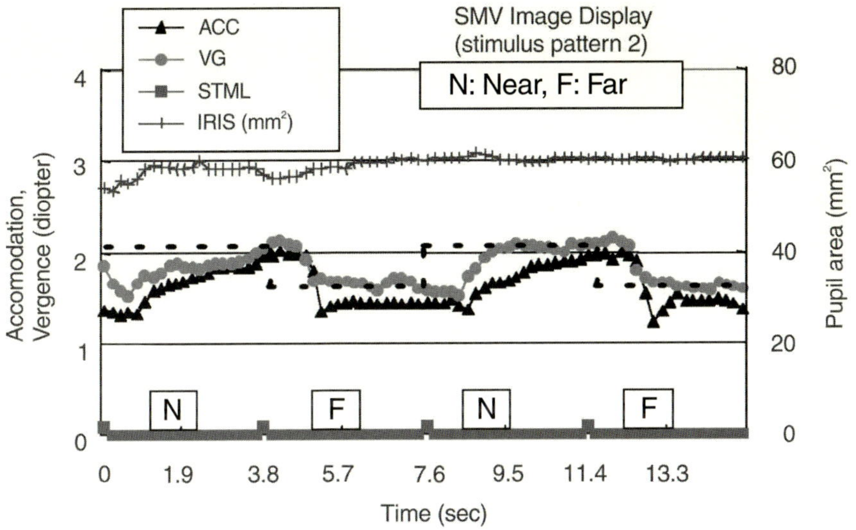

Fig. 19.24. Vergence and accommodation of KM to SMV images

vergence or accommodation distance was mutually different. The values in bold in the tables had no statistical significance.

When the stability of vergence was analyzed focusing only on the statistically significant differences, the data on viewing position were reproducible in only two or three of the eight trials under any of the display conditions. However, in many of the trials, the t-values obtained under the SMV conditions were smaller than those obtained under the other conditions, indicating that the deviation in the vergence distance is relatively smaller under SMV conditions.

A statistically significant difference in accommodation was observed, at the 1% significance level, in seven of the eight trials under the monocular and stereoscopic conditions. Under the SMV conditions, no significant difference

Table 19.2. Results of t-test on vergence

Stimuli	Trial	Position	Mono	Stereo	SMV
Pattern 1	1	Near	8.00	17.74	**0.60**
	1	Far	4.03	14.57	9.12
	2	Near	**0.49**	4.36	**1.08**
	2	Far	9.10	**1.88**	**2.32**
Pattern 2	1	Near	7.64	**1.78**	10.15
	1	Far	16.09	9.03	12.80
	2	Near	8.70	17.60	6.52
	2	Far	**1.34**	13.29	3.73

Table 19.3. Results of t-test on accommodation

Stimuli	Trial	Position	Mono	Stereo	SMV
Pattern 1	1	Near	10.96	8.43	3.96
	1	Far	29.30	5.35	**0.78**
	2	Near	5.65	5.89	**2.12**
	2	Far	6.65	19.30	**1.48**
Pattern 2	1	Near	18.10	58.00	9.38
	1	Far	4.05	**1.00**	3.72
	2	Near	**0.57**	5.42	4.49
	2	Far	8.48	9.33	**0.21**

was observed in half the trials. In other words, under the monocular and stereoscopic conditions, there was no reproducibility of the accommodation distance data in most of the trials, whereas a similar accommodation distance was observed in half the trials under the SMV conditions. In particular, stability tended to be better in the trials with the longer distances. This suggests that the variation in accommodation is larger under monocular and stereoscopic conditions and that accommodation under SMV conditions is relatively stable under all conditions. In particular, accommodative stability is better under SMV conditions.

19.6 Concluding Remarks

While the results presented in this chapter are insightful, they were obtained with only one subject. To generalize them, further studies are required with a larger number of subjects, additional conditions, and some modification of the measuring procedure for stable accommodation.

There is still a long way to go in achieving a perfectly natural, 3-D display. The "Advanced 3-D Tele-Vision Project" was started in 1992, by the Telecommunication Advancement Organization (TAO) of Japan with the objective of shortening the distance to this goal. The distance has been shortened sightly. To shorten it a bit more, we are focusing our research on three areas:

- developing a 3-D display
- measuring the eye control of viewers watching displayed 3-D images
- reducing the amount of multiview point image data by increasing the video speed.

This paper reports the results of our work as of summer 2000. Our work since then will be reported in papers to be published.

References

1. Y. Kajiki, H. Yoshikawa, T. Honda: 'Three-dimensional Display with Focused Light Array', Proc. SPIE **2652**, 106–116 (1996)
2. Y. Kajiki, H. Yoshikawa, T. Honda: 'Hologram-Like Video Images by 45-view Stereoscopic Display', Proc. SPIE **3012**, 154–166 (1997)
3. Y. Kajiki, H. Yoshikawa, T. Honda: 'Ocular Accommodation by Super-multiview Stereogram and 45-view Stereoscopic Display', Proc. 3rd Int. Display Workshop (IDW'96), 2, pp. 489–492 (1996)
4. T. Endo, Y. Kajiki, T. Honda, M. Sato: 'A Cylindrical 3-D Video Display Observable from All Directions', Proc. SPIE **3957**, 225–233 (2000)
5. T. Sudo, H. Morishita, T. Osaka, N. Taniguchi: '3-D Display Using Intersection of Light Beams', Proc. SPIE **3957**, 215–224 (2000)
6. V.V. Krishnan, S. Phillips, L. Stark: 'Frequency Analysis of Accommodation, Accommodative Vergence and Disparity Vergence', Vision Res. **13**, 1545–1554 (1973)

Index

Printing: Druckhaus Berlin-Mitte
Binding: Buchbinderei Stein & Lehmann, Berlin